**Tailored Functional Oxide Nanomaterials**

# Tailored Functional Oxide Nanomaterials

From Design to Multi-Purpose Applications

*Edited by Chiara Maccato and Davide Barreca*

**Editors**

*Prof. Chiara Maccato*
Padova University and INSTM
Department of Chemical Sciences
via F. Marzolo, 1
35131 Padova
Italy

*Dr. Davide Barreca*
ICMATE-CNR and INSTM
c/o Department of Chemical Sciences
Padova University
via F. Marzolo, 1
35131 Padova
Italy

**Cover Image:** © Yurchanka
Siarhei/Shutterstock

All books published by **WILEY-VCH** are carefully produced. Nevertheless, authors, editors, and publisher do not warrant the information contained in these books, including this book, to be free of errors. Readers are advised to keep in mind that statements, data, illustrations, procedural details or other items may inadvertently be inaccurate.

**Library of Congress Card No.:** applied for

**British Library Cataloguing-in-Publication Data**
A catalogue record for this book is available from the British Library.

**Bibliographic information published by the Deutsche Nationalbibliothek** The Deutsche Nationalbibliothek lists this publication in the Deutsche Nationalbibliografie; detailed bibliographic data are available on the Internet at <http://dnb.d-nb.de>.

© 2022 WILEY-VCH GmbH, Boschstr. 12, 69469 Weinheim, Germany

All rights reserved (including those of translation into other languages). No part of this book may be reproduced in any form – by photoprinting, microfilm, or any other means – nor transmitted or translated into a machine language without written permission from the publishers. Registered names, trademarks, etc. used in this book, even when not specifically marked as such, are not to be considered unprotected by law.

**Print ISBN:** 978-3-527-34759-9
**ePDF ISBN:** 978-3-527-82692-6
**ePub ISBN:** 978-3-527-82693-3
**oBook ISBN:** 978-3-527-82694-0

**Typesetting**  Straive, Chennai, India
**Printing and binding**  CPI Group (UK) Ltd, Croydon CR0 4YY

Printed on acid-free paper

# Contents

**Preface**  *xiii*

**1  Vapor Phase Growth of Metal-Oxide Thin Films and Nanostructures**  *1*
*Lynette Keeney and Ian M. Povey*
1.1  Introduction to Vapor Phase Deposition  *1*
1.2  Vapor Phase Deposition Methodologies  *1*
1.2.1  Chemical Vapor Deposition  *2*
1.2.2  Atomic Layer Deposition  *2*
1.3  Precursors and Chemistry  *3*
1.4  Applications of Metal-Oxide Vapor Phase Deposition  *4*
1.4.1  Case Study 1: Ferroelectric Oxide Materials  *4*
1.4.1.1  Ferroic Thin Films  *5*
1.4.2  Case Study 2: Dielectric Oxide Materials  *18*
1.5  Conclusions  *27*
References  *28*

**2  Addressing Complex Transition Metal Oxides at the Nanoscale: Bottom-Up Syntheses of Nano-objects and Properties**  *43*
*David Portehault, Francisco Gonell, and Isabel Gómez-Recio*
2.1  Introduction  *43*
2.2  Multicationic Oxides  *45*
2.2.1  Layered Oxide-Based Materials  *45*
2.2.2  Oxidation States Stable in Organic Media: The Case of Perovskites  *50*
2.2.3  Oxidation States Poorly Stable in Organic Media: The Case of Perovskites  *54*
2.3  Oxides with Uncommon Metal Oxidation States: The Case of Titanium(III) in Oxides and Extension to Tungsten Oxides  *58*
2.3.1  Crystal Structures and Requirements for the Synthesis of Oxides Bearing Titanium(III) Species  *59*
2.3.2  $Ti_2O_3$ Nanostructures  *61*
2.3.3  Mixed Valence Ti(III)/Ti(IV) Oxides: Magnéli Phases  *63*

| | | |
|---|---|---|
| 2.3.4 | Comparison to Metal Oxidation States Stable in Organic Media: Mixed W(V)/W(VI) Oxides  *68* | |
| 2.4 | Stabilization of New Crystal Structures at the Nanoscale  *73* | |
| 2.4.1 | Hard Templating to Isolate Bulk Metastable Oxides at High Temperatures  *74* | |
| 2.4.2 | Beyond Hard Templating for Isolating Nanostructures of Metastable Oxides  *75* | |
| 2.4.3 | Colloidal Syntheses  *75* | |
| 2.5 | Concluding Remarks  *76* | |
| | References  *77* | |
| | | |
| **3** | **Nanosized Oxides Supported on Arrays of Carbon Nanotubes: Synthesis Strategies and Performances of TiO$_2$/CNT Systems**  *89* | |
| | *Maria Letizia Terranova and Emanuela Tamburri* | |
| 3.1 | Introduction  *89* | |
| 3.2 | Synthesis Strategies for Preparation of CNT Arrays  *90* | |
| 3.3 | Selected Examples of Supported Nano-oxides  *91* | |
| 3.4 | A Focus on the TiO$_2$/CNT Systems  *93* | |
| 3.4.1 | Synthesis of TiO$_2$ on CNT  *99* | |
| 3.4.1.1 | Wet Chemistry  *100* | |
| 3.4.1.2 | Vacuum Techniques  *103* | |
| 3.5 | Concluding Remarks  *107* | |
| | References  *108* | |
| | | |
| **4** | **Computational Approaches to the Study of Oxide Nanomaterials and Nanoporous Oxides**  *111* | |
| | *Ettore Fois and Gloria Tabacchi* | |
| 4.1 | Introduction  *111* | |
| 4.2 | Overview of Theoretical Approaches  *113* | |
| 4.3 | Molecular Behavior at Nanomaterials Surfaces  *114* | |
| 4.3.1 | Molecular Interactions on Manganese Oxide Nanomaterials  *114* | |
| 4.3.2 | Insight on Molecule-to-Material Conversion in Chemical Vapor Deposition  *116* | |
| 4.4 | Oxide Porous Materials  *121* | |
| 4.4.1 | Structural Properties  *121* | |
| 4.4.2 | Behavior Under High-Pressure Conditions  *124* | |
| 4.4.3 | Hybrid Microporous Functional Materials  *127* | |
| 4.5 | Outlook and Perspectives  *131* | |
| | References  *133* | |
| | | |
| **5** | **Functional Spinel Oxide Nanomaterials: Tailored Synthesis and Applications**  *137* | |
| | *Zheng Fu and Mark T. Swihart* | |
| 5.1 | Introduction and Topic Overview  *137* | |

| 5.2 | Syntheses  *138* |
|---|---|
| 5.2.1 | Vapor Phase  *138* |
| 5.2.1.1 | Chemical Vapor Deposition  *138* |
| 5.2.1.2 | Atomic Layer Deposition  *138* |
| 5.2.1.3 | Spray Pyrolysis  *140* |
| 5.2.1.4 | Laser Pyrolysis  *141* |
| 5.2.1.5 | Plasma Methods  *142* |
| 5.2.2 | Solution Phase  *143* |
| 5.2.2.1 | Sol–Gel Methods  *143* |
| 5.2.2.2 | Hydrothermal Methods  *143* |
| 5.2.2.3 | Thermal Decomposition  *143* |
| 5.2.2.4 | Solvothermal Methods  *145* |
| 5.2.3 | Solid Phase  *146* |
| 5.2.3.1 | Solid-State Thermal Decomposition  *146* |
| 5.2.3.2 | Combustion  *147* |
| 5.2.3.3 | Ball Milling  *148* |
| 5.2.3.4 | High-Temperature Solid Solution Method  *148* |
| 5.3 | Structure–Effect Applications  *150* |
| 5.3.1 | One-Dimensional (1D) Structures  *151* |
| 5.3.1.1 | Nanorods  *151* |
| 5.3.1.2 | Nanowires  *154* |
| 5.3.1.3 | Nanotubes  *154* |
| 5.3.2 | Two-Dimensional (2D) Structures  *159* |
| 5.3.2.1 | Nanofilms  *159* |
| 5.3.2.2 | Nanosheets  *159* |
| 5.3.2.3 | Nanoplatelets  *163* |
| 5.3.3 | Three-Dimensional (3D) Structures  *165* |
| 5.3.4 | One- and Two-Dimensional (1&2D) Structure  *170* |
| 5.3.5 | One- and Three-Dimensional (1&3D) Structures  *171* |
| 5.3.6 | Two- and Three-Dimensional (2&3D) Structure  *173* |
| 5.4 | Self-Assembled Structures  *175* |
| 5.5 | Conclusions and Future Perspectives  *180* |
|  | References  *184* |
|  |  |
| **6** | **Photoinduced Processes in Metal Oxide Nanomaterials**  *193* |
|  | *Nikolai V. Tkachenko and Ramsha Khan* |
| 6.1 | Introduction  *193* |
| 6.2 | Photophysics of Bulk MOs  *195* |
| 6.2.1 | Energy-Level Structure and Steady-State Spectra  *195* |
| 6.2.2 | Photoexcitation and Relaxation Dynamics  *201* |
| 6.2.3 | Emission Decay Kinetics, Time-Resolved PL  *203* |
| 6.2.4 | Transient Absorption (TA) Spectroscopy  *205* |
| 6.3 | Nanostructures  *208* |
| 6.3.1 | Quantum Confinement  *208* |
| 6.3.2 | Surfaces and Interfaces  *211* |

| 6.4 | Photophysical Aspects of MO Applications 218 |
| 6.4.1 | Solar Cells 218 |
| 6.4.2 | Light Emitting Devices 219 |
| 6.4.3 | Photocatalysis 219 |
| 6.4.4 | Photodegradation 219 |
| 6.4.5 | Solar Driven Chemistry 220 |
| 6.5 | Conclusions 220 |
| | References 221 |

**7 Metal Oxide Nanomaterials for Nitrogen Oxides Removal in Urban Environments** 229
*M. Cruz-Yusta, M. Sánchez, and L. Sánchez*

| 7.1 | Introduction: Photocatalytic Removal of Nitrogen Oxides Gases 229 |
| 7.2 | $TiO_2$-Based Materials 230 |
| 7.2.1 | Tailoring the Energy Band Gap and Edges' Potentials 231 |
| 7.2.2 | Dopant Elements and Quantum Dots 234 |
| 7.2.3 | Defects, Vacancies, and Crystal Facets in the $TiO_2$ Nanostructure 235 |
| 7.2.4 | Composites/Substrates 236 |
| 7.2.5 | Titanium-Based Oxides 237 |
| 7.3 | Alternative Advanced Photocatalysts 238 |
| 7.3.1 | Bismuth Oxides 238 |
| 7.3.2 | Tin- and Zinc-Based Oxides 242 |
| 7.3.3 | Transition Metal Oxides 247 |
| 7.4 | New Insights into the NOx Gases Photochemical Oxidation Mechanism 251 |
| 7.5 | Field Studies in Urban Areas 253 |
| 7.5.1 | Photocatalytic Construction Materials 253 |
| 7.5.2 | Field Studies of NOx Abatement in Real Environments 254 |
| 7.6 | Conclusions and Perspectives 256 |
| | References 259 |

**8 Synthesis and Characterization of Oxide Photocatalysts for $CO_2$ Reduction** 277
*Fernando Fresno and Patricia García-Muñoz*

| 8.1 | Introduction 277 |
| 8.2 | Fundamentals of Heterogeneous Photocatalysis 279 |
| 8.3 | Applications of Heterogeneous Photocatalysis 281 |
| 8.4 | Photocatalytic $CO_2$ Reduction: State of the Art and Main Current Issues 283 |
| 8.4.1 | $TiO_2$-Based Photocatalysts for $CO_2$ Reduction 286 |
| 8.4.2 | Other Oxide Photocatalysts 291 |
| 8.5 | Oxide-Based Heterojunctions and Z-Scheme Photocatalytic Systems 295 |
| 8.5.1 | Cocatalysts for $CO_2$ Reduction: Metal-Oxide Synergies 299 |
| 8.6 | Conclusions and Future Perspectives 303 |
| | References 303 |

| 9 | **Functionalized Titania Coatings for Photocatalytic Air and Water Cleaning** *317* |
|---|---|
| | Ksenija Maver, Andraž Šuligoj, Urška Lavrenčič Štangar, and Nataša Novak Tušar |
| 9.1 | Introduction *317* |
| 9.1.1 | Titania as a Photocatalyst for Air and Water Cleaning *317* |
| 9.1.2 | Titania Functionalization *319* |
| 9.1.3 | Fabrication of Titania-Based Coatings *320* |
| 9.1.4 | Characterization of Titania-Based Materials *321* |
| 9.2 | Case Studies *323* |
| 9.2.1 | $SiO_2$-Supported $TiO_2$ for Removal of Volatile Organic Pollutants from Indoor Air Under UV Light *323* |
| 9.2.2 | Sn-Functionalized $TiO_2$ as a Photocatalytic Thin Coating for Removal of Organic Pollutants from Water Under UV Light *325* |
| 9.2.3 | $SiO_2$-Supported $TiO_2$ Functionalized with Transition Metals for Removal of Organic Pollutants from Water Under Visible Light *329* |
| 9.3 | Conclusion and Further Outlook *335* |
| | References *335* |
| | |
| 10 | **Metal Oxides for Photoelectrochemical Fuel Production** *339* |
| | Gian Andrea Rizzi and Leonardo Girardi |
| 10.1 | Introduction to Photoelectrochemical Cells *339* |
| 10.1.1 | The Photoelectrochemistry Approach *344* |
| 10.2 | Metal Oxides Photoelectrode Candidate Materials *347* |
| 10.2.1 | Photoanodes *349* |
| 10.2.2 | Photocathodes *349* |
| 10.3 | Tailoring Surface Catalytic Sites and Catalyst Use *350* |
| 10.4 | Metal Oxide Heterostructures *353* |
| 10.5 | Metal Oxides as a Protective Anti-corrosion Layer in Photoelectrodes *354* |
| 10.6 | Evaluation of Photoelectrode Efficiencies *359* |
| 10.7 | Conclusions and Perspectives *365* |
| | References *367* |
| | |
| 11 | **Tailoring Porous Electrode Structures by Materials Chemistry and 3D Printing for Electrochemical Energy Storage** *379* |
| | Sally O'Hanlon and Colm O'Dwyer |
| 11.1 | Strategies for Functional Porosity in Electrochemical Systems *379* |
| 11.2 | Benefits and Limitations of Structural Engineering for Electrochemical Performance *382* |
| 11.3 | Tailoring the Pore Structure of Metal Oxides for Li-ion Battery Cathodes and Anodes *383* |
| 11.4 | Developments in 3D Printing of Porous Electrodes for Electrochemical Energy Storage *389* |
| 11.5 | Porous Current Collectors by 3D Printing *390* |

| 11.6 | Battery and Supercapacitor Materials from 3D Printing  *392* |
| 11.7 | Conclusions and Outlook  *394* |
| | References  *396* |

## 12  Ferroic Transition Metal Oxide Nano-heterostructures: From Fundamentals to Applications  *405*
*G. Varvaro, A. Omelyanchik, and D. Peddis*

| 12.1 | Introduction  *405* |
| 12.2 | Ferroic Properties of Complex Transition Metal Oxides  *408* |
| 12.2.1 | Spinel Ferrites  *408* |
| 12.2.2 | Perovskites  *411* |
| 12.2.3 | Other Magnetic Oxides  *412* |
| 12.3 | Magnetic Oxide Heterostructures  *413* |
| 12.3.1 | Hard/Soft Exchange-Coupled Systems  *413* |
| 12.3.2 | Ferro(i)magnetic/Antiferromagnetic Systems  *416* |
| 12.3.3 | All-Oxide Synthetic Antiferromagnets  *419* |
| 12.4 | Artificial Multiferroic Oxide Heterostructures  *421* |
| 12.4.1 | Strain Transfer Mechanism  *423* |
| 12.4.2 | Charge Modulation Mechanism  *426* |
| 12.4.3 | Exchange Interaction Mechanism  *427* |
| 12.5 | All-Oxide Spintronic Heterostructures  *427* |
| 12.6 | Conclusion and Perspectives  *430* |
| | References  *431* |

## 13  Metal-Oxide Nanomaterials for Gas-Sensing Applications  *439*
*Pritamkumar V. Shinde, Nanasaheb M. Shinde, Shoyebmohamad F. Shaikh, and Rajaram S. Mane*

| 13.1 | Introduction  *439* |
| 13.2 | Types of Gas Sensors  *442* |
| 13.3 | Metal-Oxide Nanomaterial-Based Gas Sensors  *443* |
| 13.4 | Preparation of Metal-Oxide Gas Sensors  *446* |
| 13.4.1 | Operation Mechanism  *446* |
| 13.4.2 | Morphology-Related Structural Parameters  *448* |
| 13.4.2.1 | Grain Size  *448* |
| 13.4.2.2 | Pore Size  *449* |
| 13.4.3 | Crystallographic Defective and Heterointerface Structures  *453* |
| 13.4.3.1 | Defect Structure  *453* |
| 13.4.3.2 | Heterointerface Structure  *455* |
| 13.4.4 | Chemical Composition  *458* |
| 13.4.5 | Addition of Noble Metal Particles  *458* |
| 13.4.6 | Humidity and Temperature  *461* |
| 13.5 | Gas-Sensing Mechanisms  *462* |
| 13.5.1 | Adsorption/Desorption Model  *462* |
| 13.5.1.1 | Oxygen Adsorption Model  *464* |

| | | |
|---|---|---|
| 13.5.1.2 | Chemical Adsorption/Desorption | *467* |
| 13.5.1.3 | Physical Adsorption/Desorption | *470* |
| 13.5.2 | Bulk Resistance Control Mechanism | *471* |
| 13.5.3 | Gas Diffusion Control Mechanism | *472* |
| 13.6 | Conclusions and Future Perspectives | *474* |
| | References | *475* |

**Index** *487*

# Preface

Metal-oxide-based materials represent an extremely appealing class of functional platforms, thanks to the multiformity of their structures, stoichiometries, and chemico-physical characteristics. This broad variety, in conjunction with the modulation of their structure and morphology at the nanoscale, empowers with a precious toolkit toward the implementation of nanosystems featuring intelligent properties and specific functions. The latter, ranging from size-dependent optical properties to novel chemical reactivity, have triggered a significant attention for various applications, including, among others, heterogeneous (photo)catalysts for water/air purification and fuel production, solid state gas sensors, batteries, supercapacitors, and magnetic devices. Significant advancements in this regard have been enabled by the fast growth and implementation of innovative fabrication routes and analytical tools, as well as by the progress of computational modeling, whose synergism enables the study of an impressive variety of physico-chemical phenomena relevant from both a fundamental and an applied point of view.

In this context, the present book offers a unique combination of fundamentals and selected key applications for metal oxide nanomaterials, ranging from their preparation up to the mastering of their characteristics in an interdisciplinary view. The volume provides basic concepts related to material design through vapor- and liquid-phase approaches, addressing various issues currently under way from preparation, to theoretical research, up to a broad scenario of advanced technological utilizations. Throughout the text, the authors illustrate various aspects for the understanding and real-world end uses of the most important class of inorganic materials, elucidating the interplay between fabrication, chemico-physical characterization, and functional behavior in selected fields. The latter comprise the following:
- Photo-activated pollutant degradation.
- (Photo)electrochemical systems and devices for sustainable energy production and storage.
- Current trends regarding ferromagnetic/ferroelectric and spintronic heterostructures, batteries, electrochromics, and miniaturized gas sensors for day-to-day life.

Beside an overview of the key topics, users will also find actual perspectives and emerging research directions, in an attempt to answer to the question "what's next?" in nano-oxide development for science and technology of the future.

The key features of the present book can be summarized as follows:

- Provides a fundamental understanding of interrelations between synthesis, chemico-physical properties, and functional behavior of a variety of single and multicomponent metal oxide-based nanomaterials.
- Presents an advanced and broad perspective on metal oxide nanomaterials for various technological end uses through the contribution of internationally renowned experts.
- Offers a multi-purpose coverage of recent research advancements in synthesis, modeling, and applications in electrochemical devices for energy production/storage, magnetic materials, gas sensors, photo-assisted water/air purification, and fuel production.
- Illustrates various case studies promoting the critical thinking of students and researchers toward the development of novel oxide-based functional materials and devices.

In fact, the present volume stands as a versatile instructive text for undergraduate/postgraduate fellows to boost their interest in designing advanced oxide nanomaterials with tailored characteristics for the next-generation functional systems. In addition, the book will also serve as a powerful reference for chemists, physicists, engineers, and materials scientists working in universities and research institutions, as well as for professionals involved in the mastering and technology transfer of oxide nanosystems toward cutting-edge applications.

We believe that this book will develop an important framework for the beginners and researchers working in the target fields, exploring various aspects from basic background up to material properties and their eventual uses. It will be also a very valuable reference book for trained scientists in both academic and industrial contexts.

A brief overview of each chapter included in this book is provided in the following, in order to give the readers a general and synoptic idea of the volume content.

In **Chapter 1**, I. Povey and L. Keeney introduce the basic concepts of vapor phase growth of metal oxides, with particular regard to chemical vapor deposition (CVD) and atomic layer deposition (ALD) processes. The authors dedicate attention to selected material classes (such as ferroelectrics and dielectrics) where such kind of routes have made, or have the potential to make, significant advances in technology that could arise over the next decade.

In **Chapter 2** by D. Portehault, F. Gonell, and I. Gómez-Recio, the text focuses on chemical routes to complex oxide nanomaterials. The systems dealt with include multicationic oxides, oxides of metals with uncommon oxidation states, and new oxide crystal structures accessible only at the nanoscale. Selected case studies are proposed to provide a survey on the role of processing parameters, drawing some guidelines for material fabrication by design.

In **Chapter 3**, E. Tamburri and M.L. Terranova pay attention to the coupling of nanosized oxides with carbon nanotubes, the latter acting as supports for the dispersion of metal oxides, with particular reference to $TiO_2$-containing systems. The authors illustrate preparative methodologies and selected applications of the target hybrid materials, with special attention on the interplay between functional properties and material morphology.

**Chapter 4**, by E. Fois and G. Tabacchi, outlines key concepts related to the simulation of oxide materials and the theoretical approaches to investigate these systems. The authors propose representative studies on materials that take active part to chemical phenomena exploited in several technological applications, highlighting the potential of computational approaches and proposing a viewpoint to meet the key open challenges for future progress in these areas.

In **Chapter 5**, Z. Fu and M. Swihart offer a comprehensive review on functional nanosystems based on oxide spinels, with regard to their preparation, characterization, and functional applications. The authors perform a survey on various case studies, summarizing synthetic strategies and providing an insight on which nanostructures are possible and on how structure affects performances, providing guidance for the preparation of future spinel materials.

In **Chapter 6** by N.V. Tkachenko and R. Khan, the topic in focus is light interaction with metal oxide nanostructures. Beside fundamentals, the chapter covers the photo-physics of oxide nanostructures and light-activated reactions in the target materials and at their interfaces. The pertaining aspects are considered in view of both existing and perspective applications of metal oxide nanosystems in devices powered by solar light, which is of importance in different fields.

**Chapter 7**, by M. Cruz-Yusta, M. Sánchez, and L. Sánchez, focuses on the use of metal oxide nanomaterials for the removal of nitrogen oxides in urban environments, toward the development of air purification technologies. The authors provide relevant examples ranging from titanium oxide up to other less explored materials, with attention on the property–performance interrelations and the selectivity toward the obtainment of harmless products.

In **Chapter 8**, F. Fresno and P. García-Muñoz cover in overview functional oxide nanostructures for photocatalytic $CO_2$ reduction, with particular emphasis on the used synthetic methods and the characteristics of the resulting materials. After a summary on the basics of photocatalytic processes for the target end uses, the pertaining state of the art and main current issues are presented with regard to both pure oxides and heterostructured systems.

In **Chapter 9** by K. Maver, A. Šuligoj, U. Lavrenčič Štangar, and N. Novak Tušar, attention is devoted to photocatalytic air and water purification by means of coatings based on titanium dioxide, the most widely exploited photocatalyst so far. The text first introduces the synthesis, functionalization, and chemico-physical characterization of titania-based systems, presenting subsequently case studies on the photocatalytic degradation of volatile organic pollutants in indoor air and water under different illumination conditions.

In **Chapter 10**, L. Girardi and G.A. Rizzi summarize recent developments on photoelectrochemical fuel production using metal oxide nanosystems. The authors examine various aspects related to photocathodes, photoanodes, and their property tailoring, along with the evaluation of photoelectrode efficiency, before presenting the main bottlenecks for the development of a solar-based economy through bias-free water splitting.

In **Chapter 11**, S. O'Hanlon and C. O'Dwyer present a comprehensive overview on functional porous materials for applications in energy storage systems. The authors provide representative examples on the modulation of pore structure in metal oxides for Li-ion battery cathodes and anodes. The chapter also contains recent developments in 3D printing technologies and the direct printing of electrodes for ultimate end uses in batteries and supercapacitors.

In **Chapter 12**, G. Varvaro, A. Omelianchik, and D. Peddis pay attention to complex transition metal oxide nanomaterials and their magnetic properties. After a brief introduction on ferroic properties, ferromagnetic/antiferromagnetic systems pertaining to heterostructures based on magnetic oxides and multiferroic magnetoelectric oxides are examined, presenting also some selected examples pertaining to spintronic heterostructures.

In **Chapter 13** by P.V. Shinde, N.M. Shinde, S.F. Shaikh, and R.S. Mane, metal oxide nanomaterials as gas sensors for the detection of toxic analytes are addressed. The authors focus on chemoresistive devices, with regard to material preparation, functional behavior, and gas sensing mechanism, as well as on strategies for property tailoring aimed at achieving increased responses, sensitivity, and selectivity for practical end uses.

This book is not an attempt to fully cover all topics relevant to metal oxide-based nanomaterial preparation and applications, but focuses rather on fundamentals and advancements of selected key pertaining aspects. The editors are convinced that the future exploitation of synthetic methods and a precise control over oxide nanosystem characteristics show a bright promise to meet the open questions related to the actual technological requirements. Disclosing general concepts and key ideas in this area can be the right path to gain a deeper knowledge of fundamental phenomena and to develop of novel materials and devices.

The editors would like to thank all the authors who contributed to this book, whose publication would not have been possible without the highly professional and excellent support from the Wiley Editorial Team (Dr. Lifen Yang, Ms. Katrina Maceda, Ms. Katherine Wong). A particular and heartfelt thanks is addressed to Ms. Katherine Wong for the invaluable help in bringing this volume to completion.

# 1

# Vapor Phase Growth of Metal-Oxide Thin Films and Nanostructures

*Lynette Keeney and Ian M. Povey*

*University College Cork, Tyndall National Institute, Lee Maltings Complex, Dyke Parade, Cork T12R5CP, Ireland*

## 1.1 Introduction to Vapor Phase Deposition

The basic concept of chemical vapor deposition (CVD) is the growth of a thin solid film by the chemical reactions of a vapor of precursor molecules. These chemical reactions can take place at the surface or in the gas phase, and there are many variants encompassing a wide array of approaches to formulate the film. The basic concept of CVD was first demonstrated in the nineteenth century for carbon and metallic thin films [1–3], and it was further developed for Si [4, 5] and III/V materials [6], but its use in oxides became more widespread after the development of oxide superconductors in the 1980s [7, 8] and the drive for Si integrated circuit technology through scaling of metal-oxide semiconductor field effect transistors (MOSFETs) and dynamic random access memory (DRAM) [9]. In this short review we will briefly introduce the basic concepts of vapor phase growth of metal oxides and selected applications where it has made, or has the potential to make, significant advances in technology.

## 1.2 Vapor Phase Deposition Methodologies

Common to all vapor deposition systems are a precursor delivery system, a reaction chamber with an energy source to control the chemistry, and an exhaust system where reaction products and excess reagents are safely disposed of. This simplistic description does little justice to the complex engineering solutions that are employed to perform these tasks; however, it is beyond the scope of this article to describe these in detail, and so the reader is directed elsewhere [10, 11]. Many vapor deposition methodologies are used to generate films [10–12], which methods are employed is determined not only by the type of film required but also by the chemical nature and availability of the precursors employed. Here, we will consider two general variants of deposition methods: CVD where reagents are introduced simultaneously

*Tailored Functional Oxide Nanomaterials: From Design to Multi-Purpose Applications*, First Edition.
Edited by Chiara Maccato and Davide Barreca.
© 2022 WILEY-VCH GmbH. Published 2022 by WILEY-VCH GmbH.

and atomic layer deposition (ALD) where the reagents introduced are separated by purges. The push for miniaturization in device technology has pushed many traditional CVD processes toward an ALD solution, where material growth is achieved at the atomic layer of precision; however it should be recognized that nanotechnology can still be achieved with traditional CVD methods and that not all materials systems lend themselves to ALD.

### 1.2.1 Chemical Vapor Deposition

CVD is a complex process that can be described in the form of a schematic (Figure 1.1) where gaseous precursors are introduced into a reactor and through a series of gas phase pyrolysis reactions; partly reacted chemicals diffuse through a boundary layer to interact with a growth surface. Such a description is very much simplified as the growth process is influenced by many parameters that determine the stoichiometry, uniformity, crystallinity, and density of defects of the resultant materials. The consequence of which is that the reactor and chemistry design is extremely complicated and must consider primary and secondary chemistry, mass transport, adsorption, surface diffusion of reactive and exhaust species, and finally nucleation and film densification. The key parameters for growth control are the choice of chemistry, the pressure, and the nature and amount of energy supplied to the system. For a more in-depth treatise on the complexities of reactor design and process control, the reader is referred elsewhere [10].

### 1.2.2 Atomic Layer Deposition

ALD is a variant of CVD that employs sequential self-limiting chemical reactions at the substrate surface. As illustrated in Figure 1.2, a key aspect of ALD is that the precursors, or their reaction products, do not meet in the gas phase through cyclic pulsing and purging. This cycle of pulses and purges is repeated to build up a thin film of the desired thickness one monolayer per full cycle.

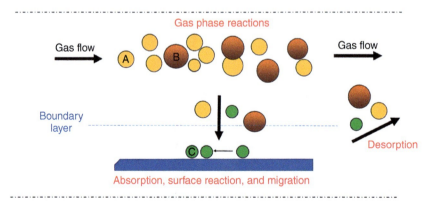

**Figure 1.1** Schematic of a generic chemical vapor deposition process. The reagents A and B are transported to the reaction zone, where they undergo a complex range of gas and surface reactions to deposit material C at the growth surface.

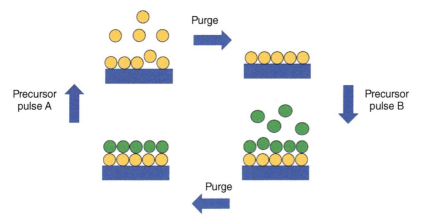

**Figure 1.2** Schematic of a typical ALD process. Two precursors A and B react at the surface to produce material AB. The two self-limiting precursor pulses are separated by purges to ensure that excess reactant is removed from the reactor between pulses to prevent gas phase reactions.

In this way the growth can be controlled at the atomic or molecular level. Another key advantage of ALD is that the self-limiting chemistry enables both large area and complex morphologies to be coated uniformly. Self-limiting behavior is a consequence of the precursor only being able to react with functional groups at the substrate surface. Hence, once all the reactive sites have undergone chemical reaction, no further reaction can take place, and excess reagent passes to the exhaust system, the resultant surface remaining in a pseudo-steady state until the co-reagent is pulsed, completing the cycle and regenerating the reactive sites for the next precursor pulse. In addition, the generally low temperature of the ALD process makes it suitable for sensitive substrates, such as polymers and biological samples. The key disadvantage of ALD is the low deposition rates relative to other methods that makes it unsuitable for thicker films. It should also be noted that the low temperature of deposition although advantageous for sensitive substrates can limit the phase selection of materials, with many material systems being limited to amorphous films. The technique of ALD has been extensively reviewed [13–15].

## 1.3 Precursors and Chemistry

Although the chemistries of variants of CVD are different, the key requirements for the chemical precursors employed are similar. Key aspects are that the precursor is stable under storage and delivery, sufficiently reactive to undergo viable reactions at or near the substrate to produce volatile by-products that can be swept from the reaction zone without contamination of the desired material. Other considerations such as ease and cost of manufacture and any safety implications are also of importance. Generally, precursors are volatile to allow easy transport to the reaction zone; however with the introduction of liquid injection methods, precursors can, in many cases, be dissolved in a suitable solvent to circumvent volatility issues [11]. The combination of precursors utilized is also vital, the production of materials requiring more than one precursor requires that their reactivity is similar under the chosen reaction conditions to ensure the desired stoichiometry. The choice of precursor used in a specific reaction is often not straightforward and is often determined by the material parameters required as well as the CVD variant being employed [10]. The basic concepts of precursor stability, reactivity, and transportability are determined by the constituent parts of the precursor molecules. ALD precursors need to be volatile enough to be easily pulsed and purged through the reaction zone at temperatures where self-limiting surface chemistry is preserved, but reactive enough so that facile reactions with the surface species and co-reagents are efficient. CVD sources require high stability at evaporation temperatures and clean decomposition at elevated temperatures. These properties are often in conflict; for example, volatility is determined by the intermolecular forces of a source, and this can be improved by increasing the stearic nature of ligands, but this in turn can reduce the reactivity. Thus, what might be a stable, volatile precursor for CVD may not undergo facile reactions with oxygen containing co-reagents that would make it suitable for ALD. Another complication is that trace contaminants that might not be relevant for one application may be seen as catastrophic for others; for example, F substitution of β-diketonates is an excellent method of improving volatility, but F is a serious contaminant, even at low levels, in silicon-based electronics. The chemistry of precursor development for CVD and its variants is complex but is well reviewed elsewhere [10–15].

## 1.4 Applications of Metal-Oxide Vapor Phase Deposition

Vapor deposition methods are well established and widespread and have been employed to generate metal-oxide-based materials for many application spaces, notably transparent conducting oxides, catalysts, dielectric materials, and protective and decorative coatings. These and many other applications are described in detail in many excellent reviews on the subject [10–15]. In this review, however, due to the wide range of vapor deposition processes and established applications, we will focus on materials and methods, on which our own personal research is centered, namely, dielectric metal oxides, where nanometer scale engineering is essential for realizing many applications. With respect to dielectrics, we will present case studies on two classes of materials: firstly, ferroelectric that displays hysteresis in their electric field response and, secondly, insulating with a linear polarization response to electric field.

### 1.4.1 Case Study 1: Ferroelectric Oxide Materials

Ferroelectric materials are a subclass of ferroic materials, which also include ferromagnetic and ferroelastic materials. Ferroelectrics are dielectrics that have a net permanent dipole moment due to their polar nature and non-centrosymmetric structure. This spontaneous dipole moment can be switched between two states using an applied electric field, and a remanent polarization remains when the electric field is removed. All ferroelectrics exhibit pyroelectric and piezoelectric effects, meaning that they can transform thermal energy into electrical energy, and they can create electric charge when mechanically stressed. Ferroelectrics are thus widely used in sensing, actuating, infrared, optoelectronic, surface acoustic wave (SAW), ultrasound, and nonvolatile data storage applications [16].

Oxide ferroelectrics are being pursued due to their diverse properties, their robust performances, and the intriguing phenomena they exhibit. Their structures and compositions can be tuned to control the functionality of the targeted application. Perovskite $KNbO_3$ and ilmenite $LiNbO_3$ have received much attention due to their large piezoelectric constants and electro-optic coefficients (e.g. $d_{33}$ of 121 pC/N for $KNbO_3$ [17] and 6–70 pC/N for $LiNbO_3$, depending on the orientation [18]) and have been used in SAW and electro-optic devices [19]. $LiNbO_3$ has recently emerged as a promising platform for photonic integrated circuits [20]. Piezoelectric, pyroelectric, and electro-optic coefficients can be optimized by forming solid solutions between end-member perovskites. One of the most technologically important and commercially successful ferroelectric systems is the perovskite solid solution system between $PbZrO_3$ and $PbTiO_3$ ($PbZr_xTi_{1-x}O_3$; often referred to as PZT), and excellent properties can be tailored by selecting ceramic compositions in different parts of the phase diagram. Properties can be further improved by partial substitution with selected cations (e.g. A-site substitution by $Sr^{2+}$, $La^{3+}$, K+, or B-site substitution with $Sn^{4+}$, $Nb^{5+}$, $Ni^{3+}$). PZT has been widely used as an active component in microelectromechanical systems (MEMS), actuators, pressure sensors, pyroelectric sensors, medical ultrasound, ultrasonic transducers, and piezoelectric micro-pumps [21, 22].

Layered structured ferroelectrics offer very flexible frameworks for a wide variety of applications, given that differing cations can be accommodated both at the A- and B-sites. Aurivillius bismuth-based materials are naturally two-dimensional (2D) nanostructured, where $(Bi_2O_2)^{2+}$ fluorite-type layers are interleaved with $m$ numbers of perovskite-type units (Figure 1.3). Ferroelectricity arises from polar displacements of cations relative to anions, which stabilizes the polar ground state. This has been exploited for commercial use in ferroelectric random-access memory (Fe-RAM) devices [25, 26, 28]. Recently, it has been shown that ferroelectricity can persist down to thicknesses of only 2.4 nm, which equates to one-half of the normal crystal unit cell, demonstrating that wide-band gap oxide ferroelectrics can be added to the growing class of 2D materials [29]. Room temperature multiferroic (ferroelectric and ferromagnetic) behavior has also been demonstrated [30, 31], offering the prospect of producing eight-state information storage based on simultaneous magnetoresistance and electroresistance [32].

### 1.4.1.1 Ferroic Thin Films

Thin films of ferroelectric and multiferroic materials are being widely used or are currently being developed for commercial applications including memories, microwave electronic components, and micro-devices with pyroelectric and piezoelectric micro-sensors and actuators [33]. Successful thin film synthesis to satisfy

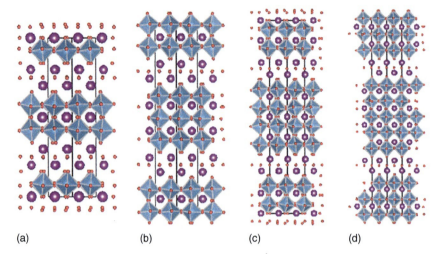

**Figure 1.3** Schematics of Aurivillius phase crystal structures projected down the <110> orientation produced using visualization for electronic structural analysis (VESTA) [23] software. (a) SrBi$_2$Ta$_2$O$_9$ ($m = 2$; space group *A21am* [24]); (b) Bi$_4$Ti$_3$O$_{12}$ ($m = 3$; space group *B1a1* [25]); (c) Bi$_5$Ti$_3$Fe$_2$O$_{15}$ ($m = 4$; space group *A21am* [26]); and (d) Sr$_2$Bi$_4$Ti$_5$O$_{18}$ ($m = 5$; space group *B2eb* [27]). Bismuth or *A*-site atoms are represented by purple spheres; oxygen atoms are represented by smaller red spheres; and $BO_6$ octahedra are represented by blue polyhedra. Source: (a) Modified from Perez-Mato et al. [24], (b) Modified from Guo et al. [25], (c) Modified from Hervoches et al. [26], (d) Modified from Ismunandar et al. [27].

the requirements of a particular application is demanding [16]. Depending on the targeted technology, it may be necessary that films are uniform, conformal, device compatible, epitaxial, have high orientation control, and are strain controlled. Furthermore, it may be a condition that films are ultrathin, have atomic-level precision, and have atomically sharp interfaces. Aspects to consider during successful processing of the fabricated film into specific device structures may involve integration with heterostructures and multilayers, integration with metallic and conductive oxide electrode layers, integration with silicon, the ability to produce patterned and nano-patterned structures, the use of comparatively low processing temperatures, and the use of reasonably inexpensive processes as well as processes that yield high reproducibility [22].

Over the past three decades, there have been significant improvements in the development of instrumentation and growth processes to enable the synthesis of high quality, reliably produced, and ferroic thin films with technological value. Methods for successful thin film fabrication of piezoelectric, ferroelectric, and multiferroic oxides include chemical solution deposition (CSD) [32], sputter deposition [34], pulsed laser deposition (PLD) [27, 35], ALD [36, 37], molecular beam epitaxy (MBE) [27, 38], and CVD [22, 39]. CVD has been a pioneering method to produce films used in high density nonvolatile Fe-RAM technologies [40]. Within this contribution, we focus on key factors involved in the CVD fabrication of high-quality ferroic thin films. We discuss the specific advantages of the CVD technique toward the growth of ferroic perovskite and layered materials.

Ferroelectrics and multiferroics are often comprised of elements with high volatility such as bismuth or lead. PLD, sputter deposition, and MBE growth techniques operate at relatively low oxygen partial pressures, where issues arising from migration of such volatile elements are intensified. Especially during epitaxial growth, the combination of high crystalline anisotropy and unidirectional growth fronts mean that structural rearrangement is limited to the surface layers of a growing film. As bismuth and lead diffuse out of the films, inadequate activation energies exist to remedy the losses within the film and the structures crystallographically shear to form microstructural defects and secondary impurity phases [41]. Microstructural defects can influence domain wall activity in ferroelectric and ferromagnetic thin films. If defects serve as nucleation sites for reverse domains, coercivity may be decreased. Alternatively, if defects contribute to "pinning effects," these can resist domain wall motion and may lead to increased coercivity [42]. Defects such as oxygen vacancies often lead to degradation of ferroelectric properties, including decreased piezoelectric coefficients and decreased remanent polarization, and can result in increased leakage currents and polarization fatigue [43]. Impurity phases such as non-ferroelectric pyrochlore $Pb_2Ti_2O_7$ and $Bi_2Ti_2O_7$ can also diminish ferroelectric properties [44]. Whereas the presence of ferrimagnetic spinel impurity phases (such as $Fe_3O_4$ and $CoFe_2O_4$) can complicate the interpretation of an intrinsic magnetic response [45, 46]. Formation of defects and impurity phases can be inhibited by increasing the oxygen partial pressure. However, for techniques such as PLD, sputtering, and MBE, it is often found that this can only be achieved using relatively narrow ranges of deposition conditions. Oxygen loss tends to occur during sputtering deposition of thin film oxides, leading to oxygen vacancy defects and associated ferroelectric fatigue. Furthermore, the use of high-energy particles during deposition can lead to physical damage and defects [47, 48]. Oxygen stoichiometry is a significant issue in MBE, since the ultra-high vacuum conditions (typically $10^{-10}$ to $10^{-5}$ Torr) required for MBE growth oppose an oxidizing atmosphere. Accordingly, either molecular or atomic oxygen sources are commonly necessary in MBE to provide sufficient oxygen for growth of thin film oxides [49]. Even so, incomplete metallic oxidation during MBE growth can lead to the formation of oxygen vacancies, n-type conductivity, and electrical leakage issues [50].

Conversely, CVD with its relatively high gas pressure (typically 0.6–70 mbar) and oxygen-rich environments (typically 100–2000 sccm; 5–70% $O_2$) compared with high vacuum techniques can prevent the re-evaporation of PbO or $Bi_2O_3$ over wider ranges of growth parameters [22]. Funakubo et al. [51] synthesized epitaxially grown PZT on MgO (100) substrates by CVD over wide deposition conditions (deposition temperature ranged from 600 to 700 °C, with oxygen partial pressures of between 64 Pa and 1.1 kPa and with total gas pressures of between 1.1 and 6.7 kPa). Whereas Pb/(Pb + Zr + Ti) deposition rates notably decreased below oxygen partial pressures of 64 Pa, no significant changes to deposition rates were observed above 64 Pa. The successful preparation of pyrochlore-free, stoichiometric PZT films at lower temperatures and over a wider temperature range compared with sputtering [52] and PLD [53] methods was attributed to the prevented revaporization of lead under the relatively high total gas and partial gas pressures enabled by this CVD method [51]. Bartasyte et al. [52] employed a pulsed liquid injection organic chemical vapor deposition (MOCVD) technique for the synthesis of $PbTiO_3$ on $SrTiO_3$ (001) and $LaO_3$ (001) substrates at 650 °C and explored the influence of the oxygen percentage (7%, 15%, 37.5%, and 50%) in the gas flow during deposition. A maximum Pb/(Pb + Ti) ratio and a maximum growth rate were achieved at 37.5% oxygen, due to the promoted formation of PbO compared with lower oxygen concentrations. However, when the oxygen concentration was too high (50%), lead deficient films were achieved. This was explained by the formation of high valency $Pb_3O_4$ and $PbO_2$ at higher oxygen partial pressures. By utilizing an increased partial pressure of 0.4 atm oxygen during cooling from high temperatures (800 °C), the lead desorption rate can be reduced. Thus, higher oxygen pressures during cooling stabilize the $PbTiO_3$ phase and enabled an increase in deposition temperature [52]. Dormans et al. [53] report on the organometallic CVD growth kinetics of $PbTiO_3$ on $SrTiO_3$ (001). A PbO partial pressure of at least $10^{-3}$ mTorr was necessary to form stoichiometric $PbTiO_3$, which was easily achieved by this CVD process. At temperatures of 700 °C and above, the rate of formation of $PbTiO_3$ from PbO and $TiO_2$ was faster than the rate of desorption of PbO. It was found that a relatively large process window of parameters can be used and stoichiometric $PbTiO_3$ was grown independent of the deposition temperature and ratio of the precursor partial pressures. The growth rate was almost independent of oxygen partial pressure; however, it was noted that the epitaxial quality of the grown $PbTiO_3$ is optimum at an oxygen partial pressure of 2.1 Torr. At higher oxygen partial pressure, it is believed that diffusion of adsorbed species is hindered by adsorbed oxygen, while at low oxygen partial pressures, the films deteriorate due to oxygen deficiency [53].

Furthermore, Micard et al. [54] explain that although good quality $BiFeO_3$ films on $IrO_2$/Si substrates in terms of structure and composition can be achieved using initial MOCVD parameters (temperatures of 650–750 °C, argon flow of 150 sccm, oxygen flow of 150 sccm, and duration of one hour), it was observed that inclusion of higher oxygen gas flow step had a marked influence on optimizing the film morphology and quality. The initial high oxygen step (900 sccm oxygen gas flow for 10 minutes) encouraged the formation of several $BiFeO_3$ sites, whereas the second growth step (150 sccm for 50 minutes) produced the growth of denser films compared to the previous conditions [54]. Similarly, Tohma et al. [55] report that morphology and grain size of $BaTiO_3$ films prepared on (100)Pt/(100)MgO substrates by MOCVD were influenced by the oxygen partial pressure used during growth, with a maximum grain size of 130 nm achieved at oxygen partial pressure of 66 Pa at deposition temperature of 973 K. The dielectric properties of the $BaTiO_3$ films were in turn influenced by the deposition conditions, with dielectric constant increasing with increasing grain size from 20 to 130 nm.

The ferroelectric and multiferroic properties of complex oxide materials are sensitively affected by their chemical composition. In addition, even slight alterations to composition can lead to abrupt changes to crystal structure and can result in phase transitions between tetragonal and rhombohedral ferroelectric phases across a morphotropic phase boundary [56]. CSD has the ability to vary stoichiometry in order to discover new phases and to investigate physical properties as a function of composition, using facile and relatively low-cost methods [30, 44, 45, 57]. ALD also enables excellent control over film composition. However, the comparatively low deposition temperatures employed by these techniques tend to produce amorphous films. A subsequent higher temperature post-anneal step is often required for the crystallization of complex oxide phases by CSD and ALD [58]. Progress in the design of CVD process technologies has enabled simple fine-tuning of stoichiometry and reproducibly produced compositions during the direct production of crystalline, epitaxial complex oxide ferroics [59].

Many of the metalorganic precursors available for the growth of ferroelectric and multiferroic oxides are relatively nonvolatile at room temperature, meaning that during conventional CVD, it is necessary to heat the precursors within their bubblers in order to sublime or evaporate the precursors. Not only does this accelerate chemical degradation of the precursors, but it is also often difficult to precisely control the stoichiometry of complex oxide materials during the conventional delivery of low vapor pressure precursors. In 1995, Van Buskirk et al. [60] describe how these issues spurred the use of liquid delivery MOCVD techniques, where multiple liquid metalorganic precursors, even those with low thermal stability and low vapor pressure, are kept at ambient temperature and are maintained under pressure [61]. This allows for the transport of several precursors that can be vaporized instantaneously. Film composition is precisely controlled with process reproducibility by real-time volumetric mixing of the component solutions [60]. Liquid delivery MOCVD studies where the Ba/Sr ratio (70/30) was kept constant, but the Ti flow varied, enabled stoichiometric adjustment of $BaSrTiO_3$ thin films on $Pt/ZrO_2/Si$ substrates. A linear relationship between film composition and precursor concentration was obtained with run-to-run repeatability. The need for stringent process control was illustrated for 90 nm films, where small variations in film composition (approx. 3 Ti at%) yielded a considerable difference (approx. 100%) in dielectric constant. Liquid source MOCVD delivery of high crystalline quality PZT films was also investigated in this work. A wide range of compositions were explored at a deposition temperature of 550 °C, and again a linear relationship between precursor composition and film composition was observed. Measured dielectric constants varied monotonically with composition, with the highest values close to the PZT morphotropic phase boundary [60]. Excellent control of stoichiometry with high compositional uniformity has also been reported for CVD-grown epitaxial $LiNbO_3$ films on $LaTiO_3$ substrates for SAW and optical device applications [59, 62].

The ability to include 5 nm thick $Bi_2O_3$ buffer layers during liquid delivery MOCVD of $SrBi_2Ta_2O_9$ on $Pt/Ti/SiO_2/Si$ substrates prevents the diffusion of Ti from the substrate and minimizes Bi migration away from the films. In addition to preventing secondary phases at the interface, the bottom $Bi_2O_3$ buffer layer also acts as a seeding layer, improving the crystallinity of the films $SrBi_2Ta_2O_9$. It was found that the remanent polarization ($2P_r$) of $SrBi_2Ta_2O_9$ samples with a buffer layer (22.5 $\mu Ccm^{-2}$) significantly improved compared with samples without buffer layers (14.7 $\mu Ccm^{-2}$) [63]. By increasing the thickness of the MOCVD-grown $BiO_x$ buffer

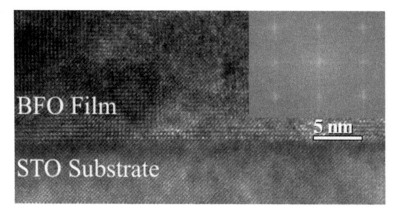

**Figure 1.4** A high-resolution transmission electron microscopy (HRTEM) bright field image showing the crystalline nature of the BFO film and a sharp interface with the substrate. The inset image shows the fast Fourier transform of the BFO film obtained from the HRTEM image. Source: Deepak et al. [65]/Reprinted with permission of American Chemical Society.

layer from 5 to 30 nm, it was possible to lower the MOCVD process temperature for 90 nm thick $Bi_4Ti_3O_{12}$ films (on Pt-coated Si substrates) from 550 to 400 °C [64].

Deepak et al. [65] used direct liquid injection chemical vapor deposition (DLI-CVD) to carefully control the concentration of the iron precursor and achieved bismuth self-limiting growth of epitaxial $BiFeO_3$ (BFO) films on $SrTiO_3$ substrates (Figure 1.4). Here, the precursor injection rates were an important parameter controlling stoichiometry. The growth window was defined in this work as the range of relative Bi-to-Fe precursor injection ratios, in which phase pure, stoichiometric BFO films can be prepared. CVD growth of BFO is generally challenged by the volatile nature of bismuth. The growth window is often very narrow, requiring strict control of bismuth volumes and with Bi/Fe precursor injection ratios of between 2.33 and 2.55 [66, 67]. Instead, Deepak et al. [65] exploited the volatile nature of bismuth oxide, to develop a bismuth self-limiting process at a growth temperature of 650 °C, pressure of 10 mbar, and oxygen flow of 1000 sccm (out of a total gas flow of 3000 sccm). Utilizing a significantly higher concentration of volatile Bi and the correct amount of the nonvolatile Fe component, an extension of the CVD growth window was enabled with Bi/Fe ratios of 1.33 to 1.81, without leading to the formation of impurity phases and without the need for stringent control of bismuth injection volumes during growth (Figure 1.4). At the decreased Bi/Fe ratio, the film surface becomes atomically smooth, with a root-mean-square (rms) roughness below 0.4 nm. Figure 1.5a,b demonstrates the amplitude strain butterfly loop and phase loops measured for ultrathin BFO films, respectively. A 180° phase switching can be observed in Figure 1.5b, which is an indication of the retention of ferroelectric switching behavior at sub-10 nm thickness and potential for application in nonvolatile memory storage applications.

Faraz et al. [30] developed DLI-CVD growth processes to synthesize thin films of the complex Aurivillius oxide, $Bi_6Ti_{2.99}Fe_{1.46}Mn_{0.68}O_{18}$. This material system is a rare example of a single-phase room temperature multiferroic to challenge

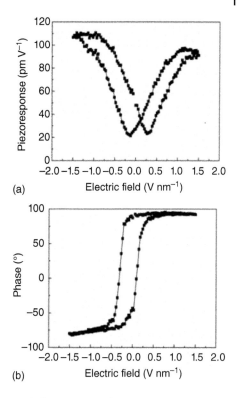

**Figure 1.5** Piezoresponse force microscopy (a) amplitude butterfly loop (piezoresponse) and (b) phase hysteresis loop for BiFeO$_3$ thin film deposited on Nb-doped SrTiO$_3$. Source: Reprinted with permission from Deepak et al. [65], Copyright 2015 American Chemical Society.

BFO. Stoichiometric control of manganese is key to ensuring room temperature ferromagnetic behavior [68], and higher magnetization values ($M_S$ 24.25 emu g$^{-1}$) (Figure 1.6) are achieved using DLI-CVD-grown films compared with CSD-grown films ($M_S$ 0.74 emu g$^{-1}$) [30].

Increasing levels of integration of ferroic thin films into devices requires the capacity to conformally coat miniature features having challenging aspect ratios at the submicron scales and ultimately at the nanoscale. A chief advantage of CVD over other thin film growth methods is the capability to deposit uniform and conformal thin films of high crystallinity over three-dimensional (3D) structures and within trenches using scalable growth processes. During reaction-rate-limited growth, the deposition rate can be described by the Arrhenius equation [69] with rate constant increasing exponentially with increase in process temperature. Due to the dominating temperature dependence, the deposition rate is only weakly dependent on the homogeneity of precursor flow; therefore, growth in the reaction-rate-limited regime is highly suited to conformal deposition [70].

Burgess et al. [71] demonstrated highly uniform ($3\sigma = 2.25\%$ at 180 nm thickness) planar films of MOCVD-deposited strontium bismuth tantalate and near 100% step-coverage of several patterned structures and recessed features with a 2 : 1 aspect ratio (0.5 µm). Excellent run-to-run reproducibility and conformality of approximately 90% for growth of SrBi$_2$Ta$_2$O$_9$ on Pt/Ti/SiO$_2$/Si on 1 : 1 aspect ratio structures of 0.5–0.6 µm size is also reported [72–74]. By optimizing MOCVD growth conditions, Goux et al. [75] improved the sidewall compositions of SrBi$_2$Ta$_2$O$_9$ 3D

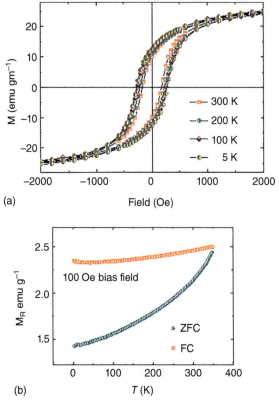

**Figure 1.6** (a) Magnetization (M) vs. magnetic field (Oe) of the $Bi_6Ti_{2.99}Fe_{1.46}Mn_{0.55}O_{18}$ sample. (b) Zero field cooled (ZFC) and field cooled (FC) measurements ($M_R$ vs. T). The saturated hysteresis loops and clear split between ZFC-FC curves demonstrate the ferromagnetic nature of the DLI-CVD-grown $Bi_6Ti_{2.99}Fe_{1.46}Mn_{0.55}O_{18}$ at room temperature. Source: With permission from Faraz et al. [30], Copyright 2017 Wiley.

ferroelectric capacitors and decreased the segregation of bismuth impurities and defects at the sidewalls. The ferroelectric contribution of the sidewalls to the overall 3D capacitor was found to be highly contingent on bismuth segregation at the sidewalls. Deposition at 440 °C followed a mass transport-limited regime, where bismuth segregation occurred at nucleation sites at the sidewall due to the increased mobility of bismuth at that temperature. Under these deposition conditions, the sidewall region did not contribute anything to the ferroelectric response and a value of 2.6 µC cm$^{-2}$ was measured for the remanent polarization ($P_r$) of the 3D capacitor. Optimized growth at 405 °C followed a reaction-rate-limited deposition regime, where bismuth segregation at the sidewalls was reduced. The ferroelectric and leakage properties were notably enhanced, with an increase in sidewall efficiency of up to 50% and $P_r$ for the 3D capacitor increasing to 7.5 µC cm$^{-2}$.

Close to 100% step-coverage can also be achieved for MOCVD-grown PZT ($PbZr_xTi_{1-x}O_3$) in 3D Ir/PZT/Ir stacks with 2/2 µm (width/distance) trench lines with 1 µm depth (aspect ratio = 0.5) [76]. The PZT films were found to be conformal on trench lines of these dimensions; however, the step-coverage decreased to 66% for narrower trench lines of width/distance/depth of 0.7/0.7/1 µm. Shin et al. [77] demonstrate how PZT (PbZr$_x$Ti$_{1-x}$O$_3$, where $x = 0.35$) thin films can be deposited conformally on 3D nanoscale trenches by liquid delivery MOCVD for high density Fe-RAM applications. Almost 100% step-coverage of Ir/PZT/Ir (20/60/20 nm) was obtained for trench capacitors of diameter 250 nm and height of 400 nm. A comparison of trench sizes having differing diameters indicated that the proportion of columnar-type grains increases with increasing trench width. It was observed that trench capacitors having a higher proportion of columnar PZT grains demonstrated superior ferroelectric properties. At 2.1 V, the remanent polarization ($2P_r$) for the 250 nm-diameter trench capacitors was 19 µC cm$^{-2}$, increasing to 24 µC cm$^{-2}$ for the 320 nm-diameter trench capacitors.

The benefits of conformal CVD growth have been extended to the fabrication of 3D composite magnetoelectric structures. Here, two different materials, one with ferroelectric order and one possessing magnetic order are combined, so when the order parameters couple, polarization can be manipulated by a magnetic field and magnetization can be influenced by an electric field, enabling low-power memories, spintronics, and sensing technologies. Reduced clamping effects from the substrate can be achieved for 3D composites compared with laminate composites, in principle allowing for higher magnetoelectric coupling response. Migita et al. [78] produced microplates of ferroelectric (Bi$_{3.25}$Nd$_{0.65}$Eu$_{0.10}$)Ti$_3$O$_{12}$ (BNEuT) ($a$-axis orientated) and deposited ferromagnetic CoFe$_2$O$_4$ (CFO) by MOCVD at 550 °C within the plate gaps with pitches of 5 µm to form composite micro-pillars. The deposition time was varied from 90 to 150 minutes, and a step-coverage of up to 66% was obtained, implying that the deposition progressed via the reaction-rate-limited regime. The CFO structures were single phase with polycrystalline grains, and a preferred orientation along (222) and (511) was reported to promote magnetic anisotropy. For a deposition time of 120 minutes, CFO/BNEuT($h$00) composite micro-pillars displayed $M_r$ (remanent magnetization) of 25.7 emu g$^{-1}$ and $H_c$ (magnetic coercivity) of 1.0 kOe. The effect of leakage current was also lowest for this deposition time and ferroelectric hysteresis loops displayed $P_r$ (remanent polarization) of 49 µC cm$^{-2}$ with $E_c$ (electric field) of 264 kV cm$^{-1}$.

Ultrathin ferroelectrics (typically thinner than 100 nm) are highly desirable due to their technological prospects in data storage applications [59, 79, 80]. For instance, in purely electronic memristors based on tunnel junctions of an ultrathin ferroelectric, the junction resistance sensitively depends on the relative fraction of ferroelectric domains having their polarization pointing toward one or the other electrode. Large on/off ratios between resistance states can be achieved. Furthermore, because ferroelectric switching involves domain nucleation and growth, artificial nano-synapses can be created where synaptic strengths evolve depending on the switching pulse amplitude and duration. This multivalued logic will allow data processing using not only "yes" and "no" but also "either yes or no" or "maybe" operations and thus will serve as unique memory elements in the quest for neuromorphic computing systems [81–83]. A challenge to commercialization is the formation of polarization induced surface charges induced by the out-of-plane polarization which can produce a depolarization field of magnitude inversely proportional to the dimensions of the ferroelectric [84]. This means that the production of high-quality films, with stable polarization, is essential for effective ferroelectric device performance and reproducibility when scaling down to miniaturized dimensions.

When MOCVD processes follow Frank–Van der Merwe-type growth, in principle it is possible to deposit films in atomic, layer-by-layer deposition modes to achieve films of atomic layer thickness [85]. This is exemplified by $BaTiO_3/SrTiO_3$ superlattices grown epitaxially on $LaAlO_3$ (012) substrates having individual layer thicknesses of 2 nm each (approximately five lattice constants) [86].

Nonumura et al. [87] discuss that for MOCVD growth of ultrathin epitaxial $Pb(ZrTiO)_3$ (PZT) films, the growth mode, film surface flatness, and ferroelectric characteristics are substantially affected by the top surface plane and surface flatness of the underlying substrate. When $SrTiO_3$ substrates were surface treated to a buffered acid etch and anneal, this enabled deposition of $SrRuO_3$ bottom electrodes with uniform terrace ledges. This in turn enabled MOCVD of highly crystalline 20 and 15 nm PZT films by the Stranski–Krastanov growth mode, having sharp and clear substrate–film interfaces and with terrace ledges of similar height to the $SrRuO_3$ terrace ledges. PZT films of 20 nm thick having terrace ledges displayed saturated ferroelectric hysteresis loops (remanent polarizations of 29–33 $\mu C\,cm^{-2}$ and coercive fields of 340–370 $kV\,cm^{-1}$). In contrast, 20 nm thick PZT films without terrace ledges exhibited unsaturated, leaky loops. Ferroelectricity was also observed for the 15 nm PZT films, although large leakage currents at this thickness prevented saturated hysteresis loops.

Fujisawa et al. [88] also used buffered-acid etches and pre-anneals of substrates before MOCVD of epitaxial $PbTiO_3$ films on $SrTiO_3$(100), La-doped $SrTiO_3$(100), and $SrRuO_3/SrTiO_3$(100). High angle annular dark field–scanning transmission electron microscopy (HAADF-STEM) determined an epitaxial relationship and sharp interface between the film and substrate layers and that $PbTiO_3$ films as thin as 1–20 monolayers (0.4–8 nm) were achieved. Contact-resonance piezoresponse force microscopy (CR-PFM) was used to demonstrate piezoelectric response and ferroelectric polarization switching in seven monolayer thick (2.7 nm) $PbTiO_3$ on $SrRuO_3/SrTiO_3$(100).

It has been emphasized [85] that a critical requirement for future data-storage technologies based on multiferroic materials is the demonstration of room temperature magnetoelectric coupling at the sub-10 nm length scale. Although this is recognized within the multiferroics community as a challenging task, Keeney et al. [89] have made progress in the optimization of multiferroic $Bi_6Ti_xFe_yMn_zO_{18}$ (B6TFMO) toward achieving this goal. Previous films grown on sapphire substrates were approximately 100 nm in thickness and required a post-growth anneal at 850 °C [29, 30]. Optimized, single-step DLI-CVD studies enabled a lowering of the crystallization temperature to 700 °C and epitaxial growth on lattice-matched substrates ($NdGaO_3$ (001), $La_{0.26}Sr_{0.76}Al_{0.61}Ta_{0.37}O_3$ and $SrTiO_3$ (100)). Epitaxial films of thickness 7 and 5 nm were achieved, which equates to 1.5 unit cell and 1 unit cell of the Aurivillius structure, respectively (Figure 1.7). Since there is an enhancement in the diffusion coefficient of bismuth above 710 °C [90], a

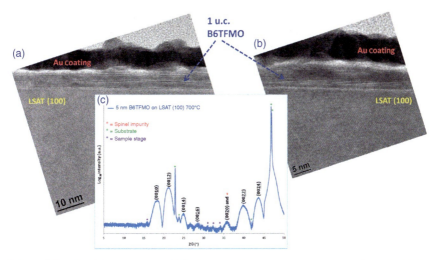

**Figure 1.7** (a), (b) Representative TEM images demonstrating continuous 5 nm Aurivillius phase films prepared by the single-step DLI-CVD process (700 °C) on LSAT (100) substrates. (c) Representative XRD pattern demonstrating that the five-layered B6TFMO Aurivillius phase is successfully achieved at 5 nm thickness (1 unit cell (u.c.) thick B6TFMO). This single-step growth process on epitaxial substrates enabled the synthesis temperature of B6TFMO to be lowered by 150 °C. Source: Keeney et al. [89]/Reprinted with permission of American Chemical Society.

lowering of the crystallization temperature by 150 °C decreased the rate of bismuth diffusion and reduced the issues associated with uncompensated stoichiometries and related structural disorder. Film quality, sample purity, and surface roughness (rms = 1.2 nm) were considerably enhanced compared with samples annealed at 850 °C; however a volume fraction of 3–4 vol% spinel magnetic impurities were identified on the film surface, which precluded measurements of magnetic properties intrinsic to the B6TFMO Aurivillius phase. Attention was focused on local ferroelectric measurements of B6TFMO by piezoresponse force microscopy (PFM) where ferroelectricity was confirmed to persist in 5 nm (1 unit cell thick) B6TFMO films, with piezoresponse increasing with increased tensile epitaxial strain. In-plane ferroelectric switching was demonstrated at 1.5 unit cell thickness (Figure 1.8). This work demonstrated that 2D ultrathin B6TFMO films can be synthesized by scalable growth methods, increasing its technological potential for utilization in novel in-plane devices enabling ferroelectric control of tunnel electron resistance. Such in-plane tunnel junction devices would have the significant advantage that performance would not be encumbered by competing depolarization fields upon scaling down to sub-10 nm dimensions.

The layered Aurivillius structures are a technologically important class of materials, demonstrating room temperature ferroelectric and multiferroic properties. Although the layering between bismuth oxide and perovskite blocks occurs naturally during most conventional growth processes, the use of a layer-by-layer growth processes offers the potential for increased control over the layering arrangements and cation site ordering, at temperatures lower than conventional growth processes.

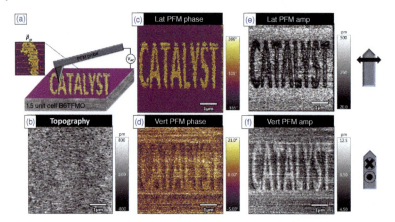

**Figure 1.8** PFM lithography studies of 7 nm BTFMO on STO (100) prepared by the one-step process (700 °C). Ferroelectric poling was performed at room temperature with application of +5 $V_{DC}$ (DC bias) to the background and by performing PFM lithography using −5 $V_{DC}$ to write "Catalyst." (a) The "read" step was performed with a probing signal ($V_{AC}$) of 3.0 V. The red and blue arrows within a close-up of a switched area represent the direction of the in-plane polarization ($P_{IP}$). (b) Unchanged sample topography after the ferroelectric poling step. (c) Lateral (Lat) PFM phase, (e) lateral PFM amplitude (amp), (d) vertical (vert) PFM phase, and (f) vertical PFM amplitude images. The direction of motion of the PFM cantilever as it scans the sample surface is indicated to the right of the figure. These studies demonstrate the stable in-plane ferroelectric switching behavior for 1.5 u.c. thick B6TFMO. Source: Keeney et al. [89]/Reprinted with permission of American Chemical Society.

Layer-by-layer deposition by ALD has proved highly successful for the growth of ferroelectric $Hf_{0.8}Zr_{0.2}O_2$ films on Si at temperatures as low as 250 °C [36, 80]. However, an extra post-anneal step is required to crystallize Aurivillius films deposited by ALD [58], whereupon the additional heat energy can alter the rearrangement of the deposited layers. Alternatively, layer-by-layer deposition using DLI-CVD enables the possible formation of multilayers and graded composition layers in addition to complex-layered oxide materials [22].

Deepak et al. [91] used a single-step, layer-by-layer, or sequenced DLI-CVD process to synthesize epitaxial ferroelectric $Bi_4Ti_3O_{12}$ films on $SrTiO_3$ (001) substrates with greater control over processing conditions and at temperatures as low as 650 °C. As opposed to simultaneous pulsing of the two liquid precursors, this layer-by-layer growth process was made possible by alternating the number of pulses of bismuth and titanium to mimic the layered Aurivillius phase structure long the c-axis. Bismuth and titanium concentrations were controlled by regulating the opening time of computer-operated injectors. The time gap between one set of pulses was kept constant at five seconds, with the total set signifying one loop, corresponding to one half of a $Bi_4Ti_3O_{12}$ unit cell. This loop was cycled 20 times to achieve films of approximately 33 nm thickness, which is equivalent to the thickness of 10 unit cells of $Bi_4Ti_3O_{12}$. The films were relatively smooth, with rms roughness of 1.4 nm. The effect of varied bismuth and titanium concentrations on $Bi_4Ti_3O_{12}$ growth was investigated, and it was found that departures from stoichiometry affected film crystallinity and encouraged the presence of out-of-phase boundary

(OPB) defects. Interestingly, it was determined that the OPB defects did not form due to reasons of bismuth migration and their presence was associated with atomic rearrangements that gave rise to an increased crystalline quality of the films [91].

This sequential, layer-by-layer growth method was extended to include iron within the Aurivillius structure, to expand its applicability toward multiferroic materials [92]. The preciseness of the liquid precursor injection system allowed insertion of monolayers of titanium oxide and iron oxide into the Aurivillius structure to investigate the growth mechanisms for $Bi_4Ti_3O_{12}$ ($m = 3$) and $Bi_5Ti_3FeO_{15}$ ($m = 4$). Structural changes as a function of miniscule variations in precursor concentrations were investigated, and it was observed that the growth mode shifts from Stranski–Krastanov growth mode to layer-by-layer (Frank–Van der Merwe) growth mode as the composition changes from $Bi_4Ti_3O_{12}$ to $Bi_5Ti_3FeO_{15}$. When the concentration of iron precursor was lower (Ti:Fe ratio of 3 : 0.3) than that required to insert a full half unit cell ($m = 3.5$) or full unit cell ($m = 4$) into the Aurivillius structure uniformly over the entire film, the only way to accommodate the extra iron and to maintain charge balance between differently charged $Fe^{3+}$ and $Ti^{4+}$ was to accommodate small regions of the four-layered Aurivillius structure, accompanied by OPB defects. As the iron concentration was increased (e.g. Ti:Fe ratio of 3 : 0.57), there was an increase in the random stacking of $m = 4$ layers between $m = 3$ layers and an increase in $c$-axis lattice parameter (Figure 1.9) until the pure, smooth (rms roughness < 0.5 nm) epitaxial $m = 4$ composition was achieved at a Ti:Fe ratio of 3 : 1.2. A precise control over structural order was demonstrated, and no bismuth oxide impurity phases were observed in this bismuth self-limiting growth method [92].

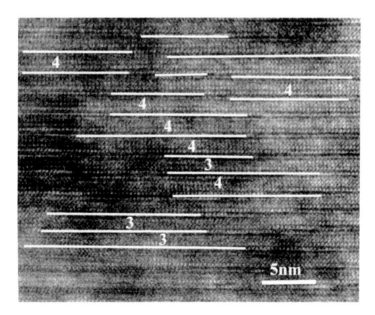

**Figure 1.9** The high resolution TEM image of the Aurivillius phase film with Ti/Fe ratio 3/0.57. Source: Deepak et al. [92]/Royal Society of Chemistry/CC BY 3.0.

The preceding sections describe how CVD is flexible and versatile, yet simple and robust and is highly compatible for the production of complex oxide ferroic thin films of technological relevance, with quality approaching that of compound semiconductor films. Specific advantages of the CVD technique toward the growth of ferroic perovskite and layered materials include relatively high gas pressures, stoichiometric control, conformal growth, and ability to produce ultrathin films and layer-by-layer growth processes. In addition, the well-understood tool design with comparatively low cost of ownership and a shower-head design allowing scalability to large wafer sizes (>300 mm) makes CVD particularly amenable to industrial processes [40].

Competitive wafer throughputs (typical deposition rates of 30 Å in 20 seconds) [93] can be obtained without compromising stoichiometry and uniformity control. Furthermore, thermal plasma spray CVD, with its relatively high deposition rate of 0.1 µm min$^{-1}$ and large area deposition (>5 in. diameter), has shown to be an effective method for the deposition of epitaxial $LiNbO_3$ films for optoelectronic device applications, typically requiring thickness of 1–100 µm [94].

Development of more cost-effective precursors is continuing and Moniz et al, [95] created a comparatively low-cost, solution-based aerosol-assisted CVD process for $BiFeO_3$ thin films using a relatively nonvolatile single-source precursor, $[CpFe(CO)_2BiCl_2]$ (Cp = cyclopentadienyl, $C_5H_5$) dissolved in tetrahydrofuran. Deposition was enabled at temperatures as low as 300 °C; however a post-deposition anneal treatment at 700 °C was required to crystallize phase pure $BiFeO_3$ films.

As miniaturization of electronic devices continues, the need for a high degree of film thickness conformality over complex device topographies will become even more important. As outlined in a recent review by Abelson and Girolami [96], various strategies exist to improve the conformality of CVD processes, including the use of growth inhibitors, the use of two co-reactants to allow a kinetic regime in which the growth is inherently super-conformal, and the use of forward-directed flux depositions.

### 1.4.2 Case Study 2: Dielectric Oxide Materials

A passive dielectric material is an insulator in which, under an applied electric field, the charges associated with the atoms are polarized and relax when the field is removed. Although such dielectric metal oxides have been used in a wide range of applications, notably stable high efficiency energy storage devices [97–100], here we focus on the requirements for high-quality dielectric material as driven by advances in the miniaturization of integrated circuits and the subsequent need to replace $SiO_2$ as the gate oxide in the MOSFET. For 30 years $SiO_2$ was considered an ideal dielectric for this role, easy to fabricate through simple thermal processing, inert, surface pacifying, and exhibiting very few electronic defects. However, the continual scaling of device dimensions reaches a limit where the tunnelling current through the $SiO_2$ is too great for effective device operation. Equation (1.1) describes the capacitive behavior of a dielectric material of dimensions $A$ and $t$, where $\varepsilon_0$ is the permittivity of free space, $k$ is the relative permittivity (dielectric constant), $A$ is the device area, and t the dielectric thickness:

$$C = \varepsilon_0 kA/t \tag{1.1}$$

From Eq. (1.1) it can be seen that to maintain the capacitance and to prevent tunnelling in a shrinking device, the dielectric material thickness and permittivity both need to increase. For comparative purposes, thickness of a high k is often referred to as the effective oxide thickness (EOT) that is equivalent to $SiO_2$ using Eq. (1.2):

$$\text{EOT} = \left(\frac{k_{SiO_2}}{k_{High-k}}\right) t_{High-k} \tag{1.2}$$

As the $k$ value of $SiO_2$ is ~3.7, there are a whole raft of materials that could be potential replacements. However, requirements for this material system are stringent; the material would need a large band offset to silicon, generate a good electrical interface, must be stable at processing temperatures, and exhibit few bulk defects, and, furthermore, to be commercially viable, it would also need to enable several generations of scaling. Many candidate systems were identified: $Al_2O_3$, $HfO_2$, $ZrO_2$, $TiO_2$, $Ta_2O_5$, $La_2O_3$, and several multicomponent oxides. The band gaps of the common alternative oxides and their $k$ values are presented in Table 1.1.

Although possessing only a moderate $k$ value, $Al_2O_3$ has been developed as a short-term solution to scaling as it has an amorphous nature, even under robust processing conditions, and has excellent thermodynamic stability at the silicon interface. Aluminum trihalides were widely used in both hydrolysis and oxidation reactions for the CVD of $Al_2O_3$ [14, 104, 105]; however the high growth temperatures (700–900 °C) are not ideal for microelectronics applications as at these growth temperatures Al will diffuse into the semiconductor material. Lower temperature processes were achieved with several alternative precursors, notably aluminum alkoxides and aluminum alkyls [106]. Although high-quality films were achieved with both precursor groups, the high stability and reactivity of trimethylaluminum (TMA) made it an ideal precursor, enabling high-quality CVD at 400–700 °C and ALD at <250 °C [14, 107–109], well within the thermal restrictions of CMOS processing. Although the pyrophoric nature of TMA is of concern for safety, a viable

**Table 1.1** Band gaps and $k$ values of alternative dielectric oxides.

| Composition | k value | Band gap | References |
|---|---|---|---|
| $Al_2O_3$ | 9 | 8.8 | [101] |
| $HfO_2$ | 25 | 5.8 | [101] |
| $ZrO_2$ | 29 | 5.8 | [101] |
| $TiO_2$ | 25–85 | 3.2 | [101] |
| $HfSiO_4$ | 15–17 | 6.5 | [102] |
| $La_2O_3$ | 30 | 6 | [103] |
| $Ta_2O_5$ | 26 | 4.4 | [103] |

alternative source has not yet been fully realized, and hence it is still the go-to source for the CVD and ALD of aluminum containing thin films [108].

Although there are a wide range of MOCVD and ALD processes for the growth of $La_2O_3$ films, their chemistry is not simple and yields films of variable quality [11]. More significantly $La_2O_3$ is not widely employed in logic devices as it tends to promote large negative flat band shifts in the capacitance voltage responses of MOS devices, hence affecting their operating voltage [109]; also, the moisture-sensitive nature renders $La_2O_3$ unsuitable for very large-scale integration (VLSI) processes as these involve frequent exposure to atmosphere.

The high thermal stability and dielectric constant of $Ta_2O_5$ coupled with its low leakage and high breakdown voltage have made it a subject of extensive research particularly in memory and off-chip capacitors devices [110]. More recently $Ta_2O_5$ has emerged as a leading candidate in memristor technology where ALD films as thin as 1–2 nm have exhibited resistive switching properties [111]. Controlling the defective nature and stoichiometry of tantalum oxides in conjunction with careful selection of electrode materials was found to strongly influence the performance of the resistive switching memory devices [112]. Furthermore, the introduction of composite oxide materials such as $ZrO_2$–$Ta_2O_5$ have also renewed interest [113, 114].

As dielectric materials, $ZrO_2$ and $HfO_2$ were deemed excellent candidates, due to their high dielectric constants and near-ideal band gaps and offsets to silicon [103], the oxides of Zr and Hf have three structural phases – monoclinic (space group $P2_1/c$), tetragonal ($P4_2/nmc$), and cubic ($Fm3m$) – with either amorphous or monoclinic materials being the dominant as-grown material phase [115]. The CVD and ALD of $ZrO_2$ were widely reported; MOCVD using the chlorides $ZrCl_4$ [116] and $HfCl_4$ [117] for $ZrO_2$ and $HfO_2$, respectively, was found to be complex, due to the low volatility of the sources and need for growth temperatures in excess of 800 °C. However, for ALD processes the high decomposition temperature of halide sources is an advantage as surface-driven reactions occur at much lower temperatures leading to lower particle formation and excellent self-limiting chemistry [118, 119]. For MOCVD a range of β-diketonates compounds were demonstrated. The β-diketonate pentane-2,4-dionate (acac) was found to produce films with very high levels of carbon contamination for both [Zr(acac)$_4$] and [Hf(acac)$_4$]. However, other β-diketonates were found to deposit high-quality films, [Zr(thd)$_4$], where thd = 2,2,6,6-tetramethylheptane-3,5-dionate; despite its low volatility and high deposition temperature (>600 °C), was one such source. The issue of low volatility of β-diketonates was addressed by differing methods; the substitution of F for H in the complexes improved volatility, but the presence of F was of concern to device manufacturers [120]. An alternative approach was to change the precursor delivery system to liquid injection using a volatile but unreactive solvent; by this method 2,7,7-trimethyl-3,5-octanedionate (tod) was employed in both [Hf(tod)$_4$] and [Zr(tod)$_4$] to grow at lower temperatures and higher growth rates than the thd β-diketonate [121]. The use of alkoxides in MOCVD has been widespread despite complications due to the formation of dimers and polymers reducing the volatility of the sources [122]. The monomeric *tert*-butoxide [(OBu$^t$)] group has been used in both $ZrO_2$ [Zr(OBu$^t$)$_4$] and $HfO_2$ [Hf(OBu$^t$)$_4$] deposition by MOCVD [123, 124]. In the case of ALD, despite high-quality films, the process was found to be not self-limiting, due to a surface β-elimination decomposition pathway being open to the [(OBu$^t$)] group [125]; consequently, it was not deemed suitable for device applications where uniformity of growth at high aspect ratios is often required. To suppress the oligomerization of alkoxides, bidentate species were added to create heteroleptic complexes that often combined the favorable properties of the differing ligands. For example, the addition of β-diketonates metal alkoxides to form a series of compounds [Zr(OBu$^t$)$_{4-x}$(acac)$_x$], where $x = 1-3$, greatly improved MOCVD process control and deposition rate [126]. The addition of a range of neutral donor functions into alkoxides produced significant advances in their viability as sources for the MOCVD of HfO$_2$ and ZrO$_2$, of note was the ligand mmp (1-methoxy-2-methyl-2-propanolate), which when used independently or in conjunction with the *tert*-butoxide group, for example, [Zr(mmp)$_4$] and [Zr(OBu$^t$)$_2$(mmp)$_2$], in liquid injection MOCVD produced capacitive structures for both metal oxides on silicon that demonstrated accumulation, depletion, and inversion (at 100 kHz). Furthermore, trapped charge and flat band shifts in the CV response were relatively small [127]. The use of the neutral donor functions in alkoxide precursors for ZrO$_2$ and HfO$_2$ ALD also produced good quality material; however, the absence of self-limiting growth was still evident [128–130].

Metal alkylamides are well-established Hf and Zr precursors for MOCVD with both [Zr(NEt$_2$)$_4$] and [Hf(NEt$_2$)$_4$] generating stoichiometric films with low levels of impurities [131, 132]. For these alkylamides, it is ALD where they have shown the greatest impact, not only maintaining the low impurity levels but also demonstrating self-limiting growth across a significant temperature window, with low impurity levels detected in the grown films [133–137]. For Hf and Zr the dimethyl, diethyl, and ethyl methyl amines have been successfully employed with ALD windows of 100–300 °C and growth rates of ~0.1 nm per cycle. Recently the dimethylamine-based precursor has become dominant, due to a higher vapor pressure and ease of carrier transport. It should be noted that many other sources have been tried in both ALD and CVD, including cyclopentadienyls [138, 139] and hydroxylamides [140]; however, despite showing promise under specific conditions, these sources have not been widely adopted, commercially or in the research community, due to either process complexities or a short fall in materials properties.

Although similar to ZrO$_2$ in many aspects, HfO$_2$ became the dominant choice as an insulating dielectric in microelectronics due to its enhanced stability at the silicon interface, leading to little or no formation of interfacial silicates [141] as opposed to ZrO$_2$ where distinct amorphous zirconium silicates were generated [142]. The greater thermodynamic stability of HfO$_2$ with Si under processing conditions, such as the 1000 °C anneal required to activate source and drain dopants, is vital as the formation of a lower $k$ interface between the semiconductor and dielectric can seriously impinge on the overall EOT of the dielectric, in addition to creating charge traps in the semiconductor band gap [103]. Significantly, CVD HfO$_2$-based high-$k$ dielectrics were introduced into the gate stack of MOSFETs at the 45 nm technology node [143].

As scaling progressed further, the search for a semiconductor replacement for silicon intensified. III/V materials had long been considered potential alternatives to silicon-based transistors due to their enhanced electron transport properties; however progress was slow relative to the successful scaling of silicon-based devices with early research moving away from vapor phase processes. The first III/V device was fabricated in 1967 based on $SiO_2$–GaAs [144]; however the limits of low $k$ materials remained. This development of alternative semiconductor materials refocused research in dielectric materials. However, in contrast to the rapid development of $Si/SiO_2$-based MOSFETs, based on simply scaling transistor design, considerably less progress has been made in the performance of the III/V transistor. Although the early gains were obtained largely with physical vapor deposition methods, high performance inversion mode III/V MOSFETs with InGaAs channel were eventually demonstrated using the vapor phase method of ALD, using high-$k$ dielectrics such as $Al_2O_3$ [145], $HfO_2$ [146], HfAlO [147], and $ZrO_2$ [148]. These advances were significant as CVD methods, despite their difficulties are a preferred commercial technology due to their demonstrable scalability. However, with CVD processes, issues with native oxides are exacerbated by the higher-pressure regime employed, as III/V surfaces readily react with atmospheric gases to form native oxides. Surface layer oxides for the $In_{0.53}Ga_{0.47}As$ system generally exist in stable oxidation states of +5, +3, and +1 and small amounts of metallic As–As [149] with stable $In_{0.53}Ga_{0.47}As$ oxides thought to include $As_2O_3$, $As_2O_5$, $Ga_2O_3$, $Ga_2O$, $In_2O_3$, $GaAsO_4$, and $InAsO_4$ [150]. Arsenic oxides have been determined to be the least stable oxides in the system, while Ga oxides the most stable [151]. GaAs systems have similar native oxides present, lacking the contribution from indium compounds and dimers, mainly consisting of $As_2O_3$, $As_2O_5$, $Ga_2O_3$, and $Ga_2O$ in addition to a small amount of elemental As [152]. The native oxide of InP has largely been shown to consist of $InPO_4$ and $In_2O_3$ with In, P, $P_2O_3$, and $P_2O_5$ [153, 154]. Antimony-based III/Vs are increasing being considered for electronics applications [155]; the stability of antimony oxides that could be formed at the interfaces has been determined by density functional theory [156].

The poor electrical quality of these mixed native oxides leads to poor devices with sub-optimal performance and high levels of device failure. Hence, to utilize vapor phase growth methodologies, these oxides need to be removed or improved in quality [157]. A wide number of studies have been performed using an array of techniques for both binary and tertiary III/V materials. These include chemical pretreatments with HF [156], $NH_4OH$ [158], or sulfides [159–165], thermal treatment methods [166–168], capping layers [169], plasma treatment [170], and combinations of these treatments [171]. Although not all successful, significant progress was made through the use of these surface treatment methods in conjunction with vapor phase deposition techniques, notably surface passivation by the use of sulfides. The use of sulfide treatments has taken several approaches; of these wet chemical methods are the most commonly employed, and of these ammonium sulfide solution, $(NH_4)_2S$, the most widespread, where a range of concentrations, exposure times, and temperatures, has been explored. For example, Peralagu et al. [165] examined $GaSb/Al_2O_3$ MOS structures applying an aqueous $(NH_4)_2S$ pretreatment prior to ALD growth of 8 nm of $Al_2O_3$. The growth process

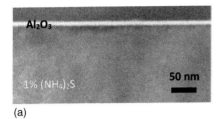 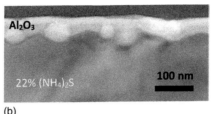

**Figure 1.10** Cross-sectional TEM micrographs of (a) 1% and (b) 22% $(NH_4)_2S$ on GaSb showing that the pre-growth etch characteristics have a significant influence on the ALD growth of oxides on II/V materials. The increased surface roughness and resulting interfacial layer are clearly evident. Source: Peralagu et al. [165]/with permission of AIP Publishing LLC.

was performed at 300 °C with TMA and water with a sub-four-minute transfer time between solution processing and reactor sample loading. Peralagu et al. studied devices with the interfaces treated with 1%, 5%, 10%, and 22% $(NH_4)_2S$ solution concentrations at 295 K. Capacitance–voltage (CV) measurements of the MOS devices obtained with the 1% solution exhibited the smallest stretch-out and flat band voltage shifts, and they also demonstrated the largest capacitance swing. Transmission electron microscopy revealed the formation of interfacial layers and increased roughness at the $Al_2O_3/p$-GaSb interface of samples treated with the higher $(NH_4)_2S$ concentration solution (22%), Figure 1.10. It should be noted that such an observation highlights that $(NH_4)_2S$, in addition to passivation, significantly etches III/V materials with the rate of etch being dependent on the III/V material in question; for example, a 10% solution of $(NH_4)_2S$ on GaSb significantly degrades the semiconductor and the interface with the dielectric, whereas for $In_{0.53}Ga_{0.47}As$ it is widely believed to exhibit the best interfacial properties after such an etch. O'Connor et al. demonstrated through multifrequency CV analysis a minority carrier response in both n-type and p-type $Au/Ni/Al_2O_3/In_{0.53}Ga_{0.47}As$ devices after such a treatment [171].

Although improvements in performance due to sulfur passivation was clear from many studies, the efficacy of processes were seen to be highly variable. Hurley et al. [172] observed that although the conditions of the passivation by $(NH_4)_2S$ were important, the key determinant of electrical performance was the length of exposure to atmosphere, between the wet chemical etching and the isolation from atmosphere in the ALD growth chamber. As illustrated in Figure 1.11, Hurley et al., employing multifrequency CV measurements on identically grown $InGaAs/Al_2O_3$ MOS structures, found that inversion was only observed if the air exposure was three minutes or less [172].

A potential solution to the short-lived passivation by sulfides was investigated by O'Connor et al., employing passivation of $In_{0.53}Ga_{0.47}As$ with $H_2S$ in situ to a metal–organic vapor phase epitaxy (MOVPE) reactor prior to transfer to an ALD reactor. $H_2S$ was found to passivate the surface in a similar manner to $(NH_4)_2S$, improving performance and suppressing the formation of arsenic oxides; however due to a short air-break between reactors, true inversion was not observed [160].

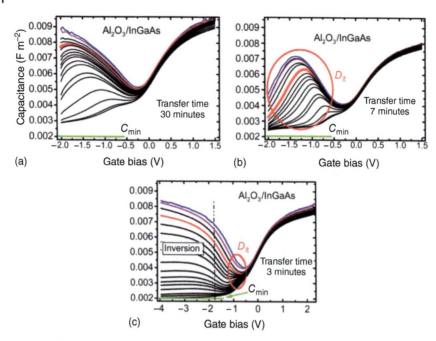

**Figure 1.11** Room temperature multifrequency C–V responses for $Al_2O_3$ (8 nm)/n – $In_{0.53}Ga_{0.47}As$/InP MOS structures with Au/Ni gates. The $In_{0.53}Ga_{0.47}As$ surface experienced a 10% $(NH_4)_2S$ surface preparation. Results are shown for varying transfer times from the 10% $(NH_4)_2S$ solution to the ALD reactor of (a) 30 minutes, (b) 7 minutes, and (c) 3 minutes. The samples illustrate the effect of a progressive reduction of $D_{it}$ and the emergence of surface inversion for the case of the 3-minute transfer time in Figure 1.9(c). Source: Hurley et al. [172]/with permission of IEEE.

Although surface treatments made significant improvements to the high k semiconductor interface, it was the growth of high-quality interface materials by ALD, often combined with surface treatments, that have made the biggest advance in electronic properties. The use of ALD was itself determined to remove surface oxides through a self-cleaning mechanism during the deposition process. This self-cleaning mechanism was influenced by the precursor and dielectric choice; consequently, the best performance characteristics of the logic devices was not determined by the dielectric's k value but the interface quality [157]. The first report of this self-cleaning by ALD was by Ye et al. [173], where a thinning of the native oxides of GaAs to 0.6 nm was observed after the ALD of $Al_2O_3$, with a corresponding reduction in the density of interface defect states ($D_{it}$) to ~$10^{12}$ cm$^{-2}$eV$^{-1}$. Frank et al. [174] also demonstrated the reduction of GaAs native oxides from 2.5 to 1 nm upon deposition of 4 nm of $Al_2O_3$. The nature of these processes for both $Al_2O_3$ and $HfO_2$ ALD were further illustrated by Brennan et al. [135], Lee et al. [175], and Milojevic et al. [176] who all reported that the initial metal precursor pulse, on GaAs and InGaAs substrates, is responsible for the majority of arsenic oxide removal with subsequent metal precursor pulses being required to minimize the presence of gallium oxides. More detailed X-ray photoelectron spectroscopy (XPS) investigations by Hinkle et al. confirmed that the metal precursor pulse removes native oxides with no reoxidation of the semiconductor surface with the subsequent oxygen containing precursor pulses. It was also revealed from this XPS study that an initial pulse of trimethylaluminum provided a reduction in the +3 surface states, whereas the TDMAH precursor provided a reduction in +5 native oxides present. This observation led to the postulation that cleanup reactions are precursor oxidation state dependent [177].

The mechanistic details for these self-cleaning reactions have been modeled; for $Al_2O_3$, Klejna et al. [178] postulated that a ligand exchange process was dominant, whereby for trimethyl aluminum, the $Al^{3+}$ preferentially replaces arsenic and gallium oxides in the 3+ oxidation state, resulting in $AlO_x$ formation and $As(CH_3)_3$ and $Ga(CH_3)_3$ as volatile reaction products. Likewise, ligand exchange was proposed to explain the removal of As—O bonds with the accompanying production of volatile elemental arsenic ($As_4$). More recently investigations of precursor dosing suggest that self-cleaning and interface modification occurs during the first two cycles of $Al_2O_3$ deposition with bulk film growth dominating after this [179]. Similarly, Suzuki et al. indicated that an interface control layer of 0.2 nm thickness may provide the same benefit as thicker depositions when seeking to improve the dielectric–semiconductor interface [180].

The self-cleaning mechanism for $HfO_2$ precursors is less well understood, the +4 oxidation of the Hf precursor provides no energetically favorable direct ligand exchange with either the +3 or +5 native oxides, implying that the prevalent cleanup reaction must take place by another mechanism. Furthermore, there are several conflicting reports of the efficacy of the process with tetrakis(dimethylamido)hafnium (TDMAHf); Shahrjerdi et al. [180] reported that no self-cleaning takes places on GaAs at 200 °C. Later Hackley et al. [181] reported interfacial oxide cleanup taking place at 275 °C, and Suri et al. [182] reported that TDMAHf requires a deposition temperature of at least 300 °C to achieve an interface almost completely free from oxides. It was suggested that this was due to the partial decomposition of TDMAHf, whereby Hf loses one out of four Hf–N linkages at 275 °C and two Hf–N linkages at 355 °C, hence cleanup reactions are facilitated by reactive intermediate species [183]. More recently, McDonnell et al. [184] reported the interface cleaning with TDMAHf at 200 °C, and after a $NH_4OH$ pretreatment etch, they attributed the differences in their observations to those of previous measurements to reoxidation of samples due the ex situ nature of the prior studies. Despite the self-cleaning mechanisms of ALD not being fully understood, it has emerged as a widely used method for device optimization, coupled with sub-nanometer thickness control of ALD and the development of interface control layers where materials having not particularly high $k$, but with excellent stability and low interface defects, have paired with higher $k$ materials to give a sufficiently high EOT values. For example, O'Mahony et al. employed thin films of either MgO and $Al_2O_3$ as interface control layers between the semiconductor and high $k$ material, to counter poor interfacial properties of $HfO_2$ and $ZrO_2$ with $In_{0.53}Ga_{0.47}As$ while maintaining a sufficiently high EOT value [185].

The drive for further scaling of complementary metal oxide semiconductor (CMOS), both in terms of physical dimension and power consumption, has created interest in replacing the channel materials with graphene and other 2D materials, for which it has been suggested that nanometer scale layers can still provide the carrier transport requirements for logic devices [186]. Current technology does not permit an all-2D material solution for device fabrication; thus the ALD of metal oxides appears to be an ideal method to enable integration of these materials; the low temperature and surface-controlled growth mechanisms of ALD enable low damage and smooth and conformal growth [15, 187]. However, 2D materials by their very nature have no out-of-plane bonding and rely on defects, dangling bonds, to initiate growth. Consequently, the higher the quality of the 2D material, the lower incidence of defects available for nucleation. The low level of active bonding sites promoted non uniform growth with selective growth along grain boundaries and step edges [188–190]. To counter this interface modification and surface activation were once again key in the vapor phase growth of dielectrics on semiconducting layers [187, 190]. The first 2D material studied was graphene, and initial studies used highly ordered pyrolytic graphite (HOPG) as a model substrate due to its ease of production and graphene-like surface with sp2 bonded carbon on terraces and defects at the step edges [191]. Uniform films of $Al_2O_3$ were produced on HOPG by dosing with $NO_2$ [192], a method that was previously employed in the ALD coating of carbon nanotubes [193]. An alternative approach was to functionalize the HOPG via a solution process prior to ALD [194]. The most widespread solution however was to replace the traditional ALD co-reagent $H_2O$ with the more reactive species $O_3$ [195–199]. This approach was tuned, through temperature and $O_3$ concentration, to prevent the formation of surface species such as epoxide, carbonyl, and carboxylic groups, which would be detrimental to the electronic properties, with the $O_3$ only being physisorbed to the graphitic surface [196, 197]. The extension of these methodologies from HOPG to graphene was not always straightforward; the ultrathin 2D nature of graphene allows significant substrate effects through the 2D material to growth interface. Unlike HOPG the underlying substrate is usually a different material; for example, graphene is often grown on a catalytic metal surface [198], or flakes are transferred from a bulk sample, potentially leaving residue. Such factors can have a significant influence on the surface chemistry and subsequent grow processes for dielectric layers [199]. Although the methods used for the vapor deposition on HOPG were employed with some degree of success, more controllable methods were deemed necessary. Methodologies to physisorb oxygen entities to promote ALD nucleation were widely studied, employing either $O_3$ or $O_2$ plasma, and it was demonstrated that by careful control of the energy of the system, these physisorbed species could initiate growth without damaging the graphene [200–203]. Chemical functionalization, with organic molecules forming π–π interactions with graphene and reactive function groups for binding to the ALD precursor, were also found to promote the formation of uniform dielectric layers. Wang et al. exploited this method with 3,4,9,10-perylenetetracarboxylic acid (PTCA), forming a uniform and continuous $Al_2O_3$ layer with TMA and water vapor at a reactor temperature of 100 °C [204]. Alaboson et al. employed 3,4,9,10-perylenetetracarboxylic dianhydride (PTCDA) with similar surface chemistry for the ALD of $Al_2O_3$, extending the process to $HfO_2$ using tetrakis(diethylamido)hafnium (TDEAHf) as the Hf source [205]. The need for functionalization of graphene has been questioned by several groups who managed to grow continuous films on graphene with a simple ALD process [206, 207]. There are many possible reasons for this, many of which are based on the premise that ALD is not a clean process, reactor pressures are such that physisorbed impurity, and precursor species will be present, especially at the lower temperatures required to prevent damage to the 2D material. In a standard self-limiting ALD process, purges are made sufficiently long to remove physisorbed species, several groups exploited this physisorption of precursor molecules to optimize growth [208–211]. It should however be recognized that this is not true ALD, and an approach based on physisorption, and not chemical bonding, would be very difficult to replicate conformally on complex 3D structures.

Although still a widely explored topic, due the relatively low level of functionality of graphene as a consequence of its metallic or semimetallic behavior [212], metal-oxide growth on 2D materials was extended to transition metal dichalcogenides (TMDs). This group of 2D materials was found to exhibit a wide range of electrical properties with band gaps that were tunable with thickness [213]. Akin to graphene, initial studies of the vapor phase growth of oxides on TMDs employed exfoliated flakes to elucidate the physics of material growth and demonstrate device performance. Several TMDs have been investigated – $WS_2$, $MoSe_2$, $WSe_2$, and $WTe_2$ – however, the most widely studied material of this class is $MoS_2$. Early work on $MoS_2$ demonstrated that physisorption could be exploited at low temperatures, to give uniform $Al_2O_3$ thin films by ALD with TMA and $H_2O$ [214]. This was extended to $HfO_2$, although uniformity was again claimed [215, 216], and it was later demonstrated to be due to the adhesion of impurities promoting $HfO_2$ growth. It was concluded that $HfO_2$ growth, directly on the TMDs, was not uniform without pretreatment, even at low temperature [217]. Further studies, with ozone pretreatments enabled uniform ALD growth of both $Al_2O_3$ and $HfO_2$ even at higher temperatures, through the formation of Mo—O bonds, an important development for device fabrication [218, 219]. The application of pretreatments to the growth of oxides on other TMDs was successfully employed for both the selenides [220] and tellurides [221]. The shift away from exfoliated flakes to directly grown TMDs further highlights the difficulties in developing schemes to enable metal-oxide growth on 2D materials. CVD grown TMDs are currently relatively poor in quality, as compared with exfoliated materials, particularly those grown at temperatures compatible with CMOS integration [222–224]. Therefore, uniform growth of oxides on the CVD-grown materials maybe misleading and, as the quality of TMD deposition improves, pretreatments and functionalization are likely to be necessary.

## 1.5 Conclusions

In this review we have introduced the concept of vapor phase growth of metal oxides. Focusing on two case studies, amorphous insulating materials, and ferroelectric oxides, we have illustrated how the nanoscale integration of these metal oxides in electronics has progressed and how the flexibility of the CVD technique has enabled the growth of highly conformal materials, at the nanometer scale, with precise stoichiometric control to produce complex-layered materials. The ability to produce ultrathin films with composition control in an industry-compatible process cannot be overstated. Although far from exhaustive this snapshot of the drive for integration has shown, the advances are huge, and it has generally been focused on exploiting the bulk properties of the selected materials. However, as scaling approaches the sub-nanometer scale, the prospect of 2D metal oxides, beyond perovskites, maybe realized [225–228]. The quantum confinement effects, due to the inherently low thickness of these materials, open them to a wide range of applications where tunable band structures can enable a range of electrical and optical properties. To date, there has been little progress in vapor phase growth of these layered materials in the 2D domain, primarily as top-down approaches, such as exfoliation from bulk, are sufficient for proof of concept and fundamental materials studies. However, as atomic scale processing methodologies improve, both in terms of addition by growth and removal by etching, the tuning of material phases and interfacial properties that enable the controlled deposition and stabilization of true 2D materials is becoming more widespread; consequently, this is an area where significant materials advances could be made to maintain metal oxides at the forefront of next-generation electronic device materials.

## References

1 Sawyer, W.E. and Mann, A. *Carbon for electric lights*, US Patent 1880, 229,335.
2 Wöhler, F. and Uslar, L. (1855). Über metallisches wolfram und molybdän. *Liebigs Ann.* 94: 255.
3 Mond, L., Langer, C., and Quinke, F. (1890). L.—Action of carbon monoxide on nickel. *J. Chem. Soc.* 57: 749.
4 Pring, J.N. and Fielding, W. (1909). CLXXI.—The preparation at high temperatures of some refractory metals from their chlorides. *J. Chem. Soc.* 95: 1497.
5 Hoelbling, R.Z. (1927). Über die Herstellung und einige Eigenschaften von reinem metallischen Silicium. *Angew. Chem.* 40: 655.
6 Jones, A.C. and O'Brien, P. (1997). *CVD of Compound Semiconductors*. Weinheim: VCH.
7 Berry, A.D., Gaskill, D.K., Holm, R.T. et al. (1988). Formation of high Tc superconducting films by organometallic chemical vapor deposition. *Appl. Phys. Lett.* 52: 1743.
8 Christian, D., Narayan, J., and Scneemeyer, L. (ed.) (1990). *High Temperature Superconductors: Fundamental Properties and Novel Materials Processing*, vol. 169. United States: Materials Research Society.
9 Chau, R., Doczy, M., Doyle, B. et al. (2004). Advanced CMOS transistors in the nanotechnology era for high-performance, low-power logic applications. In: *Proceedings of the 7th International Conference on Solid-State and Integrated Circuits Technology, 2004*, vol. 1, 26–30. IEEE, doi: 10.1109/ICSICT.2004.1434947.

10 Stringfellow, G.B. (1989). *Organometallic Vapour Phase Epitaxy*. New York: Academic Press.
11 Jones, A.C. and Hitchman, M.L. (ed.) (2009). *Chemical Vapour Deposition: Precursors, Processes and Applications*. Royal Society of Chemistry.
12 Kodas, T.T. and Hampden-Smith, M.J. (1994). *The Chemistry of Metal CVD*. Weinheim: VCH.
13 Ritala, M. and Leskela, M. (2001). *Atomic Layer Deposition. Handbook of Thin Film Materials*. San Diego, CA: Academic Press.
14 Puurunen, R.L. (2005). Surface chemistry of atomic layer deposition: a case study for the trimethylaluminum/water process. *J. Appl. Phys.* 97: 121301.
15 George, S.M. (2010). Atomic layer deposition: an overview. *Chem. Rev.* 110: 111–131.
16 Whatmore, R. (2017). Ferroelectric materials. In: *Springer Handbook of Electronic and Photonic Materials* (ed. S. Kasap and P. Capper), 597–623. Cham: Springer International Publishing.
17 Choi, J.H., Kim, J.S., Hong, S.B. et al. (2012). Crystal structure and piezoelectric properties of $KNbO_3$–$BiFeO_3$ ceramics. *J. Kor. Phys. Soc.* 61: 956–960.
18 Yue, W. and Yi-jian, J. (2003). Crystal orientation dependence of piezoelectric properties in $LiNbO_3$ and $LiTaO_3$. *Opt. Mater.* 23: 403–408.
19 Yamanouchi, K., Wagatsuma, Y., Odagawa, H., and Cho, Y. (2001). Single crystal growth of $KNbO_3$ and application to surface acoustic wave devices. *J. Eur. Ceram. Soc.* 21: 2791–2795.
20 Li, M., Ling, J., He, Y. et al. (2020). Lithium niobate photonic-crystal electro-optic modulator. *Nat. Commun.* 11: 4123.
21 Lorenz, M., Ramachandra Rao, M.S., Venkatesan, T. et al. (2016). The 2016 oxide electronic materials and oxide interfaces roadmap. *J. Phys. D: Appl. Phys.* 49: 433001.
22 Schwarzkopf, J. and Fornari, R. (2006). Epitaxial growth of ferroelectric oxide films. *Prog. Cryst. Growth Charact. Mater.* 52: 159–212.
23 Momma, K. and Izumi, F. (2011). VESTA 3 for three-dimensional visualization of crystal, volumetric and morphology data. *J. Appl. Crystallogr.* 44: 1272–1276.
24 Perez-Mato, J.M., Aroyo, M., García, A. et al. (2004). Competing structural instabilities in the ferroelectric Aurivillius compound SrBi2Ta2O9. *Phys. Rev. B* 70: 214111.
25 Guo, Y.-Y., Gibbs, A.S., Perez-Mato, J.M., and Lightfoot, P. (2019). Unexpected phase transition sequence in the ferroelectric $Bi_4Ti_3O_{12}$. *IUCrJ* 6: 438–446.
26 Hervoches, C.H., Snedden, A., Riggs, R.L. et al. (2002). Structural behavior of the four-layer Aurivillius-phase ferroelectrics $SrBi_4Ti_4O_{15}$ and $Bi_5Ti_3FeO_{15}$. *J. Solid State Chem.* 164: 280–291.
27 Ismunandar, K.T., Hoshikawa, A., Zhou, Q. et al. (2004). Structural studies of five layer Aurivillius oxides: $A_2Bi_4Ti_5O_{18}$ (A=Ca, Sr, Ba and Pb). *J. Solid State Chem.* 177: 4188–4196.
28 Birenbaum, A.Y. and Ederer, C. (2014). Potentially multiferroic Aurivillius phase $Bi_5FeTi_3O_{15}$: cation site preference, electric polarization, and magnetic coupling from first principles. *Phys. Rev. B* 90: 214109.

29 Keeney, L., Smith, R.J., Palizdar, M. et al. (2020). Ferroelectric behavior in exfoliated 2D Aurivillius oxide flakes of sub-unit cell thickness. *Adv. Electr. Mater.* 6: 1901264.
30 Faraz, A., Maity, T., Schmidt, M. et al. (2017). Direct visualization of magnetic-field-induced magnetoelectric switching in multiferroic aurivillius phase thin films. *J. Am. Ceram. Soc.* 100: 975–987.
31 Keeney, L., Maity, T., Schmidt, M. et al. (2013). Magnetic field-induced ferroelectric switching in multiferroic Aurivillius phase thin films at room temperature. *J. Am. Ceram. Soc.* 96: 2339–2357.
32 Yang, F., Tang, M.H., Ye, Z. et al. (2007). Eight logic states of tunneling magnetoelectroresistance in multiferroic tunnel junctions. *J. Appl. Phys.* 102: 044504.
33 Setter, N., Damjanovic, D., Eng, L. et al. (2006). Ferroelectric thin films: Review of materials, properties, and applications. *J. Appl. Phys.* 100: 051606.
34 Jackson, N., Keeney, L., and Mathewson, A. (2013). Flexible-CMOS and biocompatible piezoelectric AlN material for MEMS applications. *Smart Mater. Struct.* 22: 115033.
35 Gradauskaite, E., Campanini, M., Biswas, B. et al. (2020). Robust in-plane ferroelectricity in ultrathin epitaxial Aurivillius films. *Adv. Mat. Interfaces* 7: 2000202.
36 Dragoman, M., Modreanu, M., Povey, I.M. et al. (2018). Wafer-scale very large memory windows in graphene monolayer/HfZrO ferroelectric capacitors. *Nanotechnology* 29: 425204.
37 Pham, C.D., Chang, J., Zurbuchen, M.A., and Chang, J.P. (2015). Synthesis and characterization of $BiFeO_3$ thin films for multiferroic applications by radical enhanced atomic layer deposition. *Chem. Mater.* 27: 7282–7288.
38 Ihlefeld, J.F., Kumar, A., Gopalan, V. et al. (2007). Adsorption-controlled molecular-beam epitaxial growth of $BiFeO_3$. *Appl. Phys. Lett.* 91: 071922.
39 Zhang, P.F., Deepak, N., Keeney, L. et al. (2012). The structural and piezoresponse properties of c-axis-oriented Aurivillius phase $Bi_5Ti_3FeO_{15}$ thin films deposited by atomic vapor deposition. *Appl. Phys. Lett.* 101.
40 Ramesh, R., Aggarwal, S., and Auciello, O. (2001). Science and technology of ferroelectric films and heterostructures for non-volatile ferroelectric memories. *Mater. Sci. Eng. R Rep.* 32: 191–236.
41 Zurbuchen, M.A., Tian, W., Pan, X.Q. et al. (2007). Morphology, structure, and nucleation of out-of-phase boundaries (OPBs) in epitaxial films of layered oxides. *J. Mater. Res.* 22: 1439–1471.
42 Luborsky, F., Livingston, J., and Chin, G. (1996). Chapter 29 – Magnetic properties of metals and alloys. In: *Physical Metallurgy*, 4e (ed. R.W. Cahn and P. Haasen), 2501–2565. North-Holland.
43 Noguchi, Y., Matsuo, H., Kitanaka, Y., and Miyayama, M. (2019). Ferroelectrics with a controlled oxygen-vacancy distribution by design. *Sci. Rep.* 9: 4225.
44 Keeney, L., Zhang, P.F., Groh, C. et al. (2010). Piezoresponse force microscopy investigations of Aurivillius phase thin films. *J. Appl. Phys.* 108: 042004.
45 Keeney, L., Kulkarni, S., Deepak, N. et al. (2012). Room temperature ferroelectric and magnetic investigations and detailed phase analysis of Aurivillius phase $Bi_5Ti_3Fe_{0.7}Co_{0.3}O_{15}$ thin films. *J. Appl. Phys.* 112: 052010.

**46** Schmidt, M., Amann, A., Keeney, L. et al. (2014). Absence of evidence ≠ evidence of absence: statistical analysis of inclusions in multiferroic thin films. *Sci. Rep.* 4: 5712.

**47** Gifford, K.D., Auciello, O., and Kingon, A.I. (1995). Control of electrical properties of ion beam sputter-deposited PZT-based heterostructure capacitors. *Integr. Ferroelectr.* 7: 195–206.

**48** Chen, T.C., Thio, C.L., and Desu, S.B. (1997). Impedance spectroscopy of $SrBi_2Ta_2O_9$ and $SrBi_2Nb_2O_9$ ceramics correlation with fatigue behavior. *J. Mater. Res.* 12: 2628–2637.

**49** Mazet, L., Yang, S.M., Kalinin, S.V. et al. (2015). A review of molecular beam epitaxy of ferroelectric $BaTiO_3$ films on Si, Ge and GaAs substrates and their applications. *Sci. Technol. Adv. Mater.* 16: 036005.

**50** Chambers, S.A. (2008). Molecular beam epitaxial growth of doped oxide semiconductors. *J. Phys. Condens. Matter* 20: 264004.

**51** Funakubo, H., Imashita, K., Matsuyama, K. et al. (1994). Deposition condition of epitaxially grown PZT films by CVD. *J. Ceram. Soc. Jpn.* 102: 795–798.

**52** Bartasyte, A., Abrutis, A., Jimenez, C. et al. (2007). Ferroelectric $PbTiO_3$ films grown by pulsed liquid injection metalorganic chemical vapour deposition. *Ferroelectrics* 353: 104–115.

**53** Dormans, G.J.M., van Veldhoven, P.J., and de Keijser, M. (1992). Composition-controlled growth of $PbTiO_3$ on $SrTiO_3$ by organometallic chemical vapour deposition. *J. Cryst. Growth* 123: 537–544.

**54** Micard, Q., Condorelli, G.G., and Malandrino, G. (2020). Piezoelectric $BiFeO_3$ thin films: optimization of MOCVD process on Si. *Nanomaterials* 10: 630.

**55** Tohma, T., Masumoto, H., and Goto, T. (2002). Microstructure and dielectric properties of barium titanate film prepared by MOCVD. *Mater. Trans.* 43: 2880–2884.

**56** Kiat, J.M. and Dkhil, B. (2008). Chapter 14 – From the structure of relaxors to the structure of MPB systems. In: *Handbook of Advanced Dielectric, Piezoelectric and Ferroelectric Materials* (ed. Z.-G. Ye), 391–446. Woodhead Publishing.

**57** Keeney, L., Groh, C., Kulkarni, S. et al. (2012). Room temperature electromechanical and magnetic investigations of ferroelectric Aurivillius phase $Bi_5Ti_3(Fe_xMn_{1-x})O_{15}$ ($x = 1$ and 0.7) chemical solution deposited thin films. *J. Appl. Phys.* 112: 024101.

**58** Vehkamäki, M., Hatanpää, T., Ritala, M., and Leskelä, M. (2004). Bismuth precursors for atomic layer deposition of bismuth-containing oxide films. *J. Mater. Chem.* 14: 3191–3197.

**59** Wang, Y., Chen, W., Wang, B., and Zheng, Y. (2014). Ultrathin ferroelectric films: growth, characterization, physics and applications. *Materials (Basel)* 7: 6377–6485.

**60** Van Buskirk, P.C., Roeder, J.F., and Bilodeau, S. (1995). Manufacturing of perovskite thin films using liquid delivery MOCVD. *Integr. Ferroelectr.* 10: 9–22.

**61** Astié, V., Millon, C., Decams, J.-M., and Bartasyte, A. (2019). *Direct Liquid Injection Chemical Vapor Deposition*. IntechOpen.

**62** Sakashita, Y. and Segawa, H. (1995). Preparation and characterization of $LiNbO_3$ thin films produced by chemical-vapor deposition. *J. Appl. Phys.* 77: 5995–5999.

**63** Shin, W.C., Choi, K.J., and Yoon, S.G. (2002). Ferroelectric properties of SrBi$_2$Ta$_2$O$_9$ thin films with Bi$_2$O$_3$ buffer layer by liquid-delivery metalorganic chemical vapor deposition. *Thin Solid Films* 409: 133–137.

**64** Kijima, T., Ushikubo, M., and Matsunaga, H. (1999). New low-temperature processing of metalorganic chemical vapor deposition-Bi$_4$Ti$_3$O$_{12}$ thin films using BiO$_x$ buffer layer. *Jpn. J. Appl. Phys.* 38: 127–130.

**65** Deepak, N., Carolan, P., Keeney, L. et al. (2015). Bismuth self-limiting growth of ultrathin BiFeO$_3$ films. *Chem. Mater.* 27: 6508–6515.

**66** Yang, S.Y., Zavaliche, F., Mohaddes-Ardabili, L. et al. (2005). Metalorganic chemical vapor deposition of lead-free ferroelectric BiFeO$_3$ films for memory applications. *Appl. Phys. Lett.* 87: 102903.

**67** Thery, J., Dubourdieu, C., Baron, T. et al. (2007). MOCVD of BiFeO$_3$ thin films on SrTiO$_3$. *Chem. Vap. Deposition* 13: 232–238.

**68** Keeney, L., Downing, C., Schmidt, M. et al. (2017). Direct atomic scale determination of magnetic ion partition in a room temperature multiferroic material. *Sci. Rep.* 7: 1737.

**69** Logan, S.R. (1982). The origin and status of the Arrhenius equation. *J. Chem. Educ.* 59: 279.

**70** Waser, R., Schneller, T., Hoffmann-eifert, S., and Ehrhart, P. (2001). Advanced chemical deposition techniques – from research to production. *Integr. Ferroelectr.* 36: 3–20.

**71** Burgess, D., Schienle, F., Lindner, J. et al. (2000). Metal organic chemical vapor deposition and characterization of strontium bismuth tantalate (SBT) thin films. *Jpn. J. Appl. Phys.* 39: 5485–5488.

**72** Hintermaier, F., Hendrix, B., Desrochers, D. et al. (1998). Properties of SrBi$_2$Ta$_2$O$_9$ thin films grown by MOCVD for high density FeRAM applications. *Integr. Ferroelectr.* 21: 367–379.

**73** Roeder, J.F., Hendrix, B.C., Hintermaier, F. et al. (1999). Ferroelectric strontium bismuth tantalate thin films deposited by metalorganic chemical vapour deposition (MOCVD). *J. Eur. Ceram. Soc.* 19: 1463–1466.

**74** Hendrix, B.C., Hintermaier, F., Desrocherst, D.A. et al. (2011). MOCVD of SrBi$_2$Ta$_2$O$_9$ for integrated ferroelectric capacitors. *MRS Proc.* 493: 225.

**75** Goux, L., Lisoni, J.G., Schwitters, M. et al. (2005). Composition control and ferroelectric properties of sidewalls in integrated three-dimensional SrBi$_2$Ta$_2$O$_9$-based ferroelectric capacitors. *J. Appl. Phys.* 98: 054507.

**76** Yeh, C.-P., Lisker, M., Kalkofen, B., and Burte, E.P. (2016). Fabrication and investigation of three-dimensional ferroelectric capacitors for the application of FeRAM. *AIP Adv.* 6: 035128.

**77** Shin, S., Koo, J.-M., Kim, S. et al. (2006). Fabrication of 3-dimensional PbZr$_{1-x}$Ti$_x$O$_3$ nanoscale thin film capacitors for high density ferroelectric random access memory devices. *J. Nanosci. Nanotechnol.* 6: 3333–3337.

**78** Migita, T., Kobune, M., Ito, R. et al. (2020). Fabrication and characterization of micropillar-type multiferroic composite thin films by metal organic chemical vapor deposition using a ferroelectric microplate structure. *Jpn. J. Appl. Phys.* 59: SCCB08.

79 Yuan, S., Luo, X., Chan, H.L. et al. (2019). Room-temperature ferroelectricity in MoTe$_2$ down to the atomic monolayer limit. *Nat. Commun.* 10: 1775.

80 Cheema, S.S., Kwon, D., Shanker, N. et al. (2020). Enhanced ferroelectricity in ultrathin films grown directly on silicon. *Nature* 580: 478–482.

81 Boyn, S., Grollier, J., Lecerf, G. et al. (2017). Learning through ferroelectric domain dynamics in solid-state synapses. *Nat. Commun.* 8: 14736.

82 Pantel, D., Goetze, S., Hesse, D., and Alexe, M. (2012). Reversible electrical switching of spin polarization in multiferroic tunnel junctions. *Nat. Mater.* 11: 289.

83 Baudry, L., Lukyanchuk, I., and Vinokur, V.M. (2017). Ferroelectric symmetry-protected multibit memory cell. *Sci. Rep.* 7: 42196.

84 Nordlander, J., De Luca, G., Strkalj, N. et al. (2018). Probing ferroic states in oxide thin films using optical second harmonic generation. *Appl. Sci.* 8: 570.

85 Spaldin, N.A. and Ramesh, R. (2019). Advances in magnetoelectric multiferroics. *Nat. Mater.* 18: 203–212.

86 Lindner, J., Weiss, F., Senateur, J.P. et al. (2000). Growth of BaTiO$_3$/SrTiO$_3$ superlattices by injection MOCVD. *Integr. Ferroelectr.* 30: 53–59.

87 Nonomura, H., Fujisawa, H., Shimizu, M. et al. (2003). Ferroelectric properties of 15–20 nm-thick PZT ultrathin films prepared by MOCVD. *MRS Online Proc. Lib.* 748: 29.

88 Fujisawa, H., Horii, T., Takashima, Y. et al. (2006). Microstructure and ferroelectric properties of ultrathin PbTiO$_3$ films by MOCVD. *MRS Online Proc. Lib.* 902: 331.

89 Keeney, L., Saghi, Z., O'Sullivan, M. et al. (2020). Persistence of ferroelectricity close to unit-cell thickness in structurally disordered Aurivillius phases. *Chem. Mater.* 32: 10511–10523.

90 Palkar, G.D., Sitharamarao, D.N., and Dasgupta, A.K. (1963). Self-diffusion of bismuth in bismuth oxide. *Trans. Faraday Soc.* 59: 2634–2638.

91 Deepak, N., Zhang, P.F., Keeney, L. et al. (2013). Atomic vapor deposition of bismuth titanate thin films. *J. Appl. Phys.* 113: 187207.

92 Deepak, N., Carolan, P., Keeney, L. et al. (2015). Tunable nanoscale structural disorder in Aurivillius phase, $n = 3$ Bi$_4$Ti$_3$O$_{12}$ thin films and their role in the transformation to $n = 4$, Bi$_5$Ti$_3$FeO$_{15}$ phase. *J. Mater. Chem. C* 3: 5727–5732.

93 Schumacher, M., Baumann, P.K., and Seidel, T. (2006). AVD and ALD as two complementary technology solutions for next generation dielectric and conductive thin-film processing. *Chem. Vap. Deposition* 12: 99–108.

94 Yamaguchi, N., Hattori, T., Terashima, K., and Yoshida, T. (1998). High-rate deposition of LiNbO$_3$ films by thermal plasma spray CVD. *Thin Solid Films* 316: 185–188.

95 Moniz, S.J.A., Quesada-Cabrera, R., Blackman, C.S. et al. (2014). A simple, low-cost CVD route to thin films of BiFeO$_3$ for efficient water photo-oxidation. *J. Mater. Chem. A* 2: 2922–2927.

**96** Abelson, J.R. and Girolami, G.S. (2020). New strategies for conformal, superconformal, and ultrasmooth films by low temperature chemical vapor deposition. *J. Vacuum Sci. Technol. A* 38: 030802.

**97** Chu, B., Zhou, X., Ren, K. et al. (2006). A Dielectric Polymer with High Electric Energy Density and Fast Discharge Speed. *Science* 313: 334–336.

**98** Sherrill, S.A., Banerjee, P., Rubloff, G.W., and Lee, S.B. (2011). High to ultra-high power electrical energy storage. *Phys. Chem. Chem. Phys.* 13: 20714–20723.

**99** Pan, H., Kursumovic, A., Lin, Y.-H. et al. (2020). Dielectric films for high performance capacitive energy storage: multiscale engineering. *Nanoscale* 12: 19582–19591.

**100** Teranishi, T., Yoshikawa, Y., Yoneda, M. et al. (2018). Aluminum interdiffusion into $LiCoO_2$ using atomic layer deposition for high-rate lithium ion batteries. *ACS Appl. Energy Mater.* 1 (7): 3277–3282F.

**101** McPherson, J.W., Kim, J., Shanware, A. et al. (2003). Trends in the ultimate breakdown strength of high dielectric-constant materials. *IEEE Trans. Electron Devices* 50: 1771–1778.

**102** Zhu, H., Tang, C., Fonseca, L.R.C., and Ramprasad, R. (2012). Recent progress in ab initio simulations of hafnia-based gate stacks. *J. Mater. Sci.* 47: 7399–7416.

**103** Wilk, G.D. and Wallace, R.M. (1999). Electrical properties of hafnium silicate gate dielectrics deposited directly on silicon. *Appl. Phys. Lett.* 74: 2854.

**104** Kamoshida, M., Mitchell, I.V., and Mayer, J.W. (1971). Influence of deposition temperature on properties of hydrolytically grown aluminum oxide films. *Appl. Phys. Lett.* 18: 292.

**105** Mehta, D.A., Butler, S.R., and Feigl, F.J. (1973). Effects of postdeposition annealing treatments on charge trapping in CVD $Al_2O_3$ films on Si. *J. Electrochem. Soc.* 120: 1707.

**106** Jones, A.C., Houlton, D.J., Rushworth, S.A., and Critchlow, G.W. (1995). A new route to the deposition of $Al_2O_3$ by MOCVD. *J. Phys. IV* 5: 557.

**107** Yom, S.S., Kang, W.N., Yoon, Y.S. et al. (1992). Growth of $\gamma$-$Al_2O_3$ thin films on silicon by low pressure metal-organic chemical vapour deposition. *Thin Solid Films* 213: 72.

**108** Mai, L., Boysen, N., Zanders, D. et al. (2019). Potential precursor alternatives to the pyrophoric trimethylaluminium for the atomic layer deposition of aluminium oxide. *Chem. Eur. J.* 25: 7406–7406.

**109** Guha, S., Cartier, E., Gribelyuk, M.A. et al. (2000). Atomic beam deposition of lanthanum and yttrium based oxide thin films for gate dielectrics. *Appl. Phys. Lett.* 77: 2710–2712.

**110** Chaneliere, C., Autran, J.L., Devine, R.A.B., and Balland, B. (1988). Tantalum pentoxide ($Ta_2O_5$) thin films for advanced dielectric applications. *Mater. Sci. Eng., A* R22: 269–322.

**111** Kim, T., Son, H., Kim, I. et al. (2020). Reversible switching mode change in $Ta_2O_5$-based resistive switching memory (ReRAM). *Sci. Rep.* 10: 11247.

112 Song, S.J., Park, T., Yoon, K.J. et al. (2017). Comparison of the atomic layer deposition of tantalum oxide thin films using Ta(N$^t$Bu)(NEt$_2$)$_3$, Ta(N$^t$Bu)(NEt$_2$)$_2$Cp, and H$_2$O. *ACS Appl. Mater. Interfaces* 9 (1): 537–547.

113 Kukli, K., Kemell, M., Vehkamäki, M. et al. (2017). Atomic layer deposition and properties of mixed Ta$_2$O$_5$ and ZrO$_2$ films. *AIP Adv.* 7: 025001.

114 Cho, H., Park, K.-W., Park, C.H. et al. (2015). Abnormally enhanced dielectric constant in ZrO$_2$/Ta$_2$O$_5$ multi-laminate structures by metallic Ta formation. *Mater. Lett.* 154: 148–151.

115 Falkowski, M., Künneth, C., Materlik, R., and Kers, A. (2018). Unexpectedly large energy variations from dopant interactions in ferroelectric HfO$_2$ from high-throughput ab initio calculations. *NPJ Comput. Mater.* 4: 73.

116 Tauber, R.N., Dumbri, A.C., and Caffrey, R.E. (1971). Preparation and properties of pyrolytic Zirconium Dioxide films. *J. Electrochem. Soc.* 118: 747.

117 Powell, C.F. (1966). *Chemically Deposited Nonmetals*. New York: Wiley.

118 Ritala, M. and Leskelä, M. (1994). Zirconium dioxide thin films deposited by ALE using zirconium tetrachloride as precursor. *Appl. Surf. Sci.* 75: 333.

119 Ritala, M., Leskelä, M., Niinistö, L. et al. (1994). Development of crystallinity and morphology in hafnium dioxide thin films grown by atomic layer epitaxy. *Thin Solid Films* 250: 72.

120 Balog, M., Scheiber, M., Michman, M., and Patai, S. (1972). Thin films of metal oxides on silicon by chemical vapor deposition with organometallic compounds. I. *J. Cryst. Growth* 17: 298.

121 Pasko, S.V., Hubert-Pfalzgraf, L.G., Abrutis, A. et al. (2004). New sterically hindered Hf, Zr and Y β-diketonates as MOCVD precursors for oxide films. *J. Mater. Chem.* 14: 1245.

122 Bradley, D.C. (1989). Metal alkoxides as precursors for electronic and ceramic materials. *Chem. Rev.* 89: 1317.

123 Gould, B.J., Povey, I.M., Pemble, M.E., and Flavell, W.R. (1994). Chemical vapour deposition of ZrO$_2$ thin films monitored by IR spectroscopy. *J. Mater. Chem.* 4: 1815–1819.

124 Takahashi, Y., Kawae, T., and Nasu, M. (1986). Chemical vapour deposition of undoped and spinel-doped cubic zirconia film using organometallic process. *J. Cryst. Growth* 74: 409.

125 Kukli, K., Ritala, M., and Leskelä, M. (2000). Low-Temperature Deposition of Zirconium Oxide–Based Nanocrystalline Films by Alternate Supply of Zr[OC(CH$_3$)$_3$]4 and H$_2$O. *Chem. Vap. Deposition* 6: 297.

126 Kim, D.-Y., Lee, C.-H., and Park, S.J. (1996). Preparation of zirconia thin films by metalorganic chemical vapor deposition using ultrasonic nebulization. *J. Mater. Res.* 11: 2583.

127 Taylor, S., Williams, P.A., Roberts, J.L. et al. (2002). HfO$_2$ and ZrO$_2$ alternative gate dielectrics for silicon devices by liquid injection chemical vapour deposition. *Electron. Lett* 38: 1285.

128 Jones, A.C., Williams, P.A., Roberts, J.L. et al. (2002). Atomic-Layer Deposition of $ZrO_2$ Thin Films Using New Alkoxide Precursors. *Mater. Res. Soc. Symp. Proc.* 716: 145.

129 Kukli, K., Ritala, M., Leskelä, M. et al. (2003). Atomic Layer Deposition of Hafnium Dioxide Films Using Hafnium Bis(2-butanolate)bis(1-methoxy-2-methyl-2-propanolate) and Water. *Chem. Vap. Deposition* 9: 315.

130 Kukli, K., Ritala, M., Leskelä, M. et al. (2003). Atomic Layer Deposition of Hafnium Dioxide Films from 1-Methoxy-2-methyl-2-propanolate Complex of Hafnium. *Chem. Mater.* 15: 1722.

131 Bastianini, A., Battiston, G.A., Gerbasi, R. et al. (1995). Chemical Vapor Deposition of $ZrO_2$ Thin Films Using $Zr(NEt_2)4$ as Precursor. *J. Phys. IV* 5: 525.

132 Ohshita, Y., Ogura, A., Hoshino, A. et al. (2001). $HfO_2$ growth by low-pressure chemical vapor deposition using the $Hf(N(C_2H_5)_2)_4/O_2$ gas system. *J. Cryst. Growth* 233: 292.

133 Hausmann, D.M., Kim, E., Becker, J., and Gordon, R.G. (2002). Atomic Layer Deposition of Hafnium and Zirconium Oxides Using Metal Amide Precursors. *Chem. Mater.* 14: 4350.

134 Hausmann, D.M. and Gordon, R.G. (2003). Surface morphology and crystallinity control in the atomic layer deposition (ALD) of hafnium and zirconium oxide thin films. *J. Cryst. Growth* 249: 251.

135 Brennan, B., Milojevic, M., Kim, H.C. et al. (2009). Half-Cycle Atomic Layer Deposition Reaction Study Using $O_3$ and $H_2O$ Oxidation of $Al_2O_3$ on $In_{0.53}Ga_{0.47}As$. *Electrochem. Solid-State Lett.* 12 (6): H205–H207.

136 Chen, W., Sun, Q.-Q., Xu, M. et al. (2007). Atomic Layer Deposition of Hafnium Oxide from Tetrakis(ethylmethylamino)hafnium and Water Precursors. *J. Phys. Chem. C* 111: 6495–6499.

137 Hurley, P.K., O'Connor, E., Monaghan, S. et al. (2009). Structural and Electrical Properties of $HfO_2/n$-$In_xGa_{1-x}As$ structures ($x$: 0, 0.15, 0.3 and 0.53). *Electrochem. Soc. Trans.* 25: 113.

138 Putkonen, M., Niinistö, J., Kukli, K. et al. (2003). $ZrO_2$ Thin Films Grown on Silicon Substrates by Atomic Layer Deposition with $Cp_2Zr(CH_3)_2$ and Water as Precursors. *Chem. Vap. Deposition* 9: 207.

139 Niinistö, J., Putkonen, M., Niinistö, L. et al. (2005). Controlled growth of $HfO_2$ thin films by atomic layer deposition from cyclopentadienyl-type precursor and water. *J. Mater. Chem.* 15: 2271.

140 Williams, P.A., Jones, A.C., Tobin, N.L. et al. (2003). Growth of Hafnium Dioxide Thin Films by Liquid-Injection MOCVD Using Alkylamide and Hydroxylamide Precursors. *Chem. Vap. Deposition* 9: 309.

141 Gutowski, M., Jaffe, J.E., Liu, C.-L. et al. (2002). Thermodynamic stability of high-K dielectric metal oxides $ZrO_2$ and $HfO_2$ in contact with Si and $SiO_2$. *Appl. Phys. Lett.* 80: 1897.

142 Schlom, D.G. and Haeni, H.J. (2002). A Thermodynamic Approach to Selecting Alternative Gate Dielectrics. *MRS Bull.* 27: 198–204.

**143** Mistry, K. et al. (2007). A 45 nm logic technology with high-$k$+metal gate transistors, strained silicon, 9 Cu interconnect layers, 193 nm dry patterning, and 100% Pb-free packaging. *IEEE Int. Electron Devices Meeting* 2007: 247–250.

**144** Becke, H.W. and White, J.P. (1967). GaAs FET's Outperform Conventional Silicon MOS Devices. *Electronics* 40: 82.

**145** Lin, D., Waldron, N., Brammertz, G. et al. (2010). Exploring the ALD $Al_2O_3/In_{0.53}Ga_{0.47}As$ and $Al_2O_3$/Ge Interface Properties: A Common Gate Stack Approach for Advanced III-V/Ge CMOS. *ECS Trans.* 28 (5): 173–183.

**146** Xuan, Y., Wu, Y.Q., Shen, T., Yang, T., Ye, P.D. (2007) High performance sub-micron inversion-type enhancement-mode InGaAs MOSFETs with ALD $Al_2O_3$, $HfO_2$ and HfAlO as gate dielectrics, *2007 IEEE International Electron Devices Meeting*, 637–640.

**147** Lin, J.Q., Lee, S.J., Oh, H.J. et al. (2008). Inversion-Mode Self-Aligned $In_{0.53}Ga_{0.47}As$ N-Channel Metal-Oxide-Semiconductor Field-Effect Transistor With HfAlO Gate Dielectric and TaN Metal Gate. *IEEE Electron Device Lett.* 29 (9): 977.

**148** Goel, N., Heh, D.; Koveshnikov, S., Ok, I. et al. (2008). Addressing the gate stack challenge for high mobility InxGa1-xAs channels for NFETs, *2008 IEEE International Electron Devices Meeting*, 1–4. https://doi.org/10.1109/IEDM.2008.479669.

**149** Hollinger, G., Skheyta-Kabbani, R., and Gendry, M. (1994). Oxides on GaAs and InAs surfaces: An x-ray-photoelectron-spectroscopy study of reference compounds and thin oxide layers. *Phys. Rev. B* 49: 11159.

**150** Brennan, B. and Hughes, G. (2010). Identification and thermal stability of the native oxides on InGaAs using synchrotron radiation-based photoemission. *J. Appl. Phys.* 108: 053516.

**151** Lin, L. and Robertson, J. (2011). Defect states at III–V semiconductor oxide interfaces. *Appl. Phys. Lett.* 98: 082903.

**152** Hinkle, C.L., Milojevic, M., Sonnet, A.M. et al. (2009). Surface Studies of III-V Materials: Oxidation Control and Device Implications. *ECS Trans.* 19: 387.

**153** Cabrera, W., Halls, M.D., Povey, I.M., and Chabal, Y.J. (2014). Role of interfacial aluminum silicate and silicon as barrier layers for atomic layer deposition of $Al_2O_3$ films on chemically cleaned InP(100) surfaces. *J. Phys. Chem. C* 118 (50): 29164–29179.

**154** Cabrera, W., Halls, M.D., Povey, I.M., and Chabal, Y.J. (2014). Surface oxide characterization and interface evolution in atomic layer deposition of $Al_2O_3$ on InP(100) studied by in situ infrared spectroscopy. *J. Phys. Chem. C* 118 (11): 5862–5871.

**155** Bennett, B.R., Magno, R., Boos, J.B. et al. (2005). Antimonide-based compound semiconductors for electronic devices: a review. *Solid-State Electron.* 49: 1875–1895.

**156** Allen, J.P., Carey, J.J., Walsh, A. et al. (2013). Electronic structures of antimony oxides. *J. Phys. Chem. C* 117 (28): 14759–14769.

**157** Hinkle, C.L., Vogel, E.M., Ye, P.D., and Wallace, R.M. (2011). Interfacial chemistry of oxides on $In_xGa_{(1-x)}As$ and implications for MOSFET applications. *Curr. Opin. Solid State Mater. Sci.* 15: 188.

**158** Lebedev, M.V., Ensling, D., Hunger, R. et al. (2004). Synchrotron photoemission spectroscopy study of ammonium hydroxide etching to prepare well-ordered GaAs (100) surfaces. *Appl. Surf. Sci.* 229: 226–232.

**159** Shin, J., Geib, K.M., Wilmsen, C.W., and Lilliental-Weber, Z. (1894). The chemistry of sulfur passivation of GaAs surfaces. *J. Vacuum Sci. Technol. A* 1990: 8.

**160** O'Connor, E., Long, R.D., Cherkaoui, K. et al. (2008). In situ $H_2S$ passivation of $In_{0.53}Ga_{0.47}As$/InP metal-oxide-semiconductor capacitors with atomic-layer deposited $HfO_2$ gate dielectric'. *Appl. Phys. Lett.* 92: 022902–022902.

**161** Hasegawa, H., Akazawa, M., Domanowska, A., and Adamowicz, B. (2010). Surface passivation of III–V semiconductors for future CMOS devices—Past research, present status and key issues for future. *Appl. Surf. Sci.* 256: 5698–5707.

**162** Lim, H., Carraro, C., Maboudian, R. et al. (2004). Chemical and thermal stability of alkanethiol and sulfur passivated InP(100). *Langmuir* 20: 743–747.

**163** O'Connor, É., Monaghan, S., Long, R.D. et al. (2009). Temperature and frequency dependent electrical characterization of $HfO_2/In_xGa_{1-x}As$ interfaces using capacitance-voltage and conductance methods. *Appl. Phys. Lett.* 94: 102902.

**164** Hou, C.H., Chen, M.C., Wu, T.B., and Chiang, C.D. (2008). Interfacial Cleaning Effects in Passivating InSb with $Al_2O_3$ by Atomic Layer Deposition. *Electrochem. Solid-State Lett.* 11 (6): D60–D63.

**165** Peralagu, U., Povey, I.M., Carolan, P. et al. (2014). Electrical and physical characterization of the $Al_2O_3$/*p*-GaSb interface for 1%, 5%, 10%, and 22% $(NH_4)_2S$ surface treatments. *Appl. Phys. Lett.* 105: 162907.

**166** Fu, Y.-C., Peralagu, U., Millar, D.A.J. et al. (2017). The impact of forming gas annealing on the electrical characteristics of sulfur passivated $Al_2O_3/In_{0.53}Ga_{0.47}As$ (110)metal-oxide-semiconductor capacitors. *Appl. Phys. Lett.* 110: 142905.

**167** Cabrera, W., Brennan, B., Dong, H. et al. (2014). Diffusion of $In_{0.53}Ga_{0.47}As$ elements through hafnium oxide during post deposition annealing. *Appl. Phys. Lett.* 104: 011601.

**168** Djara, V., Cherkaoui, K., Schmidt, M. et al. (2012). Impact of forming gas annealing on the performance of surface-channel $In_{0.53}Ga_{0.47}As$ MOSFETs with an ALD $Al_2O_3$ gate dielectric. *IEEE Trans. Electron Devices* 59: 108.

**169** Zhernokletov, D.M., Dong, H., Barry Brennan, B. et al. (2013). Investigation of arsenic and antimony capping layers, and half cycle reactions during atomic layer deposition of $Al_2O_3$ on GaSb(100). *J. Vacuum Sci. Technol. A* 31: 060602.

**170** Losurdo, M., Capezzuto, P., Bruno, G. et al. (2002). $N_2$–$H_2$ remote plasma nitridation for GaAs surface passivation. *Appl. Phys. Lett.* 81: 16–18.

**171** O'Connor, É., Monaghan, S., Cherkaoui, K. et al. (2011). Analysis of the minority carrier response of n-type and p-type Au/Ni/$Al_2O_3$/$In_{0.53}Ga_{0.47}As$ capacitors following optimized $(NH_4)_2S$ treatment. *Appl. Phys. Lett. A* 99: 212901.

172 Hurley, P.K., O'Connor, É., Djara, V. et al. (2013). The Characterization and Passivation of Fixed Oxide Charges and Interface States in the $Al_2O_3$/InGaAs MOS System. *IEEE Trans. Device Mater. Reliab.* 13: 429.

173 Ye, P.D., Wilk, G.D., Yang, B. et al. (2003). GaAs metal–oxide–semiconductor field-effect transistor withnanometer-thin dielectric grown by atomic layer deposition. *Appl. Phys. Lett.* 2003 (83): 180.

174 Frank, M.M., Wilk, G.D., Starodub, D. et al. (2004). $HfO_2$ and $Al_2O_3$ gate-dielectrics on GaAs grown by atomic layer deposition. *Appl. Phys. Lett.* 86: 152904.

175 Lee, K.Y., Lee, Y.J., Chang, P. et al. (2008). Achieving 1nm capacitive effective thickness in atomic layerdeposited $HfO_2$ on $In_{0.53}Ga_{0.47}As$. *Appl. Phys. Lett.* 92: 252908.

176 Milojevic, M., Aguirre-Tostado, F.S., Hinkle, C.L. et al. (2008). Half-cycle atomic layer deposition reaction studies of $Al_2O_3$ on $In_{0.2}Ga_{0.8}As$ (100) surfaces. *Appl. Phys. Lett.* 93: 202902.

177 Hinkle, C.L., Sonnet, A.M., Vogel, E.M. et al. (2008). GaAs interfacial self-cleaning by atomic layer deposition. *Appl. Phys. Lett.* 92: 071901.

178 Klejna, S. and Elliott, S.D. (2012). First-principles modeling of the "clean-up" of native oxidesduring atomic layer deposition onto III–V substrates. *J. Phys. Chem. C* 116: 643.

179 Aguirre-Tostado, F.S., Milojevic, M., Hinkle, C.L. et al. (2008). Indium stability on InGaAs during atomic H surface cleaning. *Appl. Phys. Lett.* 92: 171906.

180 Shahrjerdi, D., Oye, M.M., Holmes, A.L., and Banerjee, S.K. (2006). Unpinned metal gate/high-κ GaAs capacitors: Fabrication and characterization. *Appl. Phys. Lett.* 89: 043501.

181 Hackley, J.C., Demaree, J.D., and Gougousi, T. (2008). Interface of atomic layer deposited $HfO_2$ films on GaAs (100) surfaces. *Appl. Phys. Lett.* 91: 162902.

182 Suri, R., Lichtenwalner, D.J., and Misra, V. (2010). Interfacial self cleaning during atomic layer deposition and annealing of $HfO_2$ films on native (100)-GaAs substrates. *Appl. Phys. Lett.* 96: 112905.

183 Kim, J.C., Cho, Y.S., and Moon, S.H. (2009). Atomic Layer Deposition of $HfO_2$ onto Si Using $Hf(NMe_2)_4$. *Jpn. J. Appl. Phys.* 48: 066515.

184 McDonnell, S., Dong, H., Hawkins, J.M. et al. (2012). Interfacial oxide re-growth in thin film metal oxide III-V semiconductor systems. *Appl. Phys. Lett.* 100: 141606.

185 O'Mahony, A., Monaghan, S., Chiodo, R. et al. (2010). Structural and electrical analysis of thin interface control layers of MgO or $Al_2O_3$ deposited by atomic layer deposition and incorporated at the high-k/III-V interface of $MO_2/In_xGa_{1-x}As$ (M = Hf|Zr, x = 0.53) gate stacks. *ECS Trans.* 33: 69.

186 Liao, L. and Duan, X. (2010). Graphene – dielectric integration for graphene transistors. *Mater. Sci. Eng., R* 70: 354–370.

187 Kim, H., Lee, H.-B.-R., and Maeng, W. (2009). Applications of atomic layer deposition to nanofabrication and emerging nanodevices. *Thin Solid Films* 517: 2563–2580.

**188** Garces, N.Y., Wheeler, V.D., and Gaskill, D.K. (2012). Graphene functionalization and seeding for dielectric deposition and device integration. *J. Vac. Sci. Technol., B* 30: 030801.

**189** Dlubak, B., Kidambi, P.R., Weatherup, R.S. et al. (2012). Substrate-assisted nucleation of ultra-thin dielectric layers on graphene by atomic layer deposition. *Appl. Phys. Lett.* 100: 173113.

**190** Choi, K., Lee, Y.T., and Im, S. (2016). Two-dimensional van Der Waals nanosheet devices for future electronics and photonics. *Nano Today* 11: 626–643.

**191** Lapshin, R.V. (1998). Automatic lateral calibration of tunneling microscope scanners. *Rev. Sci. Instrum.* 69: 3268–3276.

**192** Williams, J.R., DiCarlo, L., and Marcus, C.M. (2007). Quantum hall effect in a gate-controlled p-n junction of graphene. *Science* 317: 638–641.

**193** Farmer, D.B. and Gordon, R.G. (2006). Atomic layer deposition on suspended single-walled carbon nanotubes via gas-phase noncovalent functionalization. *Nano Lett.* 6: 699–703.

**194** Long, B., Manning, M., Burke, M. et al. (2012). Non-covalent functionalization of graphene using self-assembly of alkane-amines. *Adv. Funct. Mater.* 22: 717–725.

**195** McDonnell, S., Pirkle, A., Kim, J. et al. (2012). Trimethylaluminum and ozone interactions with graphite in atomic layer deposition of $Al_2O_3$. *J. Appl. Phys.* 112: 104110.

**196** Kim, J., Lee, B., Park, S.Y. et al. (2008). Conformal $Al_2O_3$ dielectric layer deposited by atomic layer deposition for graphene-based nanoelectronics. *Appl. Phys. Lett.* 92: 203102.

**197** Sundaram, G., Monsma, D., and Becker, J. (2008). Leading edge atomic layer deposition applications. *ECS Trans.* 16: 19–27.

**198** Speck, F., Ostler, M., Röhrl, J. et al. (2010). Atomic layer deposited aluminum oxide films on graphite and graphene studied by XPS and AFM. *Phys. Status Solidi C* 7: 398–401.

**199** Pirkle, A., McDonnell, S., Lee, B. et al. (2010). The effect of graphite surface condition on the composition of $Al_2O_3$ by atomic layer deposition. *Appl. Phys. Lett.* 97: 082901.

**200** Muñoz, R. and Gómez-Aleixandre, C. (2013). Review of CVD synthesis of graphene. *Chem. Vap. Deposition* 19: 297–322.

**201** Entani, S., Sakai, S., Matsumoto, Y. et al. (2010). Interface properties of metal/graphene heterostructures studied by micro-raman spectroscopy. *J. Phys. Chem. C* 114: 20042–20048.

**202** Jandhyala, S., Mordi, G., Lee, B. et al. (2012). Atomic layer deposition of dielectrics on graphene using reversibly physisorbed ozone. *ACS Nano* 6: 2722–2730.

**203** Jandhyala, S., Mordi, G., Lee, B., and Kim, J. (2012). In-situ electrical studies of ozone based atomic layer deposition on graphene. *ECS Trans.* 45: 39–46.

**204** Wang, X., Tabakman, S.M., and Dai, H. (2008). Atomic layer deposition of metal oxides on pristine and functionalized graphene. *J. Am. Chem. Soc.* 130: 8152–8153.

**205** Alaboson, J.M.P., Wang, Q.H., Emery, J.D. et al. (2011). Seeding atomic layer deposition of high-$k$ dielectrics on epitaxial graphene with organic self-assembled monolayers. *ACS Nano* 5: 5223–5232.

**206** Liu, G., Stillman, W., Rumyantsev, S. et al. (2009). Low-frequency electronic noise in the double-gate single-layer graphene transistors. *Appl. Phys. Lett.* 95: 033103.

**207** Meric, I., Han, M.Y., Young, A.F. et al. (2008). Current saturation in zero-bandgap, top-gated graphene field-effect transistors. *Nat. Nanotechnol.* 3: 654–659.

**208** Jeong, S.-J., Kim, H.W., Heo, J. et al. (2016). Physisorbed-precursor-assisted atomic layer deposition of reliable ultrathin dielectric films on inert graphene surfaces for low-power electronics. *2D Mater.* 3: 035027.

**209** Aria, A.I., Nakanishi, K., Xiao, L. et al. (2016). Parameter space of atomic layer deposition of ultrathin oxides on graphene. *ACS Appl. Mater. Interfaces* 8: 30564–30575.

**210** Park, Y.H., Kim, M.H., Kim, S.B. et al. (2016). Enhanced nucleation of high-$k$ dielectrics on graphene by atomic layer deposition. *Chem. Mater.* 28: 7268–7275.

**211** Zheng, L., Cheng, X.H., Cao, D. et al. (2014). Improvement of $Al_2O_3$ films on graphene grown by atomic layer deposition with pre-$H_2O$ treatment. *ACS Appl. Mater. Interfaces* 6: 7014–7019.

**212** Novoselov, K.S., Fal'ko, V.I., Colombo, L. et al. (2012). A roadmap for graphene. *Nature* 490: 192–200.

**213** Mak, K.F. and Shan, J. (2016). Photonics and optoelectronics of 2D semiconductor transition metal dichalcogenides. *Nat. Photonics* 10: 216–226.

**214** Liu, H., Xu, K., Zhang, X., and Ye, P.D. (2012). The integration of high-k dielectric on two-dimensional crystals by atomic layer deposition. *Appl. Phys. Lett.* 100: 152115.

**215** Radisavljevic, B., Radenovic, A., Brivio, J. et al. (2011). Single-layer $MoS_2$ transistors. *Nat. Nanotechnol.* 6: 147–150.

**216** Wang, H., Yu, L., Lee, Y.H. et al. (2012). Integrated circuits based on bilayer $MoS_2$ transistors. *Nano Lett.* 12: 4674–4680.

**217** Mcdonnell, S., Brennan, B., Azcatl, A. et al. (2013). $HfO_2$ on $MoS_2$ by atomic layer deposition: adsorption mechanisms and thickness scalability. *ACS Nano* 7: 10354–10361.

**218** Park, S., Kim, S.Y., Choi, Y. et al. (2016). Interface properties of atomic-layer-deposited $Al_2O_3$ thin films on ultraviolet/ozone-treated multilayer $MoS_2$ crystals. *ACS Appl. Mater. Interfaces* 8: 11189–11193.

**219** Yang, J., Kim, S., Choi, W. et al. (2013). Improved growth behavior of atomic-layer-deposited high-k dielectrics on multilayer $MoS_2$ by oxygen plasma pretreatment. *ACS Appl. Mater. Interfaces* 5: 4739–4744.

**220** Liu, W., Kang, J., Sarkar, D. et al. (2013). Role of metal contacts in designing high-performance monolayer N type $WSe_2$ field effect transistors. *Nano Lett.* 13: 1983–1990.

**221** Zhu, H., Addou, R., Wang, Q. et al. (2020). Surface and interfacial study of atomic layer deposited $Al_2O_3$ on $MoTe_2$ and $WTe_2$. *Nanotechnology* 31: 055704.

**222** Kim, T., Mun, J., Park, H. et al. (2017). Wafer-scale production of highly uniform two-dimensional $MoS_2$ by metal-organic chemical vapor deposition. *Nanotechnology* 28: 18LT01.

**223** Kang, K., Xie, S., Huang, L. et al. (2015). High-mobility three-atom-thick semiconducting films with wafer-scale homogeneity. *Nature* 520: 656–660.

**224** Lin, L., Monaghan, S., Sakhuja, N. et al. (2020). Large-area growth of $MoS_2$ at temperatures compatible with integrating back-end-of-line functionality. *2D Mater.* 8, 8025008.

**225** Hinterding, R. and Feldhoff, A. (2019). Two-dimensional oxides: recent progress in nanosheets: a retrospection on synthesis, microstructure and applications. *Z. Phys. Chem.* 233: 117–165.

**226** Wang, Z., Zhu, W., Qiu, Y. et al. (2016). Biological and environmental interactions of emerging two-dimensional nanomaterials. *Chem. Soc. Rev.* 45: 1750–1780.

**227** Sethurajaperumal, A., Ravichandran, V., Banerjee, A. et al. (2021). Two-dimensional layered nanosheets. *Micro and Nano Technol.* 465–497.

**228** Hwang, H., Iwasa, Y., Kawasaki, M. et al. (2012). Emergent phenomena at oxide interfaces. *Nat. Mater.* 11: 103–113.

# 2

# Addressing Complex Transition Metal Oxides at the Nanoscale: Bottom-Up Syntheses of Nano-objects and Properties

*David Portehault, Francisco Gonell, and Isabel Gómez-Recio*

Sorbonne Université, CNRS, Laboratoire de Chimie de la Matière Condensée de Paris (CMCP), 4 Place Jussieu, F-75005, Paris, France

## 2.1 Introduction

Metal oxides encompass a very large family of solids reported in structural databases. This extreme versatility in compositions and structures yields a wide range of properties. These properties are very often combined to stability under ambient atmosphere to set the foundations for an extended range of applications. While the family of bulk oxides is still expending, nanoscaling offers an additional dimension to tune or to discover properties.

The relevance of nanostructuration for oxides is demonstrated by many application fields pertaining to nanoparticles, e.g. nanomedicine [1], catalysis [2], energy conversion, and storage [3]. It is also well exemplified by the field of oxide epitaxial thin films. Indeed, planar deposition techniques offer a mature and versatile toolbox to deposit various oxide films and assess the role of nanoscale dimensions on magnetotransport and magnetism [4]. It also provides interesting models to study the mechanisms of catalytic and electrocatalytic [5–7] reactions over oxides, including the structural and chemical modifications occurring at the oxide surface during these reactions. Epitaxial thin film deposition can also be used to finely tune the structural features of the oxide film. For instance, the strain can be adjusted by lattice mismatch with the substrate, and the nanoscale thickness as well as the interface can be used to stabilize compositions that are very difficult to isolate otherwise or even not reported at the bulk scale [8–10].

Nano-objects as nanoparticles, nanowires, and 2D materials offer other possibilities than epitaxial thin films. Adjusting the dimensionality enables probing the effect of nanoscale sizes on properties such as surface plasmon resonance [11], and developing free-standing nano-objects that can be suspended in liquids opens the way to processing by using the nano-objects as building blocks of larger structures [12]. Growth of epitaxial thin films of complex structures and compounds, like many multicationic oxides, is well controlled because the conditions of pressure, temperature, and strain conditions enable the growth on well-chosen substrates that act as seeds,

*Tailored Functional Oxide Nanomaterials: From Design to Multi-Purpose Applications*, First Edition.
Edited by Chiara Maccato and Davide Barreca.
© 2022 WILEY-VCH GmbH. Published 2022 by WILEY-VCH GmbH.

thus limiting the energy input required for the nucleation of the compounds, which is typically achieved for bulk oxides by solid-state reactions above 1000 °C. The synthesis of free-standing nano-objects of such compounds becomes a challenging task as we cannot benefit from the advantages of epitaxial substrates. This may be one of the reasons why the range of oxides reported as nano-objects is very limited compared with the wide compositional and structural range of bulk oxides. Especially, most works on nano-objects have focused on crystal structures achievable at low temperature, and/or on simple compositions, especially single cation oxides. The other oxides, which may be qualified as "complex" in terms of structure and/or composition, have mostly been left aside. As many bulk complex oxides or oxide thin films show unique magnetic, electronic, or spin properties, it is of utmost fundamental, if not applied, interest to assess how these properties can be impacted by decreasing dimensionality in nano-objects.

Achieving the synthesis of nano-objects of oxides that have been only reported in the bulk state or as thin films calls for the use of advanced synthesis methods that can yield free-standing nano-objects. Last years have seen a tremendous boost in this effort. Techniques that could have been considered as traditional, such as the sol–gel process, aqueous precipitation, and colloidal syntheses in organic solvents, have been revisited to be applied to nanomaterials pertaining to the family of complex oxides, while new approaches, including high-temperature inorganic solvents and dedicated solid-state reactions, have been developed to reach new oxides at the nanoscale. In this chapter, we aim to provide an overview of such recent advances, with an emphasis on new or revisited synthesis approaches and on the link between the targeted compounds and the chosen synthesis conditions.

We will focus herein on three large case studies: multicationic oxides, oxides of titanium and tungsten bearing uncommon metal oxidation state or crystal structure, and crystal structures that have been discovered only in nano-objects. By addressing these wide families of solids, we deal with the following specific synthetic questions. How to obtain controlled nano-objects with low cost and environmentally friendly aqueous media? What strategies, relying on crystal structure relationships, can we use to target solids bearing transition metal oxidation states that are poorly stable in aqueous or organic media? How to drive reaction pathways toward unexplored areas of composition and structure maps? We will mostly show that answers to these questions can be found by reinvestigating aqueous precipitation processes, colloidal chemistry in organic solvents, the coupling of the sol–gel process with high-temperature treatments and hard templating and by exploring molten salts as high-temperature liquid media. We will discuss not only the underlying mechanisms that encompass dissolution–recrystallization but also a range of reactions within nano-objects, including topotactic transformations by cation exchange, exfoliation, and reduction.

## 2.2 Multicationic Oxides

### 2.2.1 Layered Oxide-Based Materials

Two-dimensional oxides have attracted considerable attention in the last years, owing to the properties derived from their layered structures. The capacity to store and/or to exchange species in the interlayer space, as well as the high accessibility to the catalytic active centers, make them interesting in energy storage and catalytic applications [13, 14]. We choose herein two illustrative examples of layered oxides with simple composition: nickel and cobalt hydroxides. Solids derived from both elements are disclosed later with the aim of searching more advanced materials with multicationic compositions.

Layered nickelates show several structures that can be obtained as a function of the nickel oxidation state as well as of the synthesis conditions. α-Ni(OH)$_2$ is a kinetic reaction intermediate obtained by precipitating Ni$^{2+}$ in basic aqueous solution, commonly NaOH or NH$_3$. This polymorph evolves to most stable β-Ni(OH)$_2$ by aging in aqueous media or heating in basic conditions [15].

Both α-Ni(OH)$_2$ and β-Ni(OH)$_2$ are composed of Ni(OH)$_2$ layers. In α-Ni(OH)$_2$, water or intercalated ions are located between these layers while no space is available in β-Ni(OH)$_2$ for accommodating these species [15]. Doping the layers with other elements, such as trivalent aluminum, has increased the α-polymorph stability [16]. In this case, the positive charge introduced in the layers is compensated by the interlayer anions that remain strongly bonded. Likewise, the ability to intercalate cations in the interlayer space is of high interest for the electrochemical properties of α-Ni(OH)$_2$ as positive electrode in alkaline batteries [17] or electrocatalyst for the oxygen evolution reaction (OER) [18, 19]. Indeed, the substitution of nickel with small amounts of iron, incorporated on purpose or from an impurity in the electrolyte, produces one of the most active materials for the OER [19, 20]. Additionally, ultrathin Ni(0)-embedded α-Ni(OH)$_2$ heterostructured nanosheets obtained by solvothermal synthesis have shown remarkable activity in hydrogen evolution reaction (HER) and OER electrocatalysis [21]. These performances have been shown to arise from the Ni(0)–Ni(II) active interface and the Ni(II) defects, respectively, making Ni(0)/α-Ni(OH)$_2$ heterostructures highly interesting as durable and precious metal-free electrocatalysts for water splitting [21].

Oxyhydroxides of Ni(III) can be obtained by precipitating a nickel(II) salt in basic and oxidizing conditions [22]. β-NiOOH and γ-NiOOH polymorphs are common positive electrodes of alkaline batteries through the redox couples β-NiOOH/β-Ni(OH)$_2$ and γ-NiOOH/γ-Ni(OH)$_2$, respectively [23].

Inspired by methodologies of metal–organic framework (MOF) crystallization, a one-pot synthesis of single-layered metal–organic nanoribbons built from nickel-oxo octahedral clusters was accomplished [24]. For this purpose, alkyl benzene monocarboxylate was used instead of rigid acrylic dicarboxylate linkers used for MOFs growth. The flexibility of the alkyl chain prevents conventional 3D growth, the monocarboxylate moieties acting as molecular spacers, pointing toward opposite directions. The inorganic clusters are bifunctionalized, so that the four equatorial positions of the octahedral clusters can undergo condensation. This produces consecutive edge-sharing $NiO_4(OH)_2$ octahedra, forming inorganic chains that are separated by alkyl benzene monocarboxylate ligands at both sides of the inorganic nodes (Figure 2.1a). The inter-cluster separation can be controlled by modifying the molecular spacer length. Post-synthesis treatment in nonpolar solvents allows the expansion of the ribbons, then the dispersion, and the exfoliation of the metal–organic 1D chains. These lamellar hybrid materials can be used to store and release, in a controlled fashion, nonpolar active drugs such as semiochemicals (Figure 2.1b).

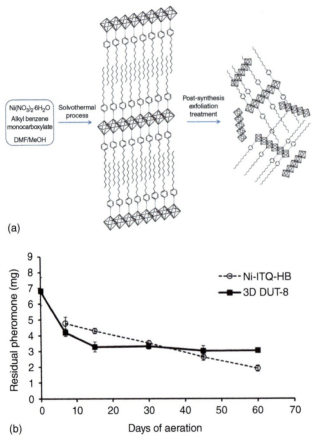

**Figure 2.1** (a) Synthesis of layered $NiO_6$-based hybrid materials obtained after condensation and then after further exfoliation in apolar solvents. (b) Kinetics of the residual pheromone (3-(S)-methyl-6-(R,S)-isopropenyl-9-decenyl acetate) in the layered $NiO_6$-based hybrid material (Ni-ITQ-HB) and 3D DUT-8(Ni) MOF. Source: Moreno et al. [25] with permission from Wiley-VCH Verlag GmbH & Co. KGaA.

As for nickel, cobalt hydroxides exhibits two polymorphs: α-Co(OH)$_2$ and β-Co(OH)$_2$. Layered cobalt oxyhydroxides show significant deviations to stoichiometry that drive various behaviors. For instance, the spontaneous transformation of α-Co(OH)$_2$ to more stable β-Co(OH)$_2$ can be hindered by introducing exchangeable anions in the interlayer space that compensate the positive charge of the Co(OH)$_{2-x}$ layers [26]. Higher oxidation state phases related to α-Co(OH)$_2$ and β-Co(OH)$_2$ show also various conduction behaviors. Especially, β-CoOOH is insulating, while γ-CoOOH is semi-metallic, thanks to the presence of higher amounts of tetravalent cobalt that is concomitant to non-stoichiometry in the form of, e.g. γ-H$_x$CoO$_2$ [27, 28]. The synthesis of cobalt oxyhydroxides is usually performed by treating cobalt(II) hydroxides under oxidizing conditions; indeed β-Co(OH)$_2$ and α-Co(OH)$_2$ evolve through topochemical transformations to β-CoOOH [29, 30] and γ-CoOOH [28], respectively. The transformation of β-Co(OH)$_2$ to γ-CoOOH does not occur, at the opposite to nickel hydroxides to oxyhydroxides transformations, probably because of the difference in iono-covalence of the M—O bond [28].

Mixed nickel–cobalt oxyhydroxides have been synthesized with interesting capacities and electrocatalytic properties [30–34]. β-Ni$_{0.89}$Co$_{0.11}$(OH)$_2$ platelets of 20–30 nm diameter obtained by coprecipitation with ammonia show higher capacities compared with the cobalt-free materials as positive electrodes in Ni-based rechargeable batteries [31].

Hierarchically structured mixed nickel–cobalt hydroxides have shown better performances than their non-structured counterparts in OER. Ni–Co (Co/Ni ratio 1/1)-layered double hydroxides, hierarchically composed by primary (smaller) and secondary (bigger) nanosheets (Figure 2.2a) obtained hydrothermally by urea hydrolysis, evolve to Ni–Co spinel through topotactic transformation by calcination. These Ni–Co oxide hierarchical nanosheets (NCO-HNSs) promote more efficiently the OER in basic media compared with Co$_3$O$_4$ and Ni–Co oxide nanorods (Co$_3$O$_4$-NRs and NCO-NRs, respectively) (Figure 2.2b). This activity enhancement arises by synergy between the large surface area provided by the 3D hierarchical structure and the promotion of the formation of nickel oxyhydroxide on the surface that acts as OER electrocatalytic active species [32]. Amorphous Ni–Co hydroxide nanocages with tunable composition (Figure 2.2c), synthesized using Cu$_2$O as template, also combine the high accessibility of the reagent with catalyst stability under *operando* conditions in OER in basic conditions, providing highly active and stable electrocatalyst (Figure 2.2d) [33]. O$_2$ bubbles formation on electrode surface during OER can hamper the electrolyte–catalyst contact, thus decreasing the catalytic activity. To overcome this limitation, electrodeposited "forests" of amorphous α-nickel–cobalt hydroxide nanodendrites (Figure 2.2e) have been developed to exhibit superhydrophilicity and ensure excellent contact between the electrode material and the electrolyte and then to provide the catalyst with high activity and durability in OER (Figure 2.2f) [34].

In an attempt to further tune electrocatalytic properties, attention has been dedicated to widening the composition range of Ni–Co-mixed-layered compounds. Nano-objects are ideal platforms for such a study because they accommodate larger strains than bulk materials. Hence, they might enable spanning a wide substitution

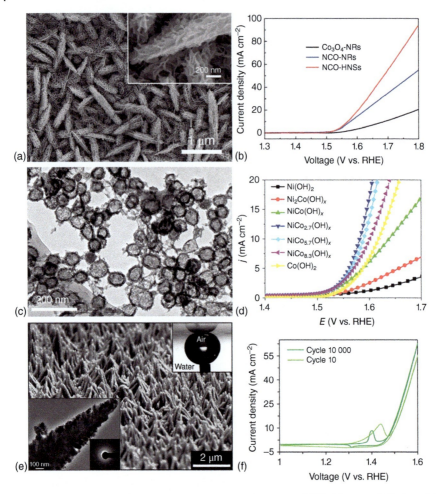

**Figure 2.2** (a) Field emission scanning electron microscopy (FESEM) image of Ni–Co oxide hierarchical nanosheets NCO-HNSs, (b) Linear sweep voltammetry cycles (LSVs) in OER conditions in basic media of NCO-HNSs, $Co_3O_4$ and Ni–Co oxide nanorods ($Co_3O_4$-NRs and NCO-NRs, respectively). (c) Transmission electron microscopy (TEM) micrograph of amorphous Ni–Co hydroxide nanocages. (d) OER LSVs of amorphous Ni–Co hydroxide nanocages with several Ni:Co ratios. (e) SEM and TEM images of amorphous α-nickel–cobalt hydroxide nanodendrite forests. The inset shows the superaerophobic behavior of the material and (f) its corresponding cyclic voltammetry (CV) before (10 cycles) and after (10 000 cycles) the accelerated stability test. Source: (a, b) Wang et al. [32], (c, d) Nai et al. [33] with permission from Wiley-VCH Verlag GmbH & Co. KgaA, and (e, f) Balram et al. [34] with permission from American Chemical Society, copyright 2017.

range in $Na_{0.6}CoO_2$-based Ni–Co mixed oxides. $Na_{0.6}Ni_{0.5}Co_{0.5}O_2$ nanoflakes of 2–10 nm basal face and 1.5–3 nm thickness have been obtained through topotactic transformation starting from Ni-substituted β-CoOOH and exchanging $H^+$ by $Na^+$ under soft conditions at 120 °C (Figure 2.3a,b). Decreasing the particle size to this level allows surpassing the Ni solubility achieved in the bulk state (Ni:Co 0.1 : 0.9 ratio) [36, 37]. The large surface area and the ability to preserve cobalt in high

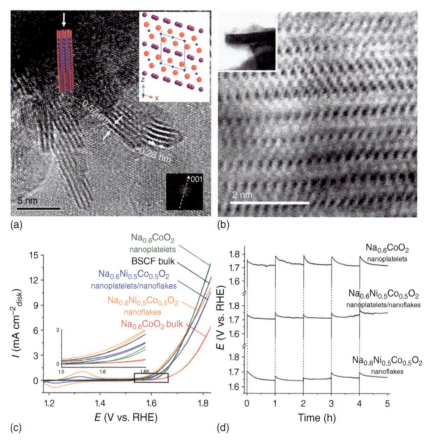

**Figure 2.3** (a) High resolution TEM (HRTEM) and (b) atomically resolved scanning TEM (STEM) images of $Na_{0.6}Ni_{0.3}Co_{0.7}O_2$. (c) CVs in OER conditions of sodium-layered Ni–Co oxides with several compositions and morphologies, compared to reference bulk $Ba_{0.5}Sr_{0.5}Co_{0.8}Fe_{0.2}O_3$ (BSCF bulk) and (d) chronopotentiometry curves at current density 5 mA cm$^{-2}$. The insets in (a) display the crystallographic orientation of the platelets. The inset in (b) show a low magnification image of a platelet. Source: Azor et al. [35] with permission from American Chemical Society, copyright 2018.

oxidation state ($Co^{3+}/Co^{4+}$) make these materials highly active and stable in OER (Figure 2.3c,d) [35].

Exfoliation of layered materials is also a good strategy for obtaining high-aspect ratio nanosheets with high surface area, ideal for applications where the active surface plays a key role. A general strategy consists in a first proton exchange of the interlayer species followed by amine intercalation. Successive intercalation of larger amines allows decreasing the interaction between layers and splitting them apart in colloidal suspensions. This approach has been successfully applied in a variety of cases, including lepidocrocite-layered titanates [38], protonated Ruddlesden–Popper phases [39], layered manganates [40], and Aurivillius phase tantalates [24]. Microwave heating fastens the exchange and intercalation steps [24], while selective centrifugation at different speeds allows the isolation of nanosheets with different levels of agglomeration [41].

Layered $Mn_{1-x}Co_xO_2$ ($x = 0.2$–$0.25$) nanosheets have been synthesized following this exfoliation strategy [42]. The $Na_{0.6}Mn_{1-x}Co_xO_2$ was acid-exchanged and then exfoliated with tetra-$n$-butylammonium cations yielding single-layered $Mn_{1-x}Co_xO_2$ nanosheets with 0.8 nm of thickness. Compared to other layered manganese and cobalt-based oxides obtained through bottom-up approaches, the material obtained by exfoliation shows larger lateral size and higher crystallinity, which is beneficial for electrochemical devices application. Indeed, these exfoliated nanosheets showed improved cycle performance compared with $MnO_2$ and higher capacitance than $Mn_{1-x}Ru_xO_2$ nanosheets in Li-ion batteries.

In this section, Ni and Co oxides have been discussed as illustrative examples of layered materials. Even showing simple composition, the corresponding hydroxides show a rich variety of structures obtained through control of the basic and oxidation–reduction synthesis conditions. Also, the versatility of organic moieties, such as monocarboxylates, allows the production of single-layered metal–organic nickel oxide nanoribbons. A good control over morphology of the mixed Ni–Co hydroxides has been accomplished in aqueous synthesis conditions, with great implications in electrocatalytic applications. Finally, the more challenging synthesis of layered materials with complex compositions is illustrated with materials obtained through soft cation exchange or chemical exfoliation. The synthesis of these materials at the nanoscale enlarges the chemical composition in a range not achievable in the bulk state.

### 2.2.2 Oxidation States Stable in Organic Media: The Case of Perovskites

$BaTiO_3$ is one of the most important perovskites because of its piezoelectric, ferroelectric, and dielectric properties, which yield important applications in electro-ceramics and materials for optics [43, 44]. The synthesis of $BaTiO_3$ by liquid-phase methods has allowed producing small and discrete nano-objects with different ferroelectric and dielectric properties compared with the bulk state, which depends on the crystal size [44].

The first class of methods are aqueous-phase syntheses. Especially hydrothermal synthesis of $BaTiO_3$ is well known, and the reaction mechanisms and kinetics have been deeply studied [45–47]. A barium precursor ($BaCl_2$ or $Ba(OH)_2$) and $TiO_2$ (usually amorphous or anatase polymorph) are mixed in a basic aqueous solution and heated under hydrothermal conditions, typically at 80–200 °C. Two synthesis pathways have been identified (Figure 2.4): in situ transformation and dissolution–reprecipitation. During in situ transformation, dissolved barium species react with suspended $TiO_2$ particles, forming a shell of $BaTiO_3$. Then $Ba^{2+}$ diffuses through the particle until $TiO_2$ has completely reacted (Figure 2.4a). In the dissolution–reprecipitation pathway, $BaTiO_3$ crystallization goes through dissolution of $TiO_2$, formation of titanium and barium species that undergo condensation yielding homogeneous nucleation of the titanate or its heterogeneous nucleation at the surface of remaining $TiO_2$ (Figure 2.4b) [47].

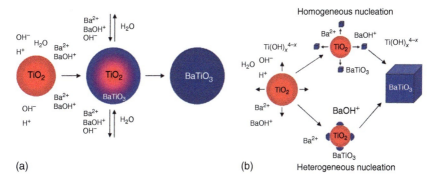

**Figure 2.4** Hydrothermal crystallization mechanisms proposed for barium titanate: (a) in situ transformation and (b) dissolution–reprecipitation pathway where homogeneous nucleation occurs in the solution or heterogeneous nucleation occurs on $TiO_2$. Source: Modeshia et al. [45] with permission from Royal Society of Chemistry.

The high versatility of hydrothermal synthesis allows the design of nanostructures with controlled morphology such as ultrathin nanowires [48] or nanotubes [49]. On one side, nanowires with high aspect ratio (more than 1300) and a diameter of less than 10 nm (Figure 2.5a–d) were synthesized using titanium alkoxide and barium hydroxide precursors dissolved in a basic aqueous ethanol solution of polyethylene glycol and treated under hydrothermal conditions. The nanowires formation occurs through a combination of Ostwald ripening and cation exchange. Under solvothermal conditions, *tert*-butyl titanate reacts with KOH producing $K_2Ti_8O_{17}$ intermediate. Then, potassium is exchanged by barium species in solution. On the other side, hydrothermal treatment of $H_2TiO_3$ exchanges $H^+$ with barium species in an aqueous ethanol solution producing $BaTiO_3$ nanotubes [49]. Both materials show high microwave radiation absorption [48, 49].

The stability of $Ba^{2+}$ and $Ti^{4+}$ species allows the synthesis of $BaTiO_3$ in organic solvents, which provide access to a wide range of surface binding molecules – solvent or ligands – to control the particle size and shape. Organic solvents also enable the use of alternative soluble precursors compared to aqueous media, especially single source barium-titanium alkoxides [50]. 8 nm barium titanate nanoparticles with narrow size distribution could be readily obtained from such a single source precursor dissolved in diphenyl ether in the presence of oleic acid. The injection of hydrogen peroxide followed by heating at 100 °C over 48 hours was used to trigger the decomposition of the single source complex (Figure 2.5e,f). By increasing or decreasing the alkoxide/oleic acid ratio, the particle size could be tuned from 10 to 5 nm, respectively [50]. Likewise, thermal decomposition of barium-titanium isopropoxide in heptadecane and in the presence of $H_2O_2$ yielded single crystalline barium titanate nanorods [52]. Solvothermal synthesis in a water–butanol mixture in basic conditions and in the presence of oleic acid also produced well-faceted barium titanate cubes easily dispersible in nonpolar solvents [53].

The high versatility of the benzyl alcohol solvent route was also instrumental in developing highly dispersible $BaTiO_3$ nanoparticles of 6 nm diameter (Figure 2.5g,h) [51, 54]. The synthesis mechanism involves Ti–O–Ti bridge formation via C–C

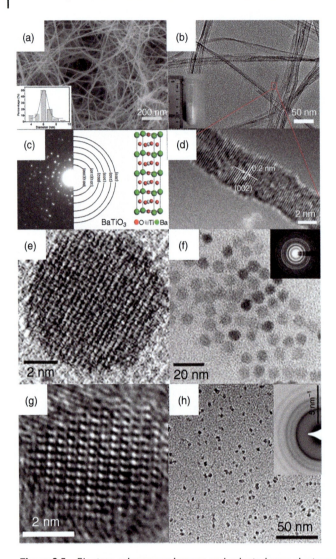

**Figure 2.5** Electron microscopy images and selected area electron diffraction (SAED) of BaTiO$_3$ (a–d) ultrathin nanowires, nanoparticles obtained in (e, f) diphenylether and oleic acid, and (g, h) benzyl alcohol synthesis. The inset in (a) shows the diameter distribution of the nanowires, the insets in (f) and (h) display corresponding SAED patterns. Source: (a–d) Yang et al. [48], copyright 2013; (e–f) from O'Brien et al. [50], copyright 2001; and (g–h) from Niederberger et al. [51], copyright 2004, with permission from American Chemical Society.

coupling between isopropoxyl ligands and benzyl alcohol, which produces 4-phenyl-2-butanol [51]. In such routes, benzyl alcohol plays the role of solvent, reagent, and surface ligand (or precursor of surface ligand). Hence, the approach offers a very simple path by limiting the number of partners to introduce in the reaction medium.

Inorganic molten salts have also been employed to enable the production of well-crystallized barium titanate with several morphologies and controlled particle size. The synthesis of nanowires was first reported by using barium oxalate and anatase $TiO_2$ in molten NaCl with nonylphenyl ether additive as ligand. By treating this mixture at high temperature (820 °C), single-crystalline nanowires of ~50–80 nm diameter and 1.5–10 µm length were obtained [55]. One may however raise the question of the role of the organic ether molecule that is decomposed at the temperature of synthesis. The local temperature raise coming from ether combustion could favor the reaction. Hence, ligand-free synthesis of $BaTiO_3$ was performed in an equimolar NaCl:KCl mixture, again from barium oxalate and anatase. After treating the medium at 950 °C, $BaTiO_3$ nanostrips were obtained [56]. Barium titanate nanostructures with controlled morphology have been described by starting from $TiO_2$ with different nanoscale morphologies and by varying the nature of the $Ba^{2+}$ precursor [57]. When spherical or rod-shaped $TiO_2$ reacted with $BaCO_3$ in NaCl:KCl eutectic mixture at 700 °C, the morphology of the titanium precursor was preserved. On the other side when rod-shaped $TiO_2$ were reacted in the same conditions with BaO, cube-shaped $BaTiO_3$ particles were produced. The authors suggest that the difference in the morphology of the resulting powders arises from the difference in barium precursor solubility. $BaCO_3$ may be more soluble in the reaction media than $TiO_2$, thus directing the synthesis through in situ transformation that results in an isomorphic transformation of the $TiO_2$ particles, keeping the morphology of the less soluble component. On the other side, the lower solubility of BaO, closer to $TiO_2$, gives the chance to break rod-shaped $TiO_2$ in several fragments before reaction with $Ba^{2+}$ species, thus producing the particles with cubic morphology. Using the same synthesis protocol in the NaCl:KCl eutectic mixture from $BaCO_3$, $BaTiO_3$ particles with controllable size were obtained as a function of the thermal treatment (600–1000 °C): the higher the temperature, the larger the particle size [58]. Well-faceted barium titanate nanocubes of 30–50 nm were obtained in softer conditions (200 °C) by using more reactive solvent and barium precursors (NaOH and $BaCl_2$, respectively) and reacting them with $TiO_2$ [59].

This section has shown with the case study of $BaTiO_3$ that the synthesis protocols to reach nanostructures of multicationic oxides with common and stable metal oxidation states span a wide range of reaction media. The stability of the aforementioned cations enables the use of different reaction media for liquid-phase synthesis. If aqueous media rely on environmentally friendly protocols, organic solvents provide the ability to further tune nucleation and growth pathways, especially by enabling the use of a wide range of surface stabilizing species, solvent molecules, or ligands. These species enable facile control of particle size and colloidal stability and can drive growth of anisotropic shapes. If organic species cannot be used in high-temperature inorganic molten salts, the latter however enable to a certain extent the control of particle faceting and shape, for instance, by tuning the relative solubility of the reagents. Nonetheless, inorganic molten salts become an essential tool when one focuses on oxides bearing high oxidation state metal cations, which are not stable in organic media.

### 2.2.3 Oxidation States Poorly Stable in Organic Media: The Case of Perovskites

Organic solvents hold reductive properties that are not compatible with obtaining high oxidation state metal cations. This is especially the case of oxide perovskites bearing $Bi^{3+}$ and $Mn^{4+}$ cations: bismuth-based ferrates and manganates perovskites.

$BiFeO_3$ is an important multiferroic material, showing both ferroelectric and ferromagnetic properties at room temperature. These properties raise interest in memory devices, spintronics, and sensors [60–62]. $BiFeO_3$ also exhibits interesting properties in photocatalysis. $BiFeO_3$ has been obtained through the Pechini process, by using tartaric acid as a complexing agent in an aqueous acid solution of bismuth(III) and iron(II) nitrates [63]. The resulting gel is then calcined under air. When citric acid is used instead of tartaric acid, $BiFeO_3$ is the minor phase accompanied by single cation iron and bismuth oxides as well as other bismuth ferrates. The authors suggest that the difference in phase selectivity arises from the ability of tartaric acid to form a heterometallic polynuclear polymeric network, which ensures a homogeneous mixture of the cations. Upon decomposition, this homogeneous gel yields the targeted phase. On the other side, citric acid forms dimeric bismuth complexes avoiding the formation of Bi–Fe heteronuclear complexes and thus yielding single-metal oxides. Similar approaches in the presence of ethylene glycol [61, 62] produces $BiFeO_3$ nanoparticles because of its ability to cross-link the gel and to enhance the homogeneity of the cationic distribution. The particle size can be tuned with the calcination temperature.

Phase selectivity in the bismuth ferrate system has been accomplished under hydrothermal conditions. $Bi_{12}Fe_{0.63}O_{18.945}$, $Bi_2Fe_4O_9$, and $BiFeO_3$ micro- and nanoparticles have been obtained with tunable morphology by adjusting the precursor ratio and the pH conditions or by adding $H_2O_2$ [64]. An equimolar $Bi^{3+}$:$Fe^{3+}$ starting solution in moderate basic (pH 8–12) media treated under hydrothermal conditions yields sillenite-type compound $Bi_{12}Fe_{0.63}O_{18.945}$ accompanied with $Fe(OH)_3$ that is easily removed by washing. Harsher conditions (pH 14, higher temperature (200 °C) and $H_2O_2$ addition) equalize cation reactivity, allowing the incorporation of both $Bi^{3+}$ and $Fe^{3+}$ in the same structure and yielding pure $BiFeO_3$. Similar results are obtained when starting with $Bi^{3+}$:$Fe^{3+}$ 1:2 ratio in harsh conditions to produce pure $Bi_2Fe_4O_9$ without the need of adding $H_2O_2$.

Phase and size control have also been achieved by the addition of several mineralizers in the reaction media under hydrothermal basic conditions. Then, pure $BiFeO_3$, $Bi_2Fe_4O_9$, and $Bi_{12}(Bi_{0.5}Fe_{0.5})O_{19.5}$ can be isolated by adding $K^+$, $Na^+$, and $Li^+$, respectively [65]. Reactant solubility differences depending on the counter cation of the mineralizer could lead to differences in the dissolution–crystallization pathway, guiding the phase selectivity. On the other side, smaller particles are obtained when the mineralizer concentration is increased. Higher ionic strength may favor the nucleation at the expense of the particle growth [65, 66]. Changing conventional heating by microwave-assisted heating yields well-faceted $BiFeO_3$ cubes of 50–200 nm [67]. Sonochemical synthesis is also convenient for the selective synthesis of 25–30 nm $BiFeO_3$ cubes, allowing doping of the structure with other elements such as $Sc^{3+}$ [60]. Molten salts have also produced $BiFeO_3$ particles, but sizes could not be downscaled below the submicron scale [68, 69].

Besides bismuth, manganese also holds oxidation states not stable in organic media, especially $Mn^{4+}$. $Mn^{4+}$ is often met in functional manganese oxide-based materials, so that its controlled incorporation in oxide nanostructures is of utmost importance to control properties of catalysis [70], electrocatalysis [71, 72], and magnetic properties [73]. Some attempts have been performed under solvothermal conditions, but they all required an additional calcination step at high temperature in order to crystallize $Mn^{4+}$-based compounds [74, 75].

In order to reach $Mn^{4+}$-based oxides through one-step liquid-phase synthesis, comproportionation of permanganate Mn(VII) and Mn(II) salts in water between room temperature and 250 °C has been proved as an efficient pathway [76]. This approach has been especially explored to synthesize rare-earth and alkali-earth manganite perovskites $Ln_{1-x}A_xMnO_{3-\delta}$ (Ln = lanthanide, A = alkali-earth metal cation, and $\delta < 0.1$). The Mn(VII)/Mn(II) reagents ratio guides the final manganese oxidation state in the perovskite and enables avoiding impurities [45]. Some studies suggest that the synthesis through aqueous-phase comproportionation at low temperature (<300 °C) with half-doped A site is only thermodynamically possible with compositions corresponding to a Goldschmidt tolerance factor close to 1 and an A-site variance (cation disorder) closest to zero [45, 77]. Other approaches able to work at high temperature for triggering perovskite crystallization should overcome this limitation in the synthesis of perovskite at nanoscale.

The sol–gel process enables the synthesis of films [78], ordered meso-/macro-structures [79], and nanoparticles [80–82], thus allowing the production of materials with relative high surface area and enhanced electrocatalytic activity, e.g. in oxygen reduction reaction (ORR). Through a related approach, poly(methylmethacrylate) colloidal crystal templates have been used to produce three-dimensional ordered macroporous $LaMnO_3$ with a surface area of $\sim 20\,m^2\,g^{-1}$. Good performances have been recorded in ORR electrocatalysis [79].

Solid-state gelation has been developed recently to increase the surface area of $LaMnO_x$ aerogel up to $74\,m^2\,g^{-1}$ [82]. Preformed amorphous lanthanum oxide/$Mn_3O_4$ (core/shell) nanoparticles of 10 nm are deposited on carbon. Carbon acts as a continuous phase matrix promoting the gelation of the reaction medium. It is then removed upon further calcination, thus yielding a perovskite–aerogel formation. The resulting material shows high mass-normalized activity in ORR, thanks to the high surface area achieved. This methodology has been extended to other perovskites compositions using various metal salts as precursors. Hence, $LaFeO_3$, $LaNiO_3$, $LaCoO_3$, $La_{0.5}Sr_{0.5}CoO_3$, and $La_{0.5}Sr_{0.5}Co_{0.5}Fe_{0.5}O_3$ could be obtained as nanostructured materials.

Synthesis in molten salts is a very interesting synthesis strategy to combine the advantages of the two previously mentioned strategies: on-step synthesis and high surface area of the resulting materials. The approach has been used to produce single crystal nano-objects with high crystallinity, controlled morphology, and particle size. By using nitrate precursors and $KNO_3$ as solvent, 20 nm $La_{0.67}Sr_{0.33}MnO_3$ well-faceted nanocubes were obtained (Figure 2.6a,b). This nanomaterial shows

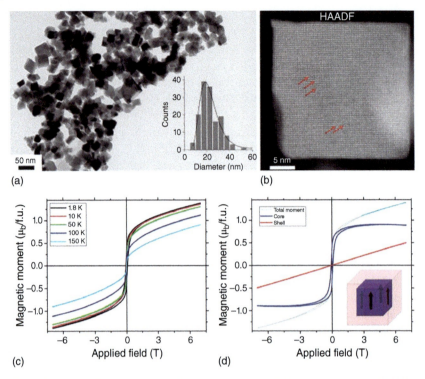

**Figure 2.6** $La_{0.67}Sr_{0.33}MnO_3$ nanocubes obtained through the molten salts method: (a) low magnification TEM. The inset in (a) shows the size distribution of the nanocubes. (b) Scanning transmission electron microscopy-high angle annular dark field (STEM-HAADF) images. (c) Magnetization curves at several temperatures and (d) decomposition of the magnetic signal. Inset: Scheme of a core–shell nanocube, with well-aligned core magnetic moments. Source: Thi N'Goc et al. [73] with permission from Wiley-VCH Verlag GmbH & Co. KGaA.

large magnetoresistance at low temperatures. This behavior has been shown to arise from a 0.8 nm thick surface layer where manganese is partially reduced to a mixed $Mn^{2+}/Mn^{3+}$ perovskite layer due to oxygen vacancies, independently of the spin polarization of the ferromagnetic core with the right oxidation state (Figure 2.6c,d) [73]. The high crystallinity mimicking epitaxial thin films, the surface free of any organic ligands that could hinder charge transfer, and the relatively large surface area (36 m² g⁻¹) provide $La_{0.67}Sr_{0.33}MnO_3$ nanocrystals with high electrocatalytic activity and remarkable stability in ORR (Figure 2.7c) [72].

The high versatility of molten salts synthesis allows tuning the composition of well-faceted nanocubes, thus yielding not only $La_{0.67}Sr_{0.33}MnO_3$ but also $La_{0.67}Ca_{0.33}MnO_3$ and $La_{0.33}Pr_{0.33}Ca_{0.3}MnO_3$ [84]. The phase selectivity could also be changed by adjusting the metal reagent ratio and by increasing the oxo-basicity of the molten salt used as solvent. Therefore, nanoparticles of the layered perovskite Ruddlesden–Popper phase $La_{0.5}Sr_{1.5}MnO_4$ could be synthesized in $NaNO_2$. The high oxo-basicity of $NaNO_2$ results in high amounts of $O^{2-}$ released in the liquid medium, allowing the synthesis of the oxygen-rich $La_{0.5}Sr_{1.5}MnO_4$ phase at the

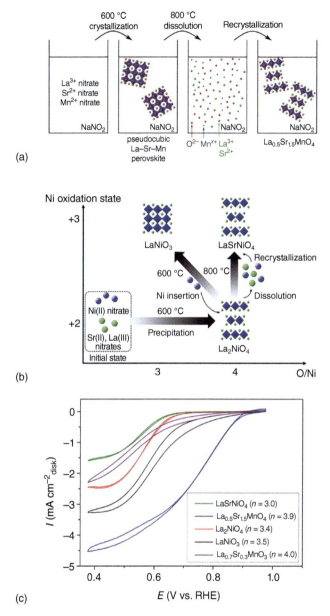

**Figure 2.7** Proposed mechanisms involved in the synthesis of (a) layered $La_{0.5}Sr_{1.5}MnO_4$, (b) nickel-based perovskites in molten $NaNO_2$, and (c) CV in ORR conditions of the nanoscaled Ni and Mn-based perovskites obtained in molten salts ($n$ = number of electrons involved in the ORR mechanism). Source: Adapted from Gonell et al. (a) [72] and (b) F. Gonell, from data in [83], with permission from American Chemical Society, copyright 2020. (c) Source: Adapted from F. Gonell, from data in [72] and [83].

nanoscale for the first time [72]. This strategy was also successfully applied to the synthesis of Sr–La–Ni perovskites. The high reactivity of nitrites compared with nitrates allows the synthesis of well-faceted $LaNiO_3$ nanoparticles at lower temperature (200 °C lower than with nitrates) as well as the synthesis of $La_2NiO_4$ and $LaSrNiO_4$-layered perovskite Ruddlesden–Popper nanoparticles [83]. The synthetic descriptors inferred in these works highlight the importance of the transition metal oxidation state, which governs the perovskite formation. The synthesis mechanism is directed by the reaction temperature; at low temperature topochemical transformation is preferred, while at high temperature the energy input is high enough to promote a dissolution–reprecipitation pathway (Figure 2.7a,b). These descriptors allow the design of nanoscaled perovskites with different activity and selectivity (number of electrons) in ORR (Figure 2.7c) [72, 83].

## 2.3 Oxides with Uncommon Metal Oxidation States: The Case of Titanium(III) in Oxides and Extension to Tungsten Oxides

Titanium oxides are being paid attention due to their technological applications on a wide range of fields such as fuel cells, batteries, photovoltaics, photocatalysis, and environmental remediation, among others [85, 86]. The potential implementation of these oxides on different devices is due to their excellent chemical stability, corrosion, and mechanical resistance, as well as their low cost, as titanium and oxygen are two of the most abundant elements on Earth [87]. The three main polymorphs of stoichiometric $TiO_2$ (anatase, rutile, and brookite) are insulators with large electronic band gap (3.0–3.2 eV), limiting light absorption to the ultraviolet range of the solar spectrum. As a consequence, their potential applications in photocatalysis and photovoltaics are restricted. Nonetheless, partial titanium reduction, from $Ti^{4+}$ to $Ti^{3+}$, gives rise to a modification of electronic, optical, and charge transport properties related to crystal structure modifications. For photovoltaics, $Ti^{4+}$ partial reduction is a promising approach for achieving narrower electronic band gap toward broader spectrum range absorption [9, 88]. Likewise, similar positive effects, related with band gap decrease on $TiO_2$ polymorphs, have been observed in other properties such as photocatalytic water splitting [87–90].

Traditionally, the most common approach for obtaining bulk reduced titanium oxides has been $TiO_2$ reduction at high temperature ($\approx$1000 °C) by hydrogen or metallic titanium for a long period of time [91, 92]. The properties and applications described earlier involve interactions between the material surface and other solution-phase or gas-phase species. As a consequence, developing synthetic approaches to decrease particle size is a suitable approach to probe the impact of the surface on the properties and possibly to enhance materials performances. A huge effort has been devoted to this trend by several groups during last years. With the aim of preventing the crystal growth, studies are focused on two kinds of approaches: encapsulating the precursors by templating to restrict particle growth and exploring less energetically demanding processes by increasing the precursor reactivity, with the aim of decreasing the reaction temperature.

### 2.3.1 Crystal Structures and Requirements for the Synthesis of Oxides Bearing Titanium(III) Species

To understand requirements in terms of synthesis conditions, it is instructive to scrutinize the crystal structures of oxides bearing titanium(III) species. Simple doping of Ti(IV) oxides is discarded from this discussion, as $TiO_2$ doped with Ti(III) as a consequence of oxygen vacancies or aliovalent cationic doping is poorly stable and readily undergoes re-oxidation to the stoichiometric compound. Likewise, related multianionic compounds, like barium titanium oxyhydride ($BaTiO_{2-x}H_x$) [93], have never been observed at the nanoscale to our knowledge and are not discussed in this chapter.

On the extreme range of oxygen stoichiometry with no Ti(III), $TiO_2$ exhibits several polymorphs. The Ti(III)-bearing compounds are all related to the rutile polymorph of $TiO_2$ (Figure 2.8). The rutile structure is built on $TiO_6$ octahedra sharing edges to form octahedral chains. These chains are linked together by sharing corners of their respective octahedra. Upon reduction of Ti(IV) by $O^{2-}$ removal, one encounters Magnéli phases of titanium with formula $Ti_{2n}O_{n-1}$. They were discovered in the 1950s by A. Magnéli [94]. It is worth noticing that this kind of structure is established not only by titanium but also by vanadium [95], molybdenum [96], and tungsten [97], although the two latter give rise to another general formula. For titanium and vanadium, the $n$ value is usually accepted as an integer number between 3 and 10; however there are studies that suggest its maximum value could be higher [98]. The structure of Magnéli phases $Ti_{2n}O_{n-1}$ can be understood by a thought experiment. Removal of $O^{2-}$ from $TiO_2$ rutile would create unstable oxygen vacancies. To stabilize the structure and eliminate these oxygen vacancies, some $TiO_6$ octahedra would shift in order to share faces and form $Ti_2O_9$ octahedra dimers. This shift is accompanied by ordering of dimers into shear planes and the formation of corundum structure blocks separating rutile domains (Figure 2.8) [86]. Further reduction yields the corundum polymorph of $Ti_2O_3$, the most common polymorph that can be prepared in not highly demanding experimental conditions [9].

Between $Ti_4O_7$ and $Ti_2O_3$, one encounters the few polymorphs of $Ti_3O_5$ (Figures 2.8 and 2.9) [99]. There are some ambiguities in the literature related with $Ti_3O_5$. Based on the general formula $Ti_nO_{2n-1}$, $n = 3$, $Ti_3O_5$ could be included as the first member of Magnéli phases. However considering the absence of crystallographic relationship involving the presence of shear planes, not all $Ti_3O_5$ polymorphs should be included in the family of Magnéli phases [99]. The first studies of A. Magnéli and coworker [100] described the two polymorphs α and β (Figure 2.9a) with anasovite-type structures, related to each other by slight atomic displacements maintaining the symmetry and similar unit cell dimensions. The absence of crystallographic shear plane in both polymorphs was interpreted as an anomaly among titanium Magnéli phases, until γ-$Ti_3O_5$ [101, 102], isostructural to $V_3O_5$, and δ-$Ti_3O_5$ [103] were obtained, which are again related to each other by

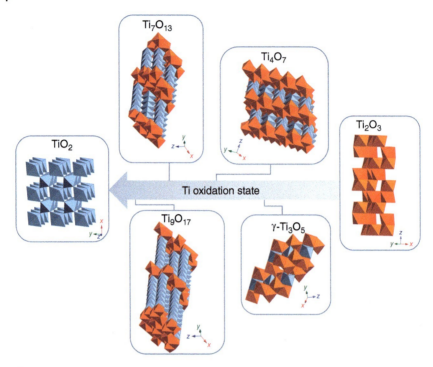

**Figure 2.8** Schemes of the crystal structures of rutile $TiO_2$, corundum $Ti_2O_3$, and titanium Magnéli phases. Face-sharing $TiO_6$ octahedra are plotted in orange. Source: Dr. Isabel Gómez-Recio.

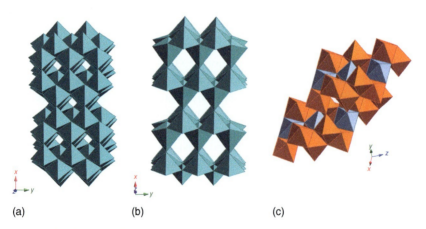

**Figure 2.9** Schemes of the crystal structures of (a) β-$Ti_3O_5$, (b) λ-$Ti_3O_5$, and (c) γ-$Ti_3O_5$. Source: Dr. Isabel Gómez-Recio.

slight atomic displacements. γ-$Ti_3O_5$ (Figure 2.8) again shows face-sharing octahedra related to shear planes, and we then ascribe it to the family of Magnéli phases. The last polymorph described, λ-$Ti_3O_5$ (Figure 2.9b) [104], shows a reversible photoinduced phase transition with β-$Ti_3O_5$. It should not be considered as a Magnéli phase. As α-$Ti_3O_5$ and δ-$Ti_3O_5$ can be considered as β-$Ti_3O_5$ and γ-$Ti_3O_5$ distortions, respectively, their structures have not been included in Figures 2.8 and 2.9, for the sake of simplicity.

Overall, the structural relationship between some $TiO_2$ and $Ti_2O_3$ polymorphs and the Magnéli phases highlights some guidelines for efficient syntheses of $Ti_2O_3$ and $Ti_nO_{2n-1}$ ($n \geq 3$) compounds, by considering most common $TiO_2$ as a starting material. First, the reduction of $TiO_2$ into Magnéli phases and $Ti_2O_3$ can only proceed through the $TiO_2$ rutile polymorph, whether it is prepared on purpose or it is formed in situ during the reaction. In other words, the topotactic transformation of $TiO_2$ rutile is the only relevant reduction pathway toward $Ti_2O_3$ corundum or Magnéli phases. This statement is supported by the outcome of reduction processes from the $TiO_2$ anatase polymorph. Indeed, the reduction of anatase does not yield the previously mentioned rutile-related phases, but other materials, especially the so-called "black anatase" [105]. Second, and as pointed out by Tominaka, Cheetham, et al. [106], the topotactic transformation from one phase to another involves collective atom displacements that are easier to trigger into nanomaterials than into bulk materials. Hence, nanomaterials are especially suited to explore such phase transformations and how they can yield full structural transformations while maintaining the overall shape of the nano-objects in the so-called isomorphic transformations. In the following parts, we explore the current reaction pathways to trigger such transformations.

### 2.3.2 $Ti_2O_3$ Nanostructures

The small band gap ($\approx 0.1$ eV) of $Ti_2O_3$ corundum enables solar light absorption in the full visible range, which is improved in nanosized particles in comparison with bulk counterparts, as light scattering is enhanced with nanoparticles [90]. This characteristic accompanied by efficient solar–thermal conversion suggests that $Ti_2O_3$ can be used as photothermal material. Under light, the material could generate steam from water, as a first step in desalination or water purification, using sunlight as power source. As a proof of concept, Wang et al. [90] performed an experiment where $Ti_2O_3$ nanoparticles were deposited on a cellulose membrane at the surface of water (Figure 2.10a). After 15 minutes under light illumination, the temperature difference between the top and bottom part was 23 °C, and water evaporation was observed (Figure 2.10b,c). In a reference experiment without nanoparticles, water was only heated by 0.5 °C, increasing water evaporation rates between 2.65 and 4.18 times higher than pure water evaporation (Figure 2.10d,e).

P. Zeng et al. [107] have coupled the sol–gel process with high-temperature reduction to synthesize $Ti_2O_3$ nano- and microstructures as cathode materials for lithium–sulfur batteries. Titanium glycolate microspheres (Figure 2.11a,b) were thermally decomposed into 1.5 µm rutile microspheres (Figure 2.11c,d) at 800 °C.

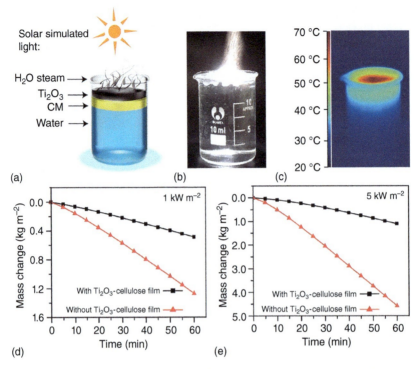

**Figure 2.10** (a) Schematic setup for proof-of-concept solar water steam production, (b) photograph of the water steam generation, and (c) IR imaging of (b). Evaporated water weights against time with and without $Ti_2O_3$ thin film device under solar illumination, under 1 kW m$^{-2}$ (d) and 5 kW m$^{-2}$ (e). Source: Wang et al. [90] with permission from Wiley-VCH Verlag GmbH & Co. KGaA.

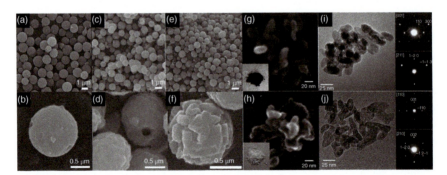

**Figure 2.11** SEM images of (a, b) titanium glycolate, (c, d) rutile, and (d, e) $Ti_2O_3$ microspheres. (g, h) SEM and (i, j) TEM images of (g, i) reduced $TiO_2$ nanoparticles and (h, j) corresponding $TiO_2$ nanoparticle precursors. Insets: (g, h) Photographs of the powders and (i, j) selected area electron diffraction patterns. Source: Adapted from (a–f) P. Zeng et al. [107] and (g–j) S. Tominaka et al. [86] with permission from American Chemical Society, copyright 2019, and Wiley-VCH Verlag GmbH & Co. KGaA, respectively.

Then, the TiO$_2$ microspheres were reduced into Ti$_2$O$_3$ microspheres (Figure 2.11e,f) by magnesium at 1000 °C. Upon reduction, the surface of the microspheres became faceted, showing clear crystal growth and microparticles as aggregates of nanoscale particles.

Although solid-state synthesis usually yields bulk phases, S. Tominaka et al. [86, 106, 108] opened an elegant route for solid-state Ti$_2$O$_3$ nanoparticles synthesis. The authors were able to successfully reduce rutile TiO$_2$ nanoparticles into ca. 25 nm Ti$_2$O$_3$ nanoparticles (Figure 2.11g-j) by using different hydrides as low-temperature reducing agents (350–420 °C): CaH$_2$ [86] and NaBH$_4$ [108]. Shifting from calcium hydride to sodium borohydride enabled shortening the reaction time from 2 weeks to 24 hours. Interestingly, it is possible to use such hydride-based reduction processes to trigger conformal transformations of TiO$_2$ nanostructures to Ti$_2$O$_3$ nanostructures, not only in nanoparticles [86, 106, 108] but also in thin films and nanorods [108]. Despite being extremely efficient in terms of morphology control, these hydride-mediated topochemical and isomorphic transformations do not enable a precise control of the composition, since an excess of hydride must always be used to ensure reduction of Ti(IV), thus yielding only the fully reduced Ti$_2$O$_3$.

### 2.3.3 Mixed Valence Ti(III)/Ti(IV) Oxides: Magnéli Phases

In Magnéli phases, the mixed valence Ti(III)/Ti(IV) ensures electronic conduction, while corundum blocks scatter phonons, which decreases thermal conductivity without impacting strongly electrical conductivity (Figure 2.12) [108]. Ti$_4$O$_7$ ($n = 4$) has a metallic behavior and is the most conductive phase of the Magnéli titanium phase family (tabulated electronic conductive 10$^3$ S cm$^{-1}$), comparable to graphite and metals [110]. When $n$ increases, the proportion of Ti(III) decreases with 2Ti$^{3+}$ for $(n-2)$Ti$^{4+}$ centers.

Low thermal conductivity and large electrical conductivity are of interest for designing thermoelectric materials for power generation by recovering waste heat (Figure 2.12) [109, 111–113]. Magnéli phases have also metal–semiconductor–insulator transitions, which can be controlled by external stimuli, as temperature, light, and pressure [87]. Magnéli phases of titanium are studied for electrochemical applications as fuel cells and batteries due to their mechanical and chemical resistance [114, 115]. Because of their high electrical conductivity and chemical inertness, Magnéli phases could indeed be alternative substrates to poorly stable carbon materials implemented in electrodes for Li–O$_2$ batteries [114]. Another field of interest arises from the expected affinity of sulfur for the Ti$_4$O$_7$ surface [116–119]. Ti$_4$O$_7$ nanostructured substrates have been designed, taking benefit from the strong adsorption of sulfur to enhance the stability of the sulfur electrode in Li–sulfur batteries [116–119]. Note however that these Ti$_4$O$_7$ materials showed surface contamination by carbon coming from the synthesis process (see following text), which may have a big role in the adsorption of sulfur as well. Magnéli phases are also studied in electrochemical wastewater treatment, as electrodes for the electrooxidation of pollutants such as dye methyl orange (Figure 2.13) [120, 121]. In all these prospective applications, increasing the surface-to-volume fraction ratio through nanostructuration is a viable approach to enhance performances.

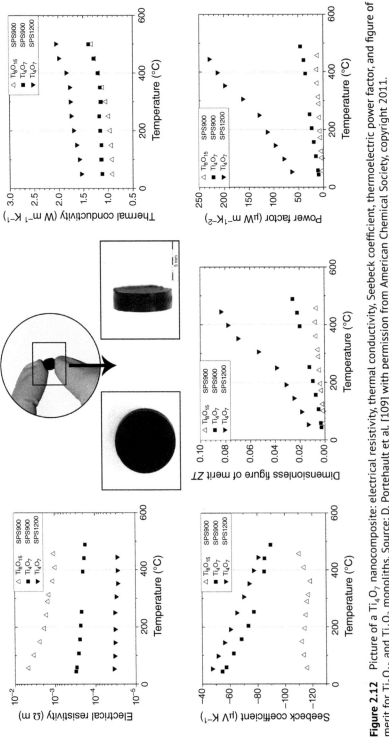

**Figure 2.12** Picture of a Ti$_4$O$_7$ nanocomposite: electrical resistivity, thermal conductivity, Seebeck coefficient, thermoelectric power factor, and figure of merit for Ti$_8$O$_{15}$ and Ti$_4$O$_7$ monoliths. Source: D. Portehault et al. [109] with permission from American Chemical Society, copyright 2011.

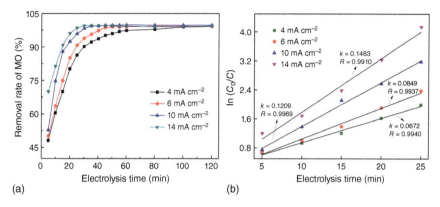

**Figure 2.13** (a) Electrochemical oxidation of azo dye methyl orange by $Ti_4O_7$ at different current densities with (b) quasi-first-order kinetics fitting. Source: G. Wang et al. [120] with permission from Elsevier Ltd.

As discussed earlier, the most efficient way to envision the synthesis of nanostructured Magnéli titanium phases is to trigger topotactic reduction of $TiO_2$ rutile. Despite the observation by Tominaka et al. [86] of $Ti_4O_7$ as an intermediate phase during the solid-state reduction of nanoparticles of $TiO_2$ rutile into nanoparticles of $Ti_2O_3$ with $CaH_2$, which exemplifies the structural relationship described in Section 3.1, no report has shown yet the production of nanostructures of titanium Magnéli phases without using templating approaches. Indeed, the two strategies available to produce titanium Magnéli nanostructures rely on templating.

The first approach for the design of nanostructured Magnéli phases is to consider the high-temperature reduction of $TiO_2$ rutile by templating the $TiO_2$ nanostructure in order to limit grain growth and maintain nanoscale features. Y. Kuroda et al. [110] described a synthetic pathway for obtaining mesoporous $Ti_6O_{11}$. A colloidal crystal made of 50 nm silica nanoparticles was infiltrated by a solution of titanium precursor, followed by inorganic polymerization of this precursor to form a silica template–$TiO_2$ material. This material was reduced under $H_2$ at 800 °C to form the Magnéli phase $Ti_6O_{11}$, before dissolution of the silica template (Figure 2.14a). Although no structural characterization of the initial $TiO_2$ material is provided to assess the presence of the rutile polymorph, it is likely that it is present in the initial stage or formed upon heating during the reduction process. In a similar manner, X. Cao et al. [114] fabricated particles of $Ti_4O_7$ with an average diameter of 300 nm, by $H_2$ reduction of $TiO_2$ nanoparticles coated by an ordered mesoporous $SiO_2$ shell. The mesoporous coating prevented $Ti_4O_7$ agglomeration and allowed diffusion of $H_2$ through the pores. Afterward the $SiO_2$ shell was dissolved by an HF solution. This method was further refined by E. Baktash et al. [118] to be applied to 50 nm nanoparticles by combining silica templating with the use of anatase particles as diluting agent for limiting particle sintering. This method produces selectively $Ti_4O_7$ and $Ti_6O_{11}$ nanoparticles of 50 nm by controlling $H_2$ reduction temperature, taking advantage of the selectivity of topotactic reduction for the rutile phase, while the anatase component did not react with $H_2$. Finally silica and anatase could be

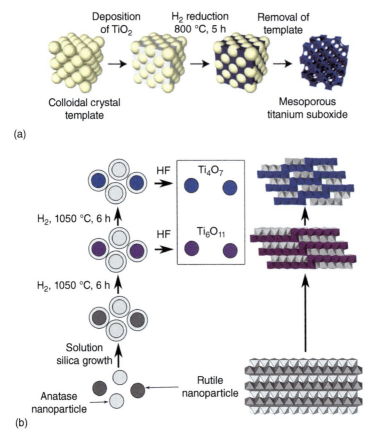

**Figure 2.14** Schemes of silica-templated syntheses of (a) mesoporous $Ti_6O_{11}$ and (b) phases selective $Ti_4O_7$ and $Ti_6O_{11}$ Magnéli phases. Source: Adapted from (a) Y. Kuroda et al. [110] and (b) E. Baktash et al. [118] with permission from Springer Science Business Media, LLC, part of Springer Nature and Wiley-VCH Verlag GmbH & Co. KGaA, respectively.

selectively dissolved by HF (Figure 2.14b) to release 50 nm free-standing Magnéli nanoparticles.

The second approach consists in using the template as a reducing agent. For this purpose, carbonaceous templates are used in order to trigger carbothermal reduction of $TiO_2$ rutile nanostructures. In these syntheses, the $TiO_2$ nanostructures and the template are produced concomitantly in a single step by using a sol–gel route (Figure 2.15). The initial reaction medium is a gel or a viscous solution made of a Ti(IV) alkoxide, a polymer, and a volatile alcohol [109]. The condensation of Ti(IV) species is triggered by heating, yielding $TiO_2$ nanoparticles. Upon temperature increase under inert atmosphere, the polymer decomposes into a carbonaceous matrix embedding the particles. The matrix confines the particles and maintains their nanoscale size. In agreement with the topotatic transformation pathway, the rutile phase appears upon further heating after anatase formation in the early stages of $TiO_2$ crystallization. $Ti_8O_{15}$ and then $Ti_4O_7$ crystallize after rutile

## 2.3 Oxides with Uncommon Metal Oxidation States | 67

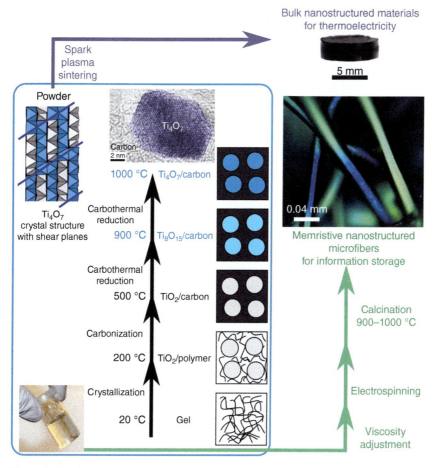

**Figure 2.15** Synthesis and nanostructured processing of Magnéli phases. Source: D. Portehault et al. [122] with permission from American Chemical Society, copyright 2018.

formation, at 900 and 1000 °C, respectively. This approach enables implementing the various shaping methods applied with the sol–gel process [12, 123]. This has been demonstrated by using electrospinning in order to produce nanostructured microfibers of different terms of Magnéli phases (Figure 2.15) [122]. These fibers show a specific internal structure, encompassing a carbon-free shell of large submicrometric particles and a core of nanoparticles still embedded into a carbon matrix. This observation highlights the need of the carbon template to limit particle growth and shows also a limit of this approach, since the reactivity of the template strongly depends on the gas flow near its surface, as the products of carbothermal reduction are gaseous CO and $CO_2$ species. The other limitation is the presence of carbon by-products at the surface of the Magnéli nanostructure, which can strongly modify the interaction of the material with the surrounding environment [119, 124].

### 2.3.4 Comparison to Metal Oxidation States Stable in Organic Media: Mixed W(V)/W(VI) Oxides

Crystal structures based on shear planes are not only encountered in titanium oxides but also with V(III)/V(IV), Mo(V)/Mo(VI), and W(V)/W(VI) oxides. The parent structure of molybdenum and tungsten Magnéli oxides is $MO_3$ (M = Mo, W) with the $ReO_3$ perovskite-related structure where the A site is empty and $MO_6$ octahedra are connected by corners [125]. Shearing leads to the formation of edge-sharing octahedra. Mo and W Magnéli oxides do not obey a general formula, although they can be described as $M_nO_{3n-1}$, $M_nO_{3n-2}$, $M_nO_{3n-3}$, $M_nO_{3n-4}$, $M_nO_{3n-5}$, etc. (Figure 2.16) [125]. They are usually described as $WO_x$ ($2.625 \leq x \leq 2.92$). For low reduction degree, ($x \geq 2.90$), shearing yields trimers of edge-sharing octahedra, while stronger reduction yields pentagonal columns of edge-sharing octahedra. Among W Magnéli oxides, only $W_{18}O_{49}$ has been isolated as a single phase. $W_{18}O_{49}$ nanostructures have also been extensively investigated [97, 125].

The gas-sensing properties of $W_{18}O_{49}$ have been widely explored for the detection of reducing gases, as $CH_4$, $NH_3$, $H_2$, or $H_2S$, and of oxidizing gases, such as $NO_x$ or CO. The redox properties of the W(V)/W(VI) oxide are the key of the detection mechanism as redox reaction with the gas modifies reversibly the W(V)/W(VI) ratio and then the electrical conductivity of the material. This change of conductivity can be measured to enable quantitative evaluation of the analyzed gas (Figure 2.17) [126]. The detection is enhanced by the high specific surface area of nanostructures, which drove efforts in their synthesis.

Owing to the strong oxidizing properties of W(VI), syntheses in organic solvents, often with reducing properties, usually do not result in stoichiometric $WO_3$, but in mixed W(V)/W(VI) phases, including $W_{18}O_{49}$. Solvothermal syntheses have been largely developed to reach nanostructures of this compound. One of the most efficient approaches is the synthesis in benzyl alcohol from $WCl_6$, which results in controlled growth of thin nanowires (Figure 2.18) with an optimized surface area to optimize the detection sensitivity [126–129].

In the presence of alkali cations, the crystallization pathway proceeds differently, as alkali tungsten oxides spontaneously form. These so-called tungsten bronzes are

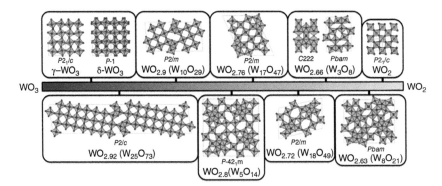

**Figure 2.16** Stoichiometric and substoichiometric tungsten oxides. Source: Adapted from Y. Lee et al. [125] with permission from American Chemical Society, copyright 2019.

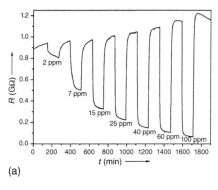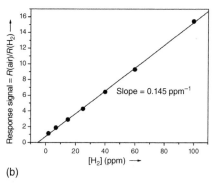

**Figure 2.17** (a) Resistance change and (b) response signal of a $W_{18}O_{49}$ nanowire thin film with respect to various concentrations of $H_2$ at room temperature. Source: Adapted from W. Cheng et al. [126] with permission from Wiley-VCH Verlag GmbH & Co. KGaA.

formed by insertion of alkali cations into the host structure of $WO_3$. Hexagonal tungsten bronzes $A_xWO_3$ (A = $Na^+$, $Cs^+$, $Rb^+$, and $NH_4^+$ or even smaller hydrated cations as $Li^+$ and $H^+$) contain $A^+$ cations inserted in the hexagonal tunnels of the parent h-$WO_3$ phase (Figure 2.19) [130], delineated by six $WO_6$ octahedra sharing corners. Cation incorporation causes tungsten reduction, which is accompanied by color change from light yellow to deep blue [132], as in Magnéli phases. This feature is largely exploited for the design of electrochromic devices, as smart windows or adjustable mirrors [130, 133, 134].

An interesting feature of nanoscaled tungsten bronzes is the emergence of localized surface plasmon resonance (LSPR), a phenomenon related to collective oscillations produced by free or conductive electrons in large carrier density materials, in response to electromagnetic radiation, which leads to a strong absorption in the near-infrared (NIR) or visible ranges, depending on the composition and morphology of the nano-objects [11, 135]. On semiconductor materials, carrier density can be tuned by chemical doping, in such a way that the dopant concentration and distribution in the crystal structure as well as the particles morphology have an impact on the LSPR response [11]. In this context, Manthiran and Alivisatos [136] described a plasmon shift and intensity decrease for $WO_{2.83}$ nanoparticles prepared by colloidal synthesis when they were treated thermally in an oxidizing environment, consistent with oxygen incorporation.

$WO_{3-\delta}$ has been mostly studied for its suitable energy band gap (2.6 eV) for interaction with visible light and for its stability in a wide range of conditions [136]. While in bulk $WO_{3-\delta}$, color changes are related with oxygen deficiency and small cations intercalation, in nanoparticles the strong absorption in the NIR region is interpreted as a consequence of small polaron hopping and/or LSPR, as it has been reported for $WO_{2.72}$ [137] and $WO_{2.83}$ [136], respectively. Polaron hopping is due to oxygen deficiencies that produce electrons trapped in 5d orbitals of tungsten atoms, followed by surrounding lattice polarization to generate polarons, which gives rise to polaron transitions between adjacent nonequivalent tungsten sites [11, 135].

To date, the studies about intense NIR absorption on $WO_{3-\delta}$ nanoparticles are still under discussion. As mentioned above, $WO_{2.72}$ color change has been interpreted

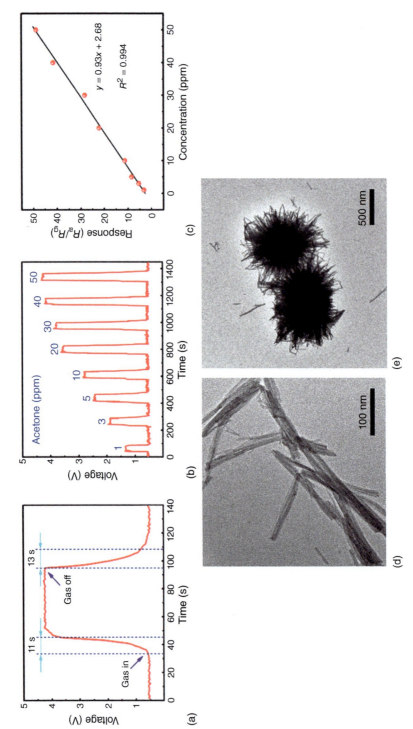

**Figure 2.18** (a) Response and recovery curve of a $W_{18}O_{49}$-based sensor when exposed to acetone, (b) $W_{18}O_{49}$ detection response to different acetone concentration, (c) calibration of a $W_{18}O_{49}$-based resistive sensor of acetone, $W_{18}O_{49}$ (d) nanowire, and (e) sea urchin-like nanoparticles. Source: W. Zhang et al. [127] with permission from American Chemical Society, copyright 2020.

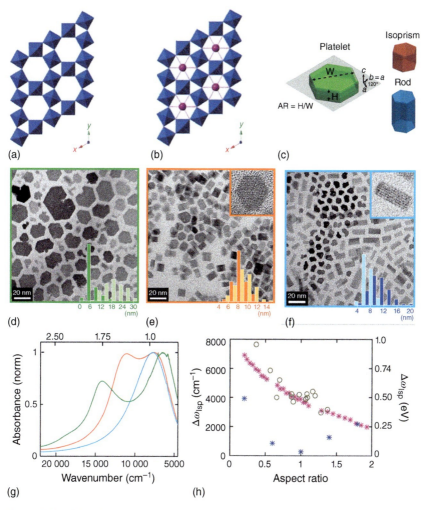

**Figure 2.19** Crystal structure of (a) h-WO$_3$ and (b) M$_x$WO$_3$. Color code: W blue and M purple. (c) Schematic drawing of particle shapes corresponding to (d–f) TEM images. Insets: The insets in (e) and (f) display HRTEM images of single nanocrystals. (g) Optical absorption spectra of h-CsWO$_3$ platelets (green), isoprisms (orange), and rods (blue) dispersed in tetrachloroethylene, showing the aspect ratio dependence of LSPR. (h) LSPR peak splitting vs. particle aspect ratio of nanocrystals (open circles) and by theoretical modeling assuming isotropic (blue stars) or anisotropic (pink stars) dielectric functions. Source: Adapted from (a, b) T. Mattox et al. [130] and (c–h) J. Kim et al. [131] with permission from American Chemical Society, copyright 2014 and 2016, respectively.

as a result of polarons hopping process, nonetheless T. Masuda and H. Yao [135] revealed an intense magneto-optical activity on WO$_{2.72}$ as consequence of circular modes of LSPR (magneto-plasmon). This magneto-optical behavior is apparent from the differential absorption of left- and right-polarized light under a magnetic field. The magneto-optical activity of WO$_{2.72}$ nanoparticles exceeds the one of typical Ag nanospheres, which are among most promising candidates for plasmonic and magneto-plasmonic applications. Therefore, $WO_{3-\delta}$ will likely attract attention as low cost and high stability potential substitute to silver nanostructures.

Again, colloidal synthesis has proved highly efficient for the precise control of morphology and particle size for tungsten bronzes. Different nanoparticles of $Cs-WO_3$ have been obtained from solutions of $WCl_4$ and CsCl in mixtures of oleylamine and oleic acid heated at 300 °C [130]. By increasing the amount of oleic acid, more faceted particles were produced, in a manner that spheres, truncated cubes, and hexagonal prisms were obtained [130]. Nanoplatelets with adjustable aspect ratio, as well as LSPR peak splitting (Figure 2.19g,h), could also be designed in a similar way by J. Kim et al. [131] by tuning reaction temperature, aging time, and tungsten precursor degassing time (Figure 2.19c–h). Likewise, aqueous soft chemistry has allowed stabilizing thin nanoplatelets of a new $H_{0.07}WO_3$ hexagonal bronze structure (Figure 2.20), firstly reported by J. Besnardiere et al. [138]. The optimal combination of nanoparticles morphology and framework preferential

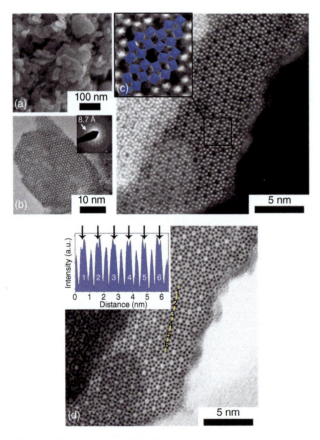

**Figure 2.20** (a) SEM and (b) TEM images of $H_{0.07}WO_3$ hexagonal bronze nanoplatelets. The inset in (b) shows a typical SAED pattern. (c) STEM-HAADF and (d) scanning transmission electron microscopy-annular bright field (STEM-ABF) micrographs showing the arrangement of tungsten octahedra (blue, inset of (c)). Source: Adapted from J. Besnadiere et al. [138] with permission from Springer Nature Licensed Under CC BY 4.0.

orientation yields promising electrochromic properties. Owing to structural tunnels perpendicularly oriented versus the nanoplatelets basal plane, cation exchange properties are optimized in such nanostructures, enabling, for instance, reversible extraction of protons and fast electrochromic switching.

A quick parallel between mixed valence titanium and tungsten oxides, especially Magnéli phases, shows that the crystallization of these structures is not occurring at the same temperature of synthesis. While tungsten Magnéli phases are readily obtained between 200 and 300 °C, to date no process has been suitable to reach titanium Magnéli phases below ca. 800 °C. The reason for this discrepancy has not been assessed to our knowledge, but one may attempt to provide several guidelines that could be useful for future investigation of these phases. A first obvious difference is the nature of the stoichiometric parent compounds. Rutile $TiO_2$ and $ReO_3$-type $WO_3$ show a different connectivity, as the latter is built only from corner-sharing octahedra while the former encompasses also edge-sharing octahedra. This difference is reflected in the relative stability of the compounds. This could explain why $WO_3$ is more prone to structural transformation than rutile $TiO_2$. Likewise, the shearing involved in the transformation of the parent phases into Magnéli phases implies the formation of edge-sharing octahedra for tungsten and of face-sharing octahedra for titanium. The latter may be related to larger activation energy, especially because it requires breaking some Ti—O bonds to enable shifting of Ti centers.

## 2.4 Stabilization of New Crystal Structures at the Nanoscale

Upon decreasing of particle size, the surface-to-volume ratio increases, resulting in the exaltation of the contribution of the surface to the total enthalpy. Because creating a surface has an energetic cost, this surface contribution to the total free enthalpy is destabilizing. Nanostructures and nanoparticles are then intrinsically metastable. An important additional consequence of this surface effect is that at a given size, the extent of the destabilization depends not only on the nature of the facets exposed but also on the crystal structure. Indeed, the surface enthalpy depends on the atomic arrangement of the surface and then on the crystal structure as well. The result is that not all solids are destabilized by the same energy increase. Therefore, inversion of the relative stability order can take place. This effect is well known in the case of polymorphism, especially for $TiO_2$ where the bulk thermodynamic stable phase is rutile, while it becomes anatase below the crossover particle size, which is situated between 10 nm and several tenth or hundreds of nanometers [139, 140]. Other examples of stability inversion at the nanoscale encompass the brookite polymorph of $TiO_2$, as well as alumina and zirconia phases [141]. This trend is driven by the relatively general situation that (i) the difference in Gibbs free energy between polymorphs is of 1–10 kJ mol$^{-1}$, which is commensurate with the surface enthalpy when the particles are small enough ($\sim$1 J m$^{-2}$ $\sim$ 8 kJ mol$^{-1}$ for $TiO_2$ of 100 m$^2$ g$^{-1}$) [140, 141], and (ii) the surface enthalpy decreases with increasing metastability, which often relates to decreasing density [141]. As a consequence, metastable phases in the bulk

## 2.4.1 Hard Templating to Isolate Bulk Metastable Oxides at High Temperatures

Perhaps the earliest example of discovery of new compounds at the nanoscale is the case of $\varepsilon$-Fe$_2$O$_3$. It has been first reported in 1934 [142], then in the 1960s [143, 144], before being reinvestigated in the 1990s [145] when it became clear that an efficient pathway to this compound required templating into silica. By a sol–gel-derived silica matrix, 10 nm maghemite $\gamma$-Fe$_2$O$_3$ nanoparticles were embedded. Upon heat treatment under air above 1400 °C, a structural transformation occurs in the nanoparticles that also slightly coalesce, yielding 30 nm $\varepsilon$-Fe$_2$O$_3$ nanoparticles (Figure 2.21). The material has been reinvestigated when its extremely high

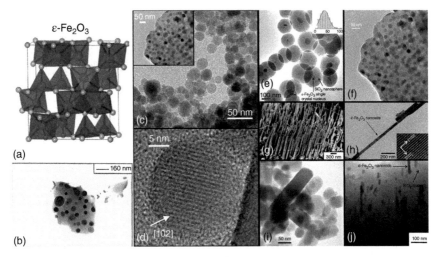

**Figure 2.21** (a) Crystal structure of $\varepsilon$-Fe$_2$O$_3$, (b, c inset) TEM images of $\varepsilon$-Fe$_2$O$_3$ nanoparticles embedded in a silica matrix, and (c) TEM and (d) HRTEM images of $\varepsilon$-Fe$_2$O$_3$ nanoparticles after removal of the silica template matrix (c inset). Spherical $\varepsilon$-Fe$_2$O$_3$ nanoparticles in silica (e) shells and (f) matrix. Insets: The inset in (e) shows the corresponding size distribution in nanometers. The inset in (h) displays an HRTEM image of the corresponding nanowire. $\varepsilon$-Fe$_2$O$_3$ (g, h) nanowires and (i, j) nanorods. Source: Adapted from (a) Sakurai et al. [146] with permission from American Chemical Society, copyright 2009; (b) Tronc et al. [145] with permission from Elsevier, Ltd., (c, d) Gich et al. [147] with permission from IOP Publishing, Ltd; (e–j) Tuček et al. [148] with permission from American Chemical Society, copyright 2010; (e) Taboada et al. [149] with permission from American Chemical Society, copyright 2009; (f) Popovici et al. [150] with permission from American Chemical Society, copyright 2004; (g) Morber et al. [151] with permission from American Chemical Society, copyright 2006; (h) Sakurai et al. [152] with permission from American Chemical Society, copyright 2008; (i) Kelm and Mader [153] with permission from Wiley-VCH Verlag GmbH & Co. KgaA; (j) Sakurai et al. [154] with permission from Elsevier, Ltd.

coercive magnetic field (2 T at room temperature) raised a large interest for data storage [150, 155]. Few years later until very recently, other properties were identified for doped or undoped ε-$Fe_2O_3$, including efficient absorption of millimeter electromagnetic waves [156], multiferroicity [157], and magneto-optical properties [158]. Further refinement of the encapsulating approach included the design of free-standing iron oxide–silica core–shell nanoparticles with 30 nm ε-$Fe_2O_3$ cores [149]. ε-$Fe_2O_3$ nanorods could also be designed [148, 155]. Investigation of the energetic landscape highlighted that ε-$Fe_2O_3$ was an intermediate in the transformation from γ-$Fe_2O_3$ to α-$Fe_2O_3$ and was stabilized for particle sizes below 30 nm, although the exact crossover size is still controversial [148].

Another example of the templating strategy to isolate new oxide phases arising at the nanoscale is λ-$Ti_3O_5$ [104]. Again, $TiO_2$ nanoparticles were produced while being encapsulated into a sol–gel-derived silica matrix. Upon reduction by heating at 1200 °C under a dihydrogen atmosphere, λ-$Ti_3O_5$ was formed as 10 nm nanoparticles and added a term to the polymorphs of $Ti_3O_5$ (Figures 2.8, 2.9, and 2.22). The crystal structure of λ-$Ti_3O_5$ is maintained up to ~500 °C before it transforms into α-$Ti_3O_5$. λ-$Ti_3O_5$ also undergoes a reversible phase transformation to β-$Ti_3O_5$ triggered by high pressures or short light pulses [104]. Interestingly both polymorphs have very different colors, blue for λ and yellow for β. The authors took benefit of this property to design optical data storage devices [104]. Later, a carbothermal reduction pathway toward λ-$Ti_3O_5$ was also explored [160], in a similar way as described earlier for Magnéli phases.

### 2.4.2 Beyond Hard Templating for Isolating Nanostructures of Metastable Oxides

Templating is not always mandatory to take benefit from the effect of the nanoscale if the growth of the nanostructures can be limited, even at high temperature. This has been demonstrated both on ε-$Fe_2O_3$ and λ-$Ti_3O_5$. ε-$Fe_2O_3$ nanowires, and nanobelts could be obtained by pulsed laser deposition (PLD) through a vapor–liquid–solid mechanism catalyzed by gold nanoparticles, which restricts lateral growth of the wires [151]. Epitaxial growth of thin films ε-$Fe_2O_3$ over several substrates, especially $AlFeO_3$, has also been reported by PLD at c. 800 °C [161]. Likewise, λ-$Ti_3O_5$ flakes made of 25 nm nanoparticles could also be obtained by treating 7 nm $TiO_2$ anatase at 1200 °C under hydrogen [104], thus suggesting that anatase is the actual $TiO_2$ precursor of λ-$Ti_3O_5$, contrary to rutile for Magnéli phases. These experiments show that hard templating is not mandatory to isolate high-temperature phases, thanks to the nanoscale, as long as nanocrystal domains remain small enough to ensure that the crystallite surface contribution can still impact the overall energetics.

### 2.4.3 Colloidal Syntheses

The structural transformations described earlier occur at high temperature. Other compounds can be formed at much lower temperature and thus do not require stringent templating or deposition strategies. This is the case of bixbyite $V_2O_3$

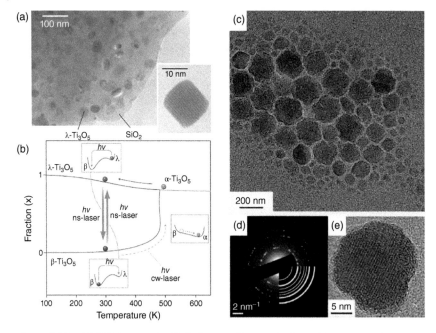

**Figure 2.22** (a) TEM image of λ-Ti$_3$O$_5$ into a silica matrix. Inset: The inset in (a) displays an HRTEM of a single λ-Ti$_3$O$_5$ nanoparticle. (b) Photo-induced phase transition between λ-Ti$_3$O$_5$ and β-Ti$_3$O$_5$. (c) TEM image of bixbyite V$_2$O$_3$, corresponding (d) SAED pattern and (e) HRTEM image. Source: (a, b) Ohkoshi et al. [104] with permission from Springer, and (c–e) Bergerud et al. [159] with permission from American Chemical Society, copyright 2013.

(Figure 2.20c–e), which was discovered recently by D. Milliron and coworkers [159]. The authors used a colloidal synthesis to form 5–30 nm nanoparticles from vanadyl acetylacetonate in squalane with oleylamine and oleic acid at 310–370 °C. These nanoparticles are able to store reversibly oxygen in the interstitial sites of bixbyite in mild conditions [162].

## 2.5 Concluding Remarks

In this chapter, we have focused on multicationic oxides and/or oxides bearing metal oxidation states that are difficult to stabilize in solids and that require specific synthetic conditions. The case studies discussed herein enable us to provide an overall survey of the synthesis strategies that have been and are currently being developed to target such materials at the nanoscale. We have stressed out that, obviously, the choice of the synthesis conditions, including the temperature, the nature of the solvent, and the use of a hard template, must be driven by the compositional and structural features of the targeted materials, and when possible, we have drawn some guidelines for choosing the right method, which we hope will be useful for the reader.

While this overview of course has not been exhaustive, we would like to highlight that, as detailed in Section 2.4, some new oxides like $\varepsilon\text{-Fe}_2O_3$ and $\lambda\text{-Ti}_2O_3$ have been isolated only at the nanoscale and sometimes with new crystal structures. To our knowledge, these new oxides have been first obtained by serendipity. Despite studies performed afterward to identify the crossover particle size in which range these oxides can be isolated, it is striking that up to now, there is no possibility to predict the emergence of a new oxide when decreasing the particle size. This can be explained by the fact that the crossover size is often in the range of 10 nm, which represents a large number of atoms and then is extremely costly in terms of computation to enable tracing (composition, structure, size) maps. In view of this lack of exploration, it is likely that many new oxides are still to be discovered at the nanoscale. Because $\varepsilon\text{-Fe}_2O_3$, $\lambda\text{-Ti}_2O_3$, and bixbyite $V_2O_3$ are showing original properties with a potentially large societal impact, it appears of utmost importance to intensify efforts in exploring the role of nanoscaling for isolating new oxides.

## References

1 Dadfar, S.M., Roemhild, K., Drude, N.I. et al. (2019). Iron oxide nanoparticles: diagnostic, therapeutic and theranostic applications. *Adv. Drug Delivery Rev.* 138: 302–325.

2 Védrine, J.C. (ed.) (2018). *Metal Oxides in Heterogeneous Catalysis*. Amsterdam: Elsevier Inc.

3 Sun, C., Alonso, J.A., and Bian, J. (2021). Recent advances in perovskite-type oxides for energy conversion and storage applications. *Adv. Energy Mater.* 11 (2): 2000459. 1–21.

4 Brahlek, M., Gupta, A.S., Lapano, J. et al. (2018). Frontiers in the growth of complex oxide thin films: past, present, and future of hybrid MBE. *Adv. Funct. Mater.* 28 (9): 1702772, 1–41.

5 Stoerzinger, K.A., Risch, M., Suntivich, J. et al. (2013). Oxygen electrocatalysis on (001)-oriented manganese perovskite films: Mn valency and charge transfer at the nanoscale. *Energy Environ. Sci.* 6 (5): 1582.

6 Lee, D., Grimaud, A., Crumlin, E.J. et al. (2013). Strain influence on the oxygen electrocatalysis of the (100)-oriented epitaxial $La_2NiO_{4+\delta}$ thin films at elevated temperatures. *J. Phys. Chem. C* 117 (100): 18789–18795.

7 Lee, D., Lee, Y.-L., Grimaud, A. et al. (2014). Strontium influence on the oxygen electrocatalysis of $La_{2-x}Sr_xNiO_{4\pm\delta}$ ($0.0 \leq x_{Sr} \leq 1.0$) thin films. *J. Mater. Chem. A* 2 (18): 6480.

8 Lee, J., Murugavel, P., Ryu, H. et al. (2006). Epitaxial stabilization of a new multiferroic hexagonal phase of $TbMnO_3$ thin films. *Adv. Mater.* 18 (23): 3125–3129.

9 Li, Y., Weng, Y., Yin, X. et al. (2018). Orthorhombic $Ti_2O_3$: a polymorph-dependent narrow-bandgap ferromagnetic oxide. *Adv. Funct. Mater.* 28 (7): 1705657.

**10** Li, Y., Yu, Z.G., Wang, L. et al. (2019). Electronic-reconstruction-enhanced hydrogen evolution catalysis in oxide polymorphs. *Nat. Commun.* 10 (1): 1–11.

**11** Agrawal, A., Cho, S.H., Zandi, O. et al. (2018). Localized surface plasmon resonance in semiconductor nanocrystals. *Chem. Rev.* 118 (6): 3121–3207.

**12** Sanchez, C., Boissiere, C., Cassaignon, S. et al. (2014). Molecular engineering or functional inorganic and hybrid materials. *Chem. Mater.* 26: 221–238.

**13** Centi, G. and Perathoner, S. (2008). Catalysis by layered materials: a review. *Microporous Mesoporous Mater.* 107 (1–2): 3–15.

**14** Han, M.H., Gonzalo, E., Singh, G., and Rojo, T. (2015). A comprehensive review of sodium layered oxides: powerful cathodes for Na-ion batteries. *Energy Environ. Sci.* 8 (1): 81–102.

**15** Hall, D.S., Lockwood, D.J., Bock, C., and MacDougall, B.R. (2015). Nickel hydroxides and related materials: a review of their structures, synthesis and properties. *Proc. R. Soc. London, Ser. A* 471 (2174): 20140792.

**16** Kamath, P.V., Dixit, M., Indira, L. et al. (1994). Stabilized α-Ni(OH)$_2$ as electrode material for alkaline secondary cells. *J. Electrochem. Soc.* 141 (11): 2956.

**17** Freitas, M.B.J.G. (2001). Nickel hydroxide powder for NiOOH/Ni(OH)$_2$ electrodes of the alkaline batteries. *J. Power Sources* 93 (1–2): 163–173.

**18** Gao, M., Sheng, W., Zhuang, Z. et al. (2014). Efficient water oxidation using nanostructured α-nickel-hydroxide as an electrocatalyst. *J. Am. Chem. Soc.* 136 (19): 7077–7084.

**19** Trotochaud, L., Young, S.L., Ranney, J.K., and Boettcher, S.W. (2014). Nickel–iron oxyhydroxide oxygen-evolution electrocatalysts: the role of intentional and incidental iron incorporation. *J. Am. Chem. Soc.* 136 (18): 6744–6753.

**20** Trotochaud, L., Ranney, J.K., Williams, K.N., and Boettcher, S.W. (2012). Solution-cast metal oxide thin film electrocatalysts for oxygen evolution. *J. Am. Chem. Soc.* 134 (41): 17253–17261.

**21** Dai, L., Chen, Z., Li, L. et al. (2020). Ultrathin Ni(0)-embedded Ni(OH)$_2$ heterostructured nanosheets with enhanced electrochemical overall water splitting. *Adv. Mater.* 32 (8): 1906915.

**22** Pan, J., Sun, Y., Wan, P. et al. (2005). Synthesis, characterization and electrochemical performance of battery grade NiOOH. *Electrochem. Commun.* 7 (8): 857–862.

**23** Oliva, P., Leonardi, J., Laurent, J.F. et al. (1982). Review of the structure and the electrochemistry of nickel hydroxides and oxy-hydroxides. *J. Power Sources* 8 (2): 229–255.

**24** Wang, Y., Delahaye, E., Leuvrey, C. et al. (2016). Efficient microwave-assisted functionalization of the Aurivillius-phase $Bi_2SrTa_2O_9$. *Inorg. Chem.* 55 (8): 4039–4046.

**25** Moreno, J.M., Navarro, I., Díaz, U. et al. (2016). Single-layered hybrid materials based on 1D associated metalorganic nanoribbons for controlled release of pheromones. *Angew. Chem. Int. Ed.* 128 (37): 11192–11196.

**26** Liu, Z., Ma, R., Osada, M. et al. (2005). Selective and controlled synthesis of α- and β-cobalt hydroxides in highly developed hexagonal platelets. *J. Am. Chem. Soc.* 127 (40): 13869–13874.

**27** Butel, M., Gautier, L., and Delmas, C. (1999). Cobalt oxyhydroxides obtained by "chimie douce" reactions: structure and electronic conductivity properties. *Solid State Ionics* 122 (1–4): 271–284.

**28** Bardé, F., Palacin, M.-R., Beaudoin, B. et al. (2004). New approaches for synthesizing γ III-CoOOH by soft chemistry. *Chem. Mater.* 16 (2): 299–306.

**29** Yang, J., Liu, H., Martens, W.N., and Frost, R.L. (2010). Synthesis and characterization of cobalt hydroxide, cobalt oxyhydroxide, and cobalt oxide nanodiscs. *J. Phys. Chem. C* 114 (1): 111–119.

**30** Pralong, V., Delahaye-Vidal, A., Beaudoin, B. et al. (1999). Oxidation mechanism of cobalt hydroxide to cobalt oxyhydroxide. *J. Mater. Chem.* 9 (4): 955–960.

**31** Audemer, A., Delahaye, R., Farhi, R. et al. (1997). Electrochemical and Raman studies of beta-type nickel hydroxides $Ni_{1-x}Co_x(OH)_2$ electrode materials. *J. Electrochem. Soc.* 144 (8): 2614–2620.

**32** Wang, H.-Y., Hsu, Y.-Y., Chen, R. et al. (2015). $Ni^{3+}$-induced formation of active NiOOH on the spinel Ni–Co oxide surface for efficient oxygen evolution reaction. *Adv. Energy Mater.* 5 (10): 1500091.

**33** Nai, J., Yin, H., You, T. et al. (2015). Efficient electrocatalytic water oxidation by using amorphous Ni–Co double hydroxides nanocages. *Adv. Energy Mater.* 5 (10): 1401880.

**34** Balram, A., Zhang, H., and Santhanagopalan, S. (2017). Enhanced oxygen evolution reaction electrocatalysis via electrodeposited amorphous α-phase nickel–cobalt hydroxide nanodendrite forests. *ACS Appl. Mater. Interfaces* 9 (34): 28355–28365.

**35** Azor, A., Ruiz-Gonzalez, M.L., Gonell, F. et al. (2018). Nickel-doped sodium cobaltite 2D nanomaterials: synthesis and electrocatalytic properties. *Chem. Mater.* 30 (15): 4986–4994.

**36** Ito, M. and Nagira, T. (2005). Effect of the polymerized complex process on doping limit of thermoelectric $Na_xCo_{1-y}M_yO_2$ (M = Mn, Ni). *Mater. Trans.* 46 (7): 1456–1461.

**37** Gayathri, N., Bharathi, A., Sastry, V.S. et al. (2006). Ground state changes induced by Ni substitution in $Na_xCoO_2$. *Solid State Commun.* 138 (10–11): 489–493.

**38** Besselink, R., Stawski, T.M., Castricum, H.L. et al. (2010). Exfoliation and restacking of lepidocrocite-type layered titanates studied by small-angle X-ray scattering. *J. Phys. Chem. C* 114 (49): 21281–21286.

**39** Schaak, R.E. and Mallouk, T.E. (2000). Prying apart Ruddlesden–Popper phases: exfoliation into sheets and nanotubes for assembly of perovskite thin films. *Chem. Mater.* 12 (11): 3427–3434.

**40** Omomo, Y., Sasaki, T., Wang, L., and Watanabe, M. (2003). Redoxable nanosheet crystallites of $MnO_2$ derived via delamination of a layered manganese oxide. *J. Am. Chem. Soc.* 125 (12): 3568–3575.

**41** Ko, J.S., Doan-Nguyen, V.V.T., Kim, H.-S. et al. (2017). $Na_2Ti_3O_7$ nanoplatelets and nanosheets derived from a modified exfoliation process for use as a high-capacity sodium-ion negative electrode. *ACS Appl. Mater. Interfaces* 9 (2): 1416–1425.

**42** Sakai, N., Fukuda, K., Ma, R., and Sasaki, T. (2018). Synthesis and substitution chemistry of redox-active manganese/cobalt oxide nanosheets. *Chem. Mater.* 30 (5): 1517–1523.

**43** Acosta, M., Novak, N., Rojas, V. et al. (2017). $BaTiO_3$-based piezoelectrics: fundamentals, current status, and perspectives. *Appl. Phys. Rev.* 4 (4): 041305.

**44** Jiang, B., Iocozzia, J., Zhao, L. et al. (2019). Barium titanate at the nanoscale: controlled synthesis and dielectric and ferroelectric properties. *Chem. Soc. Rev.* 48 (4): 1194–1228.

**45** Modeshia, D.R. and Walton, R.I. (2010). Solvothermal synthesis of perovskites and pyrochlores: crystallisation of functional oxides under mild conditions. *Chem. Soc. Rev.* 39 (11): 4303.

**46** Hertl, W. (1988). Kinetics of barium titanate synthesis. *J. Am. Ceram. Soc.* 71 (10): 879–883.

**47** Eckert, J.O. Jr., Hung-Houston, C.C., Gersten, B.L. et al. (1996). Kinetic and mechanism of hydrothermal of barium titanate. *J. Am. Ceram. Soc.* 79 (11): 2929–2939.

**48** Yang, J., Zhang, J., Liang, C. et al. (2013). Ultrathin $BaTiO_3$ nanowires with high aspect ratio: a simple one-step hydrothermal synthesis and their strong microwave absorption. *ACS Appl. Mater. Interfaces* 5 (15): 7146–7151.

**49** Zhu, Y.-F., Zhang, L., Natsuki, T. et al. (2012). Facile synthesis of $BaTiO_3$ nanotubes and their microwave absorption properties. *ACS Appl. Mater. Interfaces* 4 (4): 2101–2106.

**50** O'Brien, S., Brus, L., and Murray, C.B. (2001). Synthesis of monodisperse nanoparticles of barium titanate: toward a generalized strategy of oxide nanoparticle synthesis. *J. Am. Chem. Soc.* 123 (48): 12085–12086.

**51** Niederberger, M., Garnweitner, G., Pinna, N., and Antonietti, M. (2004). Non-aqueous and halide-free route to crystalline $BaTiO_3$, $SrTiO_3$, and $(Ba,Sr)TiO_3$ nanoparticles via a mechanism involving C—C bond formation. *J. Am. Chem. Soc.* 126 (29): 9120–9126.

**52** Urban, J.J., Yun, W.S., Gu, Q., and Park, H. (2002). Synthesis of single-crystalline perovskite nanorods composed of barium titanate and strontium titanate. *J. Am. Chem. Soc.* 124 (7): 1186–1187.

**53** Adireddy, S., Lin, C., Cao, B. et al. (2010). Solution-based growth of monodisperse cube-like $BaTiO_3$ colloidal nanocrystals. *Chem. Mater.* 22 (6): 1946–1948.

**54** Niederberger, M., Pinna, N., Polleux, J., and Antonietti, M. (2004). A general soft-chemistry route to perovskites and related materials: synthesis of $BaTiO_3$, $BaZrO_3$, and $LiNbO_3$ nanoparticles. *Angew. Chem. Int. Ed.* 43 (17): 2270–2273.

**55** Mao, Y., Banerjee, S., and Wong, S.S. (2003). Large-scale synthesis of single-crystalline perovskite nanostructures. *J. Am. Chem. Soc.* 125 (51): 15718–15719.

**56** Deng, H., Qiu, Y., and Yang, S. (2009). General surfactant-free synthesis of $MTiO_3$ (M = Ba, Sr, Pb) perovskite nanostrips. *J. Mater. Chem.* 19 (7): 976.

**57** Huang, K.-C., Huang, T.-C., and Hsieh, W.-F. (2009). Morphology-controlled synthesis of barium titanate nanostructures. *Inorg. Chem.* 48 (19): 9180–9184.

58 Fu, J., Hou, Y., Zheng, M. et al. (2015). Improving dielectric properties of PVDF composites by employing surface modified strong polarized BaTiO$_3$ particles derived by molten salt method. *ACS Appl. Mater. Interfaces* 7 (44): 24480–24491.

59 Liu, H., Hu, C., and Wang, Z.L. (2006). Composite-hydroxide-mediated approach for the synthesis of nanostructures of complex functional-oxides. *Nano Lett.* 6 (7): 1535–1540.

60 Dutta, D.P., Mandal, B.P., Naik, R. et al. (2013). Magnetic, ferroelectric, and magnetocapacitive properties of sonochemically synthesized Sc-doped BiFeO$_3$ nanoparticles. *J. Phys. Chem. C* 117 (5): 2382–2389.

61 Park, T.J., Papaefthymiou, G.C., Viescas, A.J. et al. (2007). Size-dependent magnetic properties of single-crystalline multiferroic BiFeO$_3$ nanoparticles. *Nano Lett.* 7 (3): 766–772.

62 Selbach, S.M., Tybell, T., Einarsrud, M.A., and Grande, T. (2007). Size-dependent properties of multiferroic BiFeO$_3$ nanoparticles. *Chem. Mater.* 19 (26): 6478–6484.

63 Ghosh, S., Dasgupta, S., Sen, A., and Sekhar Maiti, H. (2005). Low-temperature synthesis of nanosized bismuth ferrite by soft chemical route. *J. Am. Ceram. Soc.* 88 (5): 1349–1352.

64 Han, J.-T., Huang, Y.-H., Wu, X.-J. et al. (2006). Tunable synthesis of bismuth ferrites with various morphologies. *Adv. Mater.* 18 (16): 2145–2148.

65 Wang, Y., Xu, G., Yang, L. et al. (2007). Alkali metal ions-assisted controllable synthesis of bismuth ferrites by a hydrothermal method. *J. Am. Ceram. Soc.* 90 (11): 3673–3675.

66 Wang, Y., Xu, G., Ren, Z. et al. (2007). Mineralizer-assisted hydrothermal synthesis and characterization of BiFeO$_3$ nanoparticles. *J. Am. Ceram. Soc.* 90 (8): 2615–2617.

67 Joshi, U.A., Jang, J.S., Borse, P.H., and Lee, J.S. (2008). Microwave synthesis of single-crystalline perovskite BiFeO$_3$ nanocubes for photoelectrode and photocatalytic applications. *Appl. Phys. Lett.* 92 (24): 242106.

68 Chen, J., Xing, X., Watson, A. et al. (2007). Rapid synthesis of multiferroic BiFeO$_3$ single-crystalline nanostructures. *Chem. Mater.* 19 (15): 3598–3600.

69 He, X. and Gao, L. (2009). Synthesis of pure phase BiFeO$_3$ powders in molten alkali metal nitrates. *Ceram. Int.* 35 (3): 975–978.

70 El Hadri, A., Gómez-Recio, I., del Río, E. et al. (2017). Critical influence of redox pretreatments on the CO oxidation activity of BaFeO$_{3-\delta}$ perovskites: an in-depth atomic-scale analysis by aberration-corrected and in situ diffraction techniques. *ACS Catal.* 7: 8653–8663.

71 Stoerzinger, K.A., Lü, W., Li, C. et al. (2015). Highly active epitaxial La$_{(1-x)}$Sr$_x$MnO$_3$ surfaces for the oxygen reduction reaction: role of charge transfer. *J. Phys. Chem. Lett.* 6 (8): 1435–1440.

72 Gonell, F., Sanchez-Sanchez, C.M., Vivier, V. et al. (2020). Structure–activity relationship in manganese perovskite oxide nanocrystals from molten salts for efficient oxygen reduction reaction electrocatalysis. *Chem. Mater.* 32: 4241–4247.

73 Thi N'Goc, H.L., Mouafo, L.D.N., Etrillard, C. et al. (2017). Surface-driven magnetotransport in perovskite nanocrystals. *Adv. Mater.* 29 (9): 1604745.

74 Vázquez-Vázquez, C. and Arturo López-Quintela, M. (2006). Solvothermal synthesis and characterisation of $La_{1-x}A_xMnO_3$ nanoparticles. *J. Solid State Chem.* 179 (10): 3229–3237.

75 Hou, L., Zhang, H., Dong, L. et al. (2017). A simple non-aqueous route to nano-perovskite mixed oxides with improved catalytic properties. *Catal. Today* 287: 30–36.

76 Portehault, D., Cassaignon, S., Baudrin, E., and Jolivet, J.-P. (2008). Design of hierarchical core–corona architectures of layered manganese oxides by aqueous precipitation. *Chem. Mater.* 20 (19): 6140–6147.

77 Spooren, J. and Walton, R.I. (2005). Hydrothermal synthesis of the perovskite manganates $Pr_{0.5}Sr_{0.5}MnO_3$ and $Nd_{0.5}Sr_{0.5}MnO_3$ and alkali-earth manganese oxides $CaMn_2O_4$, $4H-SrMnO_3$, and $2H-BaMnO_3$. *J. Solid State Chem.* 178 (5): 1683–1691.

78 Shimizu, Y. and Murata, T. (2005). Sol–gel synthesis of perovskite-type lanthanum manganite thin films and fine powders using metal acetylacetonate and poly(vinyl alcohol). *J. Am. Ceram. Soc.* 80 (10): 2702–2704.

79 Lin, H., Liu, P., Wang, S. et al. (2019). A highly efficient electrocatalyst for oxygen reduction reaction: three-dimensionally ordered macroporous perovskite $LaMnO_3$. *J. Power Sources* 412: 701–709.

80 Bell, R. (2000). Influence of synthesis route on the catalytic properties of $La_{1-x}Sr_xMnO_3$. *Solid State Ionics* 131 (3–4): 211–220.

81 Taguchi, H., Matsuda, D., Nagao, M. et al. (1992). Synthesis of perovskite-type $(La_{1-x}Sr_x)MnO_3$ (0 X 0.3) at low temperature. *J. Am. Ceram. Soc.* 75 (1): 201–202.

82 Cai, B., Akkiraju, K., Mounfield, W.P. et al. (2019). Solid-state gelation for nanostructured perovskite oxide aerogels. *Chem. Mater.* 31 (22): 9422–9429.

83 Gonell, F., Sánchez-Sánchez, C.M., Vivier, V. et al. (2020). Experimental descriptors for the synthesis of multicationic nickel perovskite nanoparticles for oxygen reduction. *ACS Appl. Nano Mater.* 3 (8): 7482–7489.

84 Gonell, F., Alem, N., Dunne, P. et al. (2019). Versatile molten salt synthesis of manganite perovskite oxide nanocrystals and their magnetic properties. *ChemNanoMat* 5 (3): 358–363.

85 Chen, X., Liu, L., and Huang, F. (2015). Black titanium dioxide ($TiO_2$) nanomaterials. *Chem. Soc. Rev.* 44 (7): 1861–1885.

86 Tominaka, S., Tsujimoto, Y., Matsushita, Y., and Yamaura, K. (2011). Synthesis of nanostructured reduced titanium oxide: crystal structure transformation maintaining nanomorphology. *Angew. Chem. Int. Ed.* 50 (32): 7418–7421.

87 Jing, Y., Almassi, S., Mehraeen, S. et al. (2018). The roles of oxygen vacancies, electrolyte composition, lattice structure, and doping density on the electrochemical reactivity of Magnéli phase $TiO_2$ anodes. *J. Mater. Chem. A* 6 (46): 23828–23839.

88 Pan, Y., Li, Y.Q., Zheng, Q.H., and Xu, Y. (2019). Point defect of titanium sesquioxide $Ti_2O_3$ as the application of next generation Li-ion batteries. *J. Alloys Compd.* 786: 621–626.

89 Chen, D., Chen, C., Baiyee, Z.M. et al. (2015). Nonstoichiometric oxides as low-cost and highly-efficient oxygen reduction/evolution catalysts for low-temperature electrochemical devices. *Chem. Rev.* 115 (18): 9869–9921.

90 Wang, J., Li, Y., Deng, L. et al. (2017). High-performance photothermal conversion of narrow-bandgap $Ti_2O_3$ nanoparticles. *Adv. Mater.* 29 (3): 1603730.

91 Mercier, J. and Lakkis, S. (1973). Preparation of titanium lower oxides single crystals by chemical transport reaction. *J. Cryst. Growth* 20 (3): 195–201.

92 Strobel, P. and Le Page, Y. (1982). Crystal growth of $Ti_nO_{2n-1}$ oxides ($n = 2$ to 9). *J. Mater. Sci.* 17 (8): 2424–2430.

93 Kobayashi, Y., Hernandez, O.J., Sakaguchi, T. et al. (2012). An oxyhydride of $BaTiO_3$ exhibiting hydride exchange and electronic conductivity. *Nat. Mater.* 11 (6): 507–511.

94 Andersson, S., Collen, B., Kuylenstierna, U., and Magneli, A. (1957). Phase analysis studies on the titanium-oxygen system. *Acta Chem. Scand.* 11: 1641–1652.

95 Andersson, G. (1954). Studies on vanadium oxides. *Acta Chem. Scand.* 8: 1599–1606.

96 Magneli, A. (1948). The crystal structures of $Mo_9O_{26}$ (beta-molybdenum oxide) and $Mo_8O_{23}$ (beta-molybdenum oxide). *Acta Chem. Scand.* 2: 501–517.

97 Migas, D.B., Shaposhnikov, V.L., and Borisenko, V.E. (2010). Tungsten oxides. II. The metallic nature of Magnéli phases. *J. Appl. Phys.* 108 (9): 93714.

98 Xu, B., Sohn, H.Y., Mohassab, Y., and Lan, Y. (2016). Structures, preparation and applications of titanium suboxides. *RSC Adv.* 6 (83): 79706–79722.

99 Yoshimatsu, K., Sakata, O., and Ohtomo, A. (2017). Superconductivity in $Ti_4O_7$ and $\gamma$-$Ti_3O_5$ films. *Sci. Rep.* 7 (1): 12544.

100 Åsbrink, S. and Magnéli, A. (1959). Crystal structure studies on trititanium pentoxide, $Ti_3O_5$. *Acta Crystallogr.* 12 (8): 575–581.

101 Åsbrink, G., Åsbrink, S., Magnéli, A. et al. (1971). A $Ti_3O_5$ modification of $V_3O_5$-type structure. *Acta Chem. Scand.* 25: 3889–3890.

102 Hong, S.-H. and Åsbrink, S. (1982. https://doi.org/10.1107/S056774088200939X). The structure of $\gamma$-$Ti_3O_5$ at 297 K. *Acta Crystallogr., Sect. B: Struct. Sci.* 38 (10): 2570–2576.

103 Tanaka, K., Nasu, T., Miyamoto, Y. et al. (2015). Structural phase transition between $\gamma$-$Ti_3O_5$ and $\delta$-$Ti_3O_5$ by breaking of a one-dimensionally conducting pathway. *Cryst. Growth Des.* 15 (2): 653–657.

104 Ohkoshi, S., Tsunobuchi, Y., Matsuda, T. et al. (2010). Synthesis of a metal oxide with a room-temperature photoreversible phase transition. *Nat. Chem.* 2 (7): 539–545.

105 Chen, X., Liu, L., Yu, P.Y., and Mao, S.S. (2011). Increasing solar absorption for photocatalysis with black hydrogenated titanium dioxide nanocrystals. *Science* 331 (6018): 746–750.

106 Tominaka, S., Yoshikawa, H., Matsushita, Y., and Cheetham, A.K. (2014). Topotactic reduction of oxide nanomaterials: unique structure and electronic properties of reduced $TiO_2$ nanoparticles. *Mater. Horiz.* 1 (1): 106–110.

**107** Zeng, P., Chen, M., Jiang, S. et al. (2019). Architecture and performance of the novel sulfur host material based on $Ti_2O_3$ microspheres for lithium–sulfur batteries. *ACS Appl. Mater. Interfaces* 11 (25): 22439–22448.

**108** Tominaka, S. (2012). Facile synthesis of nanostructured reduced titanium oxides using borohydride toward the creation of oxide-based fuel cell electrodes. *Chem. Commun.* 48 (64): 7949–7951.

**109** Portehault, D., Maneeratana, V., Candolfi, C. et al. (2011). Facile general route toward tunable Magnéli nanostructures and their use as thermoelectric metal oxide/carbon nanocomposites. *ACS Nano* 5 (11): 9052–9061.

**110** Kuroda, Y., Igarashi, H., Nagai, T. et al. (2019). Templated synthesis of carbon-free mesoporous Magnéli-phase titanium suboxide. *Electrocatalysis* 10 (5): 459–465.

**111** Harada, S., Tanaka, K., and Inui, H. (2010). Thermoelectric properties and crystallographic shear structures in titanium oxides of the Magnèli phases. *J. Appl. Phys.* 108 (8): 83703.

**112** Lu, Y., Matsuda, Y., Sagara, K. et al. (2011). Fabrication and thermoelectric properties of Magneli phases by adding Ti into $TiO_2$. *Adv. Mater. Res.* 415–417: 1291–1296.

**113** Fan, Y., Feng, X., Zhou, W. et al. (2018). Preparation of monophasic titanium sub-oxides of Magnéli phase with enhanced thermoelectric performance. *J. Eur. Ceram. Soc.* 38 (2): 507–513.

**114** Cao, X., Sun, Z., Zheng, X. et al. (2017). $MnCo_2O_4$ decorated Magnéli phase titanium oxide as a carbon-free cathode for $Li–O_2$ batteries. *J. Mater. Chem. A* 5 (37): 19991–19996.

**115** Ma, M., You, S., Liu, G. et al. (2016). Macroporous monolithic Magnéli-phase titanium suboxides as anode material for effective bioelectricity generation in microbial fuel cells. *J. Mater. Chem. A* 4 (46): 18002–18007.

**116** Tao, X., Wang, J., Ying, Z. et al. (2014). Strong sulfur binding with conducting Magnéli-phase $Ti_nO_{2n-1}$ nanomaterials for improving lithium–sulfur batteries. *Nano Lett.* 14 (9): 5288–5294.

**117** Mei, S., Jafta, C.J., Lauermann, I. et al. (2017). Porous $Ti_4O_7$ particles with interconnected-pore structure as a high-efficiency polysulfide mediator for lithium–sulfur batteries. *Adv. Funct. Mater.* 27 (26): 1701176.

**118** Baktash, E., Capitolis, J., Tinat, L. et al. (2019). Different reactivity of rutile and anatase $TiO_2$ nanoparticles: synthesis and surface states of nanoparticles of mixed-valence Magnéli oxides. *Chem. Eur. J.* 25 (47): 11114–11120.

**119** Wei, H., Rodriguez, E.F., Best, A.S. et al. (2017). Chemical bonding and physical trapping of sulfur in mesoporous Magnéli $Ti_4O_7$ microspheres for high-performance Li–S battery. *Adv. Energy Mater.* 7 (4): 1601616.

**120** Wang, G., Liu, Y., Ye, J. et al. (2020). Electrochemical oxidation of methyl orange by a Magnéli phase $Ti_4O_7$ anode. *Chemosphere* 241: 125084.

**121** Huang, S.-S., Lin, Y.-H., Chuang, W. et al. (2018). Synthesis of high-performance titanium sub-oxides for electrochemical applications using combination of sol–gel and vacuum-carbothermic processes. *ACS Sustainable Chem. Eng.* 6 (3): 3162–3168.

**122** Portehault, D., Delacroix, S., Gouget, G. et al. (2018). Beyond the compositional threshold of nanoparticle-based materials. *Acc. Chem. Res.* 51 (4): 930–939.

**123** Sanchez, C., Belleville, P., Popall, M., and Nicole, L. (2011). Applications of advanced hybrid organic–inorganic nanomaterials: from laboratory to market. *Chem. Soc. Rev.* 40: 696–753.

**124** Pang, Q., Kundu, D., Cuisinier, M., and Nazar, L.F. (2014). Surface-enhanced redox chemistry of polysulphides on a metallic and polar host for lithium–sulphur batteries. *Nat. Commun.* 5 (1): 4759.

**125** Lee, Y.-J., Lee, T., and Soon, A. (2019). Phase stability diagrams of group 6 Magnéli oxides and their implications for photon-assisted applications. *Chem. Mater.* 31 (11): 4282–4290.

**126** Cheng, W., Ju, Y., Payamyar, P. et al. (2015). Large-area alignment of tungsten oxide nanowires over flat and patterned substrates for room-temperature gas sensing. *Angew. Chem. Int. Ed.* 54 (1): 340–344.

**127** Zhang, W., Fan, Y., Yuan, T. et al. (2020). Ultrafine tungsten oxide nanowires: synthesis and highly selective acetone sensing and mechanism analysis. *ACS Appl. Mater. Interfaces* 12 (3): 3755–3763.

**128** Polleux, J., Pinna, N., Antonietti, M., and Niederberger, M. (2005). Growth and assembly of crystalline tungsten oxide nanostructures assisted by bioligation. *J. Am. Chem. Soc.* 127 (44): 15595–15601.

**129** Deshmukh, R. and Niederberger, M. (2017). Mechanistic aspects in the formation, growth and surface functionalization of metal oxide nanoparticles in organic solvents. *Chem. Eur. J.* 23 (36): 8542–8570.

**130** Mattox, T.M., Bergerud, A., Agrawal, A., and Milliron, D.J. (2014). Influence of shape on the surface plasmon resonance of tungsten bronze nanocrystals. *Chem. Mater.* 26 (5): 1779–1784.

**131** Kim, J., Agrawal, A., Krieg, F. et al. (2016). The interplay of shape and crystalline anisotropies in plasmonic semiconductor nanocrystals. *Nano Lett.* 16 (6): 3879–3884.

**132** Zhou, Y., Li, N., Xin, Y. et al. (2017). $Cs_xWO_3$ nanoparticle-based organic polymer transparent foils: low haze, high near infrared-shielding ability and excellent photochromic stability. *J. Mater. Chem. C* 5 (25): 6251–6258.

**133** Runnerstrom, E.L., Llordés, A., Lounis, S.D., and Milliron, D.J. (2014). Nanostructured electrochromic smart windows: traditional materials and NIR-selective plasmonic nanocrystals. *Chem. Commun.* 50 (73): 10555–10572.

**134** Migas, D.B., Shaposhnikov, V.L., Rodin, V.N., and Borisenko, V.E. (2010). Tungsten oxides. I. Effects of oxygen vacancies and doping on electronic and optical properties of different phases of $WO_3$. *J. Appl. Phys.* 108 (9): 93713.

**135** Masuda, T. and Yao, H. (2020). Intense plasmon-induced magneto-optical activity in substoichiometric tungsten oxide ($WO_{3-x}$) nanowires/nanorods. *J. Phys. Chem. C* 124 (28): 15460–15467.

**136** Manthiram, K. and Alivisatos, A.P. (2012). Tunable localized surface plasmon resonances in tungsten oxide nanocrystals. *J. Am. Chem. Soc.* 134 (9): 3995–3998.

**137** Shi, S., Xue, X., Feng, P. et al. (2008). Low-temperature synthesis and electrical transport properties of $W_{18}O_{49}$ nanowires. *J. Cryst. Growth* 310 (2): 462–466.

**138** Besnardiere, J., Ma, B., Torres-Pardo, A. et al. (2019). Structure and electrochromism of two-dimensional octahedral molecular sieve h'-$WO_3$. *Nat. Commun.* 10 (1): 327.

**139** Ranade, M.R., Navrotsky, A., Zhang, H.Z. et al. (2002). Energetics of nanocrystalline $TiO_2$. *Proc. Natl. Acad. Sci. U.S.A.* 99: 6476-6481.

**140** Levchenko, A.A., Li, G., Boerio-Goates, J. et al. (2006). $TiO_2$ stability landscape: polymorphism, surface energy, and bound water energetics. *Chem. Mater.* 18 (26): 6324–6332.

**141** Navrotsky, A. (2004). Energetic clues to pathways to biomineralization: precursors, clusters, and nanoparticles. *Proc. Natl. Acad. Sci. U.S.A.* 101 (33): 12096–12101.

**142** Forestier, H. and Guiot-Guillain, G. (1934). Une nouvelle varieté ferromagnétique de sesquioxide de fer. *C.R. Acad. Sci.* 199: 720.

**143** Schrader, R. and Büttner, G. (1963). Eine neue Eisen (III)-oxidphase: ε-$Fe_2O_3$. *Z. Anorg. Allg. Chem.* 320 (5–6): 220–234.

**144** Walter-Lévy, L. and Quémeneu, E. (1963). Chimie minerale-sur la therolyse du sulfate ferrique basique $6Fe_2(SO_4)_3$, $Fe_2O_3 \cdot nH_2O$. *C.R. Acad. Sci.* 257 (22): 3410.

**145** Tronc, E., Chanéac, C., and Jolivet, J.P. (1998). Structural and magnetic characterization of ε-$Fe_2O_3$. *J. Solid State Chem.* 139 (139): 93–104.

**146** Sakurai, S., Namai, A., Hashimoto, K., and Ohkoshi, S.I. (2009). First observation of phase transformation of all four $Fe_2O_3$ phases ($\gamma \rightarrow \varepsilon \rightarrow \beta \rightarrow \alpha$-phase). *J. Am. Chem. Soc.* 131 (51): 18299–18303.

**147** Gich, M., Frontera, C., Roig, A. et al. (2006). Magnetoelectric coupling in ε-$Fe_2O_3$ nanoparticles. *Nanotechnology* 17 (3): 687–691.

**148** Tuček, J., Zbořil, R., Namai, A., and Ohkoshi, S.I. (2010). ε-$Fe_2O_3$: an advanced nanomaterial exhibiting giant coercive field, millimeter-wave ferromagnetic resonance, and magnetoelectric coupling. *Chem. Mater.* 22 (24): 6483–6505.

**149** Taboada, E., Gich, M., and Roig, A. (2009). Nanospheres of silica with an ε-$Fe_2O_3$ single crystal nucleus. *ACS Nano* 3 (11): 3377–3382.

**150** Popovici, M., Gich, M., Niznansky, D. et al. (2004). Optimized synthesis of the elusive ε-$Fe_2O_3$ phase via sol–gel chemistry. *Chem. Mater.* 16 (25): 5542–5548.

**151** Morber, J.R., Ding, Y., Haluska, M.S. et al. (2006). PLD-assisted VLS growth of aligned ferrite nanorods, nanowires, and nanobelts-synthesis, and properties. *J. Phys. Chem. B* 110 (43): 21672–21679.

**152** Sakurai, S., Tomita, K., Hashimoto, K. et al. (2008). Preparation of the nanowire form of ε-$Fe_2O_3$ single crystal and a study of the formation process. *J. Phys. Chem. C* 112 (51): 20212–20216.

**153** Kelm, K. and Mader, W. (2005). Synthesis and structural analysis of ε-$Fe_2O_3$. *Z. Anorg. Allg. Chem.* 631 (12): 2383–2389.

**154** Sakurai, S., Shimoyama, J., Hashimoto, K., and Ohkoshi, S. (2008). Large coercive field in magnetic-field oriented ε-$Fe_2O_3$ nanorods. *Chem. Phys. Lett.* 458 (4): 333–336.

**155** Jin, J., Ohkoshi, S.I., and Hashimoto, K. (2004). Giant coercive field of nanometer-sized iron oxide. *Adv. Mater.* 16 (1): 48–51.

**156** Ohkoshi, S.I., Kuroki, S., Sakurai, S. et al. (2007). A millimeter-wave absorber based on gallium-substituted ε-iron oxide nanomagnets. *Angew. Chem. Int. Ed.* 46 (44): 8392–8395.

**157** Guan, X., Yao, L., Rushchanskii, K.Z. et al. (2020). Unconventional ferroelectric switching via local domain wall motion in multiferroic ε-$Fe_2O_3$ films. *Adv. Electron. Mater.* 6 (4): 1901134.

**158** Ohkoshi, S., Imoto, K., Namai, A. et al. (2019). Rapid Faraday rotation on ε-iron oxide magnetic nanoparticles by visible and terahertz pulsed light. *J. Am. Chem. Soc.* 141 (4): 1775–1780.

**159** Bergerud, A., Buonsanti, R., Jordan-Sweet, J.L., and Milliron, D.J. (2013). Synthesis and phase stability of metastable bixbyite $V_2O_3$ colloidal nanocrystals. *Chem. Mater.* 25 (15): 3172–3179.

**160** Chai, G., Huang, W., Shi, Q. et al. (2015). Preparation and characterization of λ-$Ti_3O_5$ by carbothermal reduction. *J. Alloys Compd.* 621: 404–410.

**161** Gich, M., Fina, I., Morelli, A. et al. (2014). Multiferroic iron oxide thin films at room temperature. *Adv. Mater.* 26 (27): 4645–4652.

**162** Bergerud, A., Selbach, S.M., and Milliron, D.J. (2016). Oxygen incorporation and release in metastable bixbyite $V_2O_3$ nanocrystals. *ACS Nano* 10 (6): 6147–6155.

# 3

# Nanosized Oxides Supported on Arrays of Carbon Nanotubes: Synthesis Strategies and Performances of TiO$_2$/CNT Systems

*Maria Letizia Terranova and Emanuela Tamburri*

Università degli Studi di Roma "Tor Vergata", Dipartimento Scienze e Tecnologie Chimiche, Via della Ricerca Scientifica, Roma 00133, Italy

## 3.1 Introduction

In a variety of advanced technological fields, the wide class of oxides plays an outstanding role, and various applications are demanding the nanosizing of either metal- or non-metal oxides. Catalysis, energy conversion, sensors, batteries, magnetic storage media, and solar cells are all areas that strongly benefit from size reduction of materials toward the nanoscale [1, 2].

In particular, the fields of electronics and optoelectronics have significantly benefitted from the nanosizing of oxides and the consequent feasibility of device miniaturization. The task to fabricate smart nanoelectronics has been accomplished in recent years by dramatic advances in the field of materials science and by the rapid development of new functional materials.

As an example, the key technological evolution of the oxide-based gate dielectrics for advanced applications has been made possible by the improvement of industrial processes that are able to produce high-quality SiO$_2$ layers, characterized by a controlled thickness and excellent interfaces for integration in Si devices. However, the reduced SiO$_2$ layer thickness results in an increase of the gate leakage and in a decrease of the material reliability [2].

To overcome such problems, over the last few decades, some high dielectric constant ($k$) metal oxides, such as Al$_2$O$_3$, ZnO$_2$, HfO$_2$, Y$_2$O$_3$, TiO$_2$, Ta$_2$O$_5$, Gd$_2$O$_3$, and La$_2$O$_3$, have been proposed for replacing conventional SiO$_2$-based gate in SiO$_2$/Si systems and in general in ultra large-scale integration (ULSI) devices [2, 3]. The rationale for such a choice lies on the permittivity, for all such oxides higher than that of SiO$_2$, and on the subsequent feasibility to safely downscale the thickness of the gate layer. In this view high-$k$ metal oxides are now considered essential to realize future generation of nanoelectronic devices [2, 3], and recent progress in fabrication technologies are allowing to use such metal oxides also in flexible and stretchable electronics [4, 5]. However, the functional performances of the metal oxides gate layers showed an end point due to some intrinsic drawbacks of the materials and to the

*Tailored Functional Oxide Nanomaterials: From Design to Multi-Purpose Applications*, First Edition.
Edited by Chiara Maccato and Davide Barreca.
© 2022 WILEY-VCH GmbH. Published 2022 by WILEY-VCH GmbH.

difficulties experienced in integration processes. It is now well known that a critical point is the gate-insulator interfacial trap density and that a specific orientation of the bottom gate/top contact along the sweeping direction governs the charge carrier mobility, reducing therefore the operating voltage [6].

With the advent of the "nanotech era" and of the ongoing drive from micro-systems toward nano-systems, a paradigm shift is taking place. The functionalities of electronics, photonics, and photochemical systems are no more just related to the downscaling of materials but rather to a development of process technology for the tailored organization of the nanostructures. Moreover, we assist nowadays to an unrivaled rise in the design of hybrid systems formed by various nanostructures aggregated in specific architectures because it is now evident that differences in the mutual organization of the different phases dictate differences in functionality. In this context, many studies have been addressed at the beginning of 2000s to the identification of suitable nanomaterials that could act as a support for nanosized oxides, thus helping to overcome the difficulties experienced in some advanced applications of these materials. The output of research activities indicated that the carbon nanotubes (CNTs), with their exceptional mechanical, electrical, chemical, and thermal properties, could be an ideal candidate for tailoring the organization of nanostructured oxides. In this context, recent literature reports on a variety of hybrid systems coupling either single-walled carbon nanotubes (SWCNTs) or multi-walled carbon nanotubes (MWCNTs) with low-sized oxides. Some interesting examples of these systems, which find applications in several advanced technological fields, are presented in this chapter.

This chapter is organized as follows:

Section 3.2 reports on synthesis strategies for the preparation of CNT arrays.
Section 3.3 describes some selected examples of supported nano-oxides.
Section 3.4 focuses on $TiO_2$/CNT systems, illustrating preparation methodologies and proposed applications.

## 3.2 Synthesis Strategies for Preparation of CNT Arrays

As already reported in a still widely cited review paper [7], there are three main classes of techniques for production of CNTs, i.e. laser ablation, arc-discharge, and chemical vapor deposition (CVD). For the purpose to produce hybrid $TiO_2$/CNT systems, the CVD methods are especially attractive because they can be easily scaled up, are suited to growing nanotubes with a strict and tailored control of the architecture, and can enable also a "one-pot" production of the two-phases nanomaterial.

The synthesis of CNTs by CVD is accomplished starting from a source gaseous phase containing the feeding carbon (molecules, clusters) that can be activated using a variety of energy sources, giving rise to methodologies commonly named as plasma CVD (where energy for activation is released by MW, RF, or hot filament) and thermal CVD (where energy for activation is released by a rapid heating in furnaces). Both these synthesis procedures share a common key requirement, i.e. the presence of metals as process catalysts assisting the rolling up of the graphene sheets. Depending on the synthesis procedures, the metals, mainly iron, cobalt, nickel, and yttrium, can be mixed with carbonaceous solid electrodes (arc-discharge) or with the targets (laser ablation) or deposited (before the synthesis process) to form nanolayers/nanoclusters on the surface of the substrates where CNT will be deposited. Beyond the growth of nanotubes on flat surfaces, an interesting approach is the fabrication of uniform array of bundles by growing nanotubes within the nanopores of $Al_2O_3$ membranes [8]. Some schemes of plasma CVD and thermal CVD reactors suitable for nanotube production are reported in Figure 3.1.

Using the CVD technique, deposits of CNT with different structures can be obtained by tuning the process parameters, among which the appropriate metal catalyst. The nature of the metal helps in driving the process toward the preferential growth of SWCNT or MWCNT and in controlling the features of the nanotube bundles. Also, the dimensions and the chemical state of the catalyst particles are very critical for achieving CNT with regular cylindrical geometries. A specific challenge is the assembling of the bundles in ordered arrays and their orientation with respect to the substrate (Figure 3.2).

Moreover, the feasibility to produce CNT at specific locations offers a great potential for their integration in circuits and devices (Figure 3.3).

## 3.3 Selected Examples of Supported Nano-oxides

The analysis of the scientific literature shows that in the last decade a variety of hybrid materials formed by CNT arrays supporting low-dimensional metal oxides have been studied. The more common use of these systems is in the area of energetics; in particular, the amenable performances of nanostructured oxides suggested a reliable use of these materials as electrodes for supercapacitors. The highly conductive and robust CNT core, coated by a redox active oxide shell, can indeed offer a remarkable specific capacitance. Carbon nanofibers have been also employed to obtain such results. An example is reported in [9], where hybrid supercapacitor electrodes were fabricated by cathodic deposition of ultra-thin $MnO_2$ layers, coaxially coating the vertically aligned carbon nanofiber arrays (Figure 3.4). The fast kinetics of the redox reaction and the easy electrolyte access to a large volume of active electrode materials increased the total specific capacitance by adding a remarkable pseudo-capacitance to the electrical double layer capacitance [9].

On vertical CNT arrays, $IrO_2$ nanocrystals, with shapes going from particles to tubules (Figure 3.5), have been grown using $C_6H_7(C_8H_{12})Ir$ as a CVD precursor [10].

The crystalline nanoparticles of $IrO_2$ covering the CNT were found to act as a protective layer, improving both stability and lifetime of field emission (FE) with respect to the bare CNT emitters, as evidenced in Figure 3.6. Besides applications as electron emitters, the feasibility to deposit $IrO_2$ in the form of nanotubules makes the $IrO_2$/CNT nanocomposites attractive also for use as electrodes for supercapacitors.

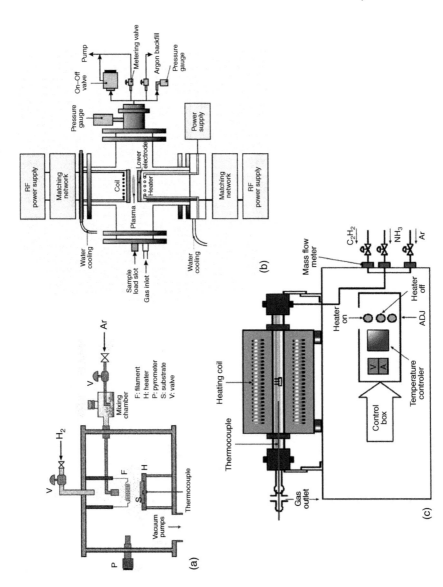

**Figure 3.1** Scheme of CVD reactors for CNT synthesis: (a) HFCVD apparatus equipped with a system for powder spraying, (b) inductively coupled plasma reactor with an independent RF power, and (c) thermal apparatus. Source: Terranova et al. [7], with permission from John Wiley and Sons.

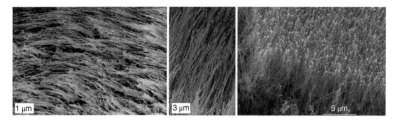

**Figure 3.2** Scanning electron microscopy (SEM) micrographs of typical CNT bundles grown with different orientation with respect to the substrate. Source: Maria Letizia Terranova and Emanuela Tamburri.

**Figure 3.3** Scanning electron microscopy (SEM) micrographs of typical CNT deposits on selected areas of Si substrates. Source: Maria Letizia Terranova and Emanuela Tamburri.

A recent interesting example of hybrid systems has been produced by the coupling of $V_2O_5$ with vertically aligned arrays of carbon nanotubes (VACNTs) [11]. Here the homogeneous decoration of VACNT arrays by a nanosized $V_2O_5$ phase has been achieved through a process of supercritical $CO_2$ impregnation and subsequent annealing. The scheme of the process is shown in Figure 3.7.

The composite $V_2O_5$/VACNT (VN) material, used as binder-free negative electrodes in aqueous asymmetric supercapacitors, exhibited a specific capacitance of 284 F g$^{-1}$ in the potential range from −1.1 to 0 V vs. SCE at 2 A g$^{-1}$, and outstanding cycling stability in $Na_2SO_4$ solution. Figure 3.8 shows the relevant electrochemical performances of the VN electrodes that enable the supercapacitor to deliver an energy density as high as 32.3 Wh kg$^{-1}$ at a power density of 118 W kg$^{-1}$. Moreover, a satisfactory cycling life with a capacitance retention of 76% after 5000 cycles has been measured. The good performances of VN electrodes can be ascribed to the in situ growth of the $V_2O_5$ nanoparticles. The resulting unique three-dimensional nanostructure ensures indeed a close contact of the active material with the CNT scaffold, whereas the spacing between nanotubes acts in relieving the internal strain within the $V_2O_5$ phase during charge/discharge cycles.

## 3.4 A Focus on the TiO$_2$/CNT Systems

Even if several hybrid structures based on CNT templates have been proposed and tested for various nanosized metal oxides, new concepts and disruptive applications diverging from the conventional technologies are stimulating research

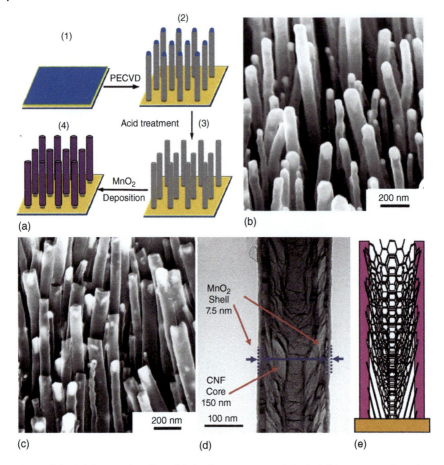

**Figure 3.4** (a) Schematics of the fabrication processes of the hybrid supercapacitors by coating a uniform manganese oxide layer on the vertically aligned carbon nanofiber array. (b) SEM image at 45° perspective view of an as-grown vertically aligned carbon nanofiber array. (c) SEM image at 45° perspective view of a vertically aligned carbon nanofiber array after 10–15 minutes of treatment in 1.0 M $HNO_3$ followed by 20 minutes of electrochemical deposition to coat manganese oxide. (d) TEM image of the same sample as that in (c), indicating the uniform coating of ~7.5 nm manganese oxide layer on the 150-diameter carbon nanofiber, presumably dominated by $MnO_2$. (e) Schematic illustration of the uniform $MnO_2$ coating on the sidewall of the cup-stacking graphitic structure of carbon nanofibers, likely associated with the active broken graphitic edges.

efforts especially in the field of high-$k$ oxides (Figure 3.9) [12]. The emergence of miniaturized, flexible, and stretchable electronics, together with the need to optimize electrical and mechanical performances, is pushing the research toward the design of highly efficient hierarchical structures.

In this context, recent advanced applications of high-$k$ nano-oxides were enabled by the settling of techniques for their organization in well-defined architectures, such as those provided by CNT arrays. In effect the CNTs do not act merely as a nanoshaped support for the oxides. Due to the feasibility to modulate the

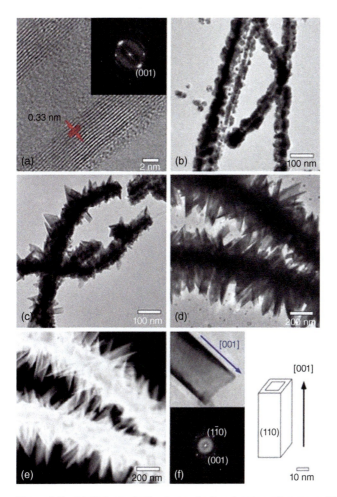

**Figure 3.5** (a) High resolution transmission electron microscopy (HRTEM) image of the pristine CNT; the inset is the FFT pattern. TEM images of IrO$_2$/CNT nanocomposites with growth time of (b) five minutes, (c) 10 minutes, and (d) 60 minutes. (e) High-angle annular dark-field (HAADF) image of IrO$_2$ nanocrystals grown on the CNT surface. (f) Upper part of left-hand side: HRTEM image focused on a IrO$_2$ nanotube; lower part of left-hand side: the FFT pattern taken from the tube side wall; right-hand side: a schematic of the tubular crystal of IrO$_2$. Source: Springer Nature.

carbon/dielectric heterojunctions by tailoring their mutual organization, such hybrid materials are nowadays finding applications in a variety of devices and technological systems like displays, sensors, printed RFID tags, bioelectronics [13]. Actually, valuable progresses in the engineering of CNT/dielectrics starting blocks guarantee the possibility to attain high-level performances, provided that the architecture of the two-phase systems is properly designed. As regards the current technological needs and the future challenges, the more relevant requirements are related to geometry and surface properties of the supporting CNT phase as well

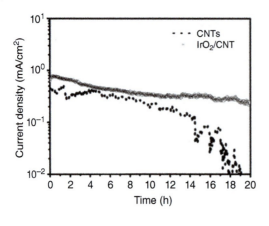

**Figure 3.6** Emission current stability of the pristine CNTs (dotted curve) and the IrO$_2$/CNT nanocomposites (open circle). Source: Reprinted with permission from Ref. [10]. Copyright © 2010 Springer.

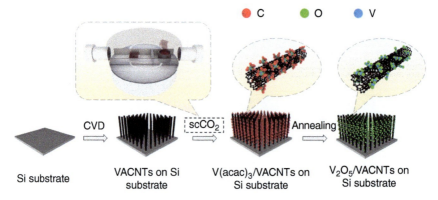

**Figure 3.7** Schematic of the synthesis of V$_2$O$_5$/VACNT composite. Source: Reprinted with permission from Ref. Sun et al. [11]. Copyright 2021 Elsevier.

as to spatial confinement and structural characteristics of the metal oxides. Such considerations influenced both fundamental and applied research and have been at the basis of materials evolution, as attested by a wide literature.

Among the class of high-$k$ materials, titanium oxide (TiO$_2$) is certainly one of the most exciting. TiO$_2$ has been from long time known as an efficient photocatalyst for its ability to degrade a wide range of organic pollutants under UV illumination. The use of TiO$_2$ as photocatalyst started in 1972 in water decomposition [14]. TiO$_2$ is a multitalented wide-gap semiconducting material [5] that can exist in three main polymorphs, *i.e.* rutile, anatase, and brookite, in each of which titanium cations are sixfold coordinated to oxygen anions, forming distorted TiO$_6$ octahedra mostly joined by sharing the octahedral edges (Figure 3.10).

These structures differ by the spatial arrangement of the TiO$_6$ octahedra building blocks. The thermodynamically stable form of bulk titania is rutile, which has tetragonal structure with a space group P42/mnm (136) (Figure 3.10a). While being meta-stable at the bulk form, anatase and brookite become stable when they are

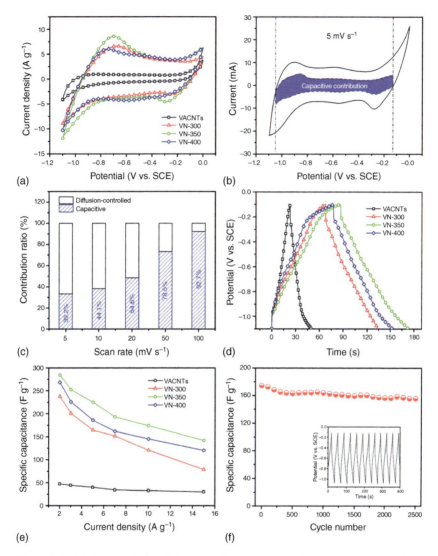

**Figure 3.8** Electrochemical performance of as-grown VACNT and the VN composites in 1 M $Na_2SO_4$. (a) CV curves at 20 mV s$^{-1}$. (b) CV curve of the VN-350 electrode with shaded area showing surface capacitive contribution. (c) Contribution of surface capacitive and diffusion-controlled components for the VN annealed at 350° at various scan rates. (d) Galvanostatic charge–discharge curves at 3 A g$^{-1}$. (e) Specific capacitance vs. current density. (f) Cycling stability of VN annealed at 350° at a current density of 10 A g$^{-1}$. Source: Reprinted with permission from Ref. Sun et al. [11]. Copyright 2021 Elsevier.

nanostructured, due to the smaller surface energy at the nanoscale. Anatase and rutile have indirect band gap, which is 3.0 eV for rutile and 3.2 eV for anatase, corresponding to UV light absorption. Anyway, it is generally accepted that anatase is the most active phase for photocatalysis applications, due to its better electronic and surface chemistry properties [15].

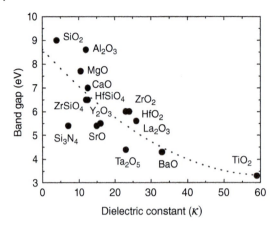

**Figure 3.9** Relation between band gap and dielectric constant for some metal oxide dielectrics. Source: Reprinted with permission from Ref. Barquinha et al. [12]. Copyright 2012 John Wiley and Sons.

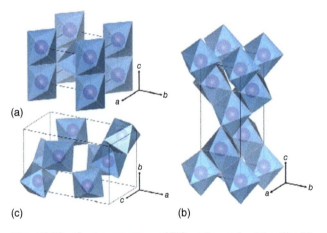

**Figure 3.10** Crystal structures of $TiO_2$ polymorphs: (a) rutile, (b) anatase, and (c) brookite. Purple spheres represent Ti atom, and the blue octahedra represent $TiO_6$ blocks. Oxygen atoms at the corner of the octahedra are omitted for clarity. Source: Zhang et al. [15], with permission from Royal Society of Chemistry.

The versatility of the $TiO_2$ in promoting the separation of the electron–hole charges generated upon irradiation has been exploited in the solar energy conversion field by developing in 1991 the dye-sensitized photoelectrochemical cells by O'Reagan and Gratzel [16].

Actually, $TiO_2$ is the focus of cutting-edge applications in many technological areas, from electronics and optoelectronics to gas sensing, photoelectrochemistry, photocatalysis, and solar-to-electric energy conversion [17, 18].

The wide applicability of bulk $TiO_2$ is even increased when the dimensions of this semiconducting oxide are downsized. In the nano-world, $TiO_2$ can be found in the form of nanoparticles, nanotubes, nanowires, nanorods, and nanoporous layers, with electronic and surface properties strictly dependent on shape and size [15, 19]. It is interesting to note that the $TiO_2$ nanosizing produces effects also in other fields, modulating, as an example, strength and durability of cement-based composite materials used for pollution degradation [20].

The outstanding possibilities opened by the nano-sizing have inspired in the last years a lot of research work, mainly addressed at the production of $TiO_2$ nano-entities ($n$-$TiO_2$) with well-defined structure, phase purity, shape, and size. The morphological control of facets and the modulation of the correlated surface energy of the $TiO_2$ nanocrystals are indeed fundamental steps for technological applications [21]. Moreover, developments of highly efficient materials for modern industry need the tailoring of nano-phase architectures. In this context, large efforts have been dedicated to $n$-$TiO_2$ immobilization on fixed supports.

A paper published in 2004 proposed for the first time the coupling, by wet chemistry approaches, of nanosized $TiO_2$ with CNT arrays [22]. From then on, the task to coat CNT, either SWCNT or MWCNT ones with $TiO_2$ at the nanoscale, has been the focus of intensive research.

The optimization of device performances guided the choice of the materials architectures and, therefore, of the procedures for the engineering of hybrid $TiO_2$/CNT systems. As a first requirement the supporting CNT must provide spatial confinement of the $TiO_2$. In this view also some features of the CNT arrays, as orientation and mutual organization with respect to the $TiO_2$ phase, must be specific for each application. However, the more intriguing task is to produce hybrid nanomaterials where the CNTs do not act only as a scaffold but provide also fast transport paths for the charges photogenerated in the semiconducting oxide [23]. It is indeed confirmed that the efficiency of charge separation and transport strongly depends on the characteristics of interfacial contacts between the two phases of the hybrid material [19]. A proper engineering of the $TiO_2$–nanocarbon interface is therefore a fundamental task to be pursued, especially in the case of environmentally friendly and renewable energy applications, such as sunlight photoconversion, $CO_2$ photoreduction, and hydrogen production from water splitting [23] (Figure 3.11).

The objective of this contribution is in particular to illustrate the state of art of the synthetic procedures that allows the immobilization of $TiO_2$ nanoentities onto geometrically controlled CNT arrays and to highlight, for some selected examples, performances of $TiO_2$/CNT systems highly compatible with some forefront technologies.

### 3.4.1 Synthesis of $TiO_2$ on CNT

$TiO_2$ at the nanoscale is obtained using a variety of already well-established techniques, such as sol–gel [24, 25], solvothermal [26], and hydrothermal [27] methodologies, direct oxidation [28], electrochemistry [29], and chemical vapor deposition [19, 30]. Some of the previously indicated chemical routes are also employed to produce $TiO_2$-on-CNT systems [31]. However, in this case, precautions must be taken to avoid the possibility that the deposition of $TiO_2$ could interrupt the network of interconnected CNT, reducing in such a way the potential of charge transport of the nanocomposite material.

**Figure 3.11** Schematic representation of $H_2$ production using renewable and non-renewable sources and its significant applications. Source: Reddy et al. [23], with permission of Elsevier.

We report in the following some successful methodologies that produced $TiO_2$-on-CNT systems characterized by properties suitable for a variety of applications. For the sake of clearness, the methodologies for the hybrid material preparation are divided in two main categories, namely, the wet chemistry and the vapor deposition ones.

#### 3.4.1.1 Wet Chemistry

Starting from the beginning of 2000s, it was discovered that $TiO_2$ anchored to CNT exhibited a significant increment of photoactivity in comparison to the pure titania. From then on, several chemical and electrochemical methodologies have been proposed to fabricate $TiO_2$/CNT hybrid materials for catalytic applications.

Among the first papers reported in literature, one can cite Jitianu et al., who used an alkoxide precursor and the classical sol–gel method to deposit anatase on MWCNT [32].

Composites formed by embedding MWCNT in nanosized $TiO_2$ were also prepared by a modified sol–gel method and used to investigate photodegradation of phenol [33]. Also, MWCNT coated by $TiO_2$ nanoparticles have been successfully tested for phenol degradation under light irradiation [34]. In this case commercial MWCNT were decorated with well-dispersed anatase nanoparticles (diameter lower than 7 nm) obtained by hydrolysis of titanium isopropoxide (TIP) in supercritical ethanol. Modulation of the $TiO_2$ decoration was obtained by changing the MWCNT:TIP ratio, and the functional test evidenced the efficiency of the $TiO_2$/MWCNT nanocomposites in the whole UV–visible region. TEM images of such materials are shown in Figure 3.12.

Also, the surface treatments of the supporting CNT exert an influence on the nanocomposite performances, as highlighted in the paper by [35]. Here the efficiency for Red X-3B dye degradation under UV irradiation has been related to the

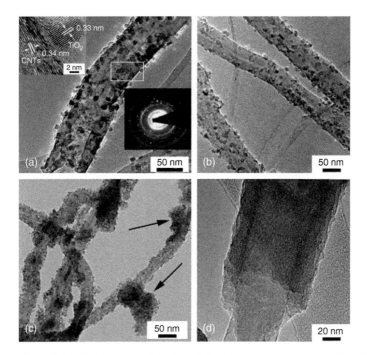

**Figure 3.12** TEM images of the TiO$_2$/MWCNT composites prepared in ethanol at 270 °C with different TIP/MWCNT mass ratios and reaction times: (a) 5 : 1, two hours, the left inset is the HRTEM image of the composite and the right one is an electron diffraction of the denoted rectangular area; (b) 1 : 1, two hours; (c) 10 : 1, two hours; and (d) 20 : 1, 0.5 hour. Source: An et al. [34], with permission of Elsevier.

degree of acid oxidization and to the presence of functional groups on the CNT surface.

The relationship between wet chemistry methodologies for the synthesis of TiO$_2$/CNT systems and their functional abilities has been analyzed and discussed in [31].

In [36], the preparation of MWCNT/TiO$_2$ nanocomposites by a "surfactant wrapping" sol–gel method is illustrated using a series of different precursors: titanium ethoxide, titanium isopropoxide, and titanium butoxide. A scheme describing the preparation of the core/shell nanocomposites by the surfactant wrapping sol–gel method is reported in Figure 3.13. Whereas all the precursors yielded uniform, continuous, and well-defined layers of nanosized anatase on individual MWCNT, differences were found in the photocatalytic and photoelectrocatalytic activity of the various mesoporous core/shell systems. The research evidenced that such properties are significantly affected by the thickness of the TiO$_2$ layer and by the conductivity of the final nanomaterial that increases with increasing the MWCNT amount.

The in situ sol–gel preparation demonstrated to be a viable synthesis path to produce TiO$_2$/MWCNT nanocomposites characterized by high functional performances [37]. These systems were found to efficiently act as photocatalyst for organic compounds degradation and moreover for photo-inactivation of bacteria.

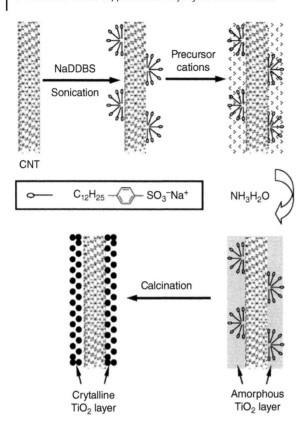

**Figure 3.13** Synthetic scheme for the preparation of CNTs/TiO$_2$ nanocomposites by the surfactant wrapping sol–gel method, in the presence of sodium dodecylbenzenesulfonate (NaDDBS). Source: Reprinted with permission from Ref. Li et al. [36]. Copyright 2011 Elsevier.

A scalable in situ sol–gel method has been also employed to produce free-standing flexible films (∼280 μm in thickness and ∼4 cm in diameter) starting from CNT super-aligned arrays with variable amount of TiO$_2$ anchored on them [38]. The scheme of the material preparation is illustrated in Figure 3.14.

The as-prepared TiO$_2$/CNT systems have been tested as anodes in fast-charging Li-ion batteries, showing an enhanced rate capability and charge/discharge reversibility due to the contribution of a pseudocapacitive storage mechanism. Such TiO$_2$/CNT films can be assembled to fabricate flexible full cells and storage devices where binders, conductive agents, and current collectors are unnecessary.

Thereafter, a series of wet chemistry routes have been exploited to synthesize CNT-based composites in simple and effective ways, avoiding the use of strong acids or of organic stabilizers [39].

Also, electrochemical routes have been proposed to obtain TiO$_2$/CNT complex architectures designed to be used as anode in Li-ion cells. Yan et al. [40] obtained a triple-layered system formed by CNT coated by a first layer of mesoporous/nanocrystalline TiO$_2$ and an external layer of carbon. The two mixed TiO$_2$ phases were prepared by electrochemistry starting from hydrous titania and resorcinol–formaldehyde resin, and the process was completed by controlled crystallization and carbonization steps. The structural characteristics of the TiO$_2$ phases coupled to carbon offered space to accommodate volume variation during the Li-ion

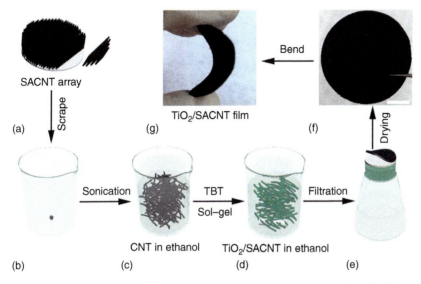

**Figure 3.14** Schematic illustration of the preparation of the free-standing flexible TiO$_2$/CNT films. Source: Reprinted with permission from Ref. Zhu et al. [38]. Copyright 2018 American Chemical Society.

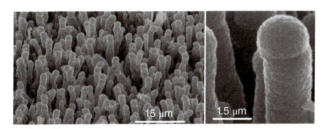

**Figure 3.15** SEM images at two different magnifications of a TiO$_2$-coated SWCNT deposit. Source: Orlanducci et al. [41], with permission of Elsevier.

intercalation stages. Such material evidenced excellent cycling stability and good performances in terms of Li-storage capacity (244 mAh g$^{-1}$ at a current density of 17 mA g$^{-1}$) and rate capability (115 mAh g$^{-1}$ at a current density of 850 mA g$^{-1}$ 5 C) [40].

### 3.4.1.2 Vacuum Techniques

The use of vacuum techniques to deposit nanosized TiO$_2$ allows to avoid the orientational disorder in the mutual aggregation of TiO$_2$ and CNT. As demonstrated by Orlanducci et al. [41], hierarchical TiO$_2$/CNT nanostructures can be obtained by the metal–organic CVD technique, using titanium tetraisopropoxide Ti(OiPr)$_4$ as a precursor. The external coating is formed by anatase grains with an average size of about 50 nm, immobilized on vertically aligned SWCNT arrays that maintain their pristine architecture. Such features make the integration of the nanocomposites in a series of devices possible (Figure 3.15).

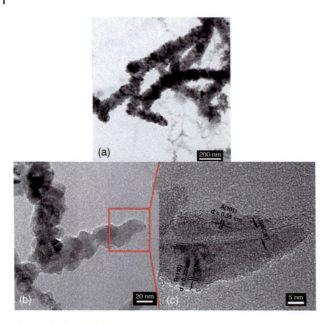

**Figure 3.16** (a) TEM image of the TiO$_2$ coated composite nanostructure at 400 °C for 60 minutes; (b) TEM image of the 20 nm thick TiO$_2$/SWCNTs composite nanostructures selected for high-resolution TEM (HRTEM); and (c) the HRTEM image observed from the box area of (b). Source: Duong et al. [42], with permission of Royal Society of Chemistry.

A low-temperature gas condensation method based on the deposition of TiO$_2$ nanoclusters onto SWCNTs has been used to produce a mixed rutile/anatase phase uniformly covering the nanotubes [42]. The high-resolution TEM images shown in Figure 3.16 evidence the morphology of the TiO$_2$ deposits.

The TiO$_2$/SWCNT system obtained in [42] demonstrated a strong adhesion to the indium tin oxide substrate, and the working electrodes assembled with the as-prepared material were characterized by favorable photo-response and photoelectrochemical effects. In particular, the photo-response was approximately 60 times higher than that of TiO$_2$ thick films.

A different methodology, namely, an in situ chemical vapor deposition, was employed by Ma et al. [43] for the direct growth of CNT in the presence of TiO$_2$ films. The CNT resulted in uniformly coated nanosized TiO$_2$ particles, giving rise to a composite material highly active for degradation of methyl orange under UV irradiation (Figure 3.17).

Di et al. [44] used continuous ultrathin (50–100 nm) sheets of highly interconnected CNT to fabricate CNT/TiO$_2$ core/shell heterostructures. The Ti-oxide deposition on the nanotube sheet was carried out by a vapor phase pyrolysis of titanium isopropoxide in a furnace kept under Ar atmosphere. This CVD method allowed to achieve a favorably oxide wrapping without altering the sheet structure formed by interconnected network of CNT. The result is a robust, flexible, and conductive film with a high photon-to-electron conversion efficiency, outperforming that of pure TiO$_2$ nanoparticles (Figure 3.18).

 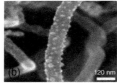 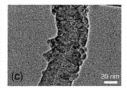

**Figure 3.17** (a) TEM image of the TiO$_2$ coated composite nanostructure at 400 °C for 60 minutes; (b) TEM image of the 20 nm thick TiO$_2$/SWCNTs composite nanostructures selected for high-resolution TEM (HRTEM); and (c) the HRTEM image observed from the box area of (b). Source: Ma et al. [43], with permission of Elsevier.

**Figure 3.18** (a) Sketch illustrating that charges transfer more smoothly between CNTs that are interconnected inside oxide coatings (right) than those isolated by oxide coatings (left); (b) a 3 cm wide CNT sheet being drawn from a 2 in. spinnable CNT array; and (c) a free-standing CNT/TiO$_2$ film. (d) Photo of a CNT/TiO$_2$ film attached to a PET film. Source: Di et al. [44], with permission from John Wiley and Sons.

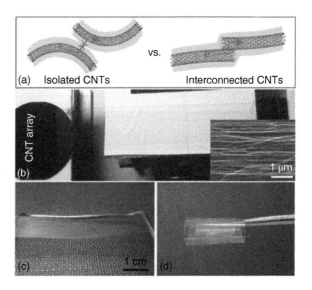

Anatase nanoparticles with various morphologies were deposited by Huang et al. on functionalized CNT using the atomic layer deposition (ALD) technique [45]. The crystalline features of the TiO$_2$ coating obtained in the $T$ range 100–300 °C, and their photocatalytic properties were found to depend on the process temperature (Figure 3.19). This result has been rationalized by considering the specific area and the electron and hole diffusion path in the hybrid nanomaterials grown at different temperatures.

The highly regular surface-controlled growth provided by the ALD technique has been exploited to obtain an excellent conformal coverage of complex CNT deposits [46]. Here hierarchical catalysts for photodegradation of organic dyes could benefit of high quality TiO$_2$/CNT heterojunctions with increased quantum efficiency.

The use of vacuum techniques to deposit nanosized TiO$_2$ allows a more detailed control of TiO$_2$/CNT heterojunctions, and therefore produces systems compatible with more sophisticated applications. In this context, also strategies mixing vacuum techniques with wet chemistry have been successfully employed to produce flexible electrodes for supercapacitors [47]. The regular coverages of CNT by TiO$_2$

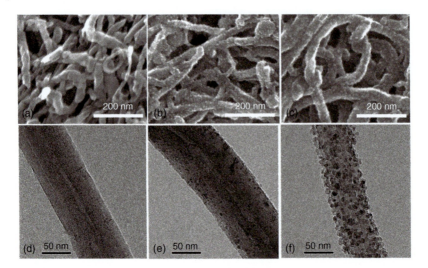

**Figure 3.19** SEM morphology of TiO$_2$@a-CNT structures formed with 50 cycles of ALD at (a) 100 °C, (b) 200 °C, and (d) 300 °C. TEM images of TiO$_2$@a-CNT structures formed various cycles of ALD at 300 °C (d) 10 cycles, (e) 25 cycles, and (f) 50 cycles. Source: Huang et al. [45], with permission of IOP Publishing Ltd.

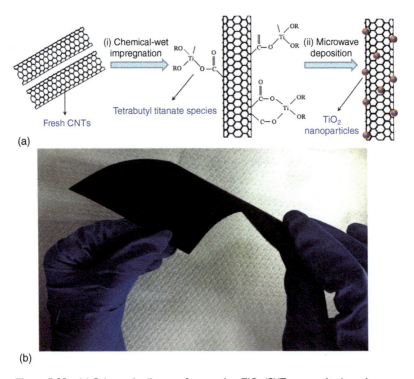

**Figure 3.20** (a) Schematic diagram for growing TiO$_2$/CNT composite by using (i) chemical-wet impregnation and (ii) pulse microwave deposition; (b) photographs of the produced TiO$_2$/CNT composite. Source: Hsieh et al. [47], with permission from Elsevier.

nanocrystals have been obtained by a wet chemistry impregnation of CVD-grown CNT mats, followed by pulse microwave irradiation. The scheme of the process along with a picture of the $TiO_2$/CNT composite is shown in Figure 3.20.

## 3.5 Concluding Remarks

According to the studies reported in the last two decades, nanoscaled metal oxides enormously enlarged their typical application fields. In particular, the use of nano-oxides allowed to make a real progress toward achievements in the ever-growing area of energetics. The performances of batteries, supercapacitors, and solar cells and of a variety of other devices for energy production and storage have been strongly influenced by the downsizing of already traditionally utilized oxides. However, in some cases the orientational disorder of the nano-oxides aggregates was found to cause incompatibility with the new standards of the microelectronic industry. The paradigm shift able to overcome such drawbacks was the design of hybrid systems where nano-oxides entities are supported by ordered arrays of foreign, compatible materials. In this context, C-$sp^2$ materials and, in particular, CNTs have shown the capability to interface effectively with the nano-oxides, assuring a good integration of the C phase with the oxides. This process, relevant for all the oxides, is of fundamental importance when dealing with high-$k$ oxides and in particular with $TiO_2$.

The outstanding intrinsic properties of $TiO_2$, combined with the additional properties provided by the nanoscale downsizing, make nanosized titania to emerge as designated high-$k$ material for future generation of nanoelectronics, especially when flexibility/stretchability are needed. Moreover, this nanomaterial is relevant for a series of strategic processes, such as photocatalysis and $CO_2$ photoreduction, and for energy-related forefront technologies such as hydrogen evolution, dye sensitized solar cells, and energy storage.

Nevertheless, the potential applications of the nanosized $TiO_2$ can be in some cases hampered by drawbacks, such as a strong tendency to aggregation and the consequent difficulty in controlling the phase arrangement. The coupling of $TiO_2$ nanoentities with CNT represents a fundamental step to overcome these obstacles, but the engineering of CNT/$TiO_2$ systems can assure much more than just a tailored arrangement of the nano-oxide.

Due to their features of charge transport and of electron–hole recombination, the supporting nanotubes can cooperate with the $TiO_2$, promoting synergetic effects that can alter the electronic behavior of the high-$k$ $TiO_2$ phase. The dependence of the electronic response of the $TiO_2$/CNT systems on the mutual organization of the two components emphasizes the need to formulate heterojunctions specifically tailored for each application.

This is the objective of an ever-increasing research field, dealing with the setting-up of routes for an efficient aggregation of such outstanding nanomaterials and of methods for their characterizations.

The final goal of the scientific efforts will be to design hierarchically organized nanostructures featuring intelligent properties and specific functions.

## References

1 Chavali, M.S. and Nikolova, M.P. (2019). Metal oxide nanoparticles and their applications in nanotechnology. *SN Appl. Sci.* 1 (6): 1–30.
2 Robertson, J. (2004). High dielectric constant oxides. *Eur. Phys. J. Appl. Phys.* 28 (3): 265–291.
3 Gusev, E.P., Cartier, E., Buchanan, D.A. et al. (2001). Ultrathin high-K metal oxides on silicon: processing, characterization and integration issues. *Microelectron. Eng.* 59 (1-4): 341–349.
4 Choi, J.H., Mao, Y., and Chang, J.P. (2011). Development of hafnium based high-k materials—A review. *Mater. Sci. Eng., R* 72 (6): 97–136.
5 Wang, B., Huang, W., Chi, L. et al. (2018). High-k gate dielectrics for emerging flexible and stretchable electronics. *Chem. Rev.* 118 (11): 5690–5754.
6 Aikawa, F., Ueno, J., Kashiwagi, T., and Itoh, E. (2020). Improvement of field-effect transistor performance with highly oriented, vertically phase separated TIPS-pentacene/polystylene blends on high-k metal oxide films by using meniscus coating. *Jpn. J. Appl. Phys.* 59 (SC): SCCA10.
7 Terranova, M.L., Sessa, V., and Rossi, M. (2006). The world of carbon nanotubes: an overview of CVD growth methodologies. *Chem. Vap. Deposition* 12 (6): 315–325.
8 Angelucci, A., Ciorba, A., Malferrari, L. et al. (2009). Field emission properties of carbon nanotube arrays grown in porous anodic alumina. *Phys. Status Solidi c* 6 (10): 2164–2169.
9 Liu, J., Essner, J., and Li, J. (2010). Hybrid supercapacitor based on coaxially coated manganese oxide on vertically aligned carbon nanofiber arrays. *Chem. Mater.* 22 (17): 5022–5030.
10 Chen, Y.M., Huang, Y.S., Lee, K.Y. et al. (2011). Characterization of $IrO_2$/CNT nanocomposites. *J. Mater. Sci. – Mater. Electron.* 22 (7): 890–894.
11 Sun, G., Ren, H., Shi, Z. et al. (2021). $V_2O_5$/vertically-aligned carbon nanotubes as negative electrode for asymmetric supercapacitor in neutral aqueous electrolyte. *J. Colloid Interface Sci.* 588: 847–856.
12 Barquinha, P., Martins, R., Pereira, L., and Fortunato, E. (2012). *Transparent Oxide Electronics: From Materials to Devices*. John Wiley & Sons.
13 Wang, B., Huang, W., Chi, L. et al. (2018). High-k gate dielectrics for emerging flexible and stretchable electronics. *Chem. Rev.* 118 (11): 5690–5754.
14 Fujishima, A. and Honda, K. (1972). Electrochemical photolysis of water at a semiconductor electrode. *Nature* 238 (5358): 37–39.
15 Zhang, Y., Jiang, Z., Huang, J. et al. (2015). Titanate and titania nanostructured materials for environmental and energy applications: a review. *RSC Adv.* 5 (97): 79479–79510.
16 O'regan, B. and Grätzel, M. (1991). A low-cost, high-efficiency solar cell based on dye-sensitized colloidal $TiO_2$ films. *Nature* 353 (6346): 737–740.
17 Dissanayake, M.A.K.L., Sarangika, H.N.M., Senadeera, G.K.R. et al. (2017). Application of a nanostructured, tri-layer $TiO_2$ photoanode for efficiency enhancement

in quasi-solid electrolyte-based dye-sensitized solar cells. *J. Appl. Electrochem.* 47 (11): 1239–1249.

18 Haider, A.J., Jameel, Z.N., and Al-Hussaini, I.H. (2019). Review on: titanium dioxide applications. *Energy Procedia* 157: 17–29.

19 Chen, X. and Mao, S.S. (2007). Titanium dioxide nanomaterials: synthesis, properties, modifications, and applications. *Chem. Rev.* 107 (7): 2891–2959.

20 Shafaei, D., Yang, S., Berlouis, L., and Minto, J. (2020). Multiscale pore structure analysis of nano titanium dioxide cement mortar composite. *Mater. Today Commun.* 22: 100779.

21 Xiao, C., Lu, B.A., Xue, P. et al. (2020). High-index-facet-and high-surface-energy nanocrystals of metals and metal oxides as highly efficient catalysts. *Joule* 4 (12): 2562–2598.

22 Jitianu, A., Cacciaguerra, T., Benoit, R. et al. (2004). Synthesis and characterization of carbon nanotubes–$TiO_2$ nanocomposites. *Carbon* 42 (5-6): 1147–1151.

23 Reddy, N.R., Bhargav, U., Kumari, M.M. et al. (2020). Review on the interface engineering in the carbonaceous titania for the improved photocatalytic hydrogen production. *Int. J. Hydrogen Energy* 45 (13): 7584–7615.

24 Del Ángel-Sanchez, K., Vázquez-Cuchillo, O., Aguilar-Elguezabal, A. et al. (2013). Photocatalytic degradation of 2, 4-dichlorophenoxyacetic acid under visible light: Effect of synthesis route. *Mater. Chem. Phys.* 139 (2-3): 423–430.

25 Pipes, R., Bhargav, A., and Manthiram, A. (2018). Nanostructured anatase titania as a cathode catalyst for Li–$CO_2$ batteries. *ACS Appl. Mater. Interfaces* 10 (43): 37119–37124.

26 Kathirvel, S., Su, C., Shiao, Y.J. et al. (2016). Solvothermal synthesis of $TiO_2$ nanorods to enhance photovoltaic performance of dye-sensitized solar cells. *Sol. Energy* 132: 310–320.

27 Novaconi, S. and Vaszilcsin, N. (2013). Inductive heating hydrothermal synthesis of titanium dioxide nanostructures. *Mater. Lett.* 95: 59–62.

28 Cai, Q., Paulose, M., Grimes, C.A., and Varghese, O.K. (2005). The effect of electrolyte composition on the fabrication of self-organized titanium oxide nanotube arrays by anodic oxidation. *J. Mater. Res.* 20 (1): 230–236.

29 Regonini, D., Bowen, C.R., Jaroenworaluck, A., and Stevens, R. (2013). A review of growth mechanism, structure and crystallinity of anodized $TiO_2$ nanotubes. *Mater. Sci. Eng., R* 74 (12): 377–406.

30 Figgemeier, E., Kylberg, W., Constable, E. et al. (2007). Titanium dioxide nanoparticles prepared by laser pyrolysis: Synthesis and photocatalytic properties. *Appl. Surf. Sci.* 254 (4): 1037–1041.

31 Langhuan, H., Houjin, W., Yingliang, L. et al. (2010). $TiO_2$/carbon nanotube composites and their synergistic effects on enhancing the photocatalysis efficiency. *Prog. Chem.* 22 (05): 86.

32 Jitianu, A., Cacciaguerra, T., Benoit, R. et al. (2004). Synthesis and characterization of carbon nanotubes–$TiO_2$ nanocomposites. *Carbon* 42 (5-6): 1147–1151.

33 Wang, W., Serp, P., Kalck, P., and Faria, J.L. (2005). Visible light photodegradation of phenol on MWNT-$TiO_2$ composite catalysts prepared by a modified sol–gel method. *J. Mol. Catal. A: Chem.* 235 (1–2): 194–199.

**34** An, G., Ma, W., Sun, Z. et al. (2007). Preparation of titania/carbon nanotube composites using supercritical ethanol and their photocatalytic activity for phenol degradation under visible light irradiation. *Carbon* 45 (9): 1795–1801.

**35** Wang, S., Ji, L., Wu, B. et al. (2008). Influence of surface treatment on preparing nanosized $TiO_2$ supported on carbon nanotubes. *Appl. Surf. Sci.* 255 (5): 3263–3266.

**36** Li, Z., Gao, B., Chen, G.Z. et al. (2011). Carbon nanotube/titanium dioxide ($CNT/TiO_2$) core–shell nanocomposites with tailored shell thickness, CNT content and photocatalytic/photoelectrocatalytic properties. *Appl. Catal., B* 110: 50–57.

**37** Koli, V.B., Dhodamani, A.G., Delekar, S.D., and Pawar, S.H. (2017). In situ sol–gel synthesis of anatase $TiO_2$-MWCNTs nanocomposites and their photocatalytic applications. *J. Photochem. Photobiol., A* 333: 40–48.

**38** Zhu, K., Luo, Y., Zhao, F. et al. (2018). Free-standing, binder-free titania/super-aligned carbon nanotube anodes for flexible and fast-charging Li-ion batteries. *ACS Sustainable Chem. Eng.* 6 (3): 3426–3433.

**39** Uysal, B.Ö., Arier, Ü.Ö.A., and Pekcan, Ö. (2020). Tailoring the electrical and optical properties of carbon nanotube reinforced transparent $TIO_2$ composites by varying nanotube concentrations. *Surf. Rev. Lett.* 27 (02): 1950103.

**40** Yan, W.W., Yuan, Y.F., Xiang, J.Y. et al. (2019). Construction of triple-layered sandwich nanotubes of carbon@ mesoporous $TiO_2$ nanocrystalline@carbon as high-performance anode materials for lithium-ion batteries. *Electrochim. Acta* 312: 119–127.

**41** Orlanducci, S., Sessa, V., Terranova, M.L. et al. (2006). Nanocrystalline $TiO_2$ on single walled carbon nanotube arrays: Towards the assembly of organized $C/TiO_2$ nanosystems. *Carbon* 44 (13): 2839–2843.

**42** Duong, T.T., Kim, D.J., Kim, C.S., and Yoon, S.G. (2011). Ultraviolet response and photoelectrochemical properties of a rutile and anatase mixture grown onto single-wall carbon nanotubes at a low temperature using nano-cluster deposition. *J. Mater. Chem.* 21 (41): 16473–16479.

**43** Ma, L., Chen, A., Zhang, Z. et al. (2013). A new fabrication method of uniformly distributed $TiO_2$/CNTs composite film by in-situ chemical vapor deposition. *Mater. Lett.* 96: 203–205.

**44** Di, J., Yong, Z., Yao, Z. et al. (2013). Robust and aligned carbon nanotube/titania core/shell films for flexible TCO-free photoelectrodes. *Small* 9 (1): 148–155.

**45** Huang, S.H., Liao, S.Y., Wang, C.C. et al. (2016). Direct formation of anatase $TiO_2$ nanoparticles on carbon nanotubes by atomic layer deposition and their photocatalytic properties. *Nanotechnology* 27 (40): 405702.

**46** Liao, S.Y., Yang, Y.C., Huang, S.H., and Gan, J.Y. (2017). Synthesis of Pt@ $TiO_2$@ CNTs hierarchical structure catalyst by atomic layer deposition and their photocatalytic and photoelectrochemical activity. *Nanomaterials* 7 (5): 97.

**47** Hsieh, C.T., Chen, Y.C., Chen, Y.F. et al. (2014). Microwave synthesis of titania-coated carbon nanotube composites for electrochemical capacitors. *J. Power Sources* 269: 526–533.

# 4

# Computational Approaches to the Study of Oxide Nanomaterials and Nanoporous Oxides

*Ettore Fois and Gloria Tabacchi*

Insubria University, Department of Science and High Technology, Via Valleggio 11, Como 22100, Italy

## 4.1 Introduction

The exceptional progress of high performance computing (HPC) in the last few years, combined with the development of techniques and algorithms in theoretical physics/chemistry, has encouraged the application of "in silico" methodologies to a wide variety of systems of great relevance for the science and technology of oxide-based (nano) materials, as documented by a number of comprehensive reviews [1, 2] and specialized books [3]. Besides being a powerful characterization tool, especially in combination with experiments, modeling approaches are particularly useful in the study of events occurring at surfaces and interfaces of oxide materials at conditions different from the ambient ones – e.g. at very high temperature or high pressure – or, more in general, in all those situations where atomistic resolution data, evidences of new complex phenomena, or mechanistic understanding are very difficult to obtain experimentally [4].

Nowadays, a vast body of literature is available on the application of computational modeling to the study of an impressive variety of physicochemical phenomena relevant for a plethora of practical technological applications encompassing, for example, sustainable energy technologies, electro-optical devices, catalysis, magnetism, and sensing [5, 6]. A number of excellent studies and various reviews devoted to the application of computational methodologies to the study of metal oxides have been published in the last two decades, which have greatly contributed to increase our microscopic-level knowledge of the inner working mechanisms of functional oxide nanomaterials. Due to the impossibility to present a complete account of the innumerable applications of modeling to nanostructured oxide systems, in this contribution, we will limit our focus to selected case studies taken from theoretical investigations from our laboratory. Importantly, these studies were mostly performed in collaboration with experimental research groups that occupy a leading position in the synthesis/characterization of nanostructured oxide materials and their technological applications. Our aim is to show how theoretical modeling may be used not only to complement experimental data but

*Tailored Functional Oxide Nanomaterials: From Design to Multi-Purpose Applications*, First Edition.
Edited by Chiara Maccato and Davide Barreca.
© 2022 WILEY-VCH GmbH. Published 2022 by WILEY-VCH GmbH.

also to provide a thorough atomistic-level understanding able to inspire the design and guide the fabrication of new nanostructured oxide materials. Moreover, the presented case studies, although limited to a few selected research areas, highlight open issues and challenges that are widespread in the general field of simulations of nanostructured oxide materials.

The present chapter is organized as follows. After a short presentation of the adopted theoretical approaches (Section 4.2), the selected case studies will be discussed in Sections 4.3 and 4.4, which constitute the main part of this work. These two sections are devoted to the use of theoretical methods to gather microscopic information, respectively, on the behavior of molecular species at the surfaces of oxide nanomaterials and within the cavities of porous oxide systems.

Section 4.3 will present computational modeling examples of a selection of relevant phenomena occurring when molecular species come in contact and interact with oxide materials surfaces. Among the immense variety of interface processes involving oxide nanosystems [7], we have chosen to focus our attention to the chemoresistive gas sensing properties of a magnetic oxide, $Mn_3O_4$, in view of the key importance of the rapid detection of chemical warfare agents (CWA) for global safety applications [8]. As a matter of fact, this manganese oxide, especially when decorated by gold nanoparticles, has been recently shown to be endowed with exceptional sensitivity in the detection of a mustard gas simulant [9]. From the theoretical point of view, this example is important because it showcases the challenges of building a simulation model for such a complex chemical system – an oxide nanomaterial–metal nanoparticle composite – and studying its interactions with the gas simulant in order to identify the molecular origin of the observed sensing behavior.

The second part of Section 4.3 will be devoted to chemical vapor deposition (CVD), a widespread bottom-up fabrication approach for oxide nanomaterials adopted in advanced systems and devices [10, 11]. In spite of the enormous successes of CVD in the synthesis of functional oxide nanosystems, the atomistic details of the involved chemical processes, starting from the fragmentation of the precursor molecules, are still far from being clarified [12]. The examples herein reported have been selected to describe the usefulness of computational modeling in deepening our atomistic-level knowledge of the mechanistic aspects of some important steps in CVD processes [12, 13] (i.e. adsorption and activation of the precursor complex, early stages of the reactivity at the substrate surface, and physisorption/bonding to the surface), which require an accurate treatment of both high-temperature effects and bond breaking/forming events.

Section 4.4 of this chapter will illustrate few selected case studies involving the theoretical modeling of an important category of porous oxide systems – zeolites. Zeolites are crystalline porous oxides mainly constituted by aluminum, silicon, and oxygen [14]. Due to their outstanding thermal stability and size–shape selectivity, zeolites have been exploited since 1960s in industrial catalysis, adsorption, and ionic exchange and are ubiquitous in detergents and desiccants [15]. More recently, they have been adopted as host matrices for nanosystems [16] or key components for energy harvesting devices [17]. Zeolites' *primary building units* are corner-sharing tetrahedra TO$_4$, where T denotes an atom – usually Si or Al – bonded to four oxygen atoms [18]. The *zeolitic frameworks* exhibit arrays of channels and cages of nanometer diameters, which can contain *extraframework* species – molecules, cations, nanoclusters, or supramolecular structures – of suitable size and shape. A wide variety of cages and channels, with different geometries and sizes, is featured among zeolite framework types. Whereas pore opening diameters normally vary between 0.3 and 0.8 nm, the inner cavities have larger diameters, ranging from 0.5 to 1.3 nm [14]. Zeolitic channels can be organized into one-dimensional (1D), two-dimensional (2D), and three-dimensional (3D) arrays. In 1D arrays, the channels do not intersect. On the other hand, 2D and 3D channel arrays present intersections between channels. From the chemical viewpoint, standard zeolites are aluminosilicates: each Al provides one negative charge to the framework, which is counterbalanced by extraframework cations coordinated by water molecules. Their interactions with framework oxygen atoms create a hydrogen-bonded network whose characteristics are dictated by the geometry of the zeolite cavity, the chemical composition of the framework, and the type and concentration of guest species. Conversely, all-silica zeolites are constituted by Si tetrahedra, have neutral frameworks, and are hydrophobic.

In particular, the computational studies described in Section 4.4 will involve both standard aluminosilicate zeolites and all-silica zeolites and will deal with (i) the structural arrangement of water molecules in zeolite channels, a general issue in zeolite science; (ii) the creation of new zeolite-based composites via high-pressure intrusion of guest molecules, an innovative approach for the fabrication of organized host–guest systems; and (iii) molecular description and electronic properties of dye–zeolite-based composites, organic–inorganic hybrid materials already adopted in electro-optical devices.

Finally, Section 4.5 will be devoted to the outlook and conclusions of this chapter, presenting some possible routes for future developments from the viewpoint of the authors.

## 4.2 Overview of Theoretical Approaches

Computational approaches used in modeling of oxide (nano) systems may be divided in three broad families, which are herein listed in increasing order of chemical accuracy: (i) geometrical approaches; (ii) classical force-field techniques, basically, Monte Carlo (MC) and molecular dynamics (MD) approaches; and (iii) quantum mechanical methodologies (a.k.a. "ab initio" or "first principles"). As all the examples discussed in this chapter required the explicit treatment of the electronic structure, this paragraph will be only limited to a short illustration of selected and widely adopted quantum mechanical approaches for the study of oxide nanomaterials based on density functional theory (DFT) [19].

The description of the electronic structure may be performed either via quantum chemistry methods – i.e. by solving the Schrödinger equation for the electronic wavefunction of the system – or via DFT approaches, where the energy as an unique functional of the electronic density. Since the exact form of the density functional is not known, approximated expressions have to be used [19]. A computationally convenient class of functionals for large-scale simulation of oxide nanomaterials include dependency from the gradient of the electronic density and are known as generalized gradient approximations (GGAs) [19]. Hybrid functionals, which combine DFT with Hartree–Fock exchange [20], have higher accuracy but are computationally more expensive than GGA approximations. Additionally, dispersion effects are also relevant, especially when the target system involves nonpolar species (e.g. hydrocarbons in zeolites), and their description may be introduced through the use of suitable dispersion corrections [20].

In particular, the computational studies described in this chapter have been mostly performed with the broadly used Perdew–Burke–Ernzerhof (PBE) approximation for DFT [21], complemented by empirical dispersion corrections [22]. Such a combination of methods (also indicated as "PBE-D2") is widespread in solid-state modeling because it provides a good balance between computational costs and accuracy in the theoretical prediction of structural properties (lattice constants, bond distances, and bond angles), as documented by benchmark studies [23].

DFT methodologies can describe the breaking and forming of chemical bonds but have a high computational overhead for large-sized simulation systems. However, crystalline materials are usually modeled by periodic-DFT, where a simulation cell – which could be e.g. the crystallographic unit cell of the target oxide system – is periodically repeated in the three dimensions.

In the cases where both a quantum mechanical approach and a description of thermal effects via the atomic motion are needed, a convenient strategy is the first-principles molecular dynamics (FPMD) method, originally formulated by Car and Parrinello [24, 25]. FPMD enables to simulate the dynamics of the target system with first-principles accuracy, because the forces on the atomic nuclei are based on a quantum mechanical description of the electrons, usually with DFT [25]. FPMD simulations, due to their significant computational load, are limited to trajectories of the order of tens of picoseconds, which are insufficient to model activated events like chemical reactions. These processes are indeed rare events at the FPMD timescale. To model reactive systems, statistical sampling techniques, e.g. the blue-moon ensemble [26], are usually adopted in combination with FPMD. These approaches, which require the definition of a suitable reaction coordinate, are able to provide quantitative information on the energy barriers of the target reaction as well as on the structure of the activated complex and represent an unvaluable tool in the study of the chemical processes involving oxide nanomaterials.

## 4.3 Molecular Behavior at Nanomaterials Surfaces

### 4.3.1 Molecular Interactions on Manganese Oxide Nanomaterials

Manganese oxide nanomaterials are appealing candidates for a broad variety of applications. The case study we consider here is $Mn_3O_4$, an environmentally friendly and low-cost multifunctional material [27]. A recent collaborative investigation [9] demonstrated that when $Mn_3O_4$ is decorated with gold nanoparticles,

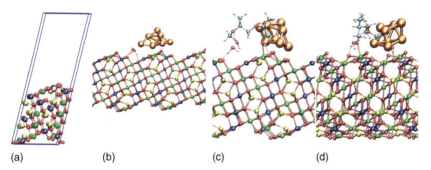

**Figure 4.1** (a) DFT-optimized slab model of the bare $Mn_3O_4$ (211) surface. The blue lines represent the (periodically repeated) simulation cell adopted in the calculations. (b) DFT-optimized Au–$Mn_3O_4$ model (orange: Au, white: H). Dashed red lines represent hydrogen bonds. (c–d) DFT-optimized DPGME + Au–$Mn_3O_4$ model in different orientations. Atom color codes: red, oxygen; yellow, tetrahedral $Mn^{2+}$; blue (up-spin) and green (down-spin), octahedral $Mn^{3+}$; orange, gold; cyan, carbon; and white, hydrogen. Source: Reproduced with permission from Bigiani et al. [9].

it acquires outstanding properties in gas sensing, especially in the detection of di(propyleneglycol) monomethyl ether (DPGME), a simulant of the vesicant nitrogen mustard gas. Specifically, the target nanomaterials, fabricated by CVD of $Mn_3O_4$ on $Al_2O_3$ followed by gold radio frequency sputtering, exhibited an exceptional sensitivity in the DPGME detection compared to the sensing of acetone and ethanol, accompanied by an excellent capacity in the selective molecular recognition of the target analyte against other warfare gas simulants [9]. In this context, a fundamental question to address, in order to extend the scope of perspective applications, was to understand the microscopic-level origin of such impressive sensing performances.

To provide molecular-level insight into the interactions between the Au–$Mn_3O_4$ nanosystems and DPGME, an extensive computational modeling effort was carried out, which allowed to identify, being responsible for the sensing mechanism, a dual-site molecular recognition based on the binding of the mustard gas simulant to both $Mn_3O_4$ and the noble metal nanoparticles [9].

The first step was the building and the geometry optimization of a slab model for the $Mn_3O_4$ (211) surfaces, which, according to X-ray diffraction data, were preferentially exposed by the fabricated manganese oxide nanomaterials. In particular, the (211) $Mn_3O_4$ surface (Figure 4.1a) was modeled by cutting a periodic slab from a $Mn_3O_4$ crystal. Due to the magnetic nature of $Mn_3O_4$, a spin-polarized DFT approach was adopted, using a Hubbard model for the Mn atoms [28].

After optimization of the bare oxide slab, a model of a gold nanoparticle was positioned on the manganese oxide surface, and a second geometry optimization was performed. Importantly, a strong interaction between gold and $Mn_3O_4$ resulted from the calculations; in particular, the binding between the noble metal cluster and the oxide surface was demonstrated by three Au–O short distances (2.10, 2.13, and 2.24 Å), which highlighted at atomistic level the presence of a close Au/$Mn_3O_4$ contact (Figure 4.1b).

The subsequent phase of the work consisted in the modeling of the adsorption of the target analyte DPGME on the Au-decorated manganese oxide slab. The interaction of the optimized Au–Mn$_3$O$_4$ (211) slab with the mustard gas simulant was modeled by randomly placing a DPGME molecule on the optimized structure of the slab and performing a geometry optimization of this new model. In order to identify distinctive features of the outstanding sensitivity and selectivity of Au–Mn$_3$O$_4$ toward DPGME from the adsorption geometry, the same procedure was then repeated with a molecule of ethanol – which, according to experimental data, exhibited a response significantly lower than DPGME.

The optimized structure of the Au–Mn$_3$O$_4$ (211) + DPGME system evidenced that both Mn$_3$O$_4$ and Au were directly interacting with the mustard gas simulant (see Figure 4.1c,d). More specifically, the DPGME hydroxyl oxygen was located in the immediate proximity (2.18 Å) of a Mn$^{3+}$ atom of the oxide surface, thus completing its octahedral coordination environment. In addition, an ethereal oxygen of DPGME was positioned close (2.35 Å) to a Au atom. Finally, the target analyte was strongly hydrogen bonded with its hydroxyl proton to a Mn$_3$O$_4$ oxygen atom (1.68 Å) and with an ethereal oxygen to a surface water molecule (1.86 Å).

By considering the Au–Mn$_3$O$_4$ + ethanol optimized model, the comparison with DPGME indicated that ethanol was steadily bound to Mn, but not to Au, its smallest separation from the noble metal cluster being 2.96 Å. This result suggested that, in line with the experimental evidences, the interaction of gold with ethanol was weaker than the gold–DPGME interaction. In other words, DPGME can establish a dual-site contact at the Au–Mn$_3$O$_4$ interface, while ethanol is not endowed with this capability. Therefore, modeling results allowed to ascribe the remarkably higher response and selectivity of DPGME with respect to ethanol to the strong dual-site interaction of DPGME with both manganese oxide and gold.

The molecular-level interaction of DPGME with Au–Mn$_3$O$_4$, occurring at the oxide–nanoparticle interface via dual anchoring to Au nanoparticle and Mn$_3$O$_4$ surface, has important practical consequences, as it enables the selective detection of DPGME mustard gas simulant among other possible interferents with outstanding responses and ultralow detection limits. As a general consideration, this example showcases the key role of theoretical modeling in understanding the fundamental working mechanisms of functional metal oxide–metal nanoparticle composites for sensing applications.

### 4.3.2 Insight on Molecule-to-Material Conversion in Chemical Vapor Deposition

Theoretical modeling is a key tool to increase our knowledge on the mechanisms underlying oxide nanomaterial fabrication methods like CVD, which, starting from molecular sources (*precursors*), allows to obtain functional nanostructured oxides exploited in advanced systems and devices [10, 29] (see Figure 4.2). Since the nature of the precursor molecule deeply affects the properties of CVD-synthesized oxide materials, theoretical studies have also aided the design of novel precursors endowed by high volatility, vaporization without undesired side reactions, and clean surface

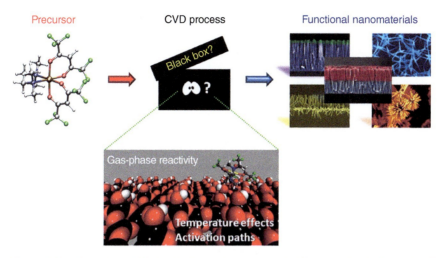

**Figure 4.2** (Top) Scheme illustrating the molecule-to-material conversion in thermal CVD. (Bottom) A pictorial representation of relevant open questions at the microscopic level, involving both gas-phase reactivity and processes at the molecule–material interface. Source: Reproduced with permission from Tabacchi et al. [12].

decomposition process [30]. It should be however noted that even the experimental conditions of the CVD process strongly influence the composition, morphology, and properties of the obtained nanomaterials [11]. In general, the molecule-to-material conversion in CVD is based on complex processes, which, from the microscopic viewpoint, may still be considered as a "black box" (Figure 4.2) that might be opened with the aid of theoretical modeling approaches [12].

In thermal-CVD, the molecular precursor – which is often a transition metal complex (see Figure 4.2) – is decomposed on a growth surface kept at a temperature considerably higher with respect to the transport gas (the operating temperatures needed for the deposition of metal oxides typically range from 500 to 750 K [10, 11, 29]). Hence, in these processes, thermal energy may be comparable with physisorption energies or vibrational stretching energies of the metal–ligand coordination bonds in the precursor molecule (at $T = 750$ K, $kT \sim 1.5$ kcal mol$^{-1}$, $kT/hc = 500$ cm$^{-1}$). These temperature effects may strongly influence the chemical and the dynamical behavior of the molecular precursor complex, initiating the fragmentation of the precursor [12]. As a consequence, activation pathways different from standard solution or gas phase chemistry may take place, leading to new nanostructured materials that can be obtained only at these harsh conditions (Figure 4.2).

The modeling of CVD processes is particularly challenging. First of all, reliable models of the growth surface need to reproduce the structure and chemical composition of the surface, also considering possible effects of the reaction atmosphere. Finite cluster models or periodically repeated slab models could both be adopted fruitfully. On one hand, cluster models of limited size may allow for the use of more accurate levels of theory, but they should be used only when the target processes take place in a spatial region of small size on insulating surfaces. Conversely, periodic slab models can provide a more realistic picture of the growth surface, which may also be of metallic or semiconducting nature.

Another important issue to consider in the theoretical modeling of CVD processes is the effect of temperature. Hence, approaches that could reproduce regions of the potential energy surface far from the (0 K) minimum energy structures obtained from geometry optimization should be adopted. In this context, all the case studies here considered employed FPMD, which explicitly takes into account both electronic structure and finite temperature effects.

In the first case study, the early stages of the behavior of Cu(hfa)$_2$TMEDA (hfa = 1,1,1,5,5,5-hexafluoro-2-4-pentanedionate, TMEDA = $N,N,N',N'$-tetramethylethylenediamine), a copper(II) diketonate–diamine precursor for the deposition of copper oxides (Figure 4.2, top left), on the CVD growth surface heated at 750 K were investigated by means of DFT structural optimization and FPMD simulations [31]. The surface reactivity of this precursor was particularly difficult to predict, because the Cu coordination sphere was completely saturated by the two diketonate and the diamine ligands.

The first step of the computational study was to set up a periodic slab model of the growth surface – which, according to experimental evidences, consisted of hydroxylated silica. Then, a Cu(hfa)$_2$TMEDA molecule was positioned on the top of the hydroxylated slab, and the model was subjected to geometry optimization. The 0 K minimum energy structure revealed that the copper precursor was physisorbed on the growth surface, with a binding energy of 5.0 kcal mol$^{-1}$ (Figure 4.2, bottom). Starting from this minimum structure, FPMD simulations were performed, and, after equilibration at the target temperature of 750 K, a production trajectory of 30 ps was collected for data analysis. Intriguingly, FPMD simulations indicated that high-temperature activation at the substrate surface could be a possible mechanism to trigger precursor decomposition. In particular, a careful analysis of the trajectory disclosed a novel phenomenon, the fast diffusional rolling motion of the Cu(hfa)$_2$TMEDA molecular complex on the model growth surface. Such rolling motion (Figure 4.3), which involved also significant oscillations of the metal—ligand bond lengths, could stimulate the liberation of the metal center by ligand detachment via highly energetic collisions between vibrationally activated molecules on the heated substrate [31].

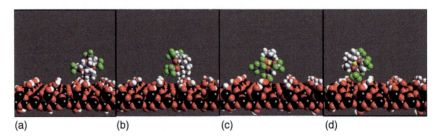

**Figure 4.3** (a–d) FPMD trajectory snapshots illustrating the rolling diffusion of Cu(hfa)$_2$TMEDA on hydroxylated silica at 750 K. All atoms in van der Waals representations, color codes as in Figure 4.2. Source: Reproduced with permission from Fois et al. [31].

The previously described rolling diffusion mechanism appeared to be general to the M(hfa)$_2$TMEDA family of precursors for the vapor deposition of metal oxides, where M is a metal of the first transition series. As a matter of fact, similar theoretical investigations were performed on the homologous Zn(hfa)$_2$TMEDA [32] and Fe(hfa)$_2$TMEDA compounds [33], highlighting the same rolling motion of the complex when in contact with a slab model of the growth surface at high-temperature conditions. However, the vibrationally excited rolling motion led to different behaviors for the three complexes, suggesting the possibility of different fragmentation mechanisms on the growth surface. Such results demonstrated the key role of the nature of the metal center in determining the behavior of the precursor on the growth substrate.

Specifically, modeling of Fe(hfa)$_2$TMEDA with the same FPMD approach at 750 K evidenced the detachment of the diamine ligand from Fe as a neutral species, generating two independently physisorbed fragments, namely, Fe(hfa)$_2$ and TMEDA [33]. In a different way, the 750 K simulation of Zn(hfa)$_2$TMEDA complex on the hydroxylated silica slab highlighted a preferential dissociation of one Zn—O bond [32], which suggested that a decrease in the metal coordination number from 6 to 5 could facilitate the precursor fragmentation.

Following this idea, in a subsequent study [34], a combined statistical sampling-FPMD approach for the simulation of reactive events was adopted to elucidate the detailed mechanism of the metal—ligand bond dissociation stage in the Zn(hfa)$_2$TMEDA complex. This theoretical approach enabled to gather microscopic information on the early precursor fragmentation steps on the CVD substrate and to quantify the energy barrier of the process.

A preliminary FPMD run of the Zn(hfa)$_2$TMEDA + hydroxylated silica model at 500 K showed that the Zn precursor, initially physisorbed and characterized by an octahedral structure, fastly moved on the model surface by a rolling diffusion stimulated by thermal exchange with the surface, in line with previous results on this family of complexes. In this motion, the Zn—O and Zn—N bonds showed significant fluctuations, leading in a few cases to the transient dissociation of the Zn—ligand bonds. Interestingly, toward the end of the simulation, a transient Zn—O bond dissociation event led to a partial detachment of one diketonate moiety, which produced a Zn(hfa)$_2$TMEDA physisorbed molecule exhibiting a five-coordinated Zn structure with square-pyramidal arrangement of the ligands. Hence, the results indicated that such unusual square-pyramidal Zn(hfa)$_2$TMEDA geometry, never found at room conditions, might however be present at high temperature on the CVD growth surface.

To identify a possible mechanism of the conversion of the octahedral Zn complex to the square pyramidal geometry, the statistical sampling simulations were started from octahedral Zn(hfa)$_2$TMEDA physisorbed on the slab. As a reaction coordinate, a Zn–O distance was selected, which was progressively increased in a series of subsequent steps until the pyramidal geometry was reached. This procedure demonstrated that the octahedral-to-pyramidal isomerization of the complex is slightly endoergic, the free energy difference between the final and the initial state being +3.6 kcal mol$^{-1}$. Remarkably, the computed energy barrier for the

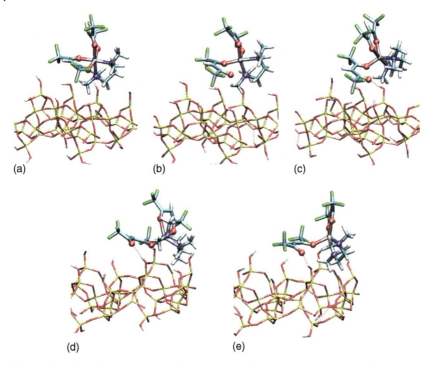

**Figure 4.4** (a–e) Five snapshots taken from the simulations corresponding to increasing values of the reaction coordinate (Zn–O distance): (a–b) octahedral Zn(hfa)$_2$TMEDA; (c) trigonal bipyramidal Zn(hfa)$_2$TMEDA (transition state); (d–e) pyramidal-Zn(hfa)$_2$TMEDA. Color codes: Zn, gray; F, green; O, red; N, blue; C, cyan; H, white; and Si: yellow. The hydrogen bond of the detached hfa oxygen O with the surface hydroxyl proton is indicated as a dotted line. Source: Barreca et al. [34]. Licensed under CC BY-4.0.

octahedral-to-pyramidal conversion resulted +5.8 kcal mol$^{-1}$, a value of the order of the thermal energy corresponding to 500 K (about 1 kcal mol$^{-1}$). Such a low activation barrier indicated that the pyramidal structure of the Zn complex can become energetically accessible when the complex is in contact with the heated growth surface.

More detailed information concerning the octahedral-to-pyramidal isomerization mechanism of the target precursor was offered by the geometry distortions showed by the complex along the reaction coordinate (i.e. the distance between Zn and one of the hfa oxygens). Figure 4.4 shows a few significant configurations taken from the simulations for increasing Zn–O distances. In particular, in the early stages of the reaction profile (Figure 4.4a,b), Zn(hfa)$_2$TMEDA maintains its standard octahedral geometry, in spite of the appreciable Zn—O bond weakening, and does not form hydrogen bonds with the surface hydroxyl groups.

Importantly, the isomerization of the precursor to the pyramidal structure was initiated by the formation of a strong hydrogen bond between the detached hfa oxygen and a surface hydroxyl group. Figure 4.4c shows the transition state, characterized by a trigonal bipyramidal structure stabilized by the aforementioned hydrogen bond.

The transition state, also exhibiting a penta-coordinated geometry, then evolved to the final state, with a square-pyramidal geometry (Figure 4.4d,e). Hence, the theoretical analysis revealed the detailed geometric changes of the Zn(hfa)$_2$TMEDA along the reaction path and highlighted a key role of hydrogen bonding with surface hydroxyl groups in the precursor isomerization process. Another important aspect disclosed by the computational analyses is that the octahedral-to-pyramidal conversion may enable the metal center to directly interact with other species on the growth surface, thus possibly triggering further reactive processes.

On the whole, these theoretical investigations devoted to the modeling of the thermal behavior of M(hfa)$_2$TMEDA complexes (M = Cu, Fe, Zn) at the CVD growth surface pinpointed the relevance of high-temperature-induced rolling diffusional motion of molecular precursors in vapor deposition fabrication of oxide nanosystems and allowed to predict the behavior of the precursor during the early stages of the CVD process.

## 4.4 Oxide Porous Materials

Regular architectures of pores and channels are a distinctive feature of several oxide materials, such as zeolites or crystalline mesoporous silicas. Here, we will focus on zeolites because these porous materials have been the prototype for the realization of a wide variety of host–guest materials holding concrete promises for applications in materials science. Numerous composites for electronics, photonics, or sensing can be obtained from the inclusion of suitable guest species inside zeolites.

Due to the presence of arrays of nanosized empty space, such frameworks may be regarded as the negative image of complex nanostructures, constituted by arrays of molecules characterized by intriguing electro-optical functionalities. Indeed, when dyes or other optically active species are confined in zeolite pores, not only they are spatially constrained by the geometry of the cavities, but they can also be remotely and individually controlled by means of an external optical input and exploited in devices, as documented by a vast body of literature in the last two decades. More recently, new methods to fabricate zeolite-based composites endowed by unique organization properties of the guest molecules have been proposed, such as high-pressure intrusion. In spite of these advances, a microscopic-level knowledge of the complex interactions between these versatile porous oxides and guest species, including water molecules, is still far from being fully established.

### 4.4.1 Structural Properties

Several powerful characterization techniques are adopted experimentally to study zeolite-based materials in order to capture microscopic structural details of the organization of guest species in zeolite channels, as well as structure–functionality relationships.

Structural analysis is commonly based on Bragg diffraction, which exploits the long-range order properties typical of periodic solids. However, guests molecules present in zeolite pores lower the crystallographic symmetry of the bare zeolite matrix, hindering structural determination with standard diffraction methods [16]. In this scenario, the use of computational approaches such as quantum chemical and MD calculations has already captured structural features otherwise difficult to gather from experiments.

In the wide body of literature concerning computational investigations of structural properties of zeolites, the study of the structural arrangement of water molecules is of general relevance, because water is normally present in the cages and cavity of aluminosilicate zeolites. The details gathered from calculations have greatly contributed to increase the microscopic-level knowledge of host–guest interactions under space confinement.

A recent interesting example of computational-guided structural resolution of an hydrated zeolite material is the elucidation, by means of dispersion-corrected DFT, of the organization of water molecules in zeolite L leading to an atomistic-level understanding of the zeolite–water and water–water interactions in the channels of this zeolite [35].

Zeolite L (Figure 4.5a) is a porous oxide material of high technological relevance since the size and shape of its 1D channels, running along the crystallographic $c$ axis, enable the incorporation of a wide variety of photoactive species, either neutral or cationic. This feature has allowed the realization of a number of host–guest composites characterized by a high degree of structural organization – such as, for example, sequenced arrays of dyes – employed in electrooptical applications (see also Section 4.4.3 and, for a recent review, Ref. [36]). Unfortunately, water is able to displace various neutral organic dyes from zeolite L channels, thus damaging the host–guest composites and hence the performances of zeolite-L-based devices [37, 38].

In this context, the theoretical study of the structural properties of hydrated zeolite L is particularly crucial for improving the stability of zeolite-L-based composites and extending their applications. As previously mentioned, structural analysis of water molecules incorporated in zeolite channels is very difficult with standard crystallographic methods because the refined structure should be interpreted as an average over many possible configurations of lower symmetry, characterized by different water arrangements [16]. As a consequence, the diffraction data provide multiple possible positions for the water oxygens, characterized by fractional occupancy. To model this system computationally, it was necessary to consider a number of water molecules corresponding to those experimentally determined by X-ray powder diffraction at standard conditions (18 water molecules per unit cell) and to select, among the water crystallographic positions with fractional occupancy, those leading to the experimental stoichiometry of the system with proper interatomic distances.

Geometry optimization of this starting model system revealed that the 1D channels of zeolite L contain a continuous supramolecular structure stabilized by the interactions of zeolitic water with potassium cations and by water–water hydrogen bonds. Evaluation of the energetics evidenced that water provided a significant energetic stabilization to the framework of zeolite L. More specifically, the minimum energy structure computed for hydrated zeolite L (Figure 4.5a) was stabilized by 326.74 kcal mol$^{-1}$ compared with the constituents (i.e. the zeolite L framework and

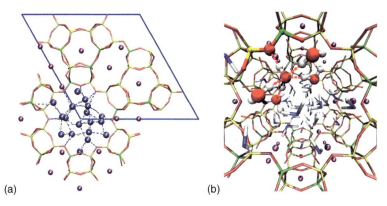

**Figure 4.5** (a) Minimum energy structure of hydrated zeolite L represented in the crystallographic *ab* plane. The blue solid lines indicate the unit cell. (b) Representation of zeolite L minimum energy structure along the channel axis showing the peculiar five-hydrogen-bonding arrangement (in ball-and-stick and dashed blue lines). Atom colors: yellow, Si; green, Al; red sticks, framework O; purple, K; blue, water O; white, H; and red spheres, oxygens participating in the five-hydrogen-bonding network. Blue dashed lines, five-hydrogen-bonding network; red dashed lines, the full hydrogen-bonding network.

18 isolated water molecules); therefore, the stabilization energy for one single water molecule was 18.15 kcal mol$^{-1}$, well in line with the values computed for water in hydrophilic zeolite frameworks. Interestingly, the stabilization energy of one water molecule, calculated for a model system containing a single water molecule per zeolite L unit cell, coordinated to a potassium cation, was found to be slightly higher, i.e. 20.42 kcal mol$^{-1}$. Therefore, the energetic analysis indicated that, by increasing the water content in zeolite L, the stabilization of a single water molecule decreases. A careful study of the geometry of the optimized structure revealed that this stabilization decrease is due to the increased competition of water molecules to bind with the potassium cations in the fully hydrated zeolite L (with 18 H$_2$O per unit cell). As a matter of fact, such a system is characterized by a larger network of hydrogen bonds, due to the higher number of water molecules per unit cell. It is important to underline that the hydrogen-bonded network, besides water and framework oxygens, involves the extraframework cations as well, which play a key role in determining the structural arrangement of water molecules in zeolites.

Intriguingly, a thorough analysis of the arrangement of the extraframework species and of the hydrogen bond connectivity indicated that the water network in zeolite L also includes a hypercoordinated structure, where a water molecule participates to five strong hydrogen bonds (Figure 4.5b). As proton transfers are generally favored in strong hydrogen-bonded networks, this five-hydrogen-bonded structure can be regarded as a water pre-dissociation complex, which could be responsible of the experimentally detected high proton activity in zeolite L nanochannels. Hence, computational modeling allowed to identify a unique local arrangement of hydrogen bonds in zeolite L nanochannels, which appear to be the origin of the propensity of water to ionization under confinement.

## 4.4.2 Behavior Under High-Pressure Conditions

Zeolites are porous oxides characterized by an impressive capability to withstand high-pressure conditions – typically, Earth scientist use gigapascal (GPa) pressures to study phase transitions in zeolite minerals, which are important constituents of the Earth crust. By compressing zeolite frameworks with molecules capable to enter the pores, new host–guest materials can be fabricated: for example, hydrocarbons can enter inside hydrophilic frameworks and water can be forced inside hydrophobic zeolites. Hence, the effect of a high applied pressure, combined with the geometric confinement of the zeolite framework, can accomplish the realization of otherwise unfeasible compounds.

Experimentally, the inclusion of guest species in a zeolite framework is enforced with a diamond anvil cell (DAC) [39, 40]. In this instrumentation, the zeolite is compressed hydrostatically using a pressure-transmitting fluid (usually a gas or a liquid) previously loaded into the DAC [41]. The pressure-transmitting fluid plays a crucial part in the experiments, because it is introduced into the zeolite pores via the application of GPa pressures, and then it is "shaped" by the zeolite framework structure. Hence, the composition of the pressure medium is very important because the penetration process depends on many factors such as e.g. the relative size of the zeolite pores and the guests, the maximum applied pressure, the rate of pressure increase, and the temperature conditions [4]. In general, empty frameworks such as hydrophobic (all-silica) zeolites are preferable to already filled ones for fabricating new host–guest compounds by high-pressure intrusion and may be considered as "molds" for pressure-assisted fabrication of new supramolecular structures.

Water was the first species to be introduced through high pressure into zeolites, via the so-called pressure-induced hydration [42]. As a matter of fact, when confined in zeolite frameworks, water molecules may form intriguing supramolecular aggregates stabilized by hydrogen bonding, like 1D chains [43, 44] or triple helices [45]. Over the years, computational studies have contributed to shed more light on the compression behavior of hydrated zeolites, suggesting that, under pressures in the GPa range, water molecules tend to enhance their hydrogen-bonding network in the zeolite pores, forming new hydrogen bonds with the oxygen atoms of the framework [46, 47]. In spite of these progresses, many atomistic details of pressure-driven water intrusion in zeolites are still far from being completely clarified.

When the pressure-transmitting fluid is a mixture, the structural properties of the guest molecules incorporated into a zeolite framework via pressure-induced intrusion become even more difficult to elucidate at a microscopic level. Understanding this aspect is particularly important, because water–alcohol mixtures of different composition perform better than water as pressure-transmitting media and are therefore often used in intrusion experiments. For these reasons, we select as a case study to illustrate the high-pressure behavior of zeolite materials the investigation of the pressure-induced intrusion of a water–ethanol mixture in the siliceous zeolite ferrierite (Si-FER) [48].

According to both experimental and computational evidences [49, 50], Si-FER is among the frameworks exhibiting highest selectivity for the separation of water/alcohol mixtures – an essential preliminary step in bioethanol production [51]. In general, the separation of mixtures of molecules of similar size is very challenging if they form hydrogen-bonding networks, such as in water/ethanol mixtures, hence the use of highly selective porous oxides – like siliceous ferrierite – is a key requirement to accomplish the separation. Nonetheless, in spite of the industrial importance of the process, the atomistic origin of the ferrierite framework selectivity was not clarified, until a joint high-pressure X-ray diffraction–computational modeling study was able to shed light on this issue [48].

In the high-pressure experiment described in Ref. [48], siliceous ferrierite was compressed in DAC up to 1.2 GPa using an ethanol:water = 1 : 3 mixture as pressure-transmitting fluid. Experimentally, the X-ray powder diffraction data indicated penetration of the fluid already at relatively low pressures, and the refined structure at 0.84 GPa suggested that, on average, four ethanol (EtOH) and about six to eight water molecules were incorporated in the Si-FER unit cell upon compression. Unfortunately, the refinement was affected by severe instabilities – as it is typical for high-pressure experiments – and was not able to provide satisfactory values for the interatomic separation between the incorporated molecules. Therefore, the diffraction data were used as a guess for building a computational model of the Si-FER·($H_2O$,EtOH) zeolite, which was then studied with a dispersion-corrected DFT approach.

The task of computational modeling was to determine the amount of water molecules penetrated in the Si-FER framework and to elucidate the structure of the hydrogen-bonding network involving water, ethanol, and framework oxygens. The strategy to accomplish this goal was to build zeolite models with four-ethanol molecules and a number of water molecules ranging from 6 to 14 per unit cell and to find the most stable of these putative models by means of DFT-based geometry optimizations. The resulting relative stabilities clearly showed that the energetically most stable zeolite composite contained eight water molecules per unit cell.

Once the unit cell stoichiometry of the pressure-created compound – Si-FER·(8$H_2O$,4EtOH) – was established, a thorough analysis of the structural features of the guest molecules was performed on the optimized model. To better understand the arrangement of the extraframework species, it is useful to describe the pore system of the zeolite host. The framework of Si-FER is characterized by a 2D architecture of empty space, composed by arrays of eight-membered ring (8MR) channels running along the $b$ crystallographic axis (as shown in Figure 4.6a), which intersect perpendicularly to two channel systems directed along the $c$ axis [14, 52]. As displayed in Figure 4.6b, the first channel system has a larger diameter (10-membered ring [10MR]), while the second one presents a smaller diameter [six-membered ring (6MR)].

The computational results unambiguously indicated that water molecules and ethanol molecules, instead of forming mixed aggregates characterized by water—ethanol hydrogen bonds, were segregated in distinct channels, as depicted in Figure 4.6c. Remarkably, whereas water molecules were confined only in the 6MR channels, the ethanol molecules were exclusively positioned in the larger 10MR channels. More specifically, the ethanol molecules were located with their

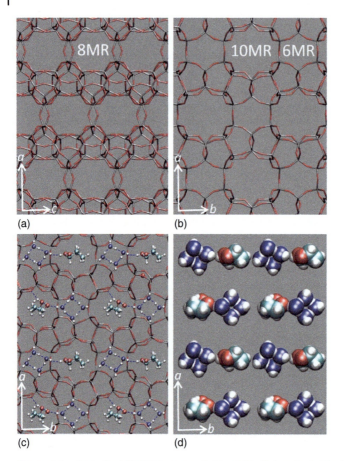

**Figure 4.6** Optimized Si-FER framework ($P = 0.84$ GPa) projected in (a) the $ac$ plane, showing the 8MR channels; (b) the $ab$ plane, showing the 10MR and 6MR channels. (c) Graphical representation of the hydrogen-bonding network (evidenced as blue dotted lines) in the minimum structure of Si-FER·4EtOH·8H$_2$O at 0.84 GPa. (d) Graphical representation of ethanol dimers and water tetramers. Source: Reproduced with permission from Arletti et al. [48].

C–C molecular axis almost perpendicular to the channel axis, giving rise to linear chains of hydrogen-bonded ethanol dimers. Although few hydrogen bonds involving ethanol and water were also present, such interactions were weaker and restricted to a limited fraction of the ethanol molecules.

Hence, the DFT results proved that compression in the GPa range induced the ethanol molecules to form a linear aggregate, which occupies only the larger channels of Si-FER and is characterized by very weak interactions with the zeolite matrix and the water molecules (located in the smaller-diameter channel). Indeed, the water molecules form a very peculiar encapsulated structure: each 6MR channel exhibits clusters of four water molecules organized in a square tetramer kept together by very strong hydrogen bonds (average hydrogen-bonding distance: ~1.7 Å), as depicted in Figure 4.6c,d. It is also important to point out that, according to the calculations, the water square tetramers incorporated in Si-FER do not form hydrogen bonds with the zeolite framework.

To establish whether such a peculiar arrangement of the water and ethanol molecules was also stable at room-temperature conditions, FPMD simulations were performed starting from the optimized model structure. The simulations confirmed that the supramolecular complexes hosted in Si-FER remained stable at room temperature as well. Most importantly, a series of calculations modeling Si-FER·($8H_2O,4EtOH$) at room pressure conditions indicated that these well-distinct clusters of water molecules and ethanol molecules remained stable even upon pressure release. In particular, the hydrogen bonds of the water square clusters evidenced only small fluctuations from the values found in the optimized structure corresponding to 0.84 GPa. This impressive stability can be explained keeping into account that water–water—hydrogen bonding governs the behavior of water in siliceous zeolites and that the square clusters of water molecules maximize inter-water—hydrogen bonding. On these bases, it can be deduced that when the pressure-transmitting fluid (the water–ethanol mixture) penetrates the channels of this hydrophobic zeolite, water molecules are energetically favored to form water square tetramers to enhance inter-water—hydrogen bonding.

Hence, theoretical modeling clarified that the driving force for the separation of water from ethanol inside this zeolite should be the formation of exceptionally stable water clusters. The tendency of water molecules to interact preferentially with each other could be the origin of the separation of water and ethanol in clusters exactly fitting into the empty spaces of siliceous ferrierite. Also importantly, structural analyses from both the DFT-optimized structures and the FPMD trajectories showed that the size of the 6MR and 10MR channels of the ferrierite framework perfectly fits, respectively, to the water square tetramers and the ethanol dimer chains. This finding indicated that even the shaping effect of the zeolite matrix is an essential requirement for the segregation of water from ethanol in siliceous ferrierite. Finally, calculations allowed to establish that, when compression was released to ambient conditions, (i) the water tetramers and the ethanol dimers maintained their structural features, and (ii) the zeolite was slightly stabilized (by 1.2 kcal mol$^{-1}$ per unit cell) with respect to high-pressure conditions.

In perspective, this latter aspect disclosed by theoretical modeling – namely, the irreversibility of the forced inclusion of guest molecules in the framework of porous oxides – may be particularly relevant for potential applications of pressure-created composite materials. On these bases, the pressure-induced intrusion of guest species in regular arrays of zeolite cavities might hold promises as a new approach to the fabrication of composite materials unattainable at normal conditions but stable upon pressure release, which could be therefore exploited in real-life technological applications.

### 4.4.3 Hybrid Microporous Functional Materials

An efficient approach to organize photoactive molecules in supramolecular aggregates endowed with appealing functionalities is to exploit a confining system to force molecules to assume a uniform orientation, which enables the encapsulated chromophores to respond to specific opto-electronic stimuli. Since the late 1980s, zeolite pores have been employed in the fabrication of confined nanosystems for electro-optical applications. To this aim, zeolites characterized by 1D arrays of nanochannels of diameter comparable to the size of organic dyes are particularly suitable to create a broad variety of organic–inorganic hybrid composites, which may be exploited e.g. in solar cells or artificial antenna devices [53]. It should however be noted that guest molecules, even when they are incorporated in 1D channel systems, do not form perfectly ordered linear arrangements, especially if the pore diameter is much larger than the size of the encapsulated molecules, thus allowing the guest molecules to rotate. Nonetheless, 1D porous matrices still remain the preferred host systems for the confinement of nanometric species. Among this family of materials, zeolite L – as already mentioned in Section 4.4.1 – is particularly convenient because of the good matching between the size and shape of its 1D channels and a number of organic chromophores of widespread use. The realization of hybrid composites upon encapsulation of dyes in zeolite L, as building units for artificial antenna systems, pioneered by the Calzaferri group [36, 54], has been widely studied with a broad variety of characterization techniques. The expertise gained has been crucial for extending the concept to the fabrication of devices for biomedical uses [55]. Nonetheless, in spite of these progresses, the microscopic details of the positioning of the dyes in the zeolite channels, as well as the understanding of the interactions between the guest dye molecules and the zeolite host, remained quite obscure for a long time. In particular, the atomistic structure of these fascinating materials is often unknown, as demonstrated by the very scarce presence in the literature of X-ray experiments on dye–zeolite L compounds. Another relevant point is that, by considering the high affinity of zeolites for water, several dye–zeolite L compounds are not stable in presence of humidity. For example, it was documented that dry p-terphenyl-zeolite L samples in presence of air of 22% relative humidity at ambient conditions release the organic dye from the zeolite matrix [53]. Fluorenone dye (Figure 4.7a) represented an interesting exception, because it is not expelled from the zeolite channels by water molecules at room conditions [57]. However, in spite of detailed spectroscopic studies and thermal analysis investigations [57], the reason of the surprising stability of fluorenone–zeolite L hybrid composite was not clarified.

Since early 2010s, our laboratory, in collaboration with Prof. Calzaferri, devoted significant efforts to the investigation of structural, dynamical, and electronic behavior of dye–zeolite L composites by means of quantum chemical approaches and FPMD simulations. The example presented herein to showcase the insight from computational modeling on dye–zeolite materials is the theoretical study of the hybrid composite of zeolite L with the neutral carbonylic dye fluorenone (Figure 4.7a), reported in Ref. [56].

The modeling strategy consisted of using FPMD- and DFT-based geometry optimizations to model both the anhydrous and the hydrated fluorenone–zeolite L compound and to compare the results obtained for the anhydrous and the hydrated

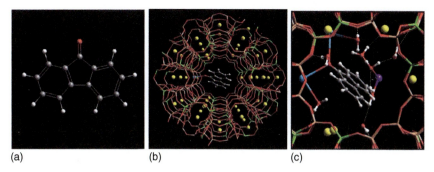

**Figure 4.7** (a) Minimum energy structure of the isolated fluorenone dye (gray, C; red, O; white, H). (b) Graphical representation of the minimum energy structure of the anhydrous fluorenone–zeolite L composite evidencing the one-dimensional channel system and the interaction of the dye carbonyl group with one of the potassium cations (highlighted in cyan). (c) Minimum energy structure of the hydrated fluorenone–zeolite L compound (gray, C; red, O; white, H; pink, Si; green, Al; yellow, K; cyan, violet, K interacting with fluorenone and/or water molecules; dotted lines, hydrogen bonds). Source: Reproduced with permission from Fois et al. [56].

model systems in order to account for the effect of the presence of water inside the nanochannels of zeolite L.

In the simulation of the dry model system, the fluorenone molecule was preferentially oriented with its carbonyl oxygen toward the channel walls, close to extraframework $K^+$ ions: the molecule only switched its carbonyl oxygen from a $K^+$ ion to a neighboring one in the course of the trajectory, suggesting the relevance of the interaction of the carbonyl group with the extraframework cations. Geometry optimizations were then carried out on various configurations sampled from the MD simulation, exhibiting different orientations of the dye. These geometry optimizations confirmed that the most stable configuration was characterized by the fluorenone strong coordination to one $K^+$, with the carbonyl oxygen at 2.59 Å from the potassium cation and the carbonyl bond of the dye appreciably elongated with respect to the gas-phase value (1.246 vs. 1.220 Å, respectively) (Figure 4.7b). Energetics analysis indicated that whereas models characterized by fluorenone–$K^+$ interactions have high stability, configurations where the coordination with $K^+$ was missing were energetically disfavored.

As regards the hydrated fluorenone–zeolite L models, during the simulation the dye molecule remained preferentially oriented as in the anhydrous system, i.e. with its carbonyl oxygen coordinated to a potassium extraframework cation. Several energy minima were obtained by geometry optimization of configurations from the MD run of the hydrated system. The lowest energy minimum, shown in Figure 4.7c, clearly evidences that, in spite of water competition in coordinating the potassium cation, the interaction of fluorenone with $K^+$ is dominant.

A careful analysis of the electronic properties of fluorenone–zeolite L composites is crucial for electrooptical applications. In experimental UV/vis spectra of fluorenone–zeolite L hybrids, the longest-wavelength signal, ascribed to the HOMO–LUMO $1\pi \to 1\pi^*$ transition of fluorenone, showed a bathochromic shift

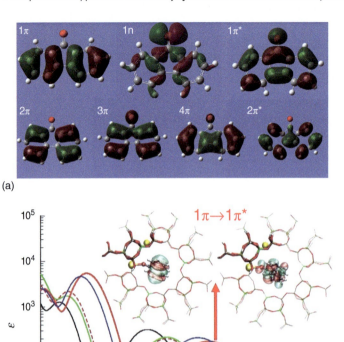

**Figure 4.8** (a) Molecular orbitals of gas-phase fluorenone involved in the electronic transitions; (b) TD-DFT calculated electronic spectra for gas-phase fluorenone (black); fluorenone in cyclohexane (green); dry fluorenone/zeolite L energy minimum (red); fluorenone molecule from the dry dye–zeolite L energy minimum (red dashed); hydrated fluorenone–zeolite L energy minimum (blue). The central inset shows the HOMO and LUMO of the dye–zeolite L composite involved in the highest-wavelength excitation. Atom colors: brown, Si; red, O; green, Al; gray, C; white, H; yellow, K. Source: Reproduced with permission from Fois et al. [56].

of 50 nm with respect to diluted cyclohexane solutions [57]. Figure 4.8 depicts the time-dependent-density functional theory (TD-DFT) spectrum computed for a cluster extracted from the minimum energy structure of the anhydrous fluorenone–zeolite L composite, along with those of gas phase fluorenone and fluorenone in cyclohexane, computed in Ref. [56]. Notably, the features of the calculated spectra were well in line with the experimental data: in particular, the $1\pi \to 1\pi^*$ HOMO–LUMO transition was computed at 402 nm in the gas phase, at 415 nm in cyclohexane, and at 456 nm in the minimum energy structure of the dry composite.

To determine that the bathochromic shift was caused directly by the fluorenone–$K^+$ interaction, electronic excitations were computed just for the fluorenone molecule from the minimum energy structure. The resulting spectrum (reported also in Figure 4.8) was similar to the spectrum of fluorenone in cyclohexane. As regards the hydrated model, the TD-DFT spectrum showed a lower redshift with respect to the anhydrous system because in the hydrated composite the interaction of the carbonyl group of the dye with the potassium cations is slightly weakened by the presence of water. Hence, the TD-DFT findings confirmed that the strong interaction of the dye with the zeolite L potassium cations – which stabilized the fluorenone LUMO, localized on the carbonyl oxygen (see Figure 4.8, top inset) – was indeed responsible for the bathochromic shift detected experimentally. Such results were also confirmed by a subsequent theoretical study based on density functional embedding methods taking into account the full zeolite system [58].

Overall, the theoretical investigation of fluorenone dye in zeolite L of Ref. [56] offered for the first time an atomistic picture of a dye–zeolite L hybrid material at conditions similar to the actual applications of zeolite L-based functional devices. Relevant features very difficult to obtain experimentally were highlighted: (i) the location of fluorenone in zeolite L nanochannels, (ii) the effect of water, and (iii) the identification of the origin of the stability of fluorenone–zeolite L materials against humidity, i.e. the strong interaction of the dye carbonyl group with the zeolite potassium cations, also tuned by the water—hydrogen bond network. From the practical viewpoint, the theoretical understanding of such stabilization mechanism, which is general to other carbonyl-containing dyes, allowed for the realization of a new series of humidity-resistant dye–zeolite composites characterized by a hierarchical organization of neutral luminescent molecules (perylene bisimide dyes) – adopted in actual energy harvesting devices [38, 59].

## 4.5 Outlook and Perspectives

The realization of oxide nanomaterials of ever-increasing complexity and the implementation of innovative approaches for the fabrication of functional nanostructured oxide composites have experienced an increasing momentum in the latest decade. In spite of the enormous progresses on the practical and technological level, the foundations of these fabrication strategies still remain largely empirical. Nonetheless, the continuous progress of computational approaches holds concrete promises to provide molecular-level bases to oxide material technologies.

The case studies discussed in the present chapter showed that theoretical modeling investigations not only complement experimental analyses but also provide molecular insight otherwise unattainable on structure and processes of oxide-based nanomaterials even at conditions very far from the ambient ones, such as elevated temperatures and pressures. Moreover, computational modeling may also help to design new fabrication strategies for oxide nanosystems. In addition, the challenging problems encountered in the study of these versatile functional materials and of their applications might also stimulate new developments in computational approaches.

In this scenario, some general theoretical advances able to further enhance the impact of modeling in oxide nanomaterials science might be (i) the decrease of the computational cost of hybrid DFT functionals to enable application to periodic systems of extensive size; (ii) the implementation of new pure-DFT approximations of improved accuracy; (iii) the development of techniques of intermediate accuracy between pure and hybrid DFT, which could treat large-sized model systems not affordable by hybrid functionals yet; and (iv) the decrease of the computational cost of wavefunction-based quantum chemical approaches, to enhance the accuracy of the theoretical modeling of transition-metal-oxide surfaces and magnetic systems.

The aforementioned advances in theoretical methodologies would be of paramount relevance to improve, for instance, the treatment of magnetic oxide nanomaterials and of their interaction with molecular species. Specifically, in the case of Au–$Mn_3O_4$ sensing, such progresses would allow, for instance, to build an improved model of the manganese oxide–gold nanoparticle interface, to extend the computational study to the dynamic aspects of the sensing action, to investigate the role of temperature, and to simulate the co-presence of target analytes and interfering molecules on the Au–$Mn_3O_4$ surface.

As far as the simulation of vapor deposition processes is concerned, in spite of the advances achieved up to date, there are still many issues to be explored, which could be of general impact. For example, the rolling diffusion of molecular precursors on the growth surface may be a plausible activation mechanism for other high-temperature surface processes besides CVD ones. In any case, the appearance of temperature-induced rolling diffusion in molecule-to-material fabrication approaches brings about further challenges for theoretical investigations. More in general, to gather atomistic understanding of the surface activation processes at high temperatures, large-sized models and longer FPMD trajectories – also complemented by statistical sampling approaches for the treatment of reactive events – would be required for an explicit simulation of bimolecular or multimolecular events at the growth surface.

Simulations have become an extremely versatile tool for addressing the complex behavior of zeolites and zeolite-based composites at both room conditions and high-pressure conditions. Regarding high-pressure phenomena in zeolites, a perspective issue to be solved is to disclose the atomistic details of the pressure-induced penetration of guest molecules in zeolite pores. This highly challenging problem would involve a careful modeling of the external surfaces of the zeolite pore opening, a task that has been already successfully performed for zeolite L-based materials at room conditions [60, 61]. In the case of electro-optical devices based on dye–zeolite hybrid composites, modeling studies have made it possible to connect the experimental optical response of the device to the structural properties of the dye–zeolite composite at atomistic-detail level. As a matter of fact, modeling has allowed to evaluate the separate effects of different variables (composition, temperature, pressure, and guest species concentration) on a macroscopic feature of the zeolitic system. However, although computational investigations have already manifested excellent predictive capabilities, theoretical models are still far from reproducing the full complexity of the real materials. To stimulate progress in this direction, in our views, experimental and computational researchers should further increase their mutual collaboration and jointly adopt integrated strategies to set up increasingly realistic model systems.

As a final consideration, computational modeling, besides aiding interpretation of experimental results, is playing a key role in opening new routes in oxide nanomaterials research. Joint theoretical–experimental strategies, which nowadays are already widely popular and successfully applied, would become even more crucial in future perspective. In this scenario, the combination of theory and experiment is expected to gain further momentum from the growing increase of computing power and faster numerical algorithms.

## References

1 Rimola, A., Costa, D., Sodupe, M. et al. (2013). Silica surface features and their role in the adsorption of biomolecules: computational modeling and experiments. *Chem. Rev.* 113 (6): 4216–4313.

2 Van Speybroeck, V., Hemelsoet, K., Joos, L. et al. (2015). Advances in theory and their application within the field of zeolite chemistry. *Chem. Soc. Rev.* 44 (20): 7044–7111.

3 Catlow, C.R.A., Smit, B., and Van Santen, R.A. (2004). *Computer Modelling of Microporous Materials*. Amsterdam, NL: Elsevier.

4 Gatta, G.D., Lotti, P., and Tabacchi, G. (2018). The effect of pressure on open-framework silicates: elastic behaviour and crystal–fluid interaction. *Phys. Chem. Miner.* 45 (2): 115–138.

5 Kühne, T.D., Iannuzzi, M., Del Ben, M. et al. (2020). CP2K: an electronic structure and molecular dynamics software package - quickstep: efficient and accurate electronic structure calculations. *J. Chem. Phys.* 152 (19): 194103.

6 Giannozzi, P., Andreussi, O., Brumme, T. et al. (2017). Advanced capabilities for materials modelling with quantum ESPRESSO. *J. Phys. Condens. Matter* 29 (46): 465901.

7 Maduraiveeran, G., Sasidharan, M., and Jin, W. (2019). Earth-abundant transition metal and metal oxide nanomaterials: synthesis and electrochemical applications. *Prog. Mater Sci.* 106: 100574.

8 Zappa, D., Galstyan, V., Kaur, N. et al. (2018). "Metal oxide-based heterostructures for gas sensors" – a review. *Anal. Chim. Acta* 1039: 1–23.

9 Bigiani, L., Zappa, D., Barreca, D. et al. (2019). Sensing nitrogen mustard gas simulant at the ppb scale via selective dual-site activation at Au/$Mn_3O_4$ interfaces. *ACS Appl. Mater. Interfaces* 11 (26): 23692–23700.

10 Devi, A. (2013). "Old chemistries" for new applications: perspectives for development of precursors for MOCVD and ALD applications. *Coord. Chem. Rev.* 257 (23–24): 3332–3384.

11 Bekermann, D., Barreca, D., Gasparotto, A., and Maccato, C. (2012). Multi-component oxide nanosystems by chemical vapor deposition and related routes: challenges and perspectives. *CrystEngComm* 14 (20): 6347–6358.

12 Tabacchi, G., Fois, E., Barreca, D., and Gasparotto, A. (2014). Opening the Pandora's jar of molecule-to-material conversion in chemical vapor deposition: insights from theory. *Int. J. Quantum Chem.* 114 (1): 1–7.

13 Pedersen, H. and Elliott, S.D. (2014). Studying chemical vapor deposition processes with theoretical chemistry. *Theor. Chem. Acc.* 133 (5): 1–10.

14 Baerlocher, C., McCusker, L.B., and Olson, D.H. (2007). *Atlas of Zeolite Framework Types*. Published on behalf of the Structure Commission of the International Zeolite Association by Elsevier.

15 Szostak, R. (1989). *Molecular Sieves*. Netherlands: Springer.

16 Tabacchi, G. (2018). Supramolecular organization in confined nanospaces. *ChemPhysChem* 19 (11): 1249–1297.

17 Calzaferri, G. and Lutkouskaya, K. (2008). Mimicking the antenna system of green plants. *Photochem. Photobiol. Sci.* 7 (8): 879–910.

18 Gottardi, G. and Galli, E. (2012). *Natural Zeolites*. Berlin, Heidelberg: Springer.

19 Parr, R.G. and Weitao, Y. (1995). *Density-Functional Theory of Atoms and Molecules*. Oxford: Oxford University Press.

20 Goerigk, L., Hansen, A., Bauer, C. et al. (2017). A look at the density functional theory zoo with the advanced GMTKN55 database for general main group thermochemistry, kinetics and noncovalent interactions. *Phys. Chem. Chem. Phys.* 19 (48): 32184–32215.

21 Perdew, J.P., Burke, K., and Ernzerhof, M. (1996). Generalized gradient approximation made simple. *Phys. Rev. Lett.* 77 (18): 3865–3868.

22 Grimme, S. (2006). Semiempirical GGA-type density functional constructed with a long-range dispersion correction. *J. Comput. Chem.* 27 (15): 1787–1799.

23 Fischer, M. and Angel, R.J. (2017). Accurate structures and energetics of neutral-framework zeotypes from dispersion-corrected DFT calculations. *J. Chem. Phys.* 146 (17): 174111.

24 Car, R. and Parrinello, M. (1985). Unified approach for molecular dynamics and density-functional theory. *Phys. Rev. Lett.* 55 (22): 2471–2474.

25 Marx, D. and Hutter, J. (2009). *Ab Initio Molecular Dynamics*. Cambridge: Cambridge University Press.

26 Ciccotti, G., Kapral, R., and Sergi, A. (2005). Simulating reactions that occur once in a blue moon. In: *Handbook of Materials Modeling*, 1597–1611. Dordrecht: Springer Netherlands.

27 Maccato, C., Bigiani, L., Carraro, G. et al. (2017). Molecular engineering of Mn II Diamine Diketonate precursors for the vapor deposition of manganese oxide nanostructures. *Chem. A Eur. J.* 23 (71): 17954–17963.

28 Cococcioni, M., and De Gironcoli, S. (2005) Linear response approach to the calculation of the effective interaction parameters in the LDA+U method, *Phys. Rev. B: Condens. Matter*, 71 (3), 035105.

29 Mishra, S. and Daniele, S. (2015). Metal-organic derivatives with fluorinated ligands as precursors for inorganic nanomaterials. *Chem. Rev.* 115 (16): 8379–8448.

30 McElwee-White, L. (2006) Design of precursors for the CVD of inorganic thin films, *J. Chem. Soc., Dalton Trans.*, (45), 5327–5333.

**31** Fois, E., Tabacchi, G., Barreca, D. et al. (2010). "Hot" surface activation of molecular complexes: insight from modeling studies. *Angew. Chem. Int. Ed.* 49 (11): 1944–1948.

**32** Tabacchi, G., Fois, E., Barreca, D., and Gasparotto, A. (2014). CVD precursors for transition metal oxide nanostructures: molecular properties, surface behavior and temperature effects. *Phys. Status Solidi* 211 (2): 251–259.

**33** Tabacchi, G., Fois, E., Barreca, D., Carraro, G., Gasparotto, A., and Maccato, C. (2016) Modeling the first activation stages of the Fe(hfa)2TMEDA CVD precursor on a heated growth surface, in *Advanced Processing and Manufacturing Technologies for Nanostructured and Multifunctional Materials II*, Ohji, T., Singh, M., and Halbig, M., Eds., John Wiley & Sons, Inc., Hoboken, NJ, USA, pp. 83–90.

**34** Barreca, D., Fois, E., Gasparotto, A. et al. (2021). The early steps of molecule-to-material conversion in chemical vapor deposition (CVD): a case study. *Molecules* 26 (7): 1988.

**35** Fois, E. and Tabacchi, G. (2019). Water in zeolite L and its MOF mimic. *Z. Kristallogr. - Cryst. Mater.* 234: 495–511.

**36** Calzaferri, G. (2020). Guests in nanochannels of zeolite L. In: *Structure and Bonding*, 1–73. Berlin, Heidelberg: Springer.

**37** Albuquerque, R.Q. and Calzaferri, G. (2007). Proton activity inside the channels of zeolite L. *Chem. A Eur. J.* 13 (32): 8939–8952.

**38** Devaux, A., Calzaferri, G., Belser, P. et al. (2014). Efficient and robust host–guest antenna composite for light harvesting. *Chem. Mater.* 26 (23): 6878–6885.

**39** Merrill, L. and Bassett, W.A. (1974). Miniature diamond anvil pressure cell for single crystal X-ray diffraction studies. *Rev. Sci. Instrum.* 45 (2): 290–294.

**40** Miletich, R., Allan, D.R., and Kuhs, W.F. (2000). High-pressure single-crystal techniques. *Rev. Mineral. Geochem.* 41 (1): 445–519.

**41** Angel, R.J., Bujak, M., Zhao, J. et al. (2007). Effective hydrostatic limits of pressure media for high-pressure crystallographic studies. *J. Appl. Crystallogr.* 40 (1): 26–32.

**42** Fraux, G., Coudert, F.X., Boutin, A., and Fuchs, A.H. (2017). Forced intrusion of water and aqueous solutions in microporous materials: from fundamental thermodynamics to energy storage devices. *Chem. Soc. Rev.* 46 (23): 7421–7437.

**43** Ståhl, K., Kvick, Å., and Ghose, S. (1989). One-dimensional water chain in the zeolite bikitaite: neutron diffraction study at 13 and 295 K. *Zeolites* 9 (4): 303–311.

**44** Quartieri, S., Sani, A., Vezzalini, G. et al. (1999). One-dimensional ice in bikitaite: single-crystal X-ray diffraction, infra-red spectroscopy and ab-initio molecular dynamics studies. *Microporous Mesoporous Mater.* 30 (1): 77–87.

**45** Fois, E., Gamba, A., and Tilocca, A. (2002). Structure and dynamics of the flexible triple helix of water inside VPI-5 molecular sieves. *J. Phys. Chem. B* 106 (18): 4806–4812.

**46** Fois, E., Gamba, A., Medici, C. et al. (2008). High pressure deformation mechanism of Li-ABW: synchrotron XRPD study and ab initio molecular dynamics simulations. *Microporous Mesoporous Mater.* 115 (3): 267–280.

**47** Ferro, O., Quartieri, S., Vezzalini, G. et al. (2002). High-pressure behavior of bikitaite: an integrated theoretical and experimental approach. *Am. Mineral.* 87 (10): 1415–1425.

**48** Arletti, R., Fois, E., Gigli, L. et al. (2017). Irreversible conversion of a water–ethanol solution into an organized two-dimensional network of alternating Supramolecular units in a hydrophobic zeolite under pressure. *Angew. Chem. Int. Ed.* 56 (8): 2105–2109.

**49** Wang, C., Bai, P., Siepmann, J.I., and Clark, A.E. (2014). Deconstructing hydrogen-bond networks in confined nanoporous materials: implications for alcohol–water separation. *J. Phys. Chem. C* 118 (34): 19723–19732.

**50** Bai, P., Jeon, M.Y., Ren, L. et al. (2015). Discovery of optimal zeolites for challenging separations and chemical transformations using predictive materials modeling. *Nat. Commun.* 6: 5912.

**51** Mittal, N., Bai, P., Siepmann, J.I. et al. (2017). Bioethanol enrichment using zeolite membranes: molecular modeling, conceptual process design and techno-economic analysis. *J. Membr. Sci.* 540: 464–476.

**52** Trudu, F., Tabacchi, G., and Fois, E. (2019). Computer modeling of apparently straight bond angles: the intriguing case of all-silica ferrierite. *Am. Mineral.* 104 (11): 1546–1555.

**53** Calzaferri, G., Huber, S., Maas, H., and Minkowski, C. (2003). Host–guest antenna materials. *Angew. Chem. Int. Ed.* 42 (32): 3732–3758.

**54** Calzaferri, G. and Gfeller, N. (1992). Thionine in the cage of zeolite L. *J. Phys. Chem.* 96 (8): 3428–3435.

**55** Popović, Z., Otter, M., Calzaferri, G., and De Cola, L. (2007). Self-assembling living systems with functional nanomaterials. *Angew. Chem. Int. Ed.* 46 (32): 6188–6191.

**56** Fois, E., Tabacchi, G., and Calzaferri, G. (2010). Interactions, behavior, and stability of fluorenone inside zeolite nanochannels. *J. Phys. Chem. C* 114 (23): 10572–10579.

**57** Devaux, A., Minkowski, C., and Calzaferri, G. (2004). Electronic and vibrational properties of fluorenone in the channels of zeolite L. *Chem. A Eur. J.* 10 (10): 2391–2408.

**58** Zhou, X., Wesolowski, T.A., Tabacchi, G. et al. (2013). First-principles simulation of the absorption bands of fluorenone in zeolite L. *Phys. Chem. Chem. Phys.* 15 (1): 159–167.

**59** Gigli, L., Arletti, R., Tabacchi, G. et al. (2018). Structure and host–guest interactions of Perylene–Diimide dyes in zeolite L nanochannels. *J. Phys. Chem. C* 122 (6): 3401–3418.

**60** Tabacchi, G., Fois, E., and Calzaferri, G. (2015). Structure of nanochannel entrances in stopcock-functionalized zeolite L composites. *Angew. Chem. Int. Ed.* 54 (38): 11112–11116.

**61** Tabacchi, G., Calzaferri, G., and Fois, E. (2016). One-dimensional self-assembly of perylene-diimide dyes by unidirectional transit of zeolite channel openings. *Chem. Commun.* 52 (75): 11195–11198.

# 5

# Functional Spinel Oxide Nanomaterials: Tailored Synthesis and Applications

*Zheng Fu and Mark T. Swihart*

*University at Buffalo (SUNY), Department of Chemical and Biological Engineering and RENEW Institute, 308 Furnas Hall, Buffalo, NY, 14260, USA*

## 5.1 Introduction and Topic Overview

In recent years, nanoscale metal oxides of the spinel crystal structure, with composition $AB_2O_4$, have been widely investigated due to their numerous useful properties and corresponding applications in electronics [1, 2], electrocatalysis [3, 4], and magnetic devices [5, 6]. These $AB_2O_4$ spinel materials are constructed from a combination of tetrahedral and octahedral units, in which A and B can be main group metals or transition metals. Nanomaterials, in general, can be classified as one-dimensional (1D) (e.g. nanorods, nanowires, nanotubes); two-dimensional (2D) (e.g. nanoplatelets, nanosheets, nanolayers); three-dimensional (3D) (e.g. nanoflowers, polyhedra, core–shell structures, nanoframes); and zero-dimensional (0D) (spherical or quasi-spherical dots whose size is <100 nm). We adopt this classification herein to frame our discussion of synthesis strategies and structure–function–application relationships in spinel nanomaterials, aiming to provide both a general background on this family of nanomaterials and specific examples and synthesis strategies. However, we also note that nanomaterials of one type may be the building blocks for another, as when 1D nanowires are interwoven or entangled to make a 2D mat or textile or when 2D nanosheets assemble into a 3D structure (often termed a "nanoflower"). We also consider the most recent advances and remaining needs and promising future directions for this class of materials.

Herein, synthesis strategies are also classified into three categories of vapor, solution, and solid-phase methods for producing these spinel nanomaterials. Then, for each major synthesis route, we classify materials based on dimensionality: 1D, 2D, and 3D structures of traditional morphology and then hierarchical one and two (1&2D), one and three (1&3D), and two and three (2&3D)-dimensional structures that assemble a lower-dimensional structure into a higher-dimensional one. We separately consider structures formed by self-assembly of small spinels, typically below 100 nm in size, which form ordered structures via reversible relatively weak interactions.

*Tailored Functional Oxide Nanomaterials: From Design to Multi-Purpose Applications*, First Edition.
Edited by Chiara Maccato and Davide Barreca.
© 2022 WILEY-VCH GmbH. Published 2022 by WILEY-VCH GmbH.

## 5.2 Syntheses

Because the preparation process controls the composition, morphology, and properties of the products, a wide array of experimental methods for synthesizing spinel nanomaterials have been developed, including both chemical and physical approaches, often with specific properties, structures, or applications in mind. Therefore, the synthesis strategies should be understood before studying structure–property–application relationships. As noted earlier, we begin by classifying methods as vapor phase, solution-phase, and solid-phase routes. Vapor-phase synthesis methods include chemical vapor deposition (CVD), atomic layer deposition (ALD), spray pyrolysis, laser pyrolysis, and plasma-driven methods. Solution-phase synthesis includes sol–gel, hydrothermal, thermal decomposition, and solvothermal methods. Solid-phase synthesis includes solid-state thermal decomposition, combustion synthesis, ball milling, and high-temperature solid solution reactions. Of course, some synthesis strategies may involve two or three phases to obtain target products, and we discuss these specific instances as they arise.

### 5.2.1 Vapor Phase

#### 5.2.1.1 Chemical Vapor Deposition

CVD, known as an efficient synthesis method to produce solid thin film materials since the 1800s [7], employs a conceptually simple strategy of reacting vaporized precursors (metal salts or organometallic compounds) on a solid substrate surface to deposit thin films of materials. Figure 5.1 A shows a simple schematic diagram of a CVD process to produce a $Co_3O_4$ catalyst applied for catalytic oxidation of hydrocarbons [11]. In this example, the CVD process is operated at low pressure (20 mbar) with a tubular horizontal hot-wall reactor to evaporate cobalt(II) acetylacetonate precursor at 160 °C, with argon as carrying gas, to mix this stream with a flow of oxygen, and to deposit $Co_3O_4$ on monolithic supports at 230 °C. Vaporizable organic salts, like the acetylacetonates, are popular reactants for the CVD method [12–14]. A key criterion for their use is that they can be vaporized without decomposing and can then react on the solid substrate. The low-pressure CVD approach allowed the uniform deposition of spinel $Co_3O_4$ throughout the catalyst monolith, at an average growth rate of 0.7 nm min$^{-1}$, to deposit a film of submicron thickness on this complex geometry.

#### 5.2.1.2 Atomic Layer Deposition

ALD shares many similarities with the CVD method. However, the layer-by-layer process of ALD can provide even more precise control of thickness of the deposition and even greater conformal coverage [15]. This control is achieved by dividing the deposition reaction into two self-limiting half-reactions. For example, a substrate can be first exposed to the metal precursor (e.g. trimethyl aluminum) that reacts with —OH groups on the substrate. Once all —OH groups are capped, no more

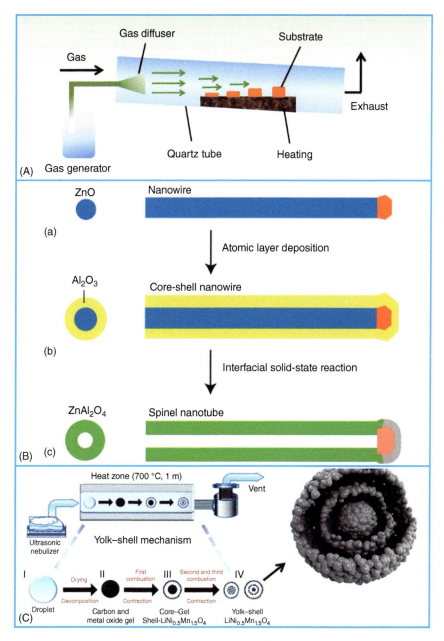

**Figure 5.1** Schematic illustrations of (A) a simple CVD reactor. Source: Adapted with permission from Zhao et al. [8], American Chemical Society. (B) The process of producing $ZnAl_2O_4$ spinel nanotubes. Source: Reproduced with permission from Jin Fan et al. [9], Springer Nature. (C) The formation mechanism of yolk–shell-structured $LiNi_{0.5}Mn_{1.5}O_4$ powders in the spray pyrolysis method. Source: Reproduced from Choi et al. [10], Royal Society of Chemistry.

precursor reacts; a monolayer of it coats the substrate. Then, the first precursor is replaced by a second precursor or reagent (e.g. water vapor) that reacts with the adsorbed metal precursor. In the example of alumina ($Al_2O_3$) ALD, the water oxidizes the adsorbed trimethyl aluminum to generate a monolayer of $Al_2O_3$ and regenerate surface —OH groups. The example in Figure 5.1B is a particular case of ALD, in which it provides a unique way to coat 1D nanostructures and convert them into hollow, single-crystalline spinel nanotubes. In that study, ZnO nanowires were first synthesized by a vapor–liquid–solid mechanism [9]. Then, nanowires were coated with a layer of $Al_2O_3$ via ALD to produce core–shell ZnO–$Al_2O_3$ nanowires. Finally, by annealing the previously mentioned structures, $ZnAl_2O_4$ spinel nanotubes were formed by solid-state reaction accompanied by the Kirkendall effect [16]. This special case demonstrates a novel use of ALD to convert nanowires to spinel nanotubes. It is also a case of processes occurring in multiple phases. The ZnO nanowires and $Al_2O_3$ are deposited from the vapor phase, but the final conversion into spinel $ZnAl_2O_4$ occurs via a solid-phase reaction.

While ALD can directly deposit many different oxides, direct deposition of spinels is relatively rare. One successful example of ALD of a spinel is the deposition of $Co_3O_4$ into preformed $TiO_2$ nanotubes [17]. As in the previous example, the ability of ALD to conformally coat even nanoscale structures was the key feature needed. Compared with $Al_2O_3$ deposition, $Co_3O_4$ ALD required both a more complex precursor (bis(cyclopentadienyl)cobalt(II)) and a stronger oxidizing co-reactant (ozone).

### 5.2.1.3 Spray Pyrolysis

Spray pyrolysis can be used to produce both thin films and powders, depending on the specific process. In this method, metal salts or other precursors are dissolved in a solvent and sprayed as small droplets in a carrier gas that delivers them to a hot surface, to deposit a film, or carries them through a heated gas region, to generate powders. When the precursor droplets, or vapor produced by evaporation of those droplets, is delivered to a hot surface, film growth occurs, similar to CVD. When the droplets react without completely evaporating, a powder is produced. Processes can be designed to produce products with internal structure that results from the dynamics of evaporation and reaction. For example, Figure 5.1C shows a mechanism of forming yolk–shell-structured $LiNi_{0.5}Mn_{1.5}O_4$ powders by spray pyrolysis in a hot-wall reactor. [10] In this example, an aqueous solution of precursor metal salts, lithium nitrate ($LiNO_3$), nickel nitrate hexahydrate [$Ni(NO_3)_2 \cdot 6H_2O$], and manganese nitrate hexahydrate [$Mn(NO_3)_2 \cdot 6H_2O$], along with sucrose, was sprayed into a tube furnace reactor at 700 °C, where the simultaneous reaction and solvent evaporation yielded a layered structure. Here, sucrose serves as a sacrificial template that decomposes to generate the yolk–shell structures. The powder was then post-treated at 700 °C for three hours in air. This simple spray pyrolysis process was able to produce $LiNi_{0.5}Mn_{1.5}O_4$ spinel powders with a large surface area and yolk–shell structure that enable them to perform extremely well as cathode materials in a lithium-ion battery, achieving a reversible discharge capacity of 99 mAh $g^{-1}$ after 500 cycles at an extremely high current of 200 C.

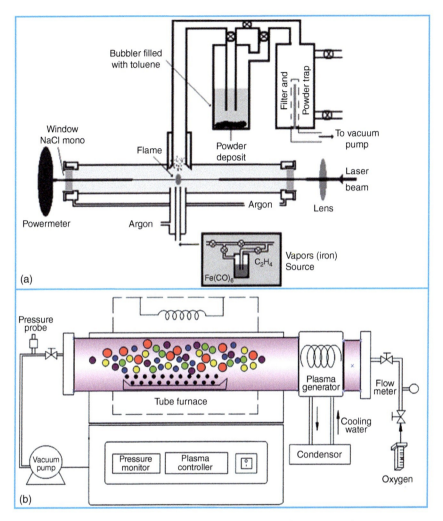

**Figure 5.2** (a) The laser pyrolysis synthesis of iron/iron oxide nanoparticles and the system for powder collection in a bubbler. Source: Reproduced with permission from Jiang et al. [19], Elsevier. (b) A plasma-enhanced tube furnace for preparing $LiMn_2O_4$. Source: Reproduced with permission from Popovici et al. [20], Royal Society of Chemistry.

### 5.2.1.4 Laser Pyrolysis

Laser pyrolysis uses $CO_2$ laser radiation as the energy source to initiate reaction upon its interaction with a reactant vapor stream. This decomposition of precursor molecules initiates the nucleation and growth of nanocrystals [18]. Figure 5.2a shows a schematic of a laser pyrolysis process for producing and collecting iron/iron oxide nanomaterials [19]. $Fe(CO)_5$ vapor as the precursor was carried by $C_2H_4$ gas into a low-pressure reactor at 3000 Pa and 25 °C. In the laser beam, $C_2H_4$ absorbed the laser energy and transferred it to the other gas molecules, driving thermal decomposition of $Fe(CO)_5$. Iron particles then nucleated and grew from the reactive decomposition products. The product particles were collected in a bubbler filled with toluene, followed by a filter to collect particles that were not collected in the bubbler. The products were then redispersed in a solvent to prepare magnetic nanofluids [21]. Particles collected on the filter were smaller than those collected in the bubbler, and with further separation by dispersion and settling of the larger particles, samples with particle sizes below 50 nm were obtained. Upon exposure to air, these particles undergo surface oxidation to produce a $Fe_3O_4/\gamma\text{-}Fe_2O_3$ shell. [19]. A similar study used both $SF_6$ and $C_2H_4$ as the photosensitizer to absorb the laser energy and compared the results for synthesis of iron nanoparticles from $Fe(CO)_5$ [22]. More generally, the laser pyrolysis approach can be used to produce metal nanoparticles that are subsequently converted to spinel metal oxides by air exposure or by annealing in air, but has not been used to directly produce spinel metal oxides. Laser pyrolysis has the advantage of producing small, high-purity materials because it can achieve very short reaction times and because only the gas is heated, and not the reactor walls. However, it requires specialized equipment and often uses toxic (e.g. $Fe(CO)_5$, $Ni(CO)_4$), pyrophoric, or otherwise hazardous or expensive precursors. Therefore, the use of laser pyrolysis for producing metal oxides, including spinels, remains relatively rare.

#### 5.2.1.5 Plasma Methods

Although plasma is often characterized as a fourth state of matter, distinct from vapor, solution, and solid phases, here we classify it in the vapor-phase category due to the vaporization processes driven by the plasma. With plasma processing, one can contrast low-pressure non-thermal plasmas, in which the electrons are at much higher effective temperature than the gas, and higher-pressure thermal plasmas, in which electrons, ions, and neutral molecules are all heated to extremely high temperature. Figure 5.2b shows a process for synthesizing $LiMn_2O_4$ by a low-pressure non-thermal plasma method [20]. In contrast to the more common approach of flowing a precursor through the reactor, this example exposes the solid precursors to the low-pressure plasma to convert them to solid nanoparticles. The small size of the product particles suggests that this involves vaporization of precursors as well as solid-state reactions. In this study, manganese dioxide was prepared from $KMnO_4$ and $MnCl_2 \cdot 4H_2O$ and mixed with lithium hydroxide by ball milling. The mixture was loaded into the tube furnace, where it was both heated to 500 °C and exposed to a plasma at a frequency of 13.56 MHz and power of 200 W. With 3 sccm flow rate of oxygen through the reactor and 66 Pa pressure in the tube, spinel $LiMn_2O_4$ nanoparticles were obtained after 30 minutes of plasma reaction. The material was smaller and much more uniform in size compared with $LiMn_2O_4$ prepared by thermal treatment alone, at higher temperature and for longer time. Therefore, the novel plasma methods produced superior $LiMn_2O_4$ for use as a cathode for lithium-ion batteries while saving energy by lowering process time and temperature.

High-pressure thermal plasma synthesis, in which electrons, ions, and neutral species are all heated to several thousand kelvin, can also be applied to produce spinel nanomaterials. Son et al. [23] employed a very simple route to obtain $NiFe_2O_4$ powders using an RF plasma system with a 50 kW to 3 MHz power supply, with NiFe permalloy powder or mixed Fe and Ni powders as metal precursors and air as the oxygen source. This robust route provides a scalable means to produce magnetic ferrites with sizes of 20–30 nm. The nickel ferrite from this process exhibited magnetization of up to 56 emu/g, decreasing with increasing nickel content, and a Neel temperature near 590 °C, almost identical to that of bulk $Fe_3O_4$ and $NiFe_2O_4$. However, this simple approach also has limitations, such as generation of by-products of unreacted Ni or Fe, and $Fe_3O_4$ along with $NiFe_2O_4$.

### 5.2.2 Solution Phase

#### 5.2.2.1 Sol–Gel Methods

Sol–gel methods are often used to produce nanoscale ceramic and glass materials [24]. Phase separation during sol–gel processing has been widely studied in recent years as a means of generating pores of controlled size and structure [25, 26]. For example, hierarchically porous $ZnAl_2O_4$ monoliths (Figure 5.3a) were prepared through a sol–gel process accompanied by phase separation, using zinc chloride and aluminum chloride as precursors dissolved in the polar solvents to form the starting sol [27]. Then, phase separation was induced by introducing polyethylene glycol to control the macrostructure formation. After aging the gel at 45 °C for 72 hours, the products were solvent exchanged and dried. Bi-continuous skeletons of $ZnAl_2O_4$ spinel monolith (with both continuous pore volume and continuous solid skeleton) were obtained after 600 °C heat treatment to achieve a high surface area of 109 $m^2\,g^{-1}$ with a pore size of around 8 nm.

#### 5.2.2.2 Hydrothermal Methods

Hydrothermal methods usually use Teflon-lined autoclaves to contain the high pressure generated by heating water above its nominal boiling point, usually operating at 100–200 °C [29]. Figure 5.3b shows a process through which magnetic iron oxide nanoparticles are prepared via a low-temperature hydrothermal method [28]. $FeSO_4 \cdot 7H_2O$ and $Fe_2(SO_4)_3 \cdot nH_2O$ solutions in a 1 : 2 ratio were mixed to reach a proper ratio to prepare the spinel iron oxide products. The mixture pH was tuned to ~10 before it was transferred into an autoclave and reacted for 18 hours. The hydrothermal methods often require relatively long reaction times and provide little ability to monitor the process. Nonetheless, they can be used to form hierarchical or hollow crystal structures, because the slow reaction process allows for kinetically controlled growth that favors or suppresses reaction on particular crystal facets. After the autoclave reaction, the obtained products usually require no more post-treatment. As in most solution-phase synthesis, the particles can be washed and separated by centrifugation.

#### 5.2.2.3 Thermal Decomposition

Thermal decomposition (or thermolysis) is a sub-category of solvothermal methods and can also occur as a solid-phase reaction. Precursor decomposition occurs when the temperature is high enough to provide the energy needed to break the chemical bonds of the precursor and initiate formation of new chemical compounds [30].

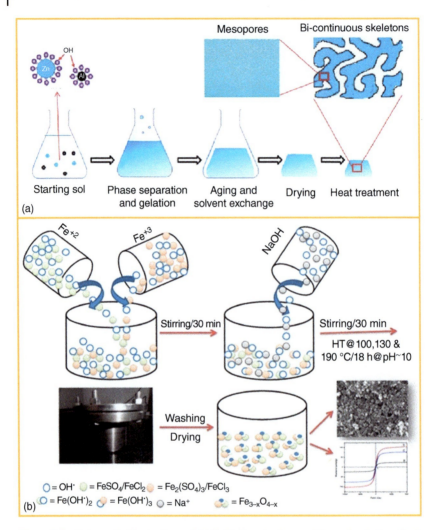

**Figure 5.3** Schematic illustrations of (a) ZnAl$_2$O$_4$ spinel monolith synthesis from ionic precursors by sol–gel method with phase separation. Source: Reproduced with permission from Guo et al. [27], Elsevier. (b) Production of magnetic iron oxide nanomaterials via low-temperature hydrothermal method. Source: Reproduced with permission from Bhavani et al. [28], Elsevier.

This process usually requires high temperature, often around 300 °C and moderate reaction times [24]. The setup in Figure 5.4A shows a thermal decomposition method setup for the synthesis of uniformly sized nanorods [31]. This method provides a stable and well-controlled reaction system to produce uniformly sized particles by controlled injection of precursors. Figure 5.4B shows TEM images of uniform Mn$_x$Fe$_{1-x}$O nanoparticles containing some spinel phase Mn$_x$Fe$_{3-x}$O$_4$, which grow as concave nanocubes [32]. The thermal decomposition of iron acetylacetonate and manganese acetate provides the metal source, reacted in a mixture of oleic acid and oleylamine as both solvent and surfactants. The shapes and sizes can be controlled by tuning the reaction parameters such as reaction time, temperature, and amount of chemicals. This method is often employed to achieve such control of morphology. However, we note that these controlled morphologies, whether produced by thermolysis or other methods, are generally unstable with respect to conversion to a sphere or low-surface-area polyhedron, particularly upon heating or transfer to water for biological applications. Therefore, more studies should focus on how to improve the stability of these spinel materials with specific morphology in future work.

### 5.2.2.4 Solvothermal Methods

Solvothermal synthesis is a wet chemical process that shares many similarities to hydrothermal methods and thermolysis, but it uses a non-aqueous solvent such

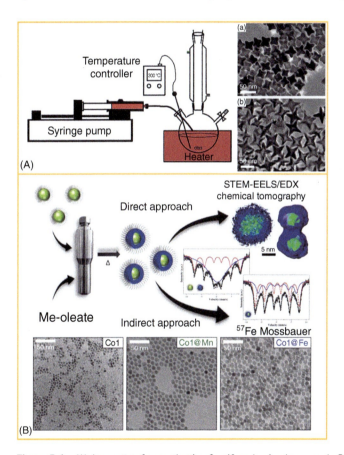

**Figure 5.4** (A) Apparatus for synthesis of uniformly sized nanorods Source: Adapted with permission from Park et al. [31], American Chemical Society) and TEM images of $Mn_xFe_{1-x}O$ concave nanocubes with different morphologies and sizes. Source: Reproduced with permission from Li et al. [32], Royal Society of Chemistry. (B) Formation of core–shell nanostructure spinel ferrite with a versatile solvothermal seed-mediated growth approach and corresponding TEM bright-field images. Source: Adapted with permission from Sanna Angotzi et al. [33], American Chemical Society.

as ethylene glycol, glycerol, or diethylene glycol [34, 35]. Figure 5.4B illustrates an example in which metal oleate precursors were used to prepare core–shell hydrophobic bimagnetic spinel cobalt ferrite and manganese ferrite nanoparticles $CoFe_2O_4$ and $MnFe_2O_4$ [33] via a low-cost and low-temperature solvothermal seed-mediated growth approach. Here, the oleate ions produced by decomposition of the precursors remains available to act as a capping ligand on the nanoparticles. This required preparing oleate precursors from their salts. The mixed Co(II) or Mn(II) and Fe(III) oleate precursors were dissolved in pentanol and used to prepare $CoFe_2O_4$ and $MnFe_2O_4$ and collected as seeds [36]. The mixture was reacted in a stainless-steel autoclave at 180–220 °C for 10 hours to produce different sizes of spinel ferrites. These seeds were then subjected to a second solvothermal growth with addition of Fe(II), Co(II), or Mn(II), and Fe(III) oleates to grow a shell of different composition on the seed particles (e.g. $CoFe_2O_4$ core with $Fe_3O_4$ or $MnFe_2O_4$ shell or $MnFe_2O_4$ core with $Fe_3O_4$ or $CoFe_2O_4$ shell). This versatile solvothermal method thus allowed production of a variety of core–shell nanostructures, allowing exploration of their magnetic properties and also suggesting this approach could be broadly applicable for magnetic and non-magnetic heterostructured nanoparticles.

### 5.2.3 Solid Phase

#### 5.2.3.1 Solid-State Thermal Decomposition

Solid-state thermal decomposition is a process that is similar to solution-phase thermal decomposition but occurring entirely in the solid state, generally without a solvent. This simplest approach uses nitrate metal salts as precursors [6]. The metal activity series (K > Na > Ca > Mg > Al > Zn > Fe > Sn > Pb > H > Cu > Ag) suggests some general principles: metals between Mg and Cu can produce metal oxides through the thermal decomposition from metal nitrates. However, metals before Mg or after Cu will remain as metal nitrates or be reduced to pure metal, respectively, upon heating to temperatures typical of thermal decomposition routes [8].

The Pechini method is a somewhat more sophisticated thermal decomposition approach, in which a precursor is prepared using citric acid to chelate metal ions and ethylene glycol or other polyol to link the citric acid moieties in a polyester structure [37]. This initial formation of the metal-chelating polyester structure is sometimes called a sol–gel route, because the citric acid–ethylene glycol polymerization leads to gelation. However, it is only an intermediate product in the synthesis. Thermal decomposition of this precursor followed by calcination can be used to prepare complex spinels with composition based on the loading of metals in the initial mixture. For example, $Co_3O_4$-coated Ni powder [38] was prepared by the Pechini method as illustrated in Figure 5.5a. Cobalt acetate was combined with citric acid and ethylene glycol in water at pH 8, and Ni powder was added. Heating at 80 °C was used to evaporate the solvent, and then the final product was obtained after 350 °C calcination for three hours to remove all the decomposable materials and organic substrate. Pechini method-synthesized spinel materials are employed

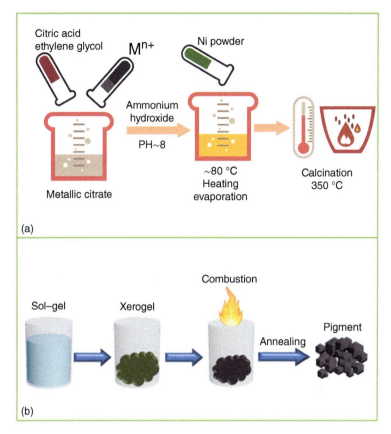

**Figure 5.5** Schematic illustrations of (a) procedure for preparing $Co_3O_4$-coated Ni powder with Pechini method [38]. Source: Zheng Fu (b) Employing sol–gel self-combustion method to fabricate $Cu_{1.5}Mn_{1.5}O_4$ spinel ceramic pigment-based TSSS paint coatings. Source: Reproduced with permission from Ma et al. [39], Royal Society of Chemistry.

as cathode materials, in applications ranging from molten carbonate fuel cells [40] to lithium-ion batteries [41], where their stability has been an advantage. Compared to solution-phase thermal decomposition, the solid-state reaction without solvents can produce more compact and stable spinel materials, more compatible for use in these energy storage applications.

### 5.2.3.2 Combustion

Combustion synthesis is a furnace-free and low-cost method to prepare oxides materials via exothermic combustion reactions [42], which can provide higher purity and smaller grain sizes than some other traditional methods[39]. Figure 5.5b shows a sol–gel self-combustion method to prepare $Cu_{1.5}Mn_{1.5}O_4$ spinel ceramic pigment [39] based on the following equation:

$$27Cu(NO_3)_2 \cdot 3H_2O + 27Mn(NO_3)_2 + 28\,C_6H_8O_7 \cdot H_2$$
$$\xrightarrow{NH_3 \cdot H_2O} 18Cu_{1.5}Mn_{1.5}O_4 + 54N_2 \uparrow + 168CO_2 \uparrow + 221H_2O$$

Copper nitrate trihydrate $Cu(NO_3)_2 \cdot 3H_2O$ and manganese nitrate $Mn(NO_3)_2$ as precursors were dissolved in water. Then citric acid was added to chelate $Cu^{2+}$ and $Mn^{2+}$, and low-molecular-weight polyethylene glycol was added to the solution as an esterifying agent, just as in the Pechini method described earlier. The mixed solution was heated and stirred to form a highly viscous gel, and then continuous heating produced Cu–Mn–citric xerogel. This xerogel was ignited in air using ethanol as initiating agent to induce the self-combustion with rapid evolution of fume, producing black powders. After annealing at 500–900 °C for one hour, the black powders were converted to $Cu_{1.5}Mn_{1.5}O_4$ spinel ceramic pigments. In this process, the annealing part also contributes to the formation of $Cu_{1.5}Mn_{1.5}O_4$ spinel materials, since the self-combustion yields both $Cu_{1.5}Mn_{1.5}O_4$ and impurity phases, which are totally transformed to the pure spinel after the annealing. This illustrates the drawback that the combustion method often produces by-products, whose removal or conversion requires additional post-treatment.

#### 5.2.3.3 Ball Milling

Ball milling as a simple grinding method is broadly used during pre- or post-treatment of materials from solution or solid-phase synthesis. However, it can also be used to drive solid-phase reactions, and here we will focus on the ball milling method to drive formation of spinel materials. The ball milling process can create oxygen defects in metal oxides, which in turn can induce lattice deformation [43]. This often improves the cathode performance of ball milling-made spinels.

Figure 5.6a shows a schematic mechanism of $LiNi_{0.5}Mn_{1.5}O_4$ (LNMO) spinel prepared by ball milling. $Li_2CO_3$, NiO, and $MnCO_3$ were combined by ball milling solid-state reaction for 16 h. The dried powders were annealed at 750 °C for 12 hours in air. Then, the ball milling process and annealing were repeated twice to yield an LNMO powder. The second round of grinding was carried out using a $ZrO_2$-lined pot and balls with LNMO powder for 6–120 hours to obtain the final LNMO spinel products. This ball milling method induced lattice deformation and introduced oxygen defects with partial $Mn^{4+}$ to increase the cathode performance. The lattice deformation within the spinel crystal should be considered as a key parameter to improve spinel cathode performances. Therefore, the relationship between ball milling and lattice deformation should be explored more deeply in future research.

#### 5.2.3.4 High-Temperature Solid Solution Method

The solid solution method is a process in which two or more solids are combined at high temperature to form a new solid material after cooling. In the following example (Figure 5.6b), microwave heating was used instead of conventional high-temperature heating, [44] and a triblock copolymer was used to template pores in the final material. Nonetheless, the spinel formation occurred via solid-state reactions. Transition metal sources including nitrates of $Ni^{2+}$, $Co^{2+}$, or $Cu^{2+}$, and a triblock copolymer were dissolved in ethanol, mixed, and stirred for four hours. Meanwhile, boehmite nanoparticles as Al source and nitric acid were mixed and heated via microwave irradiation at 70 °C for one hour. After cooling, the obtained peptized-alumina gel was mixed with the other solution and stirred for four hours.

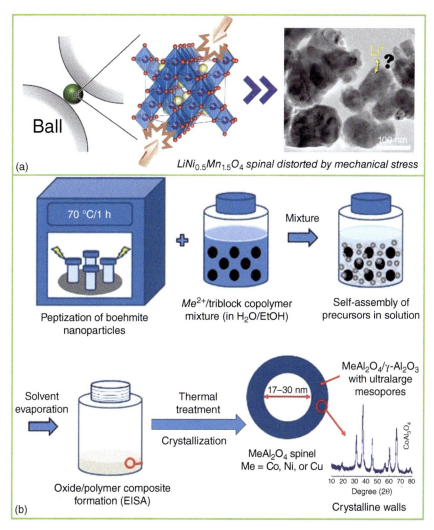

**Figure 5.6** Schematic illustrations of (a) LiNi$_{0.5}$Mn$_{1.5}$O$_4$ spinel cathode by ball milling. Source: Reproduced with permission from Kozawa [43], Elsevier. (b) The microwave-assisted synthesis to create highly porous and crystalline MeAl$_2$O$_4$ (Me = Ni, Co or Cu) supported on γ-Al$_2$O$_3$. Source: Gonçalves et al. [44], American Chemical Society.

The mixtures were then evaporated for 96 hours. Finally, calcination was carried out at 400–1100 °C to yield the final MeAl$_2$O$_4$ (Me = Ni, Co, or Cu) spinel solids. The spinel formation can be envisioned as occurring via the following processes:

$$4Al_2O_3 + 3Me^{2+} \rightarrow MeAl_2O_4 + 2Al^{3+}$$

$$2Al^{3+} + 4MeO \rightarrow MeAl_2O_4 + 3Me^{2+}$$

$$Al_2O_3 + MeO \rightarrow MeAl_2O_4$$

The multi-step process with block copolymer templating of pores allowed production of a high surface area material. Moreover, the products showed a decrease in the degree of sintering due to their high thermomechanical resistance, maintaining high surface area and large pore sizes after calcination to achieve high crystallinity.

## 5.3 Structure–Effect Applications

Morphology control of nanomaterials has long been of interest and has been a driving force for exploring growth mechanisms of crystallization. In this topic, we will discuss the various structures of spinel materials that have been reported, how they are related, and how their structures specifically relate to their application. In each case, applications derive from, and depend upon, the chemical and physical properties of materials, and these properties are determined by structure, including size, shape, and composition. At the broadest level, the morphology of nanocrystals is classified by their dimensionality. However, in many cases, only three types of structure (1D, 2D, and 3D) are discussed and reviewed, which is insufficient to describe the many reported complex spinel nanostructures. Therefore, here we classify the structure of spinel materials more finely into the following kinds (Figure 5.7): 1D (nanorods, nanowires, nanotubes); 2D (nanofilm, nanosheets, nanoplatelets); 3D (polyhedra, core–shell structure, nanoframes); 1 and 2D (textile interwoven with nanowires); 1 and 3D (nanoflowers with long-arm branches); 2 and 3D (nanoflowers comprising nanoflakes); and self-assembly-derived structures (built from spheres, cubes, or octahedra), to accurately convey the relationships between the nanostructure and corresponding applications of the spinel nanomaterials. Figure 5.7 provides a schematic overview illustrating these classifications and their interrelations.

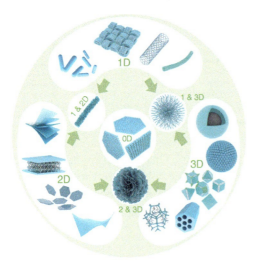

**Figure 5.7** Morphology evolution relationships among 1D, 2D, 3D, 1&2 D, 1&3 D, 2&3 D, and self-assembly-derived structures. Source: Zheng Fu.

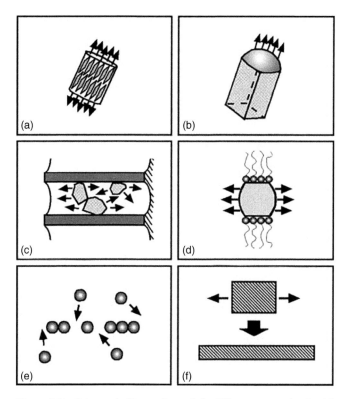

**Figure 5.8** Schematic illustrations of six different strategies for 1D growth: (a) dictation by anisotropic crystal structure, (b) vapor–liquid–solid process, (c) growth controlled by a template, (d) kinetic control via a capping reagent, (e) oriented self-assembly of monodisperse particles forming 1D structure, and (f) size reduction. Source: Reproduced with permission from Xia et al. [45], John Wiley & Sons.

### 5.3.1 One-Dimensional (1D) Structures

1D nanomaterials can be classified into three rough categories: nanorods, nanowires, and nanotubes. The most common strategies of preparing 1D nanostructures include approaches [45] for controlling the nucleation and growth of nanostructures to control their dimensions. Each of these strategies is grounded in the synthesis methods described earlier. Similar strategies are also employed to grow materials of other morphologies, with the outcome determined by specific factors, such as the crystal structure of the material or specific interactions that promote or inhibit growth in specific crystallographic directions (Figure 5.8).

#### 5.3.1.1 Nanorods

Nanorods as a well-developed crystal morphology are defined by their columnar shape with a shorter length (smaller aspect ratio) than nanowires, although there is no universally accepted length or aspect ratio that distinguishes nanorods and nanowires. Nanorod morphology control has been studied and discussed most widely in the field of pure noble metal nanocrystals, such as silver [46, 47], gold [48], platinum [49], and palladium [50] nanorods. Herein, we consider some examples of spinel nanorods to clarify the difference.

In Figure 5.9A shows $LiMn_2O_4$ (LMO) nanorods produced for use in a lithium-ion battery cathode [51]. These were prepared by a combined hydrothermal and solid-state method that yielded nanorods with an average diameter of 90 nm and length of 1.5 μm. $Li_xCoO_2$ has dominated the commercial lithium-ion battery cathode material market for many years [53]. However, the high cost, toxicity, and finite productivity of cobalt provide a major driving force for reduction or elimination of its use. Therefore, this LMO spinel as a promising cathode candidate with lower cost and better scalability is of great interest. In the cited study, the nanorod morphology played an essential role to enhance the rate capability of the spinel cathode. In comparison with commercial LMO, the LMO nanorods provided a much higher charge capacity and better cyclability. Capacity enhancement is often attributed to the large surface-to-volume ratio of nanorods. Also, selected area electron diffraction of commercial LMO and LMO nanorods, in a TEM, showed that nanorods remained single crystalline, and their cubic cell parameter decreased, during delithiation of the cubic spinel [54], which facilitates the fast kinetics of battery charging and discharging. This study illustrates one way that morphology can affect the properties from a microscopic perspective. Here, the single-crystalline structures are able to accommodate volume changes via a change in lattice parameter while maintaining their crystal structure, rather than by fragmenting or generating defects.

The $CoFe_2O_4$ (CFO) nanorod-containing nanocomposite shown in Figure 5.9B was prepared using a pulsed laser deposition process with alternating targets of CFO and $SrRuO_3$ (SRO) [55]. Phase separation led the spinel CFO to grow as vertical columns while the perovskite SRO filled in the space between columns. This strategy produced composite oxides with high interface-to-volume ratio and strong interfacial coupling [56]. This CFO spinel with nanopillar structure in the matrix can be classified in the nanorods category here. In the composite structure, large magnetoresistance (MR) is achieved by combining a magnetic material and a conductive material. Characterizing this spinel composite revealed that a large decrease in dimensions of the CFO lattice was induced by interfacial lattice coupling with the matrix, leading to large coercivity. Furthermore, EDS mapping revealed that the Co/Fe ratio in the CFO was ~0.77, higher than the value of 0.5 expected for stoichiometric CFO, suggesting escape of Fe ions from CFO to the SRO matrix. The doped $Fe^{3+}$ ions in SRO increased electron scattering [57], which enhanced the MR near the Curie temperature of SRO. The authors conclude that not only interfacial lattice coupling but also the Fe ions in the SRO helped to enhance the MR value. This is an example in which composition, morphology, and coupling of the spinel nanorods to another phase all played an important role in determining the property of interest.

Figure 5.9C shows a special example of 1D nanorods, comprising a large-scale microtube with nanorods branches on it. We include it here based upon the nanorod features, as the size of the microtube is too large to be considered a nanostructure. In Sections 5.3.4–5.3.6, we separately discuss specific multi-dimensional structures such as 1&2D, 1&3D, and 2&3D structures. This case could also be considered as

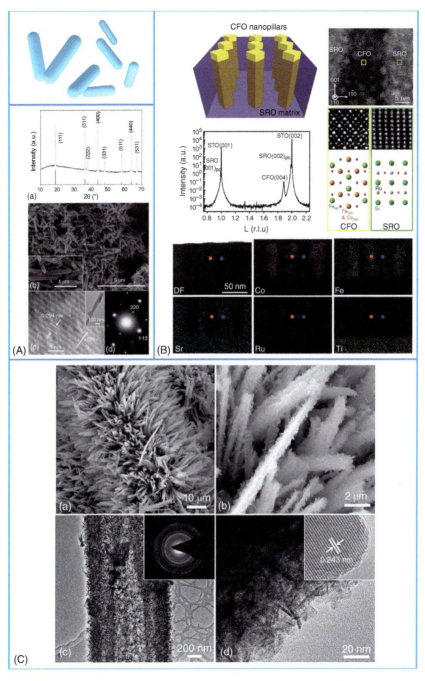

**Figure 5.9** (A) (a) XRD pattern, (b) SEM images, and (c) TEM images of LiMn$_2$O$_4$ nanorods. Source: Kim et al. [51] with permission from American Chemical Society. (B) A vertical heteroepitaxial nanocomposite of CFO nanopillars in an SRO matrix. Source: Reproduced with permission from Liu et al. [52], John Wiley & Sons. (C) SEM and TEM images, N$_2$ adsorption–desorption isotherm, and Co 2p fine XPS spectrum of Co$_3$O$_4$-MTA. Source: Reproduced with permission from Zhu et al. [3], John Wiley & Sons.

another 1&1D structure, where the 1D nanorods grows on the 1D microtube to form a hollow 3D hierarchical structure. The $Co_3O_4$ microtube arrays ($Co_3O_4$-MTA) were prepared by an electrochemical sacrificial-template method to create porosity [3], in which a single $CoHPO_4$ microrod with smooth surface and solid structure was subjected to anodic bias to drive restructuring to form this hollow MTA. The high surface area provides this spinel structure with many more electroactive sites than the smooth microtube, contributing to excellent electrocatalytic activity. Therefore, this spinel can be employed for both the hydrogen evolution reaction (HER) and oxygen evolution reaction (OER) with high catalytic activity. This simple and low-cost self-templating strategy produced a $Co_3O_4$ spinel with excellent electrocatalytic performance and may guide the design of other hollow hierarchical spinel electrode materials.

#### 5.3.1.2 Nanowires

Nanowires are characterized by nanoscale diameters and high aspect ratio and can be many microns in length, as shown in the top of Figure 5.10A. In many respects, they behave as wires and can even be bent, fabricated, or woven into various shapes. The bottom of Figure 5.10A shows highly crystalline Mn–Ni–Co ternary spinel oxide (MNCO) nanowires prepared via a facile hydrothermal method [58]. This type of ternary spinel with nanowire structure, when used as a positive electrode in a supercapacitor, delivered higher capacitance of $638\,F\,g^{-1}$ at $1\,A\,g^{-1}$ and higher capacitance retention of 93.6% after 6000 cycles compared to traditional transition-metal oxides, such as $Co_3O_4$ [61], NiO [62], and $MnO_2$ [63]. The partial replacement of Co in $Co_3O_4$ by Ni and Mn without changing the crystal structure can also lower production costs and have a positive effect on redox reactions to enhance electrocatalytic performance.

Figure 5.10B shows highly ordered mesoporous Co-containing spinel nanowires prepared using a hydrothermal method [59]. Similar to the previously mentioned nanowire case, this one also exhibits excellent CO oxidation performance due to doping with transition metals such as Cu, Fe, and Cr on the $Co^{2+}$ site of $Co_3O_4$. In this material, the $Co^{3+}$ on the catalyst surface provides the active site for CO oxidation, while $Co^{2+}$ is not active [64]. Moreover, the highly ordered mesoporous structure created by the self-aligned nanowires offered more electroactive sites than simpler structures.

Figure 5.10C illustrates another specific 1&1D structure with a hierarchical arrangement of nanowires. The mesoporous $NiCo_2O_4$ nanowire arrays (NWAs) grown on the carbon textile 1D microwire substrate were prepared via a surfactant-assisted hydrothermal method and annealing treatment [60]. As a LIB anode, it showed a reversible capacity of about $1012\,mAh\,g^{-1}$ at a current density of $0.5\,A\,g^{-1}$, retaining a capacity of $854\,mAh\,g^{-1}$ after 100 cycles.

#### 5.3.1.3 Nanotubes

Carbon nanotubes (CNTs) have been studied for many years, particularly following the landmark publication by Iijima that clearly identified them and explained what they were. In the years since, many varieties of CNTs, including single- and

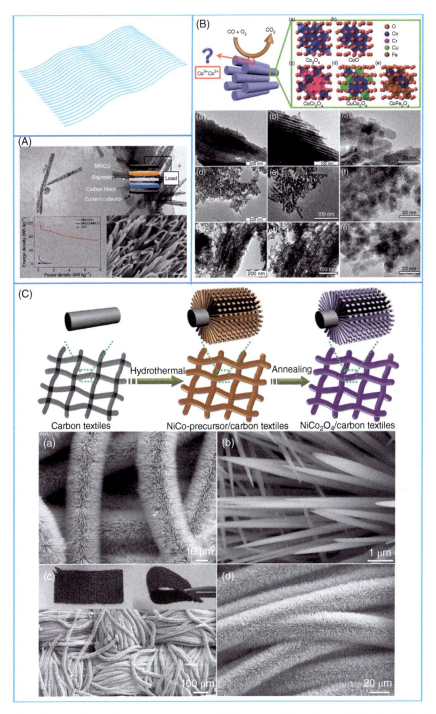

**Figure 5.10** (A) A schematic diagram and SEM images of aligned spinel Mn–Ni–Co ternary oxide nanowires. Source: Li et al. [58] with permission from American Chemical Society. (B) HR-TEM images of ordered mesoporous cobalt-based spinels. Source: Gu et al. [59] with permission from American Chemical Society. (C) A schematic illustration and SEM images of $NiCo_2O_4$ NWAs/carbon textiles composite. Source: Shen et al. [60] with permission of John Wiley & Sons, Inc.

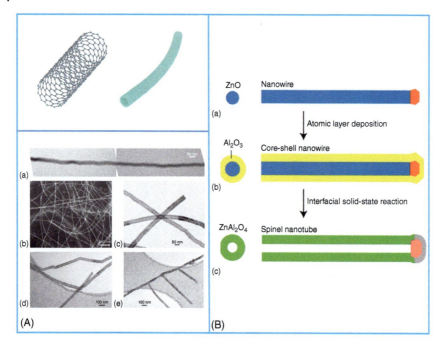

**Figure 5.11** Schematic diagram, SEM, and TEM images of $ZnAl_2O_4$ spinel nanotubes prepared by ALD coating of a ZnO nanowire with $Al_2O_3$ followed by solid-state interfacial reaction. Source: Reproduced from Jin fan et al. [9], Springer Nature.

multi-wall CNTs, helical CNTs, and branched CNTs, have been of interest for their electrical conductivity, physical strength, and thermal conductivity, allowing broad applications. In turn, CNTs have inspired the exploration of nanotubes of a wide variety of other materials. In particular spinels in nanotube structure are of interest based on the opportunities provided by separate interior and exterior surface, along with the other advantages of nanorods and nanowires already discussed.

In Figure 5.11, a $ZnAl_2O_4$ nanotube is generated by first using ALD to coat a single-crystal ZnO nanowire with $Al_2O_3$ layer, followed by interdiffusion and interfacial reaction of ZnO and $Al_2O_3$ to form $ZnAl_2O_4$, accompanied by the Kirkendall effect [9]. We have discussed this case in the ALD syntheses Section 5.2.1.2, and thus do not repeat the description of this clever synthesis. Compared with traditional methods, this robust ALD method provides a common strategy for preparing spinel nanotubes from transition metal oxide nanowires, constituting a systematic approach for fabricating a variety of materials with nanotube structure.

Figure 5.12 shows spinel materials grown on N-doped carbon nanotubes (NCNTs), which showed outstanding catalytic activity as air electrodes for the oxygen reduction reaction in zinc–air batteries [65]. As a hybrid catalyst, the $Co_2FeO_4$/NCNT spinel combines the electronic conductivity of CNT structure with the spinel high catalytic activity and surface area. Incorporation of $Fe^{3+}$ in the cobalt spinel was shown to enhance both catalytic activity and durability of the zinc–air batteries.

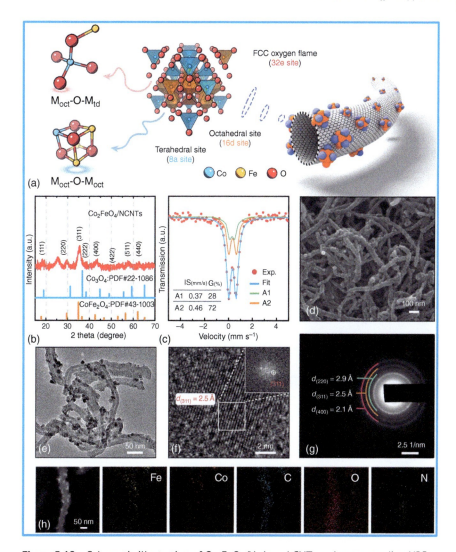

**Figure 5.12** Schematic illustration of $Co_2FeO_4$/N-doped CNTs and corresponding XRD patterns, Mossbauer spectra, SEM images, TEM images and SAED analysis, and elemental mapping. Source: Wang et al. [65] with permission of John Wiley & Sons, Inc.

Following a different approach, Figure 5.13 illustrates a combustion method to prepare mesoporous $MnCo_2O_4$ (MCO) and $CoMn_2O_4$ (CMO) spinel nanotubes [66] from electrospun polymer fibers loaded with cobalt and manganese precursors. The high-temperature calcination and exothermic oxidation of the polymer and metal precursors in air creates the mesoporous hollow structure of these spinel fibers, leading to a high density of active sites for $Li^+$ storage. The hollow fibers provided high lithium storage capability, with good cycling stability, and rate capability as anode materials in lithium-ion batteries. Beyond these examples, many other cases [67–69] show similar ability to enhance catalytic activity, lithium uptake capacity, and

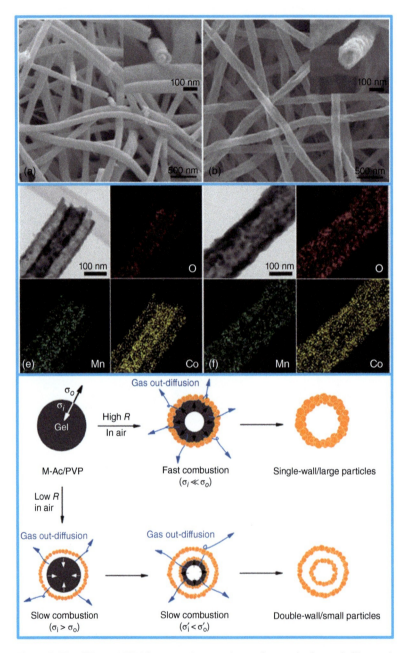

**Figure 5.13** SEM and TEM images, elemental mapping, and schematic illustration of formation of MCO nanotubes. Source: Hwang et al. [66] with permission of Royal Society of Chemistry.

similar performance measures by generating mesoporous hollow nanotube materials, and this approach will continue to be valuable for creating new energy storage materials.

In summary, 1D spinel nanomaterials including but not limited to nanorods, nanowires, and nanotubes have shown excellent electrocatalytic activities due to the inherent semiconductor properties and band structure of the specific spinel material employed as well as the ability to align or order these high-surface area structures to offer both abundant active sites and rapid transport of ions and charge carriers. Continued advances in the synthesis and applications of these 1D spinels can be expected to further impact energy storage and conversion applications.

### 5.3.2 Two-Dimensional (2D) Structures

#### 5.3.2.1 Nanofilms

Herein, we differentiate between nanofilms and nanosheets based on the synthesis methods and overall scale. We classify as nanofilms those materials fabricated by thin film deposition processes on planar or nonplanar substrates, including pulsed-spray evaporation CVD [70], ALD [71], direct pulsed-laser epitaxy [72], a combination of hydrothermal deposition and annealing [73, 74], and substrate transfer methods [75]. On the other hand, we regard as nanosheets materials produced by the direct formation of free-standing 2D materials without a substrate. Figure 5.14A shows the structure and properties of a magnetically soft epitaxial spinel NiZnAl-ferrite film prepared by pulsed laser epitaxy. This magnetoelastic thin film material can produce an excellent combination of low intrinsic damping and magnetoelastic coupling, leading to minimal extrinsic damping, soft magnetism, and large strain-induced easy-plane anisotropy [76], which was not observed in previous reports. Therefore, this work may promote both experimental and theoretical advances on the development of magnetic and spintronic spinel devices.

Figure 5.14B shows a diagram of a series of spinel ferrite $LiFe_5O_8$ (LFO) nanofilms, which were transferred onto flexible substrates after growth of an epitaxial $La_{0.67}Sr_{0.33}MnO_3$/LFO multilayer on a $SrTiO_3$ substrate [75]. In recent years, the properties such as high Curie temperature (950 K), low coercivity, and excellent thermal stability of LFO [77] have attracted attention to LFO. Here, authors successfully fabricated flexible single-crystalline LFO thin films through a simple modified transfer method. The obtained LFO thin films retained almost the same magnetic properties such as saturation magnetization, remanent magnetization, and coercivity as measured before transfer. In addition, bending tests illustrated the flexibility and stability of the transferred films, showing broader promise for application of this simple and useful transfer method to obtain single-crystal spinel thin films on flexible substrates.

#### 5.3.2.2 Nanosheets

2D nanosheets, free of any substrate, have been studied for multiple applications that benefit from their free-standing and flexible structure. Figure 5.15A shows a hierarchical nanostructure of Ni–Co oxide spinel [78], composed of secondary

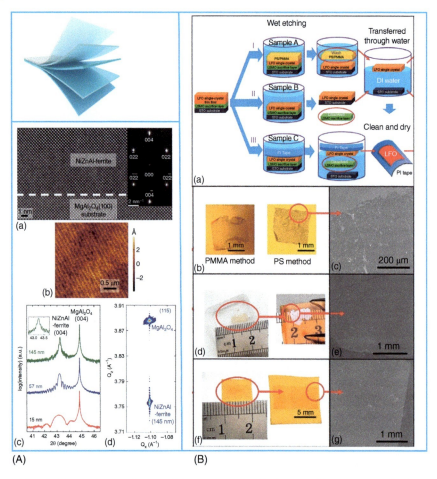

**Figure 5.14** (A) Structural properties of epitaxial NiZnAl-ferrite thin film: (a) HRTEM image and electron diffraction, (b) AFM scan, (c) XRD patterns, and (d) reciprocal lattice map of the ($\bar{1}\bar{1}5$) asymmetric diffraction. Source: Reproduced with permission from Emori et al. [72], John Wiley & Sons. (B) (a) Schematic diagram of three different transfer methods to obtain LFO films. (b, d, f) Transferred films fabricated through methods I, II, and III, respectively, showing the size of transferred films. (c, e, g) The corresponding SEM images, showing the microstructural features. Source: Reproduced with permission from Shen et al. [75], John Wiley & Sons.

nanosheets grown on primary nanosheets arrays, which can be classified to 2&2D structure according to our categories. These Ni–Co oxide hierarchical nanosheets (NCO-HNSs) were synthesized via a facile hydrothermal method followed by calcination. They provide high surface area and accessible electroactive sites due to the 2&2D nanostructure. Furthermore, in situ XANES and EXAFS data demonstrated that Ni cations were located in the tetrahedral site surrounded by coordinated oxygen molecules of spinel structure, while Co cations were in the octahedral site, allowing Ni conversion to NiOOH to decrease the overpotential and improve the OER performance, because $Ni^{3+}$ was more active than Co cations in OER.

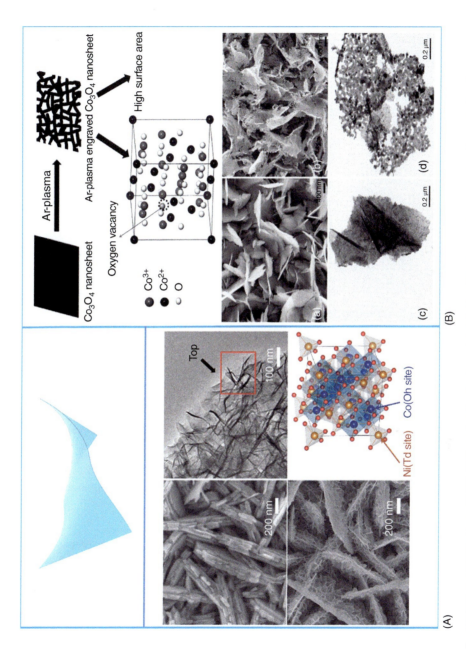

**Figure 5.15** (A) SEM and TEM images of NCO-HNSs and a structural model of NCO-HNSs. Source: Reproduced with permission from Wang et al. [78], John Wiley & Sons. (B) Illustration of the preparation strategy and SEM and TEM images of $Co_3O_4$ nanosheets. Source: Reproduced with permission from Xu et al. [79], John Wiley & Sons.

**162** | *5 Functional Spinel Oxide Nanomaterials: Tailored Synthesis and Applications*

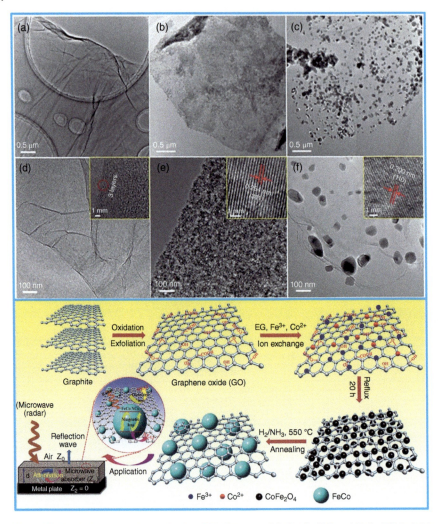

**Figure 5.16** Schematic illustration and TEM images of CoFe$_2$O$_4$/GO and FeCo/GN hybrids formation. Source: Reproduced with permission from Li et al. [80], Royal Society of Chemistry.

Similarly, the Co$_3$O$_4$ nanosheets in Figure 5.15B, which were prepared by a green and simple plasma-engraving method [79], also exhibit high surface area and good electrocatalytic performance for OER. However, this Co$_3$O$_4$ nanosheet material, without Ni$^{3+}$ at the tetrahedral site, improves the OER activity of Co$_3$O$_4$ through the presence of oxygen vacancies created by the plasma-engraving strategy.

CoFe$_2$O$_4$ in Figure 5.16 shows another application of spinel nanosheets based upon their microwave absorption properties [80]. It was prepared via a facile one-pot polyol method. Further conversion turned CoFe$_2$O$_4$/graphene oxides (GO) into FeCo/graphene nanosheet (GN) hybrids under $H_2/NH_3$ annealing. The final products showed enhanced microwave absorption properties due to the synergistic effect of high magnetocrystalline anisotropy of metallic FeCo and high conductivity of light-weight graphene. The reduction of $CoFe_2O_4$ on 2D GNs produces a metallic FeCo/GNs material, in which the spinel oxide served as an intermediate structure for creating the final metallic structure, in contrast to other cases considered earlier in which a metal alloy or other intermediate material was deposited and converted into a spinel final product. Both strategies are indeed valuable. In this particular study, the combination of magnetocrystalline anisotropy and high conductivity material provides another example of potential advantages of spinel-derived 2D materials, which are expected to have increasingly broad applications in energy conversion, lithium ion battery (LIBs), catalysts, and magnetic materials.

### 5.3.2.3 Nanoplatelets

Nanoplatelets have smaller aspect ratios (ratio of lateral dimensions to thickness) than nanosheets but still have lateral dimensions that are much greater than their thickness (aspect ratio greater than one). They often adopt specific shapes, such as triangles, rectangles, hexagons, and circles, depending on the crystalline growth direction. Unlike nanofilms and nanosheets, nanoplatelets usually have properties that depend upon the crystal facets corresponding to their faces and edges, which are determined by the crystal growth directions. The hexagonal $Co_3O_4$ nanoplatelets with exposed (112) facets shown in Figure 5.17A were synthesized by a calcination method [81]. They exhibited high activity for photocatalytic $CO_2$ reduction and were able to enhance efficient solar-driven $CO_2$ conversion with a visible-light photosensitizer. The corresponding selective area electron diffraction (SAED) pattern and HRTEM image (Figure 5.17A) reveal that (112) facets are the dominant exposed surfaces of the hexagonal platelets. Furthermore, density-functional theory (DFT) calculations supported the crystal facet dependence of photocatalytic performance for $CO_2$ reduction on $Co_3O_4$.

Figure 5.17B presents a type of core–ring-structured $NiCo_2O_4$ nanoplatelets that were synthesized by a coprecipitation hydroxide decomposition method [82]. The mechanism of core–ring formation was explained based on the evolution of the spinel crystal structure, through which the hexagonal ring formed a more compact spinel, while the core was dominated by pure Co. This specific structure showed remarkable performance for electrocatalysis of OER in alkaline media due to the large surface area of the core–ring nanostructure and the enrichment of Co atoms provided by the core part as the active sites.

Nanoplatelets can also be grown on a target substrate, by methods like those used to grow nanofilms, but without forming a continuous film, which we classify as nanoislands. For example, a cobalt ferrite in Figure 5.18, $Co_xFe_{3-x}O_4$ (CFO), was synthesized by reactive molecular beam epitaxy (MBE) on Ru (0001) as the substrate [83]. Low-energy electron microscopy (LEEM) and photoemission electron microscopy (PEEM) allowed elucidation of a comprehensive in situ growth mechanism. The produced CFO nanoislands exhibited magnetic domains several

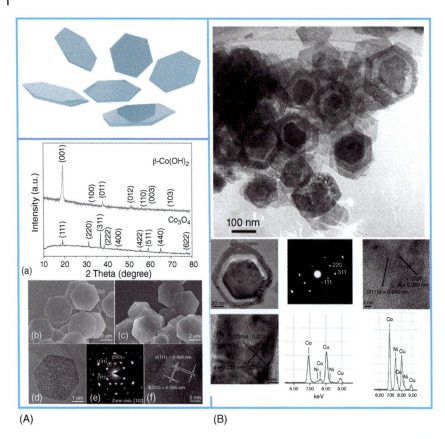

**Figure 5.17** (A) XRD patterns of β-Co(OH)$_2$ and Co$_3$O$_4$ hexagonal platelets, SEM and TEM images and SAED pattern of Co$_3$O$_4$ hexagonal platelets. Source: Reproduced with permission from Cui et al. [81], John Wiley & Sons. (B) TEM and HRTEM images and SAED pattern of core–ring NiCo$_2$O$_4$ nanoplatelets and EDS analysis of the core part and the ring part, respectively. Source: Reproduced with permission Gao et al. [82], John Wiley & Sons.

orders of magnitude wider than previous ultrathin spinel materials. Furthermore, the combination of MBE and in situ electron microscopies allowed the authors to tune the ultrathin functional spinel structure by controlling the magnetic domain size. These results can be expanded into related fields such as spintronic and nanoelectronic devices. Various combinations of the spinel nanoplatelets grown on pure metals (Ru [83], Au [84], Pt [85], and Pd [86]) provide many possibilities for these 2D nanoisland materials.

In summary, 2D materials with flexible and free-standing structure provide high surface area to expose more active sites for multiple catalytic reactions. The electrical, optical, and mechanical properties of 2D spinel materials also allow many potential applications, such as electronic and optoelectronic devices, energy storage, and sensors. Therefore, we still have a long way to go to explore the 2D spinel materials in the future.

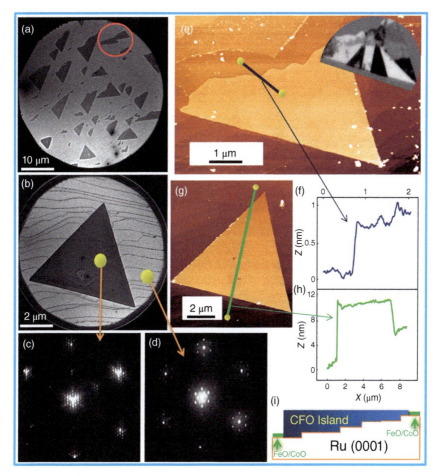

**Figure 5.18** (a) and (b) LEEM images of triangular CFO nanoislands, (c) and (d) selected area LEED patterns, (e) and (g) AFM images, and (i) a sketch of a cross-sectional view of CFO islands. Source: Reproduced with permission from Martín-García et al. [83], John Wiley & Sons.

### 5.3.3 Three-Dimensional (3D) Structures

3D spinel nanostructures have comparable size in all three dimensions, usually but not always involving hierarchical structures, which allows 3D materials to comprise a much larger family than 1D or 2D categories. Here, we present some representative examples to discuss the relations between the spinel material applications and 3D structure. Figure 5.19A shows multiple types of octahedral and cubic 3D magnetite ($Fe_3O_4$) nanocrystals (MNCs) that were synthesized using a simple one-pot solution-phase method with rational control of growth of different facets [87]. The growth mechanism was fully studied to enable a complete system for preparing specific assemblies of these monodisperse MNCs. Here, the polarity of different solvents was found to effectively control the average size of MNCs, while varying concentrations of specific additives and ligands enabled efficient tuning of the morphology of

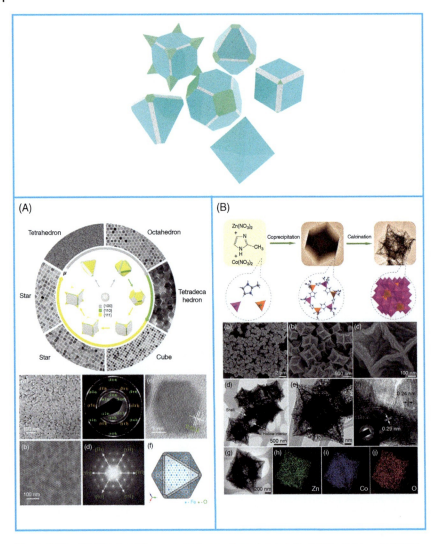

**Figure 5.19** (A) Sketch of the proposed growth mechanism for MNCs; TEM and HRTEM images and SAED patterns of MNCs. Source: Qiao et al. [87] with permission from American Chemical Society. (B) Schematic illustration, SEM, and TEM images of $Zn_xCo_{3-x}O_4$ hollow polyhedra. Source: Reproduced with permission from Wu et al. [88], American Chemical Society.

MNCs to yield cubes, stars, tetrahedra, octahedra, and tetradecahedra. The morphology of such 3D spinel materials is controlled by promotion or suppression of growth of specific crystal facets. For example, as shown as the schematic illustration, when monomer activity is high enough for deposition on (110) and (111) facets, but not on (100) facets, the octahedron grows to a tetradecahedral shape. The monomer deposition on different planes can simultaneously control the morphology, because it is governed by the order of chemical potentials of the facets and the monomer activity relative to those chemical potentials [89]. Similar cases occur in the preceding example of nanoplatelets, in which the (112) facets were controlled to produce 2D $Co_3O_4$ hexagonal platelets [81].

Figure 5.19B shows another polyhedral spinel material $Zn_xCo_{3-x}O_4$, which was prepared via a thermal decomposition process from a metal–organic framework (MOF) precursor, the cobalt-doped zeolitic imidazolate framework (ZIF)-8 [88]. The ZIF-8 template inherently grows with dodecahedral morphology based on its crystal structure. Thermal decomposition of ZIF-8 converts the dodecahedral particles to hollow and porous core–shell structures. The porous hollow dodecahedra exhibit excellent electrochemical performance with high reversible capacity and cycling stability for lithium uptake and removal. All these mesopores can contribute to provide high surface area for electrochemical reactions, combined with the ability to accommodate volume expansion during the $Li^+$ insertion or extraction processes.

Many spinel materials can be produced as 3D nanostructures by using a template to establish the 3D structure. Figure 5.20 illustrates SBA-15 and KIT-6 templating of $MnCo_2O_4$ [90]. In this approach, a soft template (self-assembled surfactant) creates a porous silica structure that serves as a hard template for growth of the spinel material. Removal of the hard template leaves the 3D spinel nanostructure. These templated materials provide remarkably high active surface area, improving catalytic performance. $MnCo_2O_4$ templated with KIT-6 and SBA-15 exhibited excellent performance as a low-temperature $NH_3$-SCR catalyst due to not only its high surface area but also its highly ordered mesopores. Following these template-based methods, many opportunities can be explored for producing and applying 3D structure spinel materials in the rational design and optimization of electrocatalysts and other functional materials that benefit from high surface area and ordered porosity.

Another type of 3D structure is a spherical hierarchical morphology. Unlike dense monodisperse spherical particles, these 3D spheres exhibit a more complex internal structure. For example, the micro-/nanospheres of Co-B nanoflakes in $ZnCo_2O_4$ (ZCO/Co-B) (Figure 5.21A) are composed of many small ZCO spherical nanocrystals with the Co-B nanoflakes as bridges connecting the ZCOs [91]. This material was produced by a simple in situ solution growth method at room temperature. As part of an electrode, ZCO/Co-B showed a superior lithium storage capacity of 843 mAh $g^{-1}$ at a high current density of 5 A $g^{-1}$. In this particle, the Co-B flakes not only act as bridges to reinforce the ZCO structure but also promote fast electron transport. DFT calculations showed that the Co–B nanoflakes can lower the lithium-ion diffusion energy barrier, thereby accelerating the $Li^+$ transfer kinetics.

Another important and widely studied 3D spherical structure of spinels is the core–shell structure. Semiconductor core–shell structures in the field of quantum dots are usually created to enhance photoluminescence quantum yield by passivating the surface trap states [93]. However, for spinel materials, the core–shell structure is usually created to enhance the surface area [9], magnetic properties [33, 92, 94], or energy storage performance [95]. Figure 5.21B shows a three-layer core–shell structure (core@shell@shell [CSS] structure), produced using a three-step thermal decomposition process [92]. In this case, both the core and outer shell comprise $Fe_{3-\delta}O_4$ spinel and an intermediate Co-doped ferrite shell is situated between them. The three steps were growth of $Fe_3O_4$

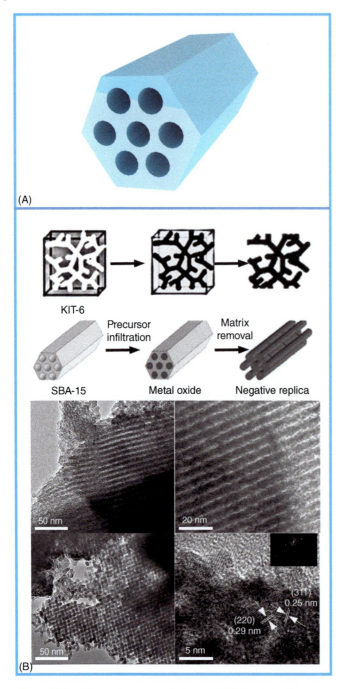

**Figure 5.20** Schematic representation and TEM images of mesoporous $MnCo_2O_4$ catalysts produced by templating with KIT-6 and SBA-15 forms of porous silica. Source: Reproduced with permission from Qiu et al. [90], Royal Society of Chemistry.

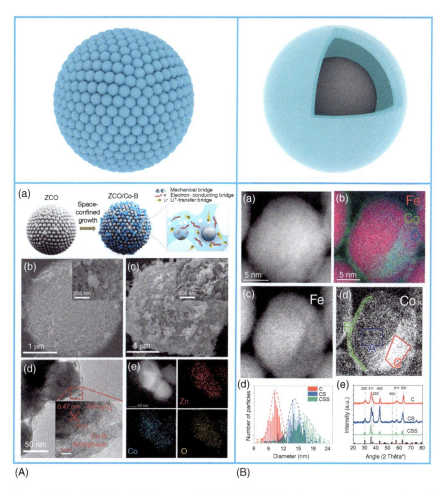

**Figure 5.21** (A) Schematic diagram of ZCO/Co-B, SEM, TEM images, and EDS mapping of ZCO/Co-B. Source: Reproduced with permission from Deng et al. [91], John Wiley & Sons. (B) Dark field STEM-HAADF micrograph and XRD patterns of CSS nanoparticles. Source: Adapted with permission from Sartori et al. [92], American Chemical Society.

cores, seed-mediated growth of a CoO shell on the cores (producing core–shell, or CS structures), and seed-mediated growth of a $Fe_3O_4$ outer shell, which was accompanied by conversion of the CoO layer to cobalt ferrite. Ultimately, the CSS structure, with varied composition but the same spinel crystal structure increased the magnetic blocking temperature and coercivity. This study provided a deeper understanding of the possibility of tuning magnetic properties by controlling the core–shell morphology of spinels.

In summary, the 3D spinel nanostructures usually grow in the following ways: (i) hierarchical morphology due to the growth only in certain unconfined directions; (ii) the connection of smaller nanoparticles with various shapes into a larger 3D structure; (iii) nanostructures grown or deposited on or in a 3D template; or (iv)

**170** | *5 Functional Spinel Oxide Nanomaterials: Tailored Synthesis and Applications*

multi-step growth to produce 3D variation in composition. The extraordinary flexibility and diversity of 3D structures expands the realm of properties and applications accessible with these spinel metal oxides.

### 5.3.4 One- and Two-Dimensional (1&2D) Structure

Aiming to improve understanding of the relationship between nanostructures of spinel materials and their applications, we further classify spinel structures here, based on combinations of their structural features, including 1&2D, 1&3D, and 2&3D structures.

1&2D structures include the following morphologies: (i) 1D nanowires arranged into 2D nanosheets or textile-like structures (Figure 5.22) [73]; (ii) 2D ultrathin nanosheets grown onto 1D nanowires, as shown Figure 5.23 [96, 97]; and (iii) 1D

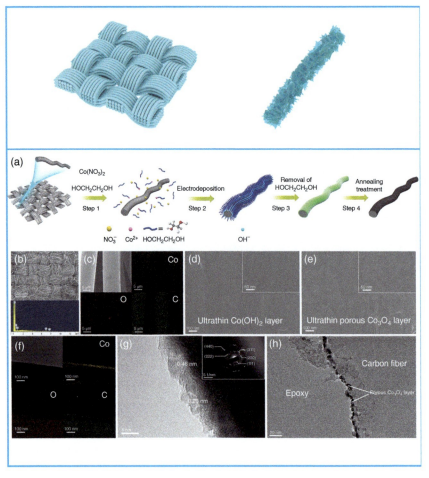

**Figure 5.22** Schematic diagram of ultrathin $Co_3O_4$/CC and the corresponding SEM and TEM images. Source: Reproduced with permission from Chen et al. [73], John Wiley & Sons.

nanowires grown on 2D nanoplatelets or substrates [98]. For the first case, we take as an example mesoporous $Co_3O_4$ layers grown on the surface of carbon cloth [73], in which the 2D nanofilms with large interfacial contact area are supported by 1D microscale carbon fibers. This $Co_3O_4$ nanofilm material exhibits remarkable electrocatalytic performance for both oxygen reduction reaction (ORR) and OER, ten times higher than the activity of a commercial $Co_3O_4$ electrode. When used in a flexible Zn-air battery, it exhibited high mechanical stability and operated without obvious performance loss under severe deformation, based on which, a prototype of an integrated device was demonstrated. This device can work under twisting, bending, and other distortions, suggesting great potential for applications in flexible and wearable optoelectronics.

In Figure 5.23, both the 1D $ZnCo_2O_4$ core and 2D $NiCo_2O_4$ sheath parts are spinels [96]. This hierarchical structure provides high surface area, from the 2D nanosheets, and enhanced conductivity through the 1D nanowires. The BET results show that this 1&2D $ZnCo_2O_4$@$NiCo_2O_4$ had more than 1.5 times the surface area of the core $ZnCo_2O_4$ nanowires, offering more active sites to benefit the electrochemical performance.

$Co_3O_4$ nanoarrays [98] were prepared via a hydrothermal method, using a 2D Ni foam as a template or substrate, with a morphology that depended on the synthesis duration (Figure 5.24). At short reaction times, nanoplates grew on the nickel substrate, but at longer hydrothermal reaction times, nanorods grew out from each nanoplate. In all cases, the hydrothermally deposited material was converted to spinel $Co_3O_4$ by calcination. The combination of 1D and 2D $Co_3O_4$ nanostructures provided higher and more accessible specific surface area (SSA) than could be provided by either morphology alone. The resulting material displayed a high catalytic activity for CO oxidation at low temperature. Each of these examples shows how the 1&2D combination provides greater possibilities for creating functional, multi-component materials.

### 5.3.5 One- and Three-Dimensional (1&3D) Structures

We consider 1&3D structure spinels to be those that present a 3D hierarchical shape with 1D branches, showing flower, urchin, star, or dandelion morphologies [99]. In Figures 5.25A and 5.25B, both $NiCo_2O_4$ micro-urchins [100] and $MCo_2O_4$ (M = Mn, Ni, Cu, and Co) nanowire flowers [101] were grown on a Ni foam substrate, while the urchin-like $Ni_xCo_{3-x}O_4$ [102] in Figure 5.26 was prepared by a facile hydrothermal method. The interesting thing here is that most of these 3D flowerlike spinels are of cobaltite compositions ($MCo_2O_4$, M = transition metals), whether they were prepared by a template method [100], hydrothermal method [103, 104], electrochemical deposition [105], or precipitation method [106, 107]. Each of these structures also exhibits high catalytic activity and/or energy storage performance resulting from the large and accessible surface area presented by the flowerlike shapes. Generally, $Co^{2+}$ is considered inactive for electrocatalysis and it is $Co^{3+}$ that provides active sites [108]. Therefore, the exchange of $Co^{2+}$ for various $M^{2+}$ ions can improve the

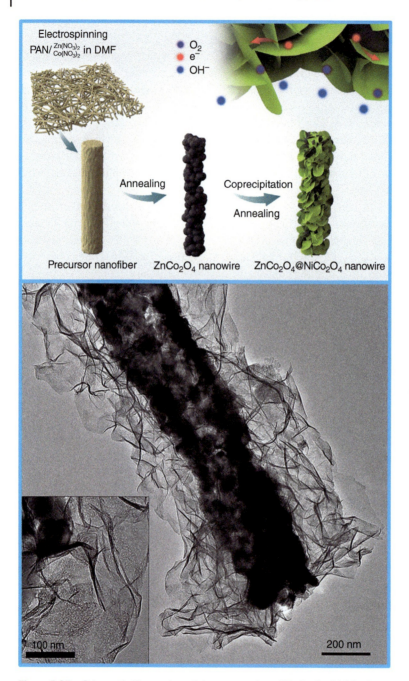

**Figure 5.23** Schematic illustration of the preparation of $ZnCo_2O_4$@$NiCo_2O_4$ core–sheath nanowires and TEM image of the product. Source: Reproduced with permission from Hwang et al. [96], John Wiley & Sons.

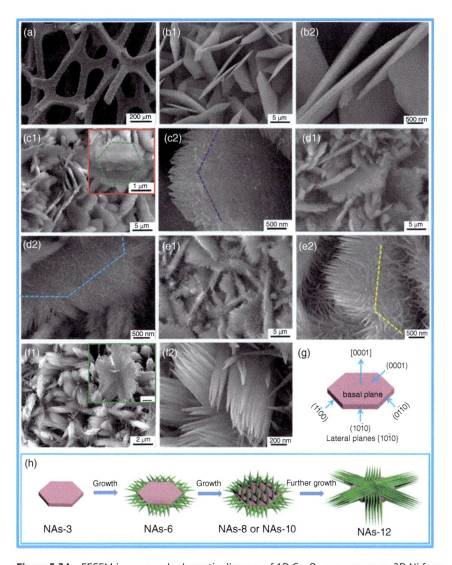

**Figure 5.24** FESEM images and schematic diagram of 1D $Co_3O_4$ nanoarrays on 2D Ni foam substrates. Source: Reproduced with permission from Mo et al. [98], Royal Society of Chemistry.

catalytic activity. These observations may guide future work on 1&3D cobaltite spinel materials in the field of electrochemistry and energy storage applications.

### 5.3.6 Two- and Three-Dimensional (2&3D) Structure

One type of 2&3D structure is the 3D flowerlike structures comprised of 2D nanosheets, which were grown on a nickel foam substrate [109]. This Zr-doped 3D $CoFe_2O_4$ on Ni foam (Figure 5.27) showed impressive bifunctional OER and HER catalytic activity. Here, the $CoFe_2O_4$ grew as flowerlike 2D nanosheet structures on

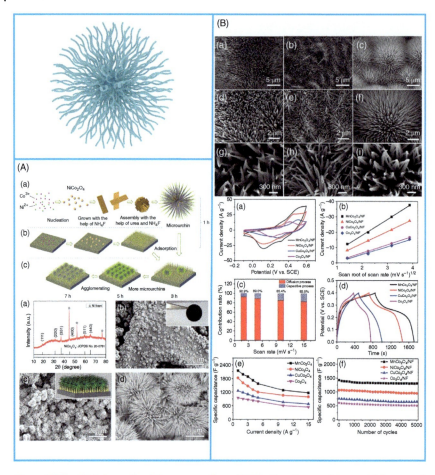

**Figure 5.25** (A) Schematic of the preparation of $NiCo_2O_4$ micro-urchins on Ni foam and the corresponding SEM images and XRD patterns. Source: Reproduced with permission from Wang et al. [100], Royal Society of Chemistry. (B) FESEM images of $MCo_2O_4$ nanowire flowers and their electrochemical performance. Source: Reproduced with permission from Liu et al. [101], Royal Society of Chemistry.

the Ni foam surface, expanding the surface area, but only when $ZrCl_4$ was added during the synthesis. Therefore, Zr-incorporation not only changed the composition but also increased the number of active sites for both OER and HER.

Another structural motif is 3D spheres or hollow structures with 2D nanosheets grown onto them. Both cabbage-like $ZnCo_2O_4$ [110] (Figure 5.28) and nanosheets assembled on hollow single-hole Ni–Co–Mn oxide (NHSNCM) spheres [111] (Figure 5.29) show a 3D spherical structure covered with 2D nanoflakes, whose formation process involves assembly of nanoflakes or nanosheets in a hydrothermal method. The purpose of synthesizing such 3D spheres with 2D nanosheets is to create more pores, voids, and interfaces at the surface, which improves the microwave absorption performance of $ZnCo_2O_4$ and the electrochemical performance of NHSNCM as an anode for lithium storage. These studies provide a novel strategy to

**Figure 5.26** SEM and TEM images of $Ni_xCo_{3-x}O_4$ hierarchical nanostructures. Source: Manivasakan et al. [102] with permission of Royal Society of Chemistry.

synthesize spheres covered with nanosheets without any templates and open a new opportunity for the 2D nanosheet-assembled spinel materials with large surface area and numerous mesopores for multiple applications.

## 5.4 Self-Assembled Structures

Self-assembly is defined as a process through which individual nanoparticles or components arrange to form organized structures based upon interactions between the individual nanoparticles. The term "0D" nanostructures was first used to describe quantum dots because of their small and uniform size. Here, we consider assembly of 0D monodisperse spinel nanoparticles with different shapes (spheres, octahedra, cubes, etc.) as the individual components. In Figures 5.30–5.32, three panels show perfectly organized layers formed from spheres [112], octahedra [113], and cubes [87]. The self-assembled MnO and $Fe_3O_4$ nanospheres in Figure 5.30 can be

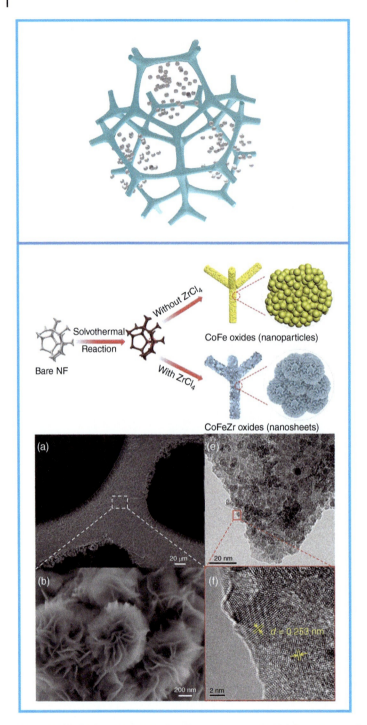

**Figure 5.27** Schematic illustration for in situ growth of CoFe oxides and CoFeZr oxides on the Ni foam and TEM images of CoFeZr oxides on Ni Foam. Source: Reproduced with permission from Huang et al. [109], John Wiley & Sons.

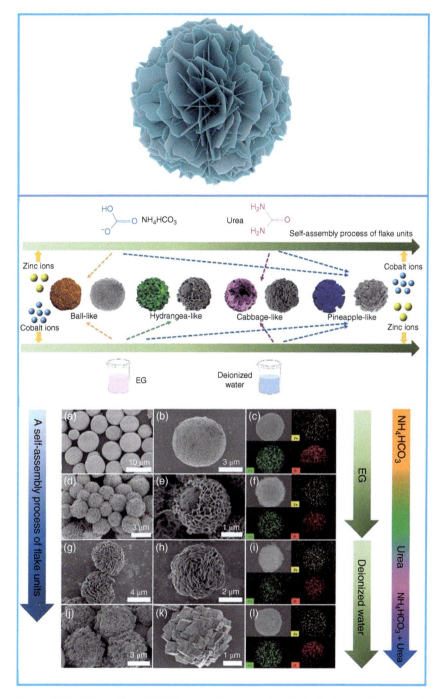

**Figure 5.28** Schematic and TEM images of cabbage-like ZnCo$_2$O$_4$. Source: Reproduced with permission from Li et al. [110], Royal Society of Chemistry.

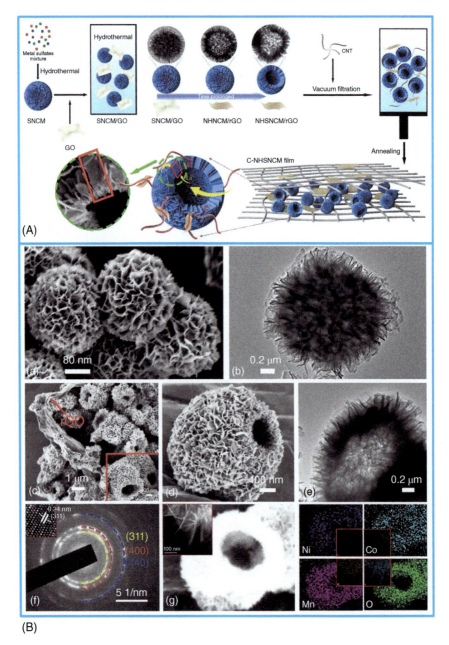

**Figure 5.29** Schematic diagram of the hollow NHSNCM spheres and the C-NHSNCM film and the corresponding TEM and SEM images, SAED pattern, and elemental mapping. Source: Reproduced with permission from Li et al. [111], John Wiley & Sons.

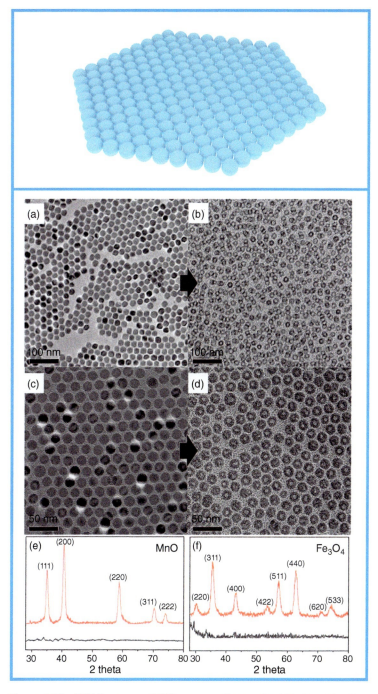

**Figure 5.30** TEM images and XRD patterns of assembled MnO and $Fe_3O_4$ monodispersed nanocrystals. Source: Adapted with permission from An et al. [112], American Chemical Society.

transformed to a core–shell hollow structure *via* acid etching and still retain the self-assembled structure.

As shown in Figure 5.31, an octahedral $Co_xFe_{3-x}O_4$ spinel [113] was synthesized using a simple solution-phase method, with excellent stability even in air up to 400 °C. Furthermore, the obtained spinel particle size was controlled from 7 to 50 nm by tuning concentration of oleic acid and oleylamine. Similar size and shape control strategy was applied on the growth of MNCs as nanocubes [87] (Figure 5.32). This nanocube spinel shows various types of cubic shapes with rounded corners, cut corners, or extended corners due to the promoted or suppressed growth on (111) plane. More importantly, the mechanism of template-free self-assembly for such magnetic spinel nanocrystals can be understood based on the following equation [114]:

$$\lambda = \frac{U}{E} = \frac{\frac{\mu^2}{D^3}}{k_B T} = \frac{\left(\frac{\pi d^3 M_s}{6}\right)^2 / D^3}{k_B T} = \frac{\pi^2 d^6 M_s^2}{36 k_B T D^3}$$

where parameter $\lambda$ is used to describe the competition of dipolar interaction and van der Waals interaction. Upon introducing all of the experimental parameters (such as surface ligands length, saturation magnetization, and critical particle size) into this equation, we can conclude that the assembly of MNCs size larger than 27 nm is dominated by magnetic dipolar interaction to form chains or rings, while the size smaller than 27 nm is dominated by van der Waals interactions, arranging in self-assembled layers. This discovery can guide future efforts in self-assembly of magnetic spinel materials with selectivity and controllability.

Most highly monodisperse spinel materials including the previously mentioned three self-assembled structure examples are synthesized via an organic solution-phase synthesis. As we mentioned in the syntheses part, the solution-phase method has the advantages of short reaction time, facile operation, and easy size and shape control. Furthermore, the organic solution-phase synthesis of the self-assembly structure spinel offers more tunable and well-understood reaction parameters (ligands, additives, solvents, etc.), not only to control the size and shape of individual components but also to promote the highly ordered and organized self-assembly structure formation.

## 5.5 Conclusions and Future Perspectives

In conclusion, we have summarized synthesis strategies for spinel nanostructures and provided detailed categorization of seven types of structures and the corresponding applications of these spinels. Morphology control is a significant factor not only for spinels but also for nearly all nanomaterials, since it affects most properties directly. This chapter provides examples and insights not only about what nanostructures are possible but also about how structure affects performance in related applications, providing guidance to the design of future spinel materials. The summarized 1D, 2D, 3D, 1&2D, 1&3D, 2&3D, and self-assembled structures

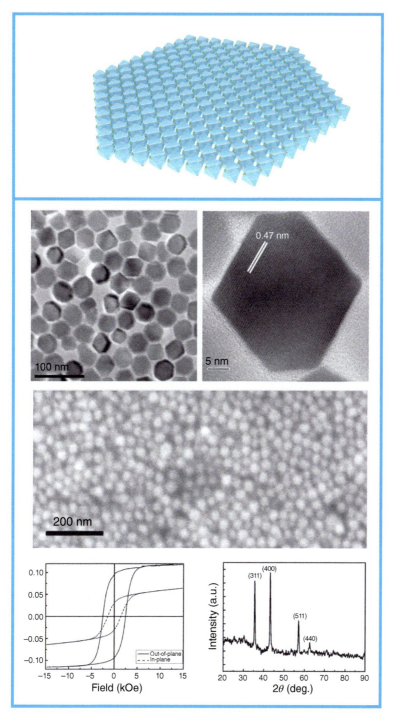

**Figure 5.31** $Co_{0.6}Fe_{2.4}O_4$ nanoparticles: TEM and SEM images, hysteresis loops, and XRD patterns. Source: Reproduced with permission from Yu et al. [113], John Wiley & Sons.

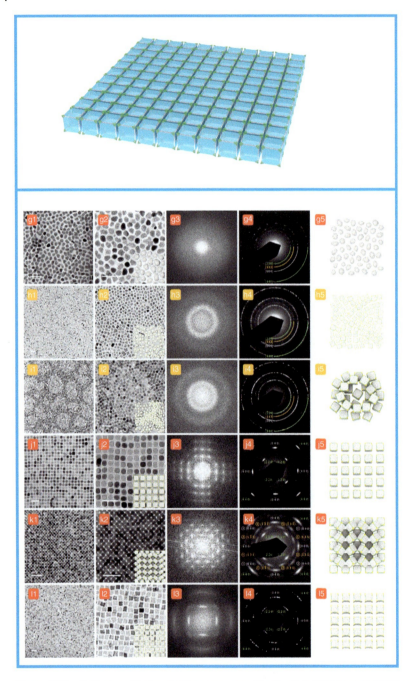

**Figure 5.32** Self-assembled monodisperse magnetite (Fe$_3$O$_4$): TEM images, SAED patterns, and 3D model illustrations. Source: Qiao et al. [87] with permission from American Chemical Society.

## 5.5 Conclusions and Future Perspectives

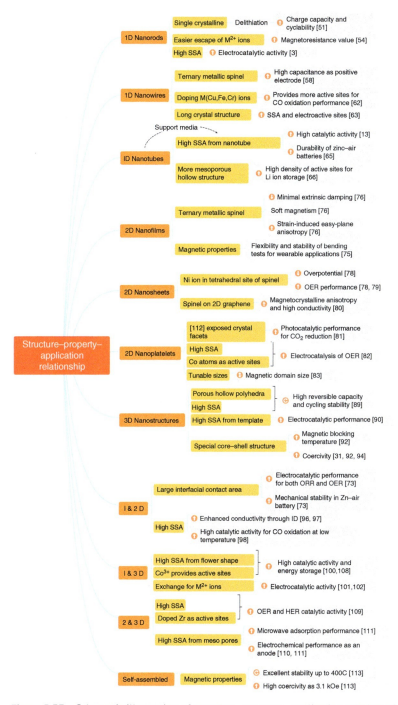

**Figure 5.33** Schematic illustration of structure–property–application relationships for functional spinel oxide nanomaterials. Source: Zheng Fu.

provide properties that benefit applications as shown schematically in Figure 5.18. These useful properties include high SSA, easy escape of doped $M^{2+}$ ions, single crystalline structure, mesoporous hollow structure useful for uptake and delivery of substances, excellent magnetic properties, the structure benefits from other nanotubes or nanosheets, dispersed metal atoms as active sites for catalysis, tunable size, and large interfacial contract area. All these useful properties from spinels can be applied to the following multiple fields: high charge capacity and cyclability for energy storage; electrocatalytic activity; CO oxidation performance; durability and mechanical stability of Zn–air batteries; wearable applications; ORR, OER, and HER electrocatalysis; photocatalytic $CO_2$ reduction; enhanced conductivity; microwave adsorption; and magnetism including minimal extrinsic damping, soft magnetism, strain-induced easy-plane anisotropy, high magnetic blocking temperature, and high coercivity. In the future, we expect more studies on spinel materials about structure–property–performance relationships for applications (Figure 5.33).

## References

1 Yao, T., Guo, X., Qin, S. et al. (2017). *Effect of rGO coating on interconnected $Co_3O_4$ nanosheets and improved supercapacitive behavior of $Co_3O_4$/rGO/NF architecture. Nano-Micro Lett.* 9 (4): 38.
2 Kushwaha, A.K., Uğur, Ş., Akbudak, S., and Uğur, G. (2017). *Investigation of structural, elastic, electronic, optical and vibrational properties of silver chromate spinels: normal ($CrAg_2O_4$) and inverse ($Ag_2CrO_4$). J. Alloys Compd.* 704: 101–108.
3 Zhu, Y.P., Ma, T.Y., Jaroniec, M., and Qiao, S.Z. (2017). *Self-templating synthesis of hollow $Co_3O_4$ microtube arrays for highly efficient water electrolysis. Angew. Chem. Int. Ed.* 56 (5): 1324–1328.
4 Yang, W., Yu, Y., Tang, Y. et al. (2017). *Enhancing electrochemical detection of dopamine via dumbbell-like FePt-$Fe_3O_4$ nanoparticles. Nanoscale* 9 (3): 1022–1027.
5 Wu, L., Jubert, P.O., Berman, D. et al. (2014). *Monolayer assembly of ferrimagnetic $Co_{(x)}Fe_{(3-x)}O_4$ nanocubes for magnetic recording. Nano Lett.* 14 (6): 3395–3399.
6 Marco, J.F., Gancedo, J.R., Gracia, M. et al. (2001). *Cation distribution and magnetic structure of the ferrimagnetic spinel $NiCo_2O_4$. J. Mater. Chem.* 11 (12): 3087–3093.
7 Park, J.-H. and Sudarshan, T. (2001). *Chemical Vapor Deposition*, vol. 2. ASM International.
8 Zhao, Q., Yan, Z., Chen, C., and Chen, J. (2017). *Spinels: Controlled preparation, oxygen reduction/evolution reaction application, and beyond. Chem. Rev.* 117 (15): 10121–10211.
9 Jin fan, H., Knez, M., Scholz, R. et al. (2006). *Monocrystalline spinel nanotube fabrication based on the Kirkendall effect. Nat. Mater.* 5 (8): 627–631.

10 Choi, S.H., Hong, Y.J., and Kang, Y.C. (2013). *Yolk-shelled cathode materials with extremely high electrochemical performances prepared by spray pyrolysis*. Nanoscale 5 (17): 7867–7871.

11 Tian, Z., Bahlawane, N., Qi, F., and Kohse-Hüinghaus K. (2009). *Catalytic oxidation of hydrocarbons over $Co_3O_4$ catalyst prepared by CVD*. Catal. Commun. 11 (2): 118–122.

12 Barreca, D., Fornasiero, P., Gasparotto, A. et al. (2010). *CVD $Co_3O_4$ nanopyramids: A nano-platform for photo-assisted $H_2$ production*. Chem. Vap. Deposition 16 (10–12): 296–300.

13 Waqas, M., El Kasmi, A., Wang, Y. et al. (2018). *CVD synthesis of Cu-doped cobalt spinel thin film catalysts for kinetic study of propene oxidation*. Colloids Surf., A 556: 195–200.

14 He, B., Chen, X., Lu, J. et al. (2016). *One-pot synthesized $Co/Co_3O_4$-N-graphene composite as electrocatalyst for oxygen reduction reaction and oxygen evolution reaction*. Electroanalysis 28 (10): 2435–2443.

15 Arifin, D., Aston, V.J., Liang, X. et al. (2012). *$CoFe_2O_4$ on a porous $Al_2O_3$ nanostructure for solar thermochemical $CO_2$ splitting*. Energy Environ. Sci. 5 (11): 9438–9443.

16 Smigelskas, A. and Kirkendall, E. (1947). *Zinc diffusion in alpha brass*. Trans. Aime 171 (1947): 130–142.

17 Huang, B., Yang, W., Wen, Y. et al. (2015). *$Co_3O_4$-modified $TiO_2$ nanotube arrays via atomic layer deposition for improved visible-light photoelectrochemical performance*. ACS Appl. Mater. Interfaces 7 (1): 422–431.

18 Tartakovskii, A. (2012). *Quantum Dots: Optics, Electron Transport and Future Applications*. Cambridge University Press.

19 Popovici, E., Dumitrache, F., Morjan, I. et al. (2007). *Iron/iron oxides core–shell nanoparticles by laser pyrolysis: Structural characterization and enhanced particle dispersion*. Appl. Surf. Sci. 254 (4): 1048–1052.

20 Jiang, Q., Zhang, H., and Wang, S. (2016). *Plasma-enhanced low-temperature solid-state synthesis of spinel $LiMn_2O_4$ with superior performance for lithium-ion batteries*. Green Chem. 18 (3): 662–666.

21 Vekas, L., Bica, D., and Avdeev, M.V. (2007). *Magnetic nanoparticles and concentrated magnetic nanofluids: Synthesis, properties and some applications*. China Particuol. 5 (1–2): 43–49.

22 He, Y., Sahoo, Y., Wang, S. et al. (2006). *Laser-driven synthesis and magnetic properties of iron nanoparticles*. J. Nanopart. Res. 8 (3): 335–342.

23 Son, S., Taheri, M., Carpenter, E. et al. (2002). *Synthesis of ferrite and nickel ferrite nanoparticles using radio-frequency thermal plasma torch*. J. Appl. Phys. 91 (10): 7589.

24 Qiao, L. and Swihart, M.T. (2016). *Solution-phase synthesis of transition metal oxide nanocrystals: Morphologies, formulae, and mechanisms*. Adv. Colloid Interface Sci.

25 Hasegawa, G., Kanamori, K., Nakanishi, K., and Hanada, T. (2010). *Facile preparation of hierarchically porous $TiO_2$ monoliths*. J. Am. Ceram. Soc. 93 (10): 3110–3115.

**26** Kido, Y., Nakanishi, K., Miyasaka, A., and Kanamori, K. (2012). Synthesis of monolithic hierarchically porous iron-based xerogels from iron(III) salts via an epoxide-mediated sol–gel process. *Chem. Mater.* 24 (11): 2071–2077.

**27** Guo, X., Yin, P., Lei, W. et al. (2017). Synthesis and characterization of monolithic $ZnAl_2O_4$ spinel with well-defined hierarchical pore structures via a sol-gel route. *J. Alloys Compd.* 727: 763–770.

**28** Bhavani, P., Rajababu, C.H., Arif, M.D. et al. (2017). Synthesis of high saturation magnetic iron oxide nanomaterials via low temperature hydrothermal method. *J. Magn. Magn. Mater.* 426: 459–466.

**29** Zhu, Z., Bai, Y., Zhang, T. et al. (2014). High-performance hole-extraction layer of sol–gel-processed NiO nanocrystals for inverted planar perovskite solar cells. *Angew. Chem.* 126 (46): 12779–12783.

**30** Yin, Y. and Alivisatos, A.P. (2005). Colloidal nanocrystal synthesis and the organic–inorganic interface. *Nature* 437 (7059): 664–670.

**31** Park, J., Koo, B., Yoon, K.Y. et al. (2005). Generalized synthesis of metal phosphide nanorods via thermal decomposition of continuously delivered metal–phosphine complexes using a syringe pump. *J. Am. Chem. Soc.* 127 (23): 8433–8440.

**32** Li, Z., Ma, Y., and Qi, L. (2014). Controlled synthesis of $Mn_xFe_{1-x}O$ concave nanocubes and highly branched cubic mesocrystals. *CrystEngComm* 16 (4): 600–608.

**33** Sanna Angotzi, M., Musinu, A., Mameli, V. et al. (2017). Spinel ferrite core-shell nanostructures by a versatile solvothermal seed-mediated growth approach and study of their nanointerfaces. *ACS Nano* 11 (8): 7889–7900.

**34** Chine, M., Sediri, F., and Gharbi, N. (2012). Solvothermal synthesis of V4O9 flake-like morphology and its photocatalytic application in the degradation of methylene blue. *Mater. Res. Bull.* 47 (11): 3422–3426.

**35** Li, C.C., Yin, X.M., Wang, T.H., and Zeng, H.C. (2009). Morphogenesis of highly uniform $CoCO_3$ submicrometer crystals and their conversion to mesoporous $Co_3O_4$ for gas-sensing applications. *Chem. Mater.* 21 (20): 4984–4992.

**36** Repko, A., Vejpravova, J., Vackova, T. et al. (2015). Oleate-based hydrothermal preparation of $CoFe_2O_4$ nanoparticles, and their magnetic properties with respect to particle size and surface coating. *J. Magn. Magn. Mater.* 390: 142–151.

**37** Pechini, M. P. (1967). Method of preparing lead and alkaline earth titanates and niobates and coating method using the same to form a capacitor. Google Patents US3330697A, filed 26 August 1963 and issued 11 July 1967.

**38** Lee, H., Hong, M., Bae, S. et al. (2003). A novel approach to preparing nano-size $Co_3O_4$-coated Ni powder by the Pechini method for MCFC cathodes. *J. Mater. Chem.* 13 (10): 2626–2632.

**39** Ma, P., Geng, Q., Gao, X. et al. (2016). Spectrally selective $Cu_{1.5}Mn_{1.5}O_4$ spinel ceramic pigments for solar thermal applications. *RSC Adv.* 6 (39): 32947–32955.

**40** Freni, S., Barone, F., and Puglisi, M. (1998). The dissolution process of the NiO cathodes for molten carbonate fuel cells: state-of-the-art. *Int. J. Energy Res.* 22 (1): 17–31.

**41** Liu, D., Zhu, W., Trottier, J. et al. (2014). *Spinel materials for high-voltage cathodes in Li-ion batteries. RSC Adv.* 4 (1): 154–167.

**42** Specchia, S., Finocchio, E., Busca, G., and Specchia, V. Combustion synthesis. In: *Handbook of Combustion* (eds. M. Lackner, F. Winter and A.K. Agarwal), 1–62. Wiley-VCH.

**43** Kozawa, T. (2019). *Lattice deformation of $LiNi_{0.5}Mn_{1.5}O_4$ spinel cathode for Li-ion batteries by ball milling. J. Power Sources* 419: 52–57.

**44** Goncalves, A.A.S., Costa, M.J.F., Zhang, L. et al. (2018). *One-pot synthesis of $MeAl_2O_4$ (Me = Ni, Co, or Cu) supported on $\gamma$-$Al_2O_3$ with ultralarge mesopores: enhancing interfacial defects in $\gamma$-$Al_2O_3$ to facilitate the formation of spinel structures at lower temperatures. Chem. Mater.* 30 (2): 436–446.

**45** Xia, Y., Yang, P., Sun, Y. et al. (2003). *One-dimensional nanostructures: synthesis, characterization, and applications. Adv. Mater.* 15 (5): 353–389.

**46** Jana, N.R., Gearheart, L., and Murphy, C.J. (2001). *Wet chemical synthesis of silver nanorods and nanowires of controllable aspect ratio. Chem. Commun.* (7): 617–618.

**47** Hu, J.-Q., Chen, Q., Xie, Z.-X. et al. (2004). *A Simple and effective route for the synthesis of crystalline silver nanorods and nanowires. Adv. Funct. Mater.* 14 (2): 183–189.

**48** Ye, X., Jin, L., Caglayan, H. et al. (2012). *Improved size-tunable synthesis of monodisperse gold nanorods through the use of aromatic additives. ACS Nano* 6 (3): 2804–2817.

**49** Yoo, S.-H. and Park, S. (2007). *Platinum-coated, nanoporous gold nanorod arrays: synthesis and characterization. Adv. Mater.* 19 (12): 1612–1615.

**50** Xiong, Y. and Xia, Y. (2007). *Shape-controlled synthesis of metal nanostructures: The case of palladium. Adv. Mater.* 19 (20): 3385–3391.

**51** Kim, D.K., Muralidharan, P., Lee, H.-W. et al. (2008). *Spinel $LiMn_2O_4$ anorods as lithium Ion battery cathodes. Nano Lett.* 8 (11): 3948–3952.

**52** Liu, H.-J., Tra, V.-T., Chen, Y.-J. et al. (2013). *Large magnetoresistance in magnetically coupled $SrRuO_3$–$CoFe_2O_4$ self-assembled nanostructures. Adv. Mater.* 25 (34): 4753–4759.

**53** Kang, K., Meng, Y.S., Bréger, J. et al. (2006). *Electrodes with high power and high capacity for rechargeable lithium batteries. Science* 311 (5763): 977–980.

**54** Ohzuku, T., Kitagawa, M., and Hirai, T. (1990). *Electrochemistry of manganese dioxide in lithium nonaqueous cell III. X-Ray diffractional study on the reduction of spinel-related manganese dioxide. J. Electrochem. Soc.* 137 (3): 769–775.

**55** Liu, H.-J., Chen, L.-Y., He, Q. et al. (2012). *Epitaxial photostriction–magnetostriction coupled self-assembled nanostructures. ACS Nano* 6 (8): 6952–6959.

**56** Mannhart, J. and Schlom, D. (2010). *Oxide interfaces—an opportunity for electronics. Science* 327 (5973): 1607–1611.

**57** Fan, J., Liao, S., Wang, W. et al. (2011). *Suppression of ferromagnetism and metal-like conductivity in lightly Fe-doped $SrRuO_3$. J. Appl. Phys.* 110 (4): 043907.

**58** Li, L., Zhang, Y., Shi, F. et al. (2014). Spinel manganese–nickel–cobalt ternary oxide nanowire array for high-performance electrochemical capacitor applications. *ACS Appl. Mater. Interfaces* 6 (20): 18040–18047.

**59** Gu, D., Jia, C.J., Weidenthaler, C. et al. (2015). Highly ordered mesoporous cobalt-containing oxides: Structure, catalytic properties, and active sites in oxidation of carbon menoxide. *J. Am. Chem. Soc.* 137 (35): 11407–11418.

**60** Shen, L., Che, Q., Li, H. et al. (2014). Mesoporous $NiCo_2O_4$ nanowire arrays grown carbon textiles as binder-free flexible electrodes for energy storage. *Adv. Funct. Mater.* 24 (18): 2630–2637.

**61** Wang, Y., Lei, Y., Li, J. et al. (2014). Synthesis of 3D-nanonet hollow structured $Co_3O_4$ for high capacity supercapacitor. *ACS Appl. Mater. Interfaces* 6 (9): 6739–6747.

**62** Meher, S.K., Justin, P., and Rao, G.R. (2011). Nanoscale morphology dependent pseudocapacitance of NiO: Influence of intercalating anions during synthesis. *Nanoscale* 3 (2): 683–692.

**63** Srivastava, M., Mishra, R.K., Singh, J. et al. (2015). Consequence of pH variation on the dielectric properties of Cr-doped lithium ferrite nanoparticles synthesized by the sol–gel method. *J. Alloys Compd.* 645: 171–177.

**64** Jansson, J., Palmqvist, A.E., Fridell, E. et al. (2002). On the catalytic activity of $Co_3O_4$ in low-temperature CO oxidation. *J. Catal.* 211 (2): 387–397.

**65** Wang, X.-T., Ouyang, T., Wang, L. et al. Redox-inert $Fe^{3+}$ ions in octahedral sites of Co–Fe spinel oxides with enhanced oxygen catalytic activity for rechargeable zinc–air batteries. *Angew. Chem. Int. Ed.*: 13291–13296.

**66** Hwang, S.M., Kim, S.Y., Kim, J.-G. et al. (2015). Electrospun manganese–cobalt oxide hollow nanofibres synthesized via combustion reactions and their lithium storage performance. *Nanoscale* 7 (18): 8351–8355.

**67** Liu, Z.-Q., Cheng, H., Li, N. et al. (2016). $ZnCo_2O_4$ qantum dots anchored on nitrogen-doped carbon nanotubes as reversible oxygen reduction/evolution electrocatalysts. *Adv. Mater.* 28 (19): 3777–3784.

**68** Lou, X.W., Deng, D., Lee, J.Y. et al. (2008). Self-supported formation of needlelike $Co_3O_4$ nanotubes and their application as lithium-Ion battery electrodes. *Adv. Mater.* 20 (2): 258–262.

**69** Aijaz, A., Masa, J., Rösler, C. et al. (2016). $Co@Co_3O_4$ encapsulated in carbon nanotube-grafted nitrogen-doped carbon polyhedra as an advanced bifunctional oxygen Electrode. *Angew. Chem. Int. Ed.* 55 (12): 4087–4091.

**70** Tian, Z.-Y., Mountapmbeme Kouotou, P., Bahlawane, N. et al. (2013). Synthesis of the catalytically active $Mn_3O_4$ spinel and its thermal properties. *J. Phys. Chem. C* 117 (12): 6218–6224.

**71** Young, M.J., Schnabel, H.-D., Holder, A.M. et al. (2016). Band diagram and rate analysis of thin film spinel $LiMn_2O_4$ formed by electrochemical conversion of ALD-grown MnO. *Adv. Funct. Mater.* 26 (43): 7895–7907.

**72** Emori, S., Gray, B.A., Jeon, H.-M. et al. (2017). Coexistence of low damping and strong magnetoelastic coupling in epitaxial spinel ferrite thin films. *Adv. Mater.* 29 (34): 1701130.

73 Chen, X., Liu, B., Zhong, C. et al. (2017). *Ultrathin $Co_3O_4$ layers with large contact area on carbon fibers as high-performance electrode for flexible zinc–air battery integrated with flexible display.* Adv. Energy Mater. 7 (18): 1700779.

74 Li, Y., Tang, F., Wang, R. et al. (2016). *Novel dual-Ion hybrid supercapacitor based on a $NiCo_2O_4$ nanowire cathode and $MoO_2$–C nanofilm anode.* ACS Appl. Mater. Interfaces 8 (44): 30232–30238.

75 Shen, L., Wu, L., Sheng, Q. et al. (2017). *Epitaxial lift-off of centimeter-scaled spinel ferrite oxide thin films for flexible electronics.* Adv. Mater. 29 (33): 1702411.

76 Hamadeh, A., Kelly, O.d.A., Hahn, C. et al. (2014). *Full control of the spin-wave damping in a magnetic insulator using spin-orbit torque.* Phys. Rev. Lett. 113 (19): 197203.

77 Pachauri, N., Khodadadi, B., Althammer, M. et al. (2015). *Study of structural and ferromagnetic resonance properties of spinel lithium ferrite ($LiFe_5O_8$) single crystals.* J. Appl. Phys. 117 (23).

78 Wang, H.-Y., Hsu, Y.-Y., Chen, R. et al. (2015). *$Ni^{3+}$-induced formation of active NiOOH on the spinel Ni–Co oxide surface for efficient oxygen evolution reaction.* Adv. Energy Mater. 5 (10): 1500091.

79 Xu, L., Jiang, Q., Xiao, Z. et al. (2016). *Plasma-engraved $Co_3O_4$ nanosheets with oxygen vacancies and high surface area for the oxygen evolution reaction.* Angew. Chem. Int. Ed. 55 (17): 5277–5281.

80 Li, X., Feng, J., Du, Y. et al. (2015). *One-pot synthesis of $CoFe_2O_4$/graphene oxide hybrids and their conversion into FeCo/graphene hybrids for lightweight and highly efficient microwave absorber.* J. Mater. Chem. A 3 (10): 5535–5546.

81 Gao, C., Meng, Q., Zhao, K. et al. (2016). *$Co_3O_4$ hexagonal platelets with controllable facets enabling highly efficient visible-light photocatalytic reduction of $CO_2$.* Adv. Mater. 28 (30): 6485–6490.

82 Cui, B., Lin, H., Li, J.-B. et al. (2008). *Core–ring structured $NiCo_2O_4$ nanoplatelets: synthesis, characterization, and electrocatalytic applications.* Adv. Funct. Mater. 18 (9): 1440–1447.

83 Martin-Garcia, L., Quesada, A., Munuera, C. et al. (2015). *Atomically flat ultrathin cobalt ferrite islands.* Adv. Mater. 27 (39): 5955–5960.

84 Fester, J., Makoveev, A., Grumelli, D. et al. (2018). *The structure of the cobalt oxide/Au catalyst interface in electrochemical water splitting.* Angew. Chem. Int. Ed. 57 (37): 11893–11897.

85 Jacobse, L., Huang, Y.-F., Koper, M.T.M., and Rost, M.J. (2018). *Correlation of surface site formation to nanoisland growth in the electrochemical roughening of Pt(111).* Nat. Mater. 17 (3): 277–282.

86 Smiljanic, M., Rakocevic, Z., Maksic, A., and Strbac S. (2014). *Hydrogen evolution reaction on platinum catalyzed by palladium and rhodium nanoislands.* Electrochim. Acta 117: 336–343.

87 Qiao, L., Fu, Z., Li, J. et al. (2017). *Standardizing size- and shape-controlled synthesis of monodisperse magnetite ($Fe_3O_4$) nanocrystals by identifying and exploiting effects of organic impurities.* ACS Nano 11: 6370–6381.

88 Wu, R., Qian, X., Zhou, K. et al. (2014). *Porous spinel $Zn_xCo_{3-x}O_4$ hollow polyhedra templated for high-rate lithium-ion batteries.* ACS Nano 8 (6): 6297–6303.

89 Zhou, K., Wang, X., Sun, X. et al. (2005). *Enhanced catalytic activity of ceria nanorods from well-defined reactive crystal planes.* J. Catal. 229 (1): 206–212.

90 Qiu, M., Zhan, S., Yu, H. et al. (2015). *Facile preparation of ordered mesoporous $MnCo_2O_4$ for low-temperature selective catalytic reduction of NO with $NH_3$.* Nanoscale 7 (6): 2568–2577.

91 Deng, J., Yu, X., Qin, X. et al. (2019). *Co–B nanoflakes as multifunctional bridges in $ZnCo_2O_4$ micro-/nanospheres for superior lithium storage with boosted kinetics and stability.* Adv. Energy Mater. 9 (14): 1803612.

92 Sartori, K., Choueikani, F., Gloter, A. et al. (2019). *Room temperature blocked magnetic nanoparticles based on ferrite promoted by a three-step thermal decomposition process.* J. Am. Chem. Soc. 141 (25): 9783–9787.

93 Reiss, P., Protiere, M., and Li, L. (2009). *Core/shell semiconductor nanocrystals.* Small 5 (2): 154–168.

94 Song, Q. and Zhang, Z.J. (2012). *Controlled synthesis and magnetic properties of bimagnetic spinel ferrite $CoFe_2O_4$ and $MnFe_2O_4$ nanocrystals with core-shell architecture.* J. Am. Chem. Soc. 134 (24): 10182–10190.

95 Peng, S., Hu, Y., Li, L. et al. (2015). *Controlled synthesis of porous spinel cobaltite core-shell microspheres as high-performance catalysts for rechargeable $Li–O_2$ batteries.* Nano Energy 13: 718–726.

96 Huang, Y., Miao, Y.E., Lu, H., and Liu, T. (2015). *Hierarchical $ZnCo_2O_4$@$NiCo_2O_4$ core-sheath nanowires: Bifunctionality towards high-performance supercapacitors and the oxygen-reduction reaction.* Chemistry 21 (28): 10100–10108.

97 Ma, F.-X., Yu, L., Xu, C.-Y., and Lou, X.W. (2016). *Self-supported formation of hierarchical $NiCo_2O_4$ tetragonal microtubes with enhanced electrochemical properties.* Energy Environ. Sci. 9 (3): 862–866.

98 Mo, S., Li, S., Ren, Q. et al. (2018). *Vertically-aligned $Co_3O_4$ arrays on Ni foam as monolithic structured catalysts for CO oxidation: effects of morphological transformation.* Nanoscale 10 (16): 7746–7758.

99 Fu, Z., Qiao, L., Liu, Y. et al. (2020). *A general hierarchical flower-shaped cobalt oxide spinel template: facile method, morphology control, and enhanced saturation magnetization.* J. Mater. Chem. C 8: 14056–14065.

100 Wang, Y., Liu, P., Zhu, K. et al. (2017). *Hierarchical bilayered hybrid nanostructural arrays of $NiCo_2O_4$ micro-urchins and nanowires as a free-standing electrode with high loading for high-performance lithium-ion batteries.* Nanoscale 9 (39): 14979–14989.

101 Liu, S., Ni, D., Li, H.-F. et al. (2018). *Effect of cation substitution on the pseudocapacitive performance of spinel cobaltite $MCo_2O_4$ (M = Mn, Ni, Cu, and Co).* J. Mater. Chem. A 6 (23): 10674–10685.

102 Manivasakan, P., Ramasamy, P., and Kim, J. (2014). *Use of urchin-like $Ni_xCo_{3-x}O_4$ hierarchical nanostructures based on non-precious metals as bifunctional electrocatalysts for anion-exchange membrane alkaline alcohol fuel cells.* Nanoscale 6 (16): 9665–9672.

**103** Kuppan, S., Shukla, A.K., Membreno, D. et al. (2017). *Revealing anisotropic spinel formation on pristine Li- and Mn-rich layered oxide surface and its impact on cathode performance. Adv. Energy Mater.* 7 (11): 1602010.

**104** Xiong, S., Chen, J.S., Lou, X.W., and Zeng, H.C. (2012). *Mesoporous $Co_3O_4$ and CoO@C topotactically transformed from chrysanthemum-like $Co(CO_3)0.5(OH)·0.11H_2O$ and their lithium-storage properties. Adv. Funct. Mater.* 22 (4): 861–871.

**105** Liu, W.-w., Lu, C., Liang, K. et al. (2014). *A three dimensional vertically aligned multiwall carbon nanotube/$NiCo_2O_4$ core/shell structure for novel high-performance supercapacitors. J. Mater. Chem. A* 2 (14): 5100–5107.

**106** Jadhav, H.S., Kalubarme, R.S., Park, C.-N. et al. (2014). *Facile and cost effective synthesis of mesoporous spinel $NiCo_2O_4$ as an anode for high lithium storage capacity. Nanoscale* 6 (17): 10071–10076.

**107** Li, J., Xiong, S., Li, X., and Qian, Y. (2013). *A facile route to synthesize multiporous $MnCo_2O_4$ and $CoMn_2O_4$ spinel quasi-hollow spheres with improved lithium storage properties. Nanoscale* 5 (5): 2045–2054.

**108** Mohamed, S.G., Tsai, Y.-Q., Chen, C.-J. et al. (2015). *Ternary spinel $MCo_2O_4$ (M = Mn, Fe, Ni, and Zn) porous nanorods as bifunctional cathode materials for lithium–$O_2$ batteries. ACS Appl. Mater. Interfaces* 7 (22): 12038–12046.

**109** Huang, L., Chen, D., Luo, G. et al. (2019). *Zirconium-regulation-induced bifunctionality in 3D cobalt–iron oxide nanosheets for overall water splitting. Adv. Mater.* 31 (28): 1901439.

**110** Li, X., Wang, L., You, W. et al. (2019). *Morphology-controlled synthesis and excellent microwave absorption performance of $ZnCo_2O_4$ nanostructures via a self-assembly process of flake units. Nanoscale* 11 (6): 2694–2702.

**111** Li, S., Zhao, X., Feng, Y. et al. (2019). *A flexible film toward high-performance lithium storage: Designing nanosheet-assembled hollow single-hole Ni–Co–Mn–O spheres with oxygen vacancy embedded in 3D carbon nanotube/graphene network. Small* 15 (27): 1901343.

**112** An, K., Kwon, S.G., Park, M. et al. (2008). *Synthesis of uniform hollow oxide nanoparticles through nanoscale acid etching. Nano Lett.* 8 (12): 4252–4258.

**113** Yu, Y., Mendoza-Garcia, A., Ning, B., and Sun, S. (2013). *Cobalt-substituted magnetite nanoparticles and their assembly into ferrimagnetic nanoparticle arrays. Adv. Mater.* 25 (22): 3090–3094.

**114** Lalatonne, Y., Richardi, J., and Pileni, M. (2004). *Van der Waals versus dipolar forces controlling mesoscopic organizations of magnetic nanocrystals. Nat. Mater.* 3 (2): 121.

# 6

# Photoinduced Processes in Metal Oxide Nanomaterials

*Nikolai V. Tkachenko and Ramsha Khan*

*Tampere University, Faculty of Engineering and Natural Science, Korkeakoulunkatu 8, 33720 Tampere, Finland*

## 6.1 Introduction

The search of greener alternatives for generating renewable energy has become an absolute necessity. Metal oxide (MO) materials are exploited and widely engineered for this purpose due to their great potential in utilizing solar energy for photocatalytic applications. These materials have diverse physical, chemical, and electronic properties, which make them suitable for photoinduced applications such as photovoltaics, water splitting, pollutant degradation, and $CO_2$ photoreduction [1–5]. These materials have become really popular because of their cost-effectiveness, low toxicity, recyclability, photo- and chemical stability, and tunabilty of electronic and optical properties. MOs are promising potential candidates to be used in combination with other materials or even to substitute some more expensive and toxic ones, such as transition metal-based complexes and dyes [6–8].

This chapter is devoted to photophysical properties of MOs, which depend essentially on electronic structures of the materials. It is common to classify materials as conductors or metals, semiconductors, and insulators or dielectric materials. Most of MOs are either insulators or semiconductors. MO insulators have a large bandgap, do not absorb visible or near UV light, and thus do not show any activity under visible light illumination. Meanwhile, well-conductive MOs are not that many and metals are more useful for conduction purposes as compared to them. However, semiconductor MOs have attracted great attention due to their unique combination of electronic, photochemical, catalytic, and mechanical properties, which can be activated or exploited by light excitation. MOs are now widely chosen on the basis of their bandgap, charge mobility, and relative ease of preparing materials with specific morphology and dimensions at nanoscale, such as materials with high specific surface area.

At the most fundamental level, these materials can be specified and distinguished by positions of the valence band (VB) and the conduction band (CB), with the VB electrons responsible for bounds between nuclei and crystal lattice and the CB electrons determining the conductivity of the materials. Within this model,

*Tailored Functional Oxide Nanomaterials: From Design to Multi-Purpose Applications*, First Edition.
Edited by Chiara Maccato and Davide Barreca.
© 2022 WILEY-VCH GmbH. Published 2022 by WILEY-VCH GmbH.

semiconductors and insulators have well separated VB and CB, and the energy difference between the top of the VB and the bottom of CB is referred to as the "bandgap." At temperatures approaching the absolute zero, the VB is completely occupied by electrons and the CB is empty. The light absorption properties of these materials are determined by the bandgap, where a photon absorption results in promoting an electron from the VB to the CB. Therefore, the photons with energy lower than the bandgap are not absorbed by the material within this simplified model.

Considering the importance of the bandgap for photophysical properties of the materials, semiconductors are typically divided into wide and narrow bandgap semiconductors. This subdivision is not strict, and the typical examples of wide bandgap semiconductors are $TiO_2$, ZnO, and $SnO_2$, which all have the bandgap larger than 3 eV and are transparent in the visible part of the solar spectrum. Since these MOs can still be fabricated with relatively high electrical conductivity, these materials are widely used as transparent electrodes but have also photorelated applications of their own, for example, in photocatalysis. Narrow bandgap semiconductors absorb visible light and thus are more efficient in solar light utilization. Examples of these MOs include $Cu_2O$, $Fe_2O_3$, and more complex materials such as $Zn_2Mo_3O_8$, which all have the bandgap close to 2 eV and absorb light effectively at wavelengths <620 nm. On the other side of the spectrum are dielectric materials that are transparent through the visible to the far UV range due to their large bandgap (>6 eV), a typical example is $Al_2O_3$. These MOs will not be considered in this chapter.

All the MO semiconductor materials considered in this chapter, such as $TiO_2$, ZnO, $Fe_2O_3$, $WO_3$, $SnO_2$, $ZrO_2$, etc. follow similar primary photoinduced reactions starting from the light absorption, which promote electrons from the VB to CB. Also these materials have common reactions following the photoexcitations and are the primary focus of this chapter. We will only mention applications that are based on MO photophysics, but aim is to overview fundamentals of photoinduced processes in MOs.

The photoinduced processes in MO materials can be observed and studied by many experimental techniques, but the most widely used are steady-state and time-resolved absorption and emission spectroscopy methods. These techniques are important to determine the electronic and optical properties of MO materials. In these methods, light from the UV–Vis region is irradiated over the MO materials, and the interaction of light with charge carriers is studied. The information one gets out of these techniques reflects on the bandgaps, defect levels, emission bands, reaction rate constants, and other photophysical properties of the material.

Widely available UV–Vis steady-state spectroscopy is used to determine the bandgap of the MO materials. Photoluminescence, PL, is one of the relaxation reaction of the excited MOs accompanied by light emission. Both, the emission spectrum and PL decay time provide important information on the fate of photogenerated carriers. Many reactions of importance, such as carrier thermal relaxation and trapping, do not emit any light but result in a change of the absorption properties of the sample and thus can be followed by the time-resolved absorption spectroscopy techniques. Most reactions of interest are ultrafast (in femtosecond time scale) and require special ultrafast laser systems to be studied. The details of spectroscopy methods go far beyond the scope of this chapter. However, there are many specialized reviews, books, and text books to refer to for practical experimental technique implementations [9–12]. The details of MOs photophysics and the phenomena that affect their properties and performance are explained in Sections 6.2–6.4.

## 6.2 Photophysics of Bulk MOs

### 6.2.1 Energy-Level Structure and Steady-State Spectra

Optical properties of semiconductors are determined by the bandgap energy, $E_{bg}$, or the gap between the top most energy of the VB, $E_{VB}$, and lower edge energy of the CB, $E_{CB}$, where $E_{bg} = E_{CB} - E_{VB}$. Photons with energies below the bandgap, $h\nu < E_{bg}$ are not absorbed by the semiconductor, and above ones, $h\nu > E_{bg}$, are absorbed, and this absorption results in promoting an electron from the VB to the CB.

A typical energy diagram for this simplified case shows relative positions (energies) of the CBs and VBs, as presented in the left side of Figure 6.1. This ignores carrier momentum. In crystals, the energy of carriers depends on the momentum and the momentum at highest possible energy of an electron in the VB may differ from the momentum of the lowest energy in CB. This leads to two types of semiconductors distinguished as direct and indirect bandgap semiconductors. For the direct bandgap semiconductors, carriers have the same momentum at the top of VB and the bottom of CB (middle case in Figure 6.1), whereas for indirect bandgap, the maximum at VB is shifted relative to the minimum in CB (right case in Figure 6.1), which reduces probability of photon absorption with the energy just above the bandgap.

The electron–hole recombination may generate a photon for direct bandgap semiconductors, whereas recombination in indirect bandgap has lower probability

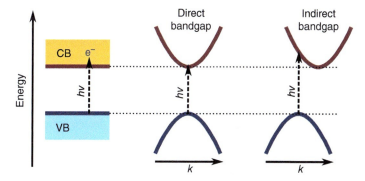

**Figure 6.1** Energy diagram: simplified conduction band (CB) and valence band (VB) presentation (left) and schematic presentation of direct (middle) and indirect (right) bandgap presentation with added momentum coordinate (k).

of photon emission and gives off heat mostly. Also experimentally, the emission of the direct bandgap semiconductor is observed close to the wavelength corresponding to the bandgap edge, $\lambda_{bg} = hc/E_{bg}$, but the emission of indirect bandgap semiconductors is typically much weaker, broader, and shifted to the red from $\lambda_{bg}$.

In general, the efficiency of photon absorption with energy above the bandgap depends on the density of electronic state of both VB and CB, and thus it depends on the type of semiconductor. The efficiency of light absorption is characterized by absorption coefficient, $\alpha$, which describes how the light intensity decays while the light propagates in an absorbing medium. Inside the absorbing medium, the light intensity decreases in the direction of the light propagation as

$$I(x) = I_0 \, e^{-\alpha x} \tag{6.1}$$

where $I_0$ is the light intensity at position $x = 0$ and $I(x)$ is at position $x$, respectively. As an example, for TiO$_2$ rutile form $\alpha \approx 10^6$ cm$^{-1}$ at 310 nm, and $\alpha$ is roughly two times lower for its anatase form [13]. It can be noted that many traditional semiconductors, such as Si, Ge, and GaAs, have much lower absorption coefficient at the wavelengths close to the bandgap. For example, at 900 nm for Si, the $\alpha$ is roughly 300 cm$^{-1}$ only. High $\alpha$ value means that already a very thin layer can be almost nontransparent. For example, 100 nm thick TiO$_2$ rutile film at 310 nm has transmittance $\exp(-\alpha x) = \exp(-10) \approx 0.000\,05$ or almost no light will pass such film at the wavelength just below the wavelength corresponding to the bandgap. However, the transmittance of a 20-nm film is expected to be 0.14 at the same wavelength.

In the example above, only light absorption inside the film was calculated. In actual measurements, the monitoring light passing through the film experiences reflection at both film interfaces in addition to the absorption in the film. To account for the reflectance at the interfaces, both transmittance, $T$, and reflectance, $R$, spectra have to be measured, and if the light interference inside the film can be neglected,[1] the sample absorbance is calculated as [14–16]

$$A = -\log\left(\frac{T}{1-R}\right) \tag{6.2}$$

In other words, referring to the example above, the actually measured transmittance of 20 nm film will be lower than that calculated using Eq. (6.1), since there will be light reflectance at both sides of the film interface. The reflectance depends on the refractive index, $n$, and for MOs it cannot be neglected since it is relatively large. For example, in the case of rutile TiO$_2$, it approaches 4 at 310 nm, and at wavelengths above the bandgap, $\lambda > 400$ nm, it is in the range of 2.4–2.8 [13]. Assuming $n = 2.5$, the reflectance at air–TiO$_2$ interface is $R = \left(\frac{n-1}{n+1}\right)^2 \approx 0.18$, which increases apparent absorbance by $A = -\log(1-R) \approx 0.06$. This is essential value since at $\lambda > 400$ nm, the absorption of crystalline and polycrystalline TiO$_2$ can be neglected at least for films with few hundreds of nanometer thickness.

Formally in optical calculations, the both transmittance and reflectance properties of films can be accounted for by introducing complex refractive index,

$$\tilde{n} = n + i\kappa \tag{6.3}$$

---

[1] For example, if the film thickness is much smaller than the wavelength.

where the real part, $n$, is responsible for the light refraction, and $\kappa$ is the extinction coefficient responsible for the light absorption and $\alpha = \frac{4\pi\kappa}{\lambda}$. Furthermore, the refractive index is directly related to dielectric constant of the material as $\tilde{n}^2 = \epsilon$. The latter is also complex value and in light absorbing media it is $\epsilon = \epsilon' + i\epsilon''$. Therefore, $\epsilon' = n^2 - \kappa^2$ and $\epsilon'' = 2n\kappa$. Nevertheless, a common parameter used to specify materials absorption property is absorption coefficient, $\alpha$, which is available from simple absorption spectra measurements.

There are no allowed electronic states inside the bandgap of an ideal monocrystalline semiconductor. However, there are many types of intraband states in real semiconductors, though the density of these states is much lower than the density of states above the CB (or below the VB). Some typical reasons for this are crystal lattice defects, impurities, intentional doping, surface defects, etc. Phenomenologically, the intraband states can be classified as shallow and deep states. The former are expected to be just a little lower than the bottom of the CB (in the case of electronic states). The electrons at the CB can be trapped by these shallow states but can also leave the states back to the CB if they get energy. This is possible if the activation energy required for the electron to jump back to the CB is not much larger than the thermal energy, $kT$. At room temperature $kT \approx 0.025$ eV and this scales to roughly 0.1–0.2 eV relative energy of the shallow states below the CB.

On the contrary, the deep intraband states are the states well separated from the CB (and the VB), and the trapped carrier cannot return back to the CB without receiving a sufficient additional energy, e.g. absorbing a photon. Nevertheless, the trapped electron can recombine with a hole in the VB and release the energy by emitting a photon. This makes PL spectroscopy a viable tool to study intraband states as discussed in Section 6.2.3.

Defects in MOs can introduce deep trap states in the crystal system. The defects can be native intrinsic defects, formed due to some missing atoms in the crystal structure or they can be extrinsic defects, formed due to introduction of dopants. The missing atoms inside the crystal lattice form vacancies [17], and the broken or dangling bonds on the surface form surface trap states. For example, in the case of $TiO_2$, there might be some oxygen vacancies in the crystal lattice, which introduce $Ti^{3+}$ trap states in the system [18]. These trap states play an important role in decelerating electron–hole recombination rate and enhancing the photocatalytic activity of $TiO_2$. These trap states also provide charges on the interfaces for chemical reactions in electrolyte environment. Another MO, $SnO_2$, shows n-type conductivity due to the presence of oxygen vacancies derived from stoichiometric defects of the material [19]. Therefore, it can be elucidated that the defect states play an important role in the trapping, diffusion, and recombination of photogenerated charge carriers [20]. In $TiO_2$, deep states promote the recombination of charge carriers, while shallow states promote diffusion of charge carriers thereby increasing their lifetimes [21]. Bauer et al. [22] studied the relaxation processes of charge carriers in ZnO nanocrystalline thin films. They observed that the charge carrier recombination is linked to the shallow trapped electron and deep trapped hole.

Doping is one of the promising ways to alter electronic properties of MOs. In this process, impurity atoms can be substituted in the crystal lattice of semiconductors to

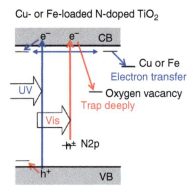

**Figure 6.2** Schematic illustration of Cu/N-TiO$_2$ powder and Fe/N-TiO$_2$ powder after excitation at 360 nm (blue arrow) and for 450 nm (orange arrow). Source: Reprinted with permission from [26]. Copyright (2013) American Chemical Society.

generate extra electrons or holes, and it affects the intrinsic carriers concentration, crystal structure, defects, and bandgap of MOs. Doping of semiconductor MOs can be done with either electron acceptors (p-type doping) or electron donors (n-type doping). Modifications of MOs can be done by adding metals such as Cu, In, Ag, Sn, Co [23], etc. or non-metals such as P, N, S, [24], etc. to influence their electronic structure.

Doping introduces new allowed energy states within the bandgap of material. The n-type impurities create allowed states near the CB, and p-type impurities create these states near the VB, and the gap between these newly created energy states and the nearest energy band is called bonding energy, $E_B$. The bonding energy must be relatively small so that even at room temperature the electrons can get to the CB or the holes to the VB, respectively. The introduction of new allowed energy levels in MO systems may increase the total recombination time, which affects the efficiency of photoinduced process.

Zhao et al. [25] prepared Zn-doped TiO$_2$ samples and their XPS results showed increased percentages of oxygen vacancies by increasing Zn doping concentration. These oxygen vacancies induce subband levels near the bottom of CB and can easily capture photoinduced electrons from TiO$_2$. Yamanaka et al. [26] prepared Cu- and Fe-loaded nitrogen-doped TiO$_2$. By this, dopant states were generated inside the MO system along with oxygen vacancy trap states and deep trap states, as shown in Figure 6.2. Charge carrier dynamics was studied by diffuse reflectance spectroscopy, and the result showed photoexcited electron transfer (ET) from TiO$_2$ to loaded Cu or Fe with long time constants under the UV light, which elucidated that metal loading in MO inhibited charge recombination between the electron and hole.

However, increasing oxygen vacancies and doping atoms beyond a certain ratio are detrimental as it impedes the efficiency of photoinduced process thereby introducing many recombination sites. Also, it should be noted that PL can be used to determine energy of the intraband states but not the density of the states since the PL intensity is relative and the emission quantum yield is usually unknown. The absorption measurements can provide quantitative information on the density of states, but the density of the intraband states is usually very low and not deducible from the

**Figure 6.3** Bandgap energy ($E_g$) determination from the Tauc plot. The linear part of the plot is extrapolated to the x-axis. Source: Reprinted with permission from [28]. Copyright (2018) American Chemical Society.

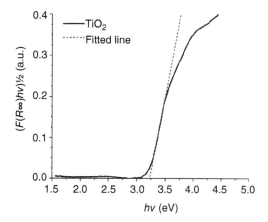

absorption spectra measurements unless the samples are intentionally doped at high level, such as few percents.

The fact that photons with energy lower than the bandgap are not absorbed by semiconductor but above the bandgap are absorbed effectively can be used to determine the bandgap by simple measurement of the semiconductor absorption spectrum. The bandgap obtained by this method is called "optical bandgap." At the qualitative level, the wavelength, $\lambda_{bg}$, at which semiconductor start absorbing the light can be used as a measure of the bandgap energy, $E_{bg} \approx \frac{hc}{\lambda_{bg}}$. More quantitative results can be obtained using "Tauc plot" presentation of the semiconductor absorption [27]. It was noted that close to the bandgap, the absorption coefficient is proportional to the photon (excess) energy as

$$h\nu\alpha \sim \left(h\nu - E_{bg}^{opt}\right)^n \tag{6.4}$$

where $E_{bg}^{opt}$ is the optical bandgap, $v$ is the light frequency, $\alpha$ is the absorption coefficient, and $n$ is the power factor which depends on the type of bandgap. For direct bandgap semiconductors, $n = \frac{1}{2}$ and for indirect it is $n = 2$. Plotting the value $(\alpha h\nu)^{1/n}$ as function of photon energy $h\nu$, approximating the initial rise at $h\nu > E_{bg}^{opt}$ and determining intersect point with abscissa gives the optical bandgap of semiconductor. As an example, Figure 6.3 shows the Tauc plot of $TiO_2$ that has indirect bandgap, where the dotted red line meeting x-axis displays its optical bandgap [28].

Simple steady-state absorption spectroscopy provides information on the semiconductor bandgap. The steady-state emission spectroscopy, or PL spectroscopy, is another technique to learn about energy structure of semiconductors. An advantage of PL is that emission spectroscopy is more sensitive method than absorption spectroscopy in general [9]. In an ideal case of defect-free sample, the only PL of a semiconductor comes out at the electron–hole recombination event at the CB and VB, respectively. This emission has a single band with maximum at the wavelength corresponding to the bandgap energy (direct bandgap semiconductor). However, different trap states may relax by emitting a photon as well, though the quantum yield of this PL is usually low. High sensitivity of the emission spectroscopy comes

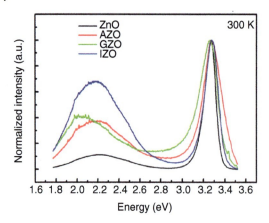

**Figure 6.4** Normalized PL spectra in energy domain (eV) of as-grown and doped ZnO nanorods, (AZO: Al-doped, GZO: Ga-doped, IZO: In-doped) [29]. Source: Kim et al. [29].

useful in this case, and the PL measurements provide information on the energies of the trap states. For example, if electron traps have energy $E_{et}$ relative to the top of the VB, the emission associated with trapped electrons recombination with the holes at the VB is expected at the wavelengths $\lambda = \frac{ch}{E_{et}}$. However, in most cases, there is a broad distribution of the trap state energies resulting in a relatively broad emission spectrum associated with these states.

An example of PL spectra of ZnO nanorods with different metals doping is shown in Figure 6.4 [29]. A relatively sharp peak close to 3.3 eV (corresponds to roughly 380 nm) arises from the CB–VB carriers recombination, and it agrees well with the optical bandgap evaluated from the Tauc plot for un-doped ZnO, which is 3.24 eV. The broad bands in the visible part of the spectrum, 1.7–2.5 eV (roughly 500–730 nm) can be attributed to PL involving doping levels in the middle of the bandgap. It can be noted that the sample without additional doping still has some PL in the visible part of the spectrum, which indicates that there are some defects in the ZnO nanorods even without doping. It is also worth noting that none of the samples has absorption in the visible part of the spectrum which would indicate a presence of defect states or doping levels.

To study the effect of doping on $TiO_2$, Wang et al. [30] doped $TiO_2$ with $In^{3+}$ by using indium chloride as precursor. Unique chemical species, $O-In-Cl_x$ was formed on the surface of $TiO_2$ which has energy states 0.3 eV below the CB of $TiO_2$. PL spectroscopy revealed two peaks around 480 and 525 nm, which were assigned to oxygen vacancies and surface states, respectively. The results showed suppression in recombination of photogenerated charge carriers in oxygen vacancies peak upon doping of $In^{3+}$. It was elucidated that electrons and holes prefer to accumulate in surface states created by $O-In-Cl_x$ and VB of $TiO_2$, respectively. Thus, emission intensity originating from the oxygen vacancies in $TiO_2$ is greatly decreased by doping, as shown in Figure 6.5. Therefore, it was concluded that surface states originated from doping allowed more efficient separation of electrons and holes.

Steady-state absorption and emission spectroscopy instruments are inexpensive and widely available, and they are efficient enough to study the key photophysical properties of semiconductor MOs, such as the bandgap energy, and present even the

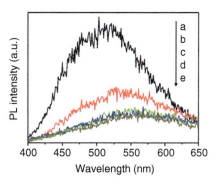

**Figure 6.5** PL spectra of (a) TiO$_2$, (b) TiO$_2$-In 3%, (c) TiO$_2$-In 7%, (d) TiO$_2$-In 10%, and (e) TiO$_2$-In 15%. Source: Reprinted with permission from [30]. Copyright (2009) American Chemical Society.

energies of the defect states. Though it is important to notice that not all defect states can be detected with these simple spectroscopy tools and much detailed analysis of the MOs electronic properties and lifetime dynamics is not possible with these instruments.

### 6.2.2 Photoexcitation and Relaxation Dynamics

Absorption of a photon with energy greater than the bandgap of the material results in the promotion of an electron from the VB to the CB. The missing electron in the VB is termed hole, and the electron–hole pair formed right after the photon absorption is called exciton. However, the act of photon absorption does not change position of the carriers, though it does change the electron localization. Formally, right after the excitation the photogenerated electrons and holes have the same position of the center of masses, or can be located at the same position. Coulomb interaction between them results in the exciton coupling stabilization energy. If the coupling energy is high, exciton may have relatively long lifetime [31]. In most MOs, coupling is weak, and the excitons dissociate quickly. For example, the exciton coupling energy in ZnO is considered to be high compared to other MOs, but it is only 60 meV [32], or only 2.4 times larger than the thermal energy $kT$ at normal conditions. As a result, the exciton dissociation occurs in a sub-picosecond time scale and after that electron and hole can move independently.

In most cases of practical importance, the excitation photon energy, $h\nu$, is larger than the bandgap energy, $E_{bg}$. This results in generation of carriers (electrons in the CB) with excess energy $\Delta E = h\nu - E_{bg}$. These carriers are called "hot carriers," and they lose the excess energy through multiple interactions with the crystal lattice producing phonons and heating the MO. Typically, this process is very fast, and it occurs in a subpicosecond time scale [33] and typically it precedes other possible relaxation reactions.

Schematically, the process of photoexcitation is presented in Figure 6.6 by a "vertical" elevation of an electron from the VB to the CB and leaving a hole in the VB as shown in reaction 1. The hot electron relaxes to the bottom of the CB, and it is shown as reaction 2 in the scheme. Considering photocarrier spatial localization, another fast reaction taking place in MOs is termed "exciton dissociation," and it

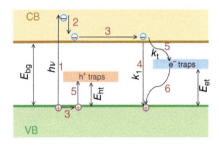

**Figure 6.6** Schematic presentation of energy levels and main carrier reaction pathways following the photoexcitation; Reaction 1: electron–hole dissociation and hot electron generation, 2: relaxation of hot electron, 3: diffusion of electrons and holes, 4: electron–hole recombination, 5: electron and hole trapping, and 6: trapped electrons relaxation.

allows electron and hole to diffuse independently [34], as was discussed above. This is reaction 3 in Figure 6.6.

The electron in the CB must recombine with the hole in the VB to return to the ground state of the system as before the excitation. The large amount of energy released in recombination process can be emitted as a photon, a process known as photoluminescence, or PL. The energy of PL photons is just below the bandgap energy. The process of the recombination depends on concentrations of both the electrons and holes. However, since the electrons and holes have different mobilities and typically, there are some excess carriers, either holes or electrons before the excitation, the recombination dynamics depends only on concentration of the minority carriers generated by photoexcitation. Within this approximation and assuming a low excitation limit, the photocarrier population is described by the first-order kinetic equation [35]

$$\frac{dn}{dt} = -k_1 n \tag{6.5}$$

where $n$ is the carrier density, electrons in the CB in most cases, and $k_1$ is the relaxation rate constant. The relaxation reaction can be the electron–hole recombination, reaction 4 in Figure 6.6, or the carrier trapping, reaction 5. If both reactions are possible, then $k_1 = k_r + k_t$, where $k_r$ is the recombination rate constant and $k_t$ is the trapping rate constant. The trapped electrons also recombine with the holes in VB, as shown in reaction 6. The solution of Eq. (6.5) is the exponential decay of the carrier density after pulsed excitation $n(t) = n_0 \exp(-k_1 t)$, where $n_0$ is the density generated by the excitation pulse.

The electron–hole recombination may results in photon emission and nonradiative recombination, which are characterized by two rate constants $k_{PL}$ and $k_{nr}$, respectively. Then, the total relaxation rate is $k_1 = k_{PL} + k_{nr} + k_t$, and the emission quantum yield is (by definition) [36]:

$$\phi_{PL} = \frac{k_{PL}}{k_1} = \frac{k_{PL}}{k_{PL} + k_{nr} + k_t} \tag{6.6}$$

Measurements of the emission decay time constant reveal the total relaxation rate constant, $k_1 = 1/\tau$.

If the excitation density is increased, the density of photogenerated carrier increases, and the interaction of carriers with each other has to be taken into account. This leads to so-called higher-order decay kinetics, such as

$$\frac{dn}{dt} = -k_1 n - k_2 n^2 + k_3 n^3 \tag{6.7}$$

where $k_2$ is the second-order interaction rate constant, and $k_3$ is the third-order process. The latter may arise from Auger effect and called Auger recombination. From the practical point of view, at a high excitation density, the relaxation dynamics for charge carriers is faster.

The kinetic equations (6.5) and (6.7) describe reasonably well the population change in homogeneous systems with constant values of $k_1$, $k_2$, and $k_3$. In many cases of practical importance and in particular in nanostructured MO, the systems are heterogeneous and sometimes the reaction rate constants are changing from spot to spot, e.g. from nanocrystal to nanocrystal. In this case, a more realistic model should be considered which accounts for some distribution of the decay rate constant. Unfortunately, there is no simple and unified mathematical description of such system. One of the simplest "solutions" in the case of distribution of the first order decay rate constants is the so-called stretched exponential decay (also known as Kohlrausch function) [37]

$$n(t) = n_0 \exp\left[-(k_s t)^\beta\right] \qquad (6.8)$$

where $k_s$ is a rate constant, and $\beta$ is called stretching parameter, which is in the range of 0 … 1. The value $\beta = 1$ corresponds to a pure exponential decay, and smaller $\beta$ results in a faster decay at the beginning and slower in the end. Although this decay model gives reasonably good approximation of the experimental data, there is no clear physical meaning associated with parameter $\beta$. The stretched exponential decay has to be used with caution since it is known to result in "unphysical" fast decay in short time scale, and the results with $\beta < 0.4$ must be carefully revised.

If the photon energy is two times higher than the bandgap or more, the excess energy of the hot carrier is sufficient to promote another electron from the VB to the CB and lose the corresponding amount of energy in a single step. This process is referred to as multiexciton generation, and it competes with the thermal relaxation of the hot carrier through generation of multiple phonons. Multiexciton generation with single photon was experimentally confirmed in quantum dots (QDs) [38], which have slower thermal relaxation of hot carriers. It was also observed in low bandgap semiconductor QDs, which makes feasible utilization of photons with energy much higher than the bandgap energy.

### 6.2.3 Emission Decay Kinetics, Time-Resolved PL

The time-resolved PL spectroscopy may be a very useful tool to gain information about intraband states. Li et al. [39] studied ZnO nanowires deposition procedure using steady-state and time-resolved PL. The presence of the defect states within the bandgap is observed as a broad emission in the visible part of the spectrum, whereas the direct recombination of the CB electrons and VB holes results in a narrow emission band in the UV part of the spectrum, similar to that shown in Figure 6.4. The steady-state PL measurements were complemented by time-resolved PL decay measurements as presented in Figure 6.7. Based on the study, the defects were attributed to oxygen vacancies in ZnO.

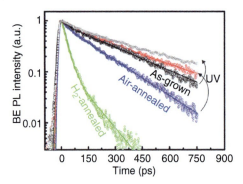

**Figure 6.7** Normalized PL decays of ZnO nanowires at the band edge. The samples were prepared and post-treated at different conditions. Source: Reproduced from Ref. [39] with permission from the Royal Society of Chemistry.

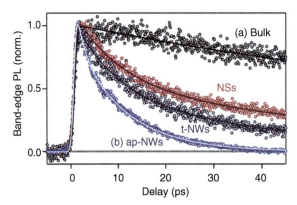

**Figure 6.8** PL decays of ZnO bulk, as deposited nanowires (ap-NWs) and heat-treated nanowires (t-NWs) at the wavelengths corresponding to the bandgap (380 nm) for (a) commercially available bulk single crystals (b) as-prepared nanowires (ap-NWs). Source: Reproduced from Ref. [41] with permission from the AIP Publishing.

Rawool et al. [40] prepared heterojunction of NiO (p-type MO) with $TiO_2$ (n-type MO) for studying photocatalytic hydrogen generation. Nanoparticles of different molar ratios of $TiO_2$ and NiO were prepared. The p- and n-type of MO semiconductor systems were designed to improve the separation of photogenerated electrons and holes. Time-resolved PL study of Ti/Ni oxides was undertaken, and it was concluded that formation of the heterojunction results in photocarrier separation with electrons collected on $TiO_2$ and holes on NiO sides, respectively, leading to a longer lifetime of the carriers and improved photocatalytic activity of the hybrids.

Appavoo et al. [41] studied PL decay of a series of ZnO bulk and nanowire samples in picosecond time domain using a Kerr-base technique, which allowed to achieve a picosecond time resolution. As shown in Figure 6.8, the longest PL lifetime was observed for bulk material. The nanowires had much shorter PL lifetime, which was only slightly improved by thermal treatment. This can be attributed to the surface involvement in the faster electron–hole recombination.

The time-resolved PL adds possibility to gain information on the lifetime of the carriers in the CB and also in the trapped states. It also retains the high sensitivity advantage of the steady-state emission spectroscopy, which makes it useful in studying thin films, nanostructures, and other objects with small amount of substance. There is a range of reasonably cost-effective instruments available for the time-resolved PL studies, which also makes it an affordable research method.

## 6.2.4 Transient Absorption (TA) Spectroscopy

Transient absorption (TA) spectroscopy is a useful tool in studying the charge carrier dynamics of semiconductor MOs in a wide temporal range from femto- to milliseconds. A distinct advantage of TA over PL is that virtually any change in electronic state of the sample has an effect on its TA response, whereas only emissive states are observed in PL. Similar to the time-resolved PL, the TA provides important information on the dynamics of the photogenerated charge carriers and helps to identify photoreactions in MO or other semiconductor samples. Furthermore, the TA spectroscopy can be used to distinguish between reactions of free and trapped holes and electrons. For example, Katoh et al. [42] studied the TA spectra of $TiO_2$ and identified TA spectra of conducting electrons, trapped electrons, and holes, as shown in Figure 6.9. The intraband transition of conducting electrons in $TiO_2$ and the excitation of electrons in the bottom of CB to higher sublevels of the CB can be seen as a broadband absorption with intensity increasing toward IR range (1000–10 000 nm). The absorption band with maximum around at 800 nm was attributed to trapped electrons which is due to d–d transition of localized $Ti^{3+}$ states, and the band with maximum close to 550 nm was assigned to trapped holes.

It must be noted that the TA response kinetics depends on the excitation density and on the density of the photocarriers in the CB. In most cases of practical importance, e.g. study of photoreactions in solar cells, the excitation photon flux creates a little excess carrier population which exclude high-order carrier relaxation kinetics discussed above (Eq. (6.7) and discussion after). The laser spectroscopy techniques can provide photon flux greater than that of the Sun by many orders of magnitude and induce phenomena that do not occur in the devices of interest. Therefore, the time-resolved spectroscopy experiments must be carried out at sufficiently low

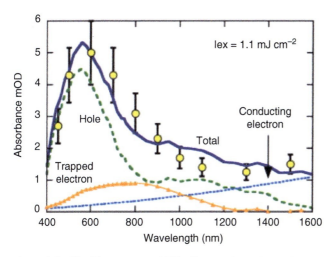

**Figure 6.9** The TA spectrum of $TiO_2$ films excited with 355 nm laser pulse decomposed on the spectra of free electrons in CB, trapped electrons, and trapped holes. Source: Reprinted with permission from [42]. Copyright (2010) Elsevier B.V.

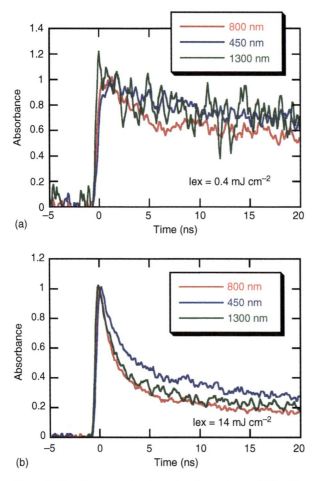

**Figure 6.10** Charge carrier dynamics dependence of $TiO_2$ films excited at 355 nm on TA excitation densities at (a) 0.4 mJ cm$^{-2}$ and (b) 14 mJ cm$^{-2}$. Source: Reprinted with permission from [42]. Copyright (2010) Elsevier B.V.

excitation densities excluding photocarrier interaction and quenching. In the above-mentioned study [42], the measurements were carried out at low (0.4 mJ cm$^{-2}$) and high (14 mJ cm$^{-2}$) excitation densities at 355 nm. The normalized responses at selected monitoring wavelengths are shown in Figure 6.10, where a much faster decay can be seen at high excitation density.

Determination of a "safe" excitation density is a common first task in TA spectroscopy experiments. It depends on the studied samples and their phenomena of interest. Yoshihara et al. [43] studied TA response of $TiO_2$ nanoparticle film prepared following Grätzel protocol in nanosecond time domain. A "safe" excitation density was determined to be close to 50 μJ cm$^{-2}$. Up to this value, the increase of the excitation density resulted in a stronger response, but the response shape was density independent. At higher excitation densities, a fast component was observed and attributed to higher-order decay reactions. Authors have also demonstrated that the characteristic intraband absorption of electrons (free electrons) follows power low $\lambda^{1.7}$ in the range of 1000–2500 nm. A broad band at 770 nm (roughly 550–1000 nm) was assigned to trapped electrons. Trapped holes have a band at 520 nm and a shoulder at 900–1200 nm.

Tamaki et al. [44] have conducted picosecond pump-probe measurements of $TiO_2$ nanocrystal film and concluded that "safe" excitation density is roughly 0.5 mJ cm$^{-2}$, and calculated spectra of holes (a broad band at 450–700 nm), trapped electrons (a broad band at 550–1000 nm), and conducting electrons (an increasing absorption toward longer wavelengths) in the range 400–1600 nm.

Sachs et al. [45] reported that photogenerated carriers recombine virtually equally fast in dense and mesoporous films of both rutile and anatase $TiO_2$ samples. The measurements were carried out in near infrared (NIR, at 1200 nm), with excitation densities in the range 0.03–0.7 mJ cm$^{-2}$ at 355 nm. At low excitation densities, the longest carrier lifetime was observed for mesoporous anatase and was estimated to be roughly 2 ns, whereas in dense anatase it was closer to 1 ns. The lifetime was close to 100 ps for rutile mesoporous film and around 40 ps for dense rutile film. It can be noted that the size of nanoparticles was relatively large (>20 nm), and no confinement effect of the size on absorption spectra was reported. The quantum confinement will be discussed below in Section 6.3.1.

It can be argued that for $TiO_2$ samples the "safe" excitation density is <0.3 mJ cm$^{-2}$ at excitation wavelengths <360 nm (shorter than the optical bandgap). Also, there are good references to distinguish spectrally the free electrons in the CB, trapped electrons, and trapped holes.

Another popular MO is ZnO. ZnO nanowires were studied by Cooper et al. [46] using fs TA. The nanowires were annealed in air followed by hydrogen treatment. Based on this series of TA measurements and accounting previous ZnO studies the energy scheme for shallow traps and different types of electron and hole intraband traps was proposed, as shown in Figure 6.11.

Cherepy et al. [47] studied ultrafast electron dynamics in γ- and α-$Fe_2O_3$ nanoparticles by femtosecond transient absorption spectroscopy, as shown in Figure 6.12. The interest to $Fe_2O_3$ is driven by its absorption in the visible region of the solar spectrum and potential use in photocatalytic water-splitting applications, and also by its abundance in nature. However, $Fe_2O_3$ conductivity is very low, showing resistivity of

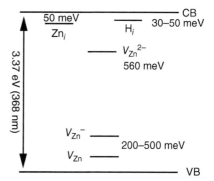

**Figure 6.11** Proposed band structure of ZnO, considering the relative energy levels estimated for its main types of defects. Source: Reprinted with permission from [46]. Copyright (2012) American Chemical Society.

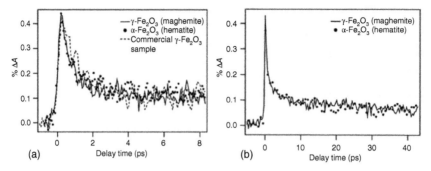

**Figure 6.12** TA decay profiles of nanoparticles of $\gamma$-$Fe_2O_3$, $\alpha$-$Fe_2O_3$, and a commercial sample of $\gamma$-$Fe_2O_3$ excited at 390 nm and probed at 720 nm from (a) 0–8 ps and (b) 0–40 ps time scales. Source: Reprinted with permission from [47]. Copyright (1998) American Chemical Society.

$10^3$–$10^6$ $\Omega$ cm for $\alpha$-$Fe_2O_3$, and the carrier mobility is also low. To study the photoreactions, the samples were excited with 390 nm pump. It was observed that the decay dynamics of both $\gamma$- and $\alpha$-$Fe_2O_3$ is fast and almost identical. Therefore, the crystal size and structure do not influence the charge carrier lifetime, and the electron–hole recombination is mediated by intrinsic mid-bandgap states and trap states.

Corby et al. [48] studied the role of oxygen vacancies in $WO_3$ for photoelectrocatalytic water splitting. Transient absorption spectroscopy was used to study the recombination dynamics of photogenerated electrons and holes. It was found that intermediate concentration of the vacancies (2% of oxygen atoms) gave the highest photoinduced charge carrier densities, and the slowest recombination kinetics across, and thus, it is preferred for the photocatalytic applications.

The transient absorption spectroscopy is a very powerful tool to study photophysics of MOs alone, in combinations with other materials and even incorporated within different photodevices. An advantage of the TA is that it covers an extremely broad time scale from femtoseconds and up to seconds. Another advantage is possibility to discriminate between different intermediate states through their distinct spectral features. The latter, however, requires additional knowledge and specially designed reference samples, which is a challenge of its own.

## 6.3 Nanostructures

### 6.3.1 Quantum Confinement

The effect of the space limiting on energy levels of nanocrystals is well understood in the frame of "the particle in the box" quantum model, which considers an elementary particle, electron in our case, placed in a potential constant in some small space and rising instantly to a higher value, or even infinite value, outside this space. The spatial restriction on the elementary particle location results in the appearance of distinct energy levels with energies inversely proportional to

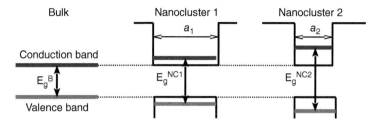

**Figure 6.13** Potential energy profiles for bulk semiconductor and two nanoclusters of bigger ($a_1$) and smaller ($a_1$) size.

the square of the "box" size. This is commonly called "quantum confinement." In the case of a semiconductor, which has a bulk bandgap $E_g$, the effect of quantum confinement is schematically presented in Figure 6.13. Limiting the size of semiconductor results in a larger bandgap, and the size effect is stronger for smaller semiconductor particles.

The quantum confinement is well studied, and the most simple system for the confinement modeling is QD, which is a spherical nanocrystal with diameter $a$. The bandgap energy for such particle depends on the bandgap of the bulk material ($E_g$), the confinement effect, and the Coulomb interaction between electron and hole after excitation [49]:

$$E = E_g + \frac{h^2}{8a^2}\left(\frac{1}{m_e} + \frac{1}{m_h}\right) - \frac{1.8 q_e^2}{4\pi\varepsilon_0 \varepsilon a} \tag{6.9}$$

where $h$ is the Plank constant, $m_e$ and $m_h$ are the effective masses of the electron and hole, $q_e$ is the electron charge, and $\varepsilon_0$ and $\varepsilon$ are the dielectric permittivities of vacuum and material, respectively. Typically, the effective mass of electron in the CB (and hole in the VB) differs greatly from the mass of resting electron. For example, the effective mass of electron in ZnO is $0.26m$, where $m$ is the free electron mass at rest [50]. This difference cannot be ignored in a reasonable quantitative estimation.

Figure 6.14 illustrates the dependence of the apparent bandgap energy, $E$, on the QD-sized $a$ for two values of dielectric constant, $\varepsilon = 10$, which is close to that of ZnO, for example, [50], and $\varepsilon = 100$. Calculations were carried out assuming $m_e = 0.26m$, $m_e \ll m_h$, and $E_{bg} = 2$ eV, which corresponds to a wavelength of 620 nm.

The effect of confinement can be already noticed when the ZnO nanocrystal size is reduced to 10 nm, and it becomes quite obvious at dimensions <8 nm [51]. The size effect depends strongly on the type of semiconductor, on effective carrier masses in the first place but also on intrinsic dielectric permittivity as well. At a qualitative level, confinement effect is observed when the nanocrystal size becomes compatible with the electronic wave function delocalization at the CB.

The most straightforward indication of the quantum confinement is the increase of the apparent bandgap energy which is available experimentally from steady-state absorption measurements. For emitting samples, one can acquire emission spectra and observe the confinement effect as a blue shift in the emission as well. It has to be noted that quantum confinement has effect on positions of both CBs and VBs, as

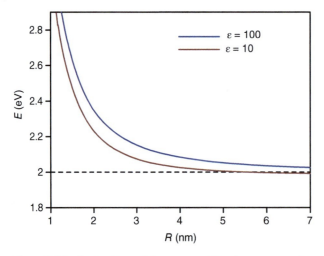

**Figure 6.14** Dependence of the apparent bandgap energy of quantum dot on its size calculated using Eq. (6.9), and $m_e = 0.26m$, $m_e \ll m_h$, and $E_{bg} = 2$ eV. The values of $\varepsilon$ are indicated in the plot.

shown in Figure 6.13. The energy of the CB increases and of the VB decreases. The absolute value of the energy shift depends on the effective masses of the electron and hole, respectively. Typically, effective mass of electron is smaller and the shift is larger.

Theoretical modeling of the size and shape effect of $TiO_2$ on carrier dynamics was undertaken by Nam et al. [52]. It was concluded that the recombination of the electron–hole pair generated by photoexcitation becomes faster as the size decreases, though the computational model was limited by 97 $TiO_2$ units, or roughly 3 nm size. Experimental study of the size effect on the carrier lifetime was carried out by Pozina et al. [53]. The PL decays of colloidal ZnO nanocrystal were studied, and a decrease in the lifetime from 22 to 6 ps was observed on the nanocrystal radius reduction from 45 to 8 nm. Samples were excited at 266 nm (third harmonic of Ti:sapphire laser).

Photocatalysis is one of the application areas that can benefit from possibility to tune the optical and photophysical properties of semiconductor nanoparticles. This motivates research interests on MO nanostructures with characteristic dimensions close to 10 ns and smaller [54].

The carrier motions are restricted in all three directions in QD leaving no possibility for the carriers to move. Therefore, QDs are often referred to as zero-dimensional (0D) quantum objects. However, the quantum confinement effect is still observed if carriers are restricted in two directions and free to move in one, thus termed as 1D objects, or restricted in only one direction and can move is a plane or in 2D. Examples of 1D structures are nanorods and nanowires. Quantum wells are examples of 2D structures. Quantum wells can be prepared using molecular beam epitaxy when a thin layer of material with relatively low bandgap is incorporated in the bulk of material with higher bandgap energy. Though any semiconductor film, e.g. $TiO_2$ or ZnO atomic layer deposited films, with thickness close to 10 nm will show the confinement effect by the blue shift of the bandgap absorption edge and emission band, and thus can be treated as 2D object.

Although the exact effect of the size on the semiconductor nanostructure depends on the dimensionality and geometry in general, in order of magnitude the scaling parameter is carrier delocalization at the CB, when one of the dimensions approaches characteristic carrier delocalization size the confinement effect is observed [54]. For one and the same material, quantum confinement starts to play the role at roughly the same size of the nanostructure independent of its other dimension such as nanoparticle, nanowire, or nanosheet. For Si, this critical size was reported to be close to 6 nm [55]. In colloidal QDs such as CdSe and PbS, the confinement effect can be observed at the size approaching 10 nm [56].

A detectable "blue shift" of the absorption band edge was observed for ZnO films with thickness close to 15 nm, and the bandgap increased by almost 0.5 eV when the layer thickness was reduced to 2.4 nm [57]. Similarly, ZnO QDs with average size of 1.9 nm show significant rise of the bandgap energy to 4.27 eV [58]. For crystals of very small size, the surface imperfections and defects result in formation of different types of defect levels within the bandgap, which was observed as broad emission of the QDs in the visible part of the spectrum (400–600 nm); however, hydroxy passivation of the surface allowed to diminish the number of defects gradually [58].

In all above mentioned cases, the quantum confinement starts to play a significant role when at least one of the dimensions of the MO structure or other semiconductor approaches 10 nm or less. A straightforward indication of the confinement is the blue shift of the bandgap, which can be observed by measuring steady state absorption and emission spectra. Also, the excited state lifetime can be affected and even new relaxation pathways can be tailored since the surface phenomena play increasing role as the dimensions decrease.

### 6.3.2 Surfaces and Interfaces

Surfaces of MOs play an important role determining how a specific material will interact with gaseous or liquid environment. The interaction of water or electrolyte with MOs is crucial and critical factor for their practical applications such as photocatalytic water splitting or $CO_2$ reduction for fuels production. MOs form interfaces with external media and can act as interpercolated high surface area junctions between electron donors and acceptors [59]. However, to control the charge transport, it is not only the MO composition but its interlayer grain boundaries, crystalline phase and even crystal planes exposed at the interface that matter. Also, the concentration of trapping sites contributes to the final efficiency of the MO in photocatalytic application. Hence, engineering the interfaces carefully and optimizing the charge carriers transport across them play a pivotal role in photocatalytic process.

Hussain et al. [60] studied the interface between rutile $TiO_2$ (110) and liquid water. They elucidated that the $TiO_2$ (110) surface has terminal hydroxyl groups in the contact layer which arise from mixed $O_2$ and $H_2O$ dissociation in the presence of

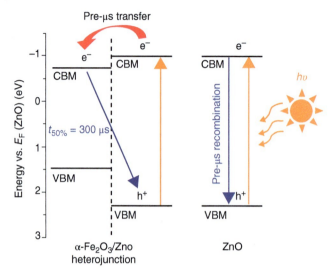

**Figure 6.15** Band alignment diagram showing charge transfer from ZnO to α-Fe$_2$O$_3$ on the excitation of ZnO but not vice versa. Source: Reprinted with permission from [61]. (Copyright (2019) Wiley-VCH)

point defects. Hence, increasing density of hydroxyl groups at the TiO$_2$ (110) interface would enhance the efficiency of photocatalytic process.

Various heterostructures of MOs have been prepared to achieve an internal charge separation which increases carrier lifetime and thus increases the efficiency of photoinduced processes. Jiamprasertboon et al. [61] studied the time-resolved TA spectra of α-Fe$_2$O$_3$/ZnO films prepared by aerosol assisted chemical vapor deposition. The carrier transfer across the interfaces was established using hole and electron scavenging agents and comparing the TA spectra. It was shown that excitation of ZnO (in UV) results in electron transfer from the ZnO CB to the CB of α-Fe$_2$O$_3$ according to the band energy alignment of the materials, as shown in Figure 6.15. However, the photoexcitation of α-Fe$_2$O$_3$ does not result in carriers transfer between two materials. The electron transfer from ZnO to α-Fe$_2$O$_3$ separates electrons in the CB of α-Fe$_2$O$_3$ from holes in the VB of ZnO and increases the photocarrier lifetimes, thus increasing the photodegradation efficiency of the system which was tested by degrading stearic acid.

Kadam et al. [62] prepared Ag-core/ZnO-shell nanostructures (CSNS) and investigated their PL spectra to study the photogenerated charge carrier transfer, separation, and recombination [62]. Bare ZnO showed emission at 378 nm, whereas in the case of Ag-ZnO CSNS, the intensity of PL was lower, as shown in Figure 6.16. This quenching was observed due to the interfacial charge transfer from ZnO shell to Ag core which acted as an electron sink. This charge transfer hampered the recombination of photogenerated charge carriers and hence can enhance the system photocatalytic performances. The prepared photocatalyst was found to be more effective for methyl orange degradation as compared to bare TiO$_2$ and ZnO.

**Figure 6.16** Photoluminescence spectra of ZnO, and Ag–ZnO core–shell nanostructures excited at 330 nm. Source: Reprinted with permission from [62]. Copyright (2018) Elsevier B.V.

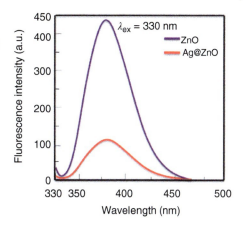

**Figure 6.17** Schematic presentation of the energy levels in $Fe_2O_3$ layer covered with ultrathin layer of $Fe_2TiO_5$ and photoinduced reactions leading to water splitting. Source: Reprinted with permission from [63]. Copyright (2015) American Chemical Society.

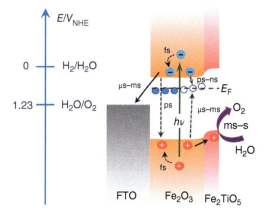

The MOs with narrow bandgap have an advantage of absorbing visible light and thus can directly utilize the solar light to power catalytic reactions or generate electric current. For example, hematite, $\alpha\text{-}Fe_2O_3$, absorbs light at $\lambda < 580$ nm and has positions of the VB and CB favorable for water splitting. But it has a disadvantage of low electron mobility and a small hole diffusion length, typically 2–20 nm. Therefore, only a thin hematite layer of particles can be used in photocatalytic applications. Furthermore, the catalytic reactions are diffusion controlled and slow, but the carriers must reach the interface and "wait" for the catalytic reaction to happen. This can be achieved by adding a very thin, few atomic layers of semiconductor on top of hematite with VB a bit above the hematite VB to trap hole at the surface, or with CB a bit lower than that of hematite to trap electrons, respectively. A similar effect can be achieved by applying a bias potential, which will force the electrons or holes to move toward the surface.

The approach to surface modification was tested by Ruoko et al. [63] by fabricating hematite layer (roughly 50 nm) with a thin $Fe_2TiO_5$ overlayer on top. Figure 6.17 presents the energy diagram of the structure and photoinduced reaction

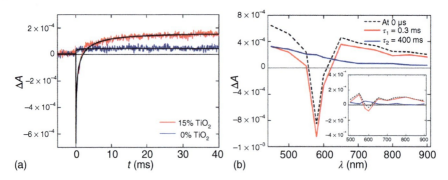

**Figure 6.18** (a) Flash-photolysis TA decays at 580 nm for the bare hematite and the hematite–titania samples at 1.6 V bias and (b) corresponding decay component spectra for the bare hematite (inset) and hematite–titania. Source: Reprinted with permission from [63]. Copyright (2015) American Chemical Society.

leading to hole trapping close to the surface and participating in the oxygen evolution cycle of the water-splitting reaction. The photoreactions were studied by two TA spectroscopy techniques, femtosecond pump-probe, and nanosecond–millisecond flash photolysis. These spectroscopy techniques allow to identify the intermediate state determine the reaction time constants and conclude on the reactions taking place in the system. In particular, a change in the sample reactivity upon addition of the $Fe_2TiO_5$ top layer is presented in Figure 6.18. The signal is stronger in the case of titania overlayer indicating that more carriers are available at long delay after excitation (up to hundreds of milliseconds). The carriers can be identified as electrons due to characteristic bleaching band close to 580 nm.

Zhang et al. [64] studied the effect of Au atoms adsorption on $TiO_2$ surface. Au was deposited on $TiO_2$ (110) surface atomically and the photostimulated desorption (PSD) study was done by desorption of $^{18}O_2$ to evaluate hole transport from $TiO_2$ bulk to surface. $^{18}O_2$ was introduced on the Au-$TiO_2$ system where it chemisorbed at the $TiO_2$ oxygen vacancies as either $O_2^-$ or $O_2^{2-}$. These species reacted with holes in the VB of $TiO_2$ and desorb as oxygen. It was seen that Au adsorption on $TiO_2$ (110) surface depressed photoinduced hole transport from bulk to $TiO_2$ surface which decreased $^{18}O_2$ PSD yield. This indicated that adsorbed Au atoms donated a fraction of its electrons to the surface which causes downward bending of $TiO_2$ VB and CB.

Shang et al. [65] prepared glass/$SnO_2$/$TiO_2$ and glass/$TiO_2$/$SnO_2$ interfacial composite semiconductor films to study their photocatalytic activity. They observed that for glass/$SnO_2$/$TiO_2$ films, due to the band energy alignment, the electrons are always collected in the CB of $SnO_2$, which has insufficient energy for the reduction reaction as shown in Figure 6.19. However, since the holes are collected in the $TiO_2$, and the hole energy is sufficient for the oxidation reaction, the oxidation reaction initiated by illumination was observed.

Semiconductor MOs such as $TiO_2$, ZnO, and $SnO_2$, etc. belong to the wide bandgap semiconductors and are very inefficient in solar light utilization since they are transparent in the visible and longer wavelength ranges. However, these MOs can act

**Figure 6.19** Schematic diagram of the charge transfer in bilayer SnO$_2$/TiO$_2$ films on glass. Source: Reprinted with permission from [65]. Copyright (2004) Elsevier B.V.

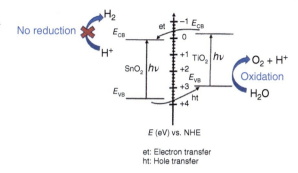

**Figure 6.20** Comparison of electron injection in RuN3-sensitized TiO$_2$, SnO$_2$, and ZnO thin films. All samples were excited at 400 nm and proved in the IR, 2150 cm$^{-1}$ (TiO$_2$), 2066 cm$^{-1}$ (TiO$_2$), and 1900 cm$^{-1}$ (ZnO). Source: Reprinted with permission from [68]. Copyright (2001) American Chemical Society.

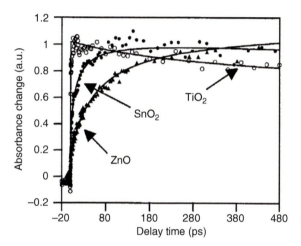

as a scaffold layer for loading dye and QDs, thus absorbing the visible and NIR light [66]. These sensitizers not only enhance the absorption of the hybrids in the visible region but also can induce efficient charge separation at the MOs/sensitizer interface to provide carriers for numerical applications, from which dye-sensitized solar cells (DSSCs) are probably the most well known [67].

Injection of electrons from a photoexcited sensitizer to the CB of MO is the primary photoreaction in DSSCs. For example, a series of MOs sensitized by RuN3 ([Ru(dcbpyH$_2$)$_2$(NCS)$_2$]) dye was studied by Asbury et al. [68] using ultrafast vibrational spectroscopy. A large difference in electron injection time constant was observed, as presented in Figure 6.20, and explained by the difference in electronic coupling, density of states at the MO side, and driving force for the electron transfer (ET) at the interface. From this point of view, comparison of TiO$_2$ and ZnO interfaces is instructive as these two materials have roughly the same energies of the CB, but the electron injection at the TiO$_2$ surface is much faster than that at the ZnO surface. This was attributed to higher density of states of TiO$_2$ and different nature of atomic orbitals involved in formation of the CB, i.e. 3d orbital in the case of TiO$_2$ and 4s orbital in ZnO [68].

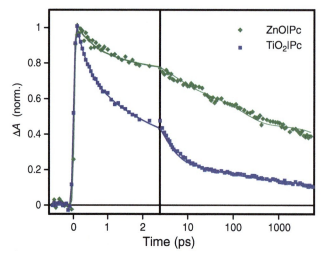

**Figure 6.21** Kinetics of the charge recombination following electron injection to TiO$_2$ and ZnO monitored by TA at wavelength specific for the sensitizer (phthalocyanine) cation absorption at 865 nm [71]. Source: Published by the Royal Society under the terms of the Creative Commons Attribution License.

It has to be noted that the photoinduced electron injection is not a simple single-step reaction. At first, an electron-cation complex is formed at the interface. This state is termed as "interfacial charge transfer complex" or ICTC. It may yield a "free electron" in the CB of the MO, or recombine in place. It is difficult to distinguish experimentally between the ICTC and free electron-sensitizer cation pair, since the binding energy is low and spectroscopic properties of these two states are very close to each other. One method successfully used in this case is time-resolved terahertz (THz) spectroscopy, which is sensitive to the conducting electron properties [69]. Also, traditional pump-probe spectroscopy at the bandgap region can be used to observe substrate-specific excitonic features to study the ET dynamics at the interface [70].

Injection of the charge carrier to the CB is the first step in the photocurrent generation in DSSC. The carrier must be delivered to the external electric circuits, which means that the charge recombination at the interface must be slower than the carrier diffusion away from the interface. This can be monitored by following the charge recombination with TA spectroscopy. It was shown that although the electron injection is faster at the TiO$_2$ interface than at ZnO, the charge recombination is also faster in TiO$_2$-dye system compared to that of ZnO-based systems [71]. The MOs were sensitized by a phthalocyanine derivative, and the cation absorption band of the dye in the NIR range was used to monitor the charge recombination (Figure 6.21). The longer carrier lifetime in ZnO was attributed to much faster carrier diffusion in ZnO compared to that in TiO$_2$.

The photoinduced charge transfer at functionalized MO surfaces and interfaces formed with other materials are the key players in many applications of practical importance, such as photocatalysis and solar cells. However, a flat MO surface with a monomolecular layer of dye molecules or QDs absorbs only a few percentages of the incident light, even for compounds with the highest molar absorption. This motivates design, fabrication, and study of materials with high specific surface area. The materials can be produced by a great variety of methods such as, by curing nanoparticles, growing nanorods or nanowires, using templates to prepare mesoporous materials or else. If at least one dimension in this material is smaller than few hundreds of nanometers, it can be classified as a nanostructure.

A common property of all these nanostructures is relatively high ratio of the material surface to its volume, which is formally specified by the specific surface area value. From the view point of material photophysics, the design goal is to prepare structures in which the surface effects dominate over the bulk ones. If the photocarriers are generated in bulk, they are expected to travel to the surface without losses and participate in the surface reactions, e.g. catalytic reactions. If the carriers are generated at the surface, as in the case of DSSC, the carriers must travel through the MO without losses to a charge-collecting electrode.

The photophysical properties of semiconductors are determined by the material electronic properties, the electronic wave functions and carrier mobilities. The free electrons in the CB are not localized at a particular atom, but delocalized over number of atoms that makes them "free" carrier and provide material conductivity. The recombination mechanisms of carriers traveling in the bulk and at the surface are different. More importantly, the carriers at the surface can interact with environment and are the main players, for example, in photocatalysis. From this point of view, the carrier diffusion length is the parameter pointing to the length scale at which photophysics of nanostructure starts to depend on the interfacial reactions rather than the bulk ones. Although, a great variation in the diffusion length can be noted for different MOs, from being few nanometers for $Fe_2O_3$ [72] and longer than hundreds of nanometers in ZnO nanocrystals [73]. An estimation of the carrier diffusion length, $L$, can be made based on the carrier diffusion coefficient, $D$, and carrier lifetime at the CB, $\tau$, as

$$L = \sqrt{D\tau} \qquad (6.10)$$

As an example, assuming carrier lifetime to be $\tau = 1$ ns and diffusion coefficient $D = 0.5 \text{ cm}^2 \text{ s}^{-1}$ estimated for bulk $TiO_2$ [50], one obtains $L \approx 200$ nm, showing that already at 100 nm size of $TiO_2$ nanomaterial the surface carrier recombination may affect the carrier lifetime.

The practical outcome determining the performance of MO materials in photophysical applications is the interplay between the bulk and surface effects, with the surface effect dominating over the bulk ones. It is because the surface is within the reach for carriers determining photophysics of these materials. Also, by changing the size of MOs, their electronic properties and surface areas can be tailored. A small change in the size can have significant effect on the adsorption, desorption, chemical bond formation, trap states, and the bandgap of the materials. Moreover, carrier's diffusion lengths of different MOs play a key role in providing charges to the interfaces for chemical reactions to occur.

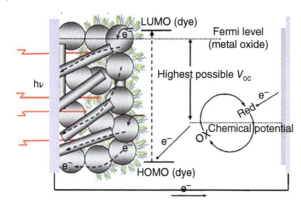

**Figure 6.22** Schematic of DSSC in which mesoporous MO acts as a photoelectrode. Source: Reprinted with permission from [77]. Copyright (2009) Wiley.

## 6.4 Photophysical Aspects of MO Applications

### 6.4.1 Solar Cells

Probably, the DSSC is the most recognized application of MOs [74, 75]. DSSCs got breakthrough when [76] high surface area mesoporous $TiO_2$ electrode was prepared with a monolayer of sensitizer. $TiO_2$ nanoparticles are deposited as a thin film and sensitized by a suitable dye, which can be an organic molecule, metal complex, QD, or else. The essential photoinduced reaction is the electron injection from the sensitizer to the CB of the $TiO_2$, as shown in Figure 6.22.

The quantum efficiency of this primary reaction, electron injection to MO is close to unity, but the overall power conversion efficiency for the best DSSC is around 13% or close to one-third of the theoretical limit for single junction solar cells [78]. This is still less than the efficiency achieved with the first- and second-generation solar cells, up to 20–30% [75]. Further improvement of the DSSCs requires mainly the development of dyes with broader absorption, finding better matching electrolytes for the dye cation regeneration and improving other factors outside the primary working MO–dye interface. MO materials in DSSCs play a significant role in efficient charge transport, reducing charge recombination and as a blocking layer (to prevent contact between redox mediator electrolyte and FTO). Although $TiO_2$ is mostly used in DSSCs, other MOs such as ZnO, $SnO_2$, $SrTiO_3$, $Zn_2SnO_4$ [79–81] can be also used reasonably well in this application.

Another strategy in solar cell design is to use a narrow bandgap semiconductor as light harvester and place it in-between electron and hole transporting layers to create a flow of the electrons and holes in opposite directions for efficient photocurrent generation. Perovskite solar cells is an example of successful applications of this approach. In this scenario, the MO must have bandgap energy close to 1 eV and be a good conductor with sufficiently long photocarriers lifetime. Unfortunately, none of the MOs studied so far can compete with perovskite with the combination of these two properties.

### 6.4.2 Light Emitting Devices

Although PL is a widely used method to study properties of MOs, the PL quantum yield is rather low, typically <6% for most MOs [82, 83]. Another obstacle in using MO to generate electroluminescence is carrier mobility and trapping. However, the wide bandgap MOs are actively used as semi-transparent electrodes in these devices [84, 85], though this used case is not related to MO photophysics and thus was not discussed here.

### 6.4.3 Photocatalysis

This is probably another best-known area of MOs applications since the demonstration of photocatalytic water splitting by $TiO_2$ and Pt electrode pair by A. Fujishima and K. Honda [2]. Since then the research in this direction keeps growing up continuously. MOs can be used directly as catalyst activated by solar light, or can be a source of photocarriers, electrons, and holes, which activate the catalyst [86]. There are a variety of photocatalytic applications of MOs, which aim for environmental remediation, energy sustainability such as in hydrogen production, and conversion of $CO_2$ to renewable energy products [87]. Generally, light absorption, electron–hole separation, hole transportation, and surface reactions are the key features in research and design, requiring materials of optimized composition, favorable structure, and morphology [88].

From the photophysical point of view, MOs must have the CB or VB at the energy level suitable for the catalyst activation. In practical applications, this can be a challenging task since the most of catalytic reactions are multistep reactions. For example, the water splitting requires only 1.23 V potential but to produce one $O_2$ (from two $H_2O$) four electrons (and four holes, respectively) are required [89]. Typical approach to the problem starts from determining energy requirements to the carrier delivering system, which are the VB and CB positions in MO, and finding an optimum photoexcitation scheme, either direct or by using sensitizer or doping. Furthermore, a combination of two or more light-harvesting/carrier-delivering systems can be designed to drive individual steps of the photocatalytic process. The optical properties and photocarrier dynamics are the key research questions then [34].

### 6.4.4 Photodegradation

Photodegradation is a wide photocatalytic application area of MOs. This application plays a vital role in the degradation of organic harmful dyes such as AZO dyes, which are carcinogenic in nature [90, 91]. Also, this application holds potential to provide antimicrobial [92, 93] and self-cleaning coatings [94, 95] for windows, walls, and textiles, which utilize, for example, a thin layer of $TiO_2$ or ZnO coating.

This application is a typical test to determine the photoactivity of different semiconductors and MOs in particular. These experiments are based on degradation (decoloration) of dyes, e.g. methyl orange, rhodamine blue, etc. but not aiming at

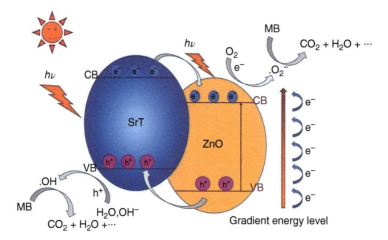

**Figure 6.23** Photodegradation of methylene blue dye by ZnO@SrTiO$_3$ nanocomposites. Source: Reprinted with permission from [97]. Copyright (2020) Wiley-VCH GmbH.

particular final product [3], which is the aim of photocatalysis. MOs can generate electrons and holes that react with water, dissolved oxygen and surface hydroxyl groups to produce hydroxyl and super oxide radical anions [96] which react with organic structure of the dyes and mineralize them into $CO_2$, $H_2O$, and ammonium and nitrate ions, as shown in Figure 6.23 [97].

### 6.4.5 Solar Driven Chemistry

The concept of solar driven chemistry is relatively new one and it has its roots in photocatalysis [98], but it formulates the problem in a more general form of powering endothermic chemical reactions by solar energy. From the photophysical side of the concept, the solar energy, photons, need to be collected and delivered to a reaction center in a form which can be consumed by the reaction center. This can be an electron or a hole with specific energy. The approach to the problem is very similar to that discussed above concerning photocatalysis, and requirements to photophysical properties of MO in this application are also similar which are, light harvesting, charge carriers generation and separation, and transport along particular energy levels.

## 6.5 Conclusions

In this chapter, we mentioned just a few applications of MO based on the solar light utilization. There are already many scenarios for building up photonic devices and many new will appear in future. They are all based on our fundamental knowledge of light–matter interaction applied to bare MOs and composite materials based on the MOs. Here we shortly reviewed photoinduced processes in MO semiconductors and their hybrids. The key aspects considered can be summarized as follows:

- photoexcitation of semiconductor MO results in promoting an electron from the VB to the CB which leaves a hole in the VB, the electron in the CB retains excess energy equal to the bandgap energy and both the electron and hole can move across the MO;
- alternatively, MOs can be sensitized by efficient dyes attached to their surface, in which case the energy of the excited electron (e.g. Lowest non-occupied molecular orbital of a molecular dye sensitizer) must be higher than that of the CB of MO to afford the electron transfer from the sensitizer to the MO, this is particularly useful for wide bandgap MOs that are not absorbing visible light by themselves;
- electrons in the CB can recombine with holes in the VB in a process that may produce a photon giving rise to PL, or both electrons and holes can be trapped by different sort of defects in the MO structure, this affects gradually the carriers migration and lifetime;
- free moving carries can reach MO surface, interact with environment, and be involved in different reactions of practical importance, such as photocatalytic water splitting or $CO_2$ reduction;
- the MO surface reactivity can be altered by chemical processing, modification of morphology and surface area, overlayer deposition, passivation and other methods, which all allow to guide the carrier migration and tune energies at the surface to optimize light-powered surface reactions.

## References

1 Riente, P. and Noël, T. (2019). Application of metal oxide semiconductors in light-driven organic transformations. *Catal. Sci. Technol.* 9 (19): 5186–5232. https://doi.org/10.1039/C9CY01170F.
2 Fujishima, A. and Honda, K. (1972). Electrochemical photolysis of water at a semiconductor electrode. *Nature* 238 (5358): 37–38. https://doi.org/10.1038/238037a0.
3 Khan, R., Riaz, A., Rabeel, M. et al. (2019). $TiO_2$@$NbSe_2$ decorated nanocomposites for efficient visible-light photocatalysis. *Appl. Nanosci.* 9 (8): 1915–1924. https://doi.org/10.1007/s13204-019-01020-6.
4 Chavali, M.S. and Nikolova, M.P. (2019). Metal oxide nanoparticles and their applications in nanotechnology. *SN Appl. Sci.* 1 (6): 607. https://doi.org/10.1007/s42452-019-0592-3.
5 Yang, Y., Niu, S., Han, D. et al. (2017). Progress in developing metal oxide nanomaterials for photoelectrochemical water splitting. *Adv. Energy Mater.* 7 (19): 1700555. https://doi.org/10.1002/aenm.201700555.
6 Gershon, T. (2011). Metal oxide applications in organic-based photovoltaics. *Mater. Sci. Technol.* 27 (9): 1357–1371. https://doi.org/10.1179/026708311X13081465539809.
7 Dalle, K.E., Warnan, J., Leung, J.J. et al. (2019). Electro- and solar-driven fuel synthesis with first row transition metal complexes. *Chem. Rev.* 119 (4): 2752–2875. https://doi.org/10.1021/acs.chemrev.8b00392.

8 Panchal, P.K. and Patel, M.N. (2005). Toxic effect of transition metal complexes on *Salmonella typhi, Escherichia coli* and *Serratia marcescens. Toxicol. Environ. Chem.* 87 (3): 407–414. https://doi.org/10.1080/02772240500254788.

9 Tkachenko, N.V. (2006). *Optical Spectroscopy: Methods and Instrumentations.* Amsterdam: Elsevier B. V. ISBN 978-0-444-52126-2. https://doi.org/10.1016/B978-0-444-52126-2.X5024-2.

10 Lemmetyinen, H., Tkachenko, N.V., Valeur, B. et al. (2014). Time-resolved fluorescence methods (IUPAC technical report). *Pure Appl. Chem.* 86 (12): 1969–1998. https://doi.org/doi:10.1515/pac-2013-0912.

11 Parson, W.W. (2015). *Modern Optical Spectroscopy*, 2e. Berlin, Heidelberg: Springer-erlag. ISBN 978-3-662-46777-0. https://doi.org/10.1007/978-3-662-46777-0.

12 Roduner, E., Krüger, T., Forbes, P., and Kress, K. (2018). *Optical Spectroscopy.* World Scientific (Europe). ISBN 978-1-78634-610-0. https://doi.org/10.1142/q0182.

13 Sbaï, N., Perriére, J., Gallas, B. et al. (2008). Structural, optical, and electrical properties of epitaxial titanium oxide thin films on $LaAlO_3$ substrate. *J. Appl. Phys.* 104 (3): 033529. https://doi.org/10.1063/1.2964114.

14 Cesaria, M., Caricato, A.P., and Martino, M. (2014). Realistic reflectance spectrum of thin films covering a transparent optically thick substrate. *Appl. Phys. Lett.* 105 (3): 031105. https://doi.org/10.1063/1.4890675.

15 Stein, H.S., Soedarmadji, E., Newhouse, P.F. et al. (2019). Synthesis, optical imaging, and absorption spectroscopy data for 179072 metal oxides. *Sci. Data* 6 (1): 1–5. https://doi.org/10.1038/s41597-019-0019-4.

16 Mayerhöfer, T.G., Mutschke, H., and Popp, J.Ã. (2016). Employing theories far beyond their limits–the case of the (Boguer-) Beer–Lambert law. *ChemPhysChem* 17 (13): 1948–1955. https://doi.org/10.1002/cphc.201600114.

17 Kofstad, P. (1996). Defect chemistry in metal oxides. *Phase Trans.* 58 (1–3): 0141–1594. https://doi.org/10.1080/01411599608242395.

18 Liu, Y., Wang, J., Yang, P., and Matras-Postolek, K. (2015). Self-modification of $TiO_2$ one-dimensional nano-materials by $Ti^{3+}$ and oxygen vacancy using $Ti_2O_3$ as precursor. *RSC Adv.* 5 (76): 61657–61663. https://doi.org/10.1039/C5RA07079A.

19 Godinho, K.G., Walsh, A., and Watson, G.W. (2009). Energetic and electronic structure analysis of intrinsic defects in $SnO_2$. *J. Phys. Chem. C* 113 (1): 439–448. https://doi.org/10.1021/jp807753t.

20 Leijtens, T., Eperon, G.E., Barker, A.J. et al. (2016). Carrier trapping and recombination: the role of defect physics in enhancing the open circuit voltage of metal halide perovskite solar cells. *Energy Environ. Sci.* 9 (11): 3472–3481. https://doi.org/10.1039/C6EE01729K.

21 Choudhury, B., Dey, M., and Choudhury, A. (2014). Shallow and deep trap emission and luminescence quenching of $TiO_2$ nanoparticles on Cu doping. *Appl. Nanosci.* 4: 499–506. https://doi.org/10.1007/s13204-013-0226-9.

**22** Bauer, C., Boschloo, G., Mukhtar, E., and Hagfeldt, A. (2004). Ultrafast relaxation dynamics of charge carriers relaxation in ZnO nanocrystalline thin films. *Chem. Phys. Lett.* 387 (1): 176–181. https://doi.org/10.1016/j.cplett.2004.01.106.

**23** Di Paola, A., Marcì, G., Palmisano, L. et al. (2002). Preparation of polycrystalline $TiO_2$ photocatalysts impregnated with various transition metal ions: characterization and photocatalytic activity for the degradation of 4-nitrophenol. *J. Phys. Chem. B* 106 (3): 637–645. https://doi.org/10.1021/jp013074l.

**24** Marschall, R. and Wang, L. (2014). Non-metal doping of transition metal oxides for visible-light photocatalysis. *Cat. Today* 225: 111–135. https://doi.org/10.1016/j.cattod.2013.10.088.

**25** Zhao, Y., Li, C., Liu, X. et al. (2008). Zn-doped $TiO_2$ nanoparticles with high photocatalytic activity synthesized by hydrogen–oxygen diffusion flame. *Appl. Catal. B: Environ.* 79 (3): 208–215. https://doi.org/10.1016/j.apcatb.2007.09.044.

**26** Yamanaka, K.-i., Ohwaki, T., and Morikawa, T. (2013). Charge-carrier dynamics in Cu- or Fe- loaded nitrogen-doped $TiO_2$ powder studied by femtosecond diffuse reflectance spectroscopy. *J. Phys. Chem. C* 117 (32): 16448–16456. https://doi.org/10.1021/jp404431z.

**27** Wood, D.L. and Tauc, J.J. (1972). Weak absorption tails in amorphous semiconductors. *Phys. Rev. B* 5 (8): 3144–3151. https://doi.org/10.1103/PhysRevB.5.3144.

**28** Makuła, P., Pacia, M., and Macyk, W. (2018). How to correctly determine the bandgap energy of modified semiconductor photocatalysts based on UV-Vis spectra. *J. Phys. Chem. Lett.* 9 (23): 6814–6817. https://doi.org/10.1021/acs.jpclett.8b02892.

**29** Kim, S., Nam, G., Park, H.-G. et al. (2013). Effects of doping with Al, Ga, and In on structural and optical properties of ZNO nanorods grown by hydrothermal method. *Bull. Korean Chem. Soc.* 34 (4): 1205–1211. https://doi.org/10.5012/BKCS.2013.34.4.1205.

**30** Wang, E., Yang, W., and Cao, Y. (2009). Unique surface chemical species on indium doped $TiO_2$ and their effect on the visible light photocatalytic activity. *J. Phys. Chem. C* 113 (49): 20912–20917. https://doi.org/10.1021/jp9041793.

**31** Scholes, G.D. (2008). Insights into excitons confined to nanoscale systems: electron-hole interaction, binding energy, and photodissociation. *ACS Nano* 2 (3): 523–537. https://doi.org/10.1021/nn700179k.

**32** Djurisić, A.B., Ng, A.M.C., and Chen, X.Y. (2010). ZnO nanostructures for optoelectronics: material properties and device applications. *Prog. Quantum Electron.* 34 (4): 191–259. https://doi.org/10.1016/j.pquantelec.2010.04.001.

**33** Sun, C.-K., Sun, S.-Z., Lin, K.-H. et al. (2005). Ultrafast carrier dynamics in ZnO nanorods. *Appl. Phys. Lett.* 87 (2): 023106. https://doi.org/10.1063/1.1989444.

**34** Le Bahers, T. and Takanabe, K. (2019). Combined theoretical and experimental characterizations of semiconductors for photoelectrocatalytic applications. *J. Photochem. Photobiol., C* 40: 212–233. https://doi.org/10.1016/j.jphotochemrev.2019.01.001.

**35** Ganguly, P., Panneri, S., Hareesh, U.S., Breen, A., Pillai, S.C. (2019). Chapter 23 - Recent advances in photocatalytic detoxification of water. In: *Micro and*

*Nano Technologies, Nanoscale Materials in Water Purification* (eds. S. Thomas, D. Pasquini, S.-Y. Leu, D.A. Gopakumar), 653–688. Elsevier, ISBN 9780128139264.

36 Lenci, F., Checcucci, G., Sgarbossa, A., Martin, M.M., Plaza, P., Angelini, N. (2005). Fluorescent biomolecules In: *Encyclopedia of Condensed Matter Physics* (eds. F. Bassani, G.L. Liedl, P. Wyder), 222–235. Elsevier, ISBN 9780123694010.

37 Berberan-Santos, M.N., Bodunov, E.N., and Valeur, B. (2005). Mathematical functions for the analysis of luminescence decays with underlying distributions 1. Kohlrausch decay function (stretched exponential). *Chem. Phys.* 315 (1): 171–182. https://doi.org/10.1016/j.chemphys.2005.04.006.

38 Klimov, V.I. (2007). Spectral and dynamical properties of multiexcitons in semiconductor nanocrystals. *Annu. Rev. Phys. Chem.* 58: 635–673. https://doi.org/10.1146/annurev.physchem.58.032806.104537.

39 Li, M., Xing, G., Foong, L. et al. (2012). Tailoring the charge carrier dynamics in ZnO nanowires: the role of surface hole/electron traps. *Phys. Chem. Chem. Phys.* 14 (9): 3075–3082. https://doi.org/10.1039/c2cp23425d.

40 Rawool, S.A., Pai, M.R., Banerjee, A.M. et al. (2018). pn heterojunctions in $NiO:TiO_2$ composites with type-II band alignment assisting sunlight driven photocatalytic $H_2$ generation. *Appl. Catal. B* 221: 443–458. https://doi.org/10.1016/j.apcatb.2017.09.004.

41 Appavoo, K., Liu, M., and Sfeir, M.Y. (2014). Role of size and defects in ultrafast broadband emission dynamics of ZnO nanostructures. *Appl. Phys. Lett.* 104 (13): 133101. https://doi.org/10.1063/1.4868534.

42 Katoh, R., Murai, M., and Furube, A. (2010). Transient absorption spectra of nanocrystalline $TiO_2$ films at high excitation density. *Chem. Phys. Lett.* 500 (4): 309–312. https://doi.org/10.1016/j.cplett.2010.10.045.

43 Yoshihara, T., Katoh, R., Furube, A. et al. (2004). Identification of reactive species in photoexcited nanocrystalline $TiO_2$ films by wide-wavelength-range (400–2500 nm) transient absorption spectroscopy. *J. Phys. Chem. B* 108 (12): 3817–3823. https://doi.org/10.1021/jp031305d.

44 Tamaki, Y., Furube, A., Katoh, R. et al. (2006). Trapping dynamics of electrons and holes in a nanocrystalline $TiO_2$ film revealed by femtosecond visible/near-infrared transient absorption spectroscopy. *C.R. Chimie* 9 (2): 268–274. https://doi.org/10.1016/j.crci.2005.05.018.

45 Sachs, M., Pastor, E., Kafizas, A., and Durrant, J.R. (2016). Evaluation of surface state mediated charge recombination in anatase and rutile $TiO_2$. *J. Phys. Chem. Lett.* 7 (19): 3742–3746. https://doi.org/10.1021/acs.jpclett.6b01501.

46 Cooper, J.K., Ling, Y., Longo, C. et al. (2012). Effects of hydrogen treatment and air annealing on ultrafast charge carrier dynamics in ZnO nanowires under in situ photoelectrochemical conditions. *J. Phys. Chem. C* 116 (33): 17360–17368. https://doi.org/10.1021/jp304428t.

47 Cherepy, N.J., Liston, D.B., Lovejoy, J.A. et al. (1998). Ultrafast studies of photoexcited electron dynamics in $\gamma$- and $\alpha$-$Fe_2O_3$ semiconductor nanoparticles. *J. Phys. Chem. B* 102 (5): 770–776. https://doi.org/10.1021/jp973149e.

**48** Corby, S., Francàs, L., Kafizas, A., and Durrant, J.R. (2020). Determining the role of oxygen vacancies in the photoelectrocatalytic performance of $WO_3$ for water oxidation. *Chem. Sci.* 11 (11): 2907–2914. https://doi.org/10.1039/C9SC06325K.

**49** Murphy, C.J. and Coffer, J.L. (2002). Quantum dot: a primer. *Appl. Spectrosc.* 56 (1): 16A–27A. https://doi.org/10.1366/0003702021954214.

**50** Zhang, Q., Dandeneau, C.S., Zhou, X., and Cao, G. (2009). ZnO nanostructures for dye-sensitized solar cells. *Adv. Mater.* 21 (4): 4087–4108. https://doi.org/10.1002/adma.200803827.

**51** Camarda, P., Vaccaro, L., Sciortino, A. et al. (2020). Synthesis of multi-color luminescent ZnO nanoparticles by ultra-short pulsed laser ablation. *Appl. Surf. Sci.* 506: 144954. https://doi.org/10.1016/j-apsusc.2019.144954.

**52** Nam, Y., Li, L., Lee, J.Y., and Prezhdo, O.V. (2018). Size and shape effects on charge recombination dynamics of $TiO_2$ nanoclusters. *J. Phys. Chem. C* 122 (9): 5201–5208. https://doi.org/10.1021/acs.jpcc.8b00691.

**53** Pozina, G., Yang, L.L., Zhao, Q.X. et al. (2010). Size dependent carrier recombination in ZnO nanocrystals. *Appl. Phys. Lett.* 97 (13): 131909. https://doi.org/10.1063/1.3494535.

**54** Edvinsson, T. (2018). Optical quantum confinement and photocatalytic properties in two-, one- and zero- dimensional nanostructures. *R. Soc. Open Sci.* 5: 180387. https://doi.org/10.1098/rsos.180387.

**55** Limpens, R., Sugimoto, H., Neale, N.R., and Fujii, M. (2018). Critical size for carrier delocalization in doped silicon nanocrystals: a study by ultrafast spectroscopy. *ACS Photonics* 5 (10): 4037–4045. https://doi.org/10.1021/acsphotonics.8b00671.

**56** Jasieniak, J., Califano, M., and Watkins, S.E. (2011). Size-dependent valence and conduction band-edge energies of semiconductor nanocrystals. *ACS Nano* 5 (7): 5888–5902. https://doi.org/10.1021/nn201681s.

**57** Barnasas, A., Kanistras, N., Ntagkas, N. et al. (2020). Quantum confinement effects of thin ZnO films by experiment and theory. *Physica E* 120: 114072. https://doi.org/10.1016/j.physe.2020.114072.

**58** Jain, G., Rocks, C., Maguire, P., and Mariotti, D. (2020). One-step synthesis of strongly confined, defect-free and hydroxy-terminated ZnO quantum dots. *Nanotechnology* 31 (21): 215707. https://doi.org/10.1088/1361-6528/ab72b5.

**59** Graetzel, M., Janssen, R.Ã.A.J., Mitzi, D.B., and Sargent, E.H. (2012). Materials interface engineering for solution-processed photovoltaics. *Nature* 488 (7411): 304–312. https://doi.org/10.1038/nature11476.

**60** Hussain, H., Tocci, G., Woolcot, T. et al. (2017). Structure of a model $TiO_2$ photocatalytic interface. *Nat. Mater.* 16 (4): 461–466. and https://doi.org/10.1038/nmat4793.

**61** Jiamprasertboon, A., Kafizas, A., Sachs, M. et al. (2019). Heterojunction $\alpha$-$Fe_2O_3$/ZnO films with enhanced photocatalytic properties grown by aerosol-assisted chemical vapour deposition. *Chem. Eur. J.* 25 (48): 11337–11345. https://doi.org/10.1002/chem.201902175.

**62** Kadam, A.N., Bhopate, D.P., Kondalkar, V.V. et al. (2018). Facile synthesis of Ag-ZnO core–shell nanostructures with enhanced photocatalytic activity. *J. Ind. Eng. Chem.* 61: 78–86. https://doi.org/10.1016/j.jiec.2017.12.003.

**63** Ruoko, T.-P., Kaunisto, K., BÃrtsch, M. et al. (2015). Subpicosecond to second time-scale charge carrier kinetics in hematite–titania nanocomposite photoanodes. *J. Phys. Chem. Lett.* 6 (15): 2859–2864. https://doi.org/10.1021/acs.jpclett.5b01128.

**64** Zhang, Z., Tang, W., Neurock, M., and Yates, J.T. (2011). Electric charge of single Au atoms adsorbed on $TiO_2$ (110) and associated band bending. *J. Phys. Chem. C* 115 (48): 23848–23853. https://doi.org/10.1021/jp2067809.

**65** Shang, J., Yao, W., Zhu, Y., and Wu, N. (2004). Structure and photocatalytic performances of glass/$SnO_2$/$TiO_2$ interface composite film. *Appl. Catal., A* 257 (1): 25–32. https://doi.org/10.1016/j.apcata.2003.07.001.

**66** Elumalai, N.K., Vijila, C., Jose, R. et al. (2015). Metal oxide semiconducting interfacial layers for photovoltaic and photocatalytic applications. *Mater. Renew. Sustain. Energy* 4 (3): https://doi.org/10.1007/s40243-015-0054-9.

**67** Hagfeldt, A., Boschloo, G., Sun, L. et al. (2010). Dye-sensitized solar cells. *Chem. Rev.* 110 (10): 6595–6663. https://doi.org/10.1021/cr900356p.

**68** Asbury, J.B., Hao, E., Wang, Y. et al. (2001). Ultrafast electron transfer dynamics from molecular adsorbates to semiconductor nanocrystalline thin films. *J. Phys. Chem. B* 105 (20): 4545–4557. https://doi.org/10.1021/jp003485m.

**69** Němec, H., Rochford, J., Taratula, O. et al. (2010). Influence of the electron-cation interaction on electron mobility in dye-sensitized ZnO and $TiO_2$ nanocrystals: a study using ultrafast terahertz spectroscopy. *Phys. Rev. Lett.* 104 (19): 197401 (1–4). https://doi.org/10.1103/PhysRevLett.104.197401.

**70** Baldini, E., Palmieri, T., Rossi, T. et al. (2017). Interfacial electron injection probed by a substrate-specific excitonic signature. *J. Am. Chem. Soc.* 139 (33): 11584–11589. https://doi.org/10.1021/jacs.7b06322.

**71** Virkki, K., Tervola, E., Ince, M. et al. (2018). Comparison of electron injection and recombination on $TiO_2$ nanoparticles and ZnO nanorods photosensitized by phthalocyanine. *R. Soc. Open Sci.* 5 (7): 180323. https://doi.org/10.1098/rsos.180323.

**72** Mulmudi, H.K., Mathews, N., Dou, X.C. et al. (2011). Controlled growth of hematite ($\alpha$-$Fe_2O_3$) nanorod array on fluorine doped tin oxide: synthesis and photoelectrochemical properties. *Electrochem. Commun.* 13 (9): 951–954. https://doi.org/10.1016/j.elecom.2011.06.008.

**73** Lin, Y., Shatkhin, M., Flitsiyan, E. et al. (2011). Minority carrier transport in p-ZnO nanowires. *J. Appl. Phys.* 109 (1): 016107. https://doi.org/10.1063/1.3530732.

**74** Hagfeldt, A., Boschloo, G., Sun, L. et al. (2010). Dye-sensitized solar cells. *Chem. Rev.* 110 (11): 6595–6663. https://doi.org/10.1021/cr900356p.

**75** Sharma, K., Sharma, V., and Sharma, S.S. (2018). Dye-sensitized solar cells: fundamentals and current status. *Nanoscale Res. Lett.* 13 (1): 381. https://doi.org/10.1186/s11671-018-2760-6.

**76** O'Regan, B. and Grätzel, M. (1991). A low-cost, high-efficiency solar cell based on dye-sensitized colloidal $TiO_2$ films. *Nature* 353 (6346): 737–740. https://doi.org/10.1038/353737a0.

**77** Jose, R., Thavasi, V., and Ramakrishna, S. (2009). Metal oxides for dye-sensitized solar cells. *J. Am. Ceram. Soc.* 92 (2): 289–301. https://doi.org/10.1111/j.1551-2916.2008.02870.x.

**78** Cao, Y., Liu, Y., Zakeeruddin, S.M. et al. (2018). Direct contact of selective charge extraction layers enables high-efficiency molecular photovoltaics. *Joule* 2 (6): 1108–1117. https://doi.org/https://doi.org/10.1016/j.joule.2018.03.017.

**79** Tan, B., Toman, E., Li, Y., and Wu, Y. (2007). Zinc stannate ($Zn_2SnO_4$) dye-sensitized solar cells. *J. Am. Chem. Soc.* 129 (14): 4162–4163. https://doi.org/10.1021/ja070804f.

**80** Anta, J.A., Guillén, E., and Tena-Zaera, R.Ã. (2012). ZnO-based dye-sensitized solar cells. *J. Phys. Chem. C* 116 (21): 11413–11425. https://doi.org/10.1021/jp3010025.

**81** DiMarco, B.N., Sampaio, R.N., James, E.M. et al. (2020). Efficiency considerations for $SnO_2$-based dye-sensitized solar cells. *ACS Appl. Mater. Interfaces* 12 (21): 23923–23930. https://doi.org/10.1021/acsami.0c04117.

**82** Kaniyankandy, S. and Ghosh, H.N. (2009). Efficient luminescence and photocatalytic behaviour in ultrafine $TiO_2$ particles synthesized by arrested precipitation. *J. Mater. Chem.* 19 (21): 3523–3528. https://doi.org/10.1039/B904589A.

**83** Paul, K.K., Jana, S., and Giri, P.K. (2018). Tunable and high photoluminescence quantum yield from self-decorated $TiO_2$ quantum dots on fluorine doped mesoporous $TiO_2$ flowers by rapid thermal annealing. *Part. Part. Syst. Charact.* 35 (9): 1800198. https://doi.org/10.1002/ppsc.201800198.

**84** Rahman, F. (2019). Zinc oxide light-emitting diodes: a review. *Opt. Eng.* 58 (1): 1–20. https://doi.org/10.1117/1.OE.58.1.010901.

**85** Yu, X., Marks, T.J., and Facchetti, A. (2016). Metal oxides for optoelectronic applications. *Nat. Mater.* 15 (4): 383–396. https://doi.org/10.1038/nmat4599.

**86** Schneider, J., Matsuoka, M., Takeuchi, M. et al. (2014). Understanding $TiO_2$ photocatalysis: mechanisms and materials. *Chem. Rev.* 114 (19): 9919–9986. https://doi.org/10.1021/cr5001892.

**87** Danish, M.S., Bhattacharya, A., Stepanova, D. et al. (2020). A systematic review of metal oxide applications for energy and environmental sustainability. *Metals* 10 (12): 1604. https://doi.org/10.3390/met10121604.

**88** Ahmed, M. and Xinxin, G. (2016). A review of metal oxynitrides for photocatalysis. *Inorg. Chem. Front.* 3 (5): 578–590. https://doi.org/10.1039/C5QI00202H.

**89** Tang, J., Durrant, J.R., and Klug, D.R. (2008). Mechanism of photocatalytic water splitting in$TiO_2$. reaction of water with photoholes, importance of charge carrier dynamics, and evidence for four-hole chemistry. *J. Am. Chem. Soc.* 130 (42): 13885–13891. https://doi.org/10.1021/ja8034637.

**90** Chung, K.-T. (2016). Azo dyes and human health: a review. *J. Environ. Sci. Health, Part C Environ. Carcinog. Ecotoxicol. Rev.* 34 (4): 233–261. https://doi.org/10.1080/10590501.2016.1236602.

**91** Lellis, B., Fávaro-Polonio, C.Z., Pamphile, J.A., and Polonio, J.C. (2019). Effects of textile dyes on health and the environment and bioremediation potential of living organisms. *Biotechnol. Res. Innovat.* 3 (2): 275–290. https://doi.org/10.1016/j.biori.2019.09.001.

**92** Soo, J.Z., Chai, L.C., Ang, B.C., and Ong, B.H. (2020). Enhancing the antibacterial performance of titanium dioxide nanofibers by coating with silver nanoparticles. *ACS Appl. Nano Mater.* 3 (6): 5743–5751. https://doi.org/10.1021/acsanm.0c00925.

**93** Dadi, R., Azouani, R., Traore, M. et al. (2019). Antibacterial activity of ZnO and CuO nanoparticles against gram positive and gram negative strains. *Mater. Sci. Eng., C* 104: 109968. https://doi.org/10.1016/j.msec.2019.109968.

**94** Shaban, M., Zayed, M., and Hamdy, H. (2019). Nanostructured ZnO thin films for self-cleaning applications. *RSC Adv.* 7 (2): 617–631. https://doi.org/10.1039/C6RA24788A.

**95** Yuranova, T., Mosteo, R., Bandara, J. et al. (2006). Self-cleaning cotton textiles surfaces modified by photoactive $SiO_2/TiO_2$ coating. *J. Mol. Catal. A: Chem.* 244 (1): 160–167. https://doi.org/10.1016/j.molcata.2005.08.059.

**96** Thomas, M., Naikoo, G.A., Sheikh, M.U.D. et al. (2016). Effective photocatalytic degradation of congo red dye using alginate/carboxymethyl cellulose/$TiO_2$ nanocomposite hydrogel under direct sunlight irradiation. *J. Photochem. Photobiol., A* 327: 33–43. https://doi.org/10.1016/j.jphotochem.2016.05.005.

**97** Zhao, W., Guo, J., and Wang, H. (2020). Synthesis and characterization of a novel rambutan-like ZnO@$SrTiO_3$/$TiO_2$ microsphere. *ChemistrySelect* 5 (32): 10029–10033. https://doi.org/10.1002/slct.202002135.

**98** Borges, M.E., Sierra, M., Cuevas, E. et al. (2016). Photocatalysis with solar energy: Sunlight-responsive photocatalyst based on $TiO_2$ loaded on a natural material for wastewater treatment. *Solar Energy* 135: 527–535. https://doi.org/10.1016/j.solener.2016.06.022.

# 7

# Metal Oxide Nanomaterials for Nitrogen Oxides Removal in Urban Environments

*M. Cruz-Yusta, M. Sánchez, and L. Sánchez*

*Córdoba University, Department of Inorganic Chemistry and Chemical Engineering, Campus de Rabanales, Córdoba 14014, Spain*

## 7.1 Introduction: Photocatalytic Removal of Nitrogen Oxides Gases

Air pollution in urban environments is a major concern for today's society. The presence of particulate matter (PM), ozone, sulfur, and nitrogen oxides is harmful to human health and the environment. Particularly, the nitrogen oxide (NOx = NO + NO$_2$) emissions come from power generation and industrial and traffic sources, the latter contributing 40% of emissions [1]. These gaseous oxides participate in the formation of PM and in the generation of ozone and, when inhaled, produce bronchitis, asthma, respiratory infections, and reduced lung function and growth [2]. In fact, because the annual limit of NOx concentration values is usually exceeded by many countries [3], around 10 000 early deaths per year in the European Union are associated with NOx emissions [4]. Therefore, it is of high interest to develop and improve methodologies to help to mitigate the emission and removal of NOx gases, known as DeNO$_x$ actions. Frequently used technologies such as selective catalytic reduction (SCR), electrical discharge, and wet/dry processes to control NOx pollution are expensive, high energy consuming, and, most importantly, cannot be applied in open urban centers [5].

Since the early 2000s, the use of photocatalysis has successfully been proved to mitigate NOx levels in certain urban places, thanks to the use of TiO$_2$-containing cementitious materials [6]. The use of a powerful photocatalytic additive, such as the TiO$_2$ semiconductor, demonstrated the capacity to remove NOx gases from air using photochemical reactions. The photochemical DeNO$_x$ action proceeds with the sole use of atmospheric water, oxygen molecules, and sunlight. Once the TiO$_2$ nanoparticles are irradiated by the UV light, the motion of charges at atomic level occurs on the surface of the semiconductor. Thus, several oxidation processes are initiated leading to the rapid conversion of NOx gases into other oxidized nongaseous forms and, therefore, being removed from the air [5]. In early years, due to successful field applications of this technology, a great interest of the scientific community arose

*Tailored Functional Oxide Nanomaterials: From Design to Multi-Purpose Applications*, First Edition.
Edited by Chiara Maccato and Davide Barreca.
© 2022 WILEY-VCH GmbH. Published 2022 by WILEY-VCH GmbH.

looking for a deeper understanding of the photochemical oxidation process (PCO) of $NO_X$ gases by using $TiO_2$-containing cementitious materials [7–15].

Titania is used as the standard photocatalyst for $DeNO_x$ applications, due to its excellent characteristics in relation to its strong oxidation power and chemical inertness. However, it fails in the whole harvesting of the sunlight, as it is only active under UV light irradiation ($\lambda < 387$ nm; 3.2 eV energy band gap) and is expensive as a raw material for large-scale applications in urban infrastructures. On the other hand, in considering the basic $NO \rightarrow NO_2^- \rightarrow NO_2 \rightarrow NO_3^-$ photocatalytic mechanism – which will be explained in depth further on – titania exhibits a low $DeNO_x$ selectivity because emissions of $NO_2$, a gas, which is more toxic than NO, occur during the PCO process. Additionally, the inhalation of $TiO_2$ nanoparticles has just recently been ascertained as a cause of cancer [16], which could determine the unfavorable use of titania-based compounds in the near future.

The issues commented previously are being addressed by the scientific community, who have turned their attention to the study of new semiconductors, modified titania, or alternative compounds, with the aim of finding low cost, sustainable, and enhanced $DeNO_x$ photocatalysts. This chapter attempts to summarize this flourishing field of research in which a vast number of publications have paid attention to different chemical actions (doping, heterojunctions, quantum dot functionalization, structural defects, etc.) and compositions (transition metal and non-transition metal oxides, oxo-halydes, oxo-hydroxides, etc.) with semiconductor nanomaterials now proposed as new $DeNO_x$ photocatalysts in a revisited or innovative study.

## 7.2 TiO₂-Based Materials

The performance of a semiconductor participating in heterogeneous photocatalytic processes is related to its chemical, physical, and electronic characteristics. It is desirable to be chemically inert and to exhibit good contact with the reactant molecules. Thus, high photocatalytic activities are found for catalysts exhibiting large surface areas, which are obtained through the suitable dispersion of the catalyst in suitable substrates or by preparing new crystalline nanostructured morphologies. On the other hand, the electronic parameters governing the photochemical process are also key factors. This process is initiated when the appropriate photonic energy of sunlight causes the promotion of electrons in the semiconductor valence band (VB) to the conduction band (CB), creating a pair of mobile charges $e^-/h^+$, which in turn initiate the redox processes. In this sense, the activity of a photocatalyst can be improved by modifying its band gap with the aim of enhancing the harvesting of the sunlight, favoring the mobility of the charge carriers, and promoting the production of radical species. Strategies such as coupling band gaps between semiconductors, element doping/co-doping, the deposition of quantum dots, and the promotion of structural defects are used to modify the electronic band structure of the photocatalyst, in order to achieve higher quantum efficiency and reaction rates.

Figure 7.1 shows a basic summary of the previously mentioned strategies and the advantages obtained for $TiO_2$ and alternative compounds to be revised in this

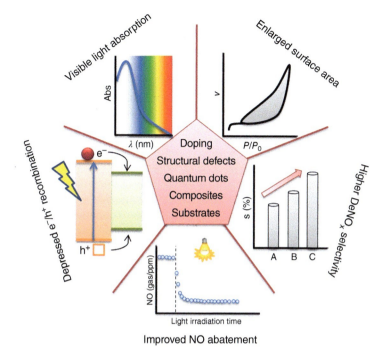

**Figure 7.1** Different strategies to modify a photocatalyst and the main improved characteristics with respect to the DeNO$_x$ action. Source: L. Sánchez.

chapter. In brief, by modifying the semiconductor structure/composition, the intention is to find a photocatalyst with a large surface area, UV and visible light absorption ability, and the lowest e$^-$/h$^+$ recombination, with the aim of increasing the efficiency and selectivity in the photochemical oxidation of NOx gases in air.

### 7.2.1 Tailoring the Energy Band Gap and Edges' Potentials

When two semiconductors are coupled, an electron heterojunction occurs. The new energy arrangement of the respective conduction and VBs permits the enlargement of the light absorption ability and enhances the mobility of the charge carriers, delaying their recombination. This strategy has been successfully used to activate TiO$_2$ under visible light and to greatly increase its photochemical performance.

Due to its narrow energy band gap (∼2.7 eV), graphitic carbon nitride is used as a coupled semiconductor to constitute composites for visible light-driven photocatalysis. Several researchers have paid attention to the study of the heterojunction formed by g-C$_3$N$_4$/TiO$_2$ photocatalysts. Trapalis et al. [17] prepared g-C$_3$N$_4$/TiO$_2$ composites obtained after the calcination of melamine/TiO$_2$ P25 mixtures with different ratios (3 : 1, 1 : 1, 2 : 3, 1 : 3, 1 : 4, and 1 : 7). The differences in composition of the photocatalysts allowed changes to be observed in both band gap energy (from 2.66 to 3.14 eV) and the CB edge (from 1.27 to 0.67 eV) values from pure g-C$_3$N$_4$ to pure TiO$_2$, respectively. These parameters seem to have no influence on the NO photocatalytic oxidation under UV irradiation, and the NO abatement efficiency was similar for $TiO_2$ and the samples, which were richer in $g-C_3N_4$. However, the photocatalyst with a ratio 1 : 4 exhibited the best photocatalytic activity under visible light, as a result of the combination of the optical and electronic properties of the composites, as explained by the authors. In a second study, this research group advanced in the knowledge of these $DeNO_x$ compounds by homogeneously incorporating calcium carbonate into $g-C_3N_4/TiO_2$ composites [18]. As a result, apart from the improved NO photooxidation under visible light, the release of $NO_2$ gas – a highly toxic intermediate – is lessened. Thus, the basic character of the $CaCO_3$ composite and the acidic character of the gaseous pollutants favor the $NO_2$ adsorption on the surface and, therefore, its oxidation. Thus, the selectivity of the $NO \rightarrow NO_3^-$ PCO process was clearly increased. Similar composites were prepared by Hu et al. in order to overtake the low quantum efficiency usually exhibited by {001} grown $TiO_2$ nanosheets [19]. The composites with visible light active response yielded higher $DeNO_x$ photochemical activity than the pure compounds. The photo-luminescence studies concluded that $g-C_3N_4/TiO_2$ heterojunction promotes the separation of the electron–hole pairs generated during the photochemical process in composites, which results in a better photocatalytic activity. The importance of building a good heterojunction is reflected in the proper interaction between both semiconductors. Thus, no improvement in the $DeNO_x$ performance is observed when $g-C_3N_4$ and $TiO_2$ samples are simply mechanically mixed [20]. The electronic interaction was studied in depth by Jiang and Zhang et al. with N-doped $TiO_2/g-C_3N_4$ composites [21]. These photocatalysts exhibited better photocatalytic NO removal efficiency, which increases with the $N-TiO_2$ content. As the N doping narrows the band gap of $TiO_2$, an effective heterojunction is formed with the $g-C_3N_4$ semiconductor. Thus, the photoexcited $e^-$ produced in $g-C_3N_4$ flows spontaneously into the CB of $N-TiO_2$, while the holes ($h^+$) generated in $N-TiO_2$ go to the VB of $g-C_3N_4$. In this manner, the charge carrier mobility is facilitated. Both charge carriers are consumed in the production of $\cdot OH$ and $\cdot O_2^-$ radicals (initiating the $DeNO_x$ PCO process), as detected by spin-trapping electron paramagnetic resonance (EPR) measurements. On the other hand, due to the high efficiency of this PCO process, the NO gas is completely mineralized into nitrate ions.

In a similar way, other interesting works on heterojunction refer to that formed by graphene/$TiO_2$ [22, 23]. N. Todorova et al. studied graphene/$TiO_2$ and graphene oxide/$TiO_2$ photocatalysts [24]. Both systems exhibited $DeNO_x$ photocatalytic activity superior to that of $TiO_2$, the enhancement being attributed to the role of graphene sheets when linked to titania particles: as an electron trap in the case of $G/TiO_2$ or photosensitizer for $GO/TiO_2$. In both cases, an enhanced visible light activity was observed and, importantly, a low emission of $NO_2$ molecules because of their affinity toward the graphene sheets. Thus, Li and Wang's research group prepared graphene embedded anatase–rutile titania mesoporous microspheres (OATMS/GP composites) constituted by radially oriented self-assembly $TiO_2$ nanowires (Figure 7.2a). As inferred from the NO concentration profile evolution under light irradiation, the incorporation of graphene nanosheets results in a highly superior NO removal ability for this composite, compared to that of pure $TiO_2$ and

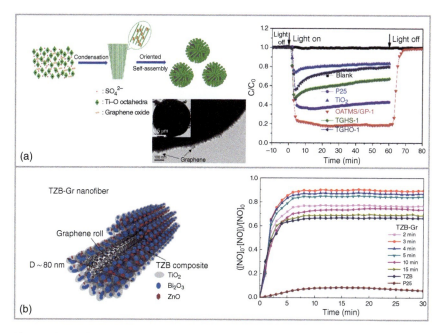

**Figure 7.2** (a) Schematic growth mechanism of the mesoporous structures of the as-prepared OATMS/GP structure, TEM image of part of mesoporous $TiO_2$ microsphere, and NO relative concentration measured under light irradiation for P25, $TiO_2$, and OATMS/GP photocatalysts. Source: Reprinted from Yang et al. [22], with permission from Elsevier. (b) Schematic diagram of as-prepared graphene/TZB nanofibers and NO removal ratio for composites with different graphene centrifuged time durations, TZB, and P25 photocatalysts. Source: Reprinted from [25], copyright 2017, with permission from Elsevier.

$TiO_2$ P25 (benchmark product widely used as a standard in the field) samples, presenting an outstanding NO removal efficiency as high as 81.4% (Figure 7.2a) [22]. An interesting advance in graphene–titania compounds is that proposed by Leung et al., preparing a graphene/TBZ where TBZ is an electrospun $TiO_2/Bi_2O_3/ZnO$ fiber [25]. The graphene sheets are rolled up into "spiral rolls" inserted in the TBZ nanofiber (Figure 7.2b). This rolling configuration is carried out with the aim of suppressing the electron–hole recombination phenomena at the edges of the graphene sheet, thus enhancing the photochemical activity. The prepared composite exhibited a large surface area and visible light activity, the latter due to the reduced band gap energy originated by the presence of zinc and bismuth oxides. This advanced photocatalyst has superior photochemical NO removal (Figure 7.2b), being 10 times more efficient than $TiO_2$ P25.

Other heterostructure types are those containing transition metal oxides: $WO_3$, $MoO_3$, and $PtO_2$ [26–29]. In this group, the use of $WO_3/TiO_2$ composites for $DeNO_x$ processes is well known. A synergic benefit is obtained by both oxides; the heterojunction lessens the inherent photocorrosion of $WO_3$ and enables the visible light response of the photocatalyst. Moreover, a significant improvement of visible light sensitivity is obtained when the composite is additionally grafted with Fe(III) nanoclusters. The iron sites trap the photoelectrons from the CB of the oxides, improving the charge carrier separation. Thus, the NO conversion is enhanced around 40% in comparison to that of $TiO_2$ P25.

The energy band gap is not only tailored by the creation of semiconductor heterojunctions. As in the case of graphene, the introduction of dopant elements or the creation of defects in the $TiO_2$ lattice, as well as the intimate contact with quantum dots, allows the creation of intermediate energy levels between CB and VB, favoring $e^-/h^+$ motion and light absorption ability. The corresponding studies are examined in the following sections.

### 7.2.2 Dopant Elements and Quantum Dots

C–$TiO_2$ mesoporous anatase nanocrystals were prepared by Li and Wang et al. [30]. By using a hydrothermal method, chitosan is used as the template and the source of carbon, resulting in a doped titania wrapped in a carbon shell. The Ti—C bond explains the presence of Ti(III) and oxygen vacancies (Ov), which promote the visible light photo-response. The high photocatalytic activity measured, about 71% NO conversion, is related to the large specific surface area (189 $m^2\,g^{-1}$) and the inhibition of $e^-/h^+$ recombination. The new electronic pathways for the photogenerated electrons are of significant importance, the superoxide being the main radical detected. Outstanding results were also obtained by Komarneni et al. [31] with a similar strategy. On the other hand, nitrogen was also studied as a dopant element [32, 33]. Kim et al. prepared N–$TiO_2$/$ZrO_2$ photocatalysts using small amounts of zirconia to stabilize the crystallization of the $TiO_2$ anatase phase [34]. The increased amount of N favors the anatase to rutile phase transformation, larger particle size, and visible light reactivity. NO removal efficiencies around 60% were measured under visible light irradiation. An advance on the N–$TiO_2$ system was performed by Yin et al. [35] combining the long afterglow phosphor $CaAl_2O_4$:(Eu, Nd) with $TiO_{2-x}N_y$. The authors obtained a persistent photocatalyst by also working at night. The combination of the two different band structures of these compounds allows $TiO_{2-x}N_y$ to absorb the long visible light afterglow emitted by $CaAl_2O_4$:(Eu, Nd) once the light is turned off. Superior $DeNO_x$ activity has also been found in co-doped B,N–$TiO_2$ [36]. The joint use of boron and nitrogen dopants improved the photocatalytic activity due to: (i) the large surface area (182 $m^2g^{-1}$), (ii) a mesoporous structure facilitating the contact with the NO reactant molecules, (iii) a decreased band gap in the developed B–O–Ti–N structure, and (iv) the creation of oxygen vacancies favoring the charge mobility. Following the route of co-doping, $Ti_{0.909}W_{0.091}O_2N_x$ was reported as a photocatalyst of outstanding $DeNO_x$ selectivity, >90%. The presence of $W^{5+}$ as trap sites for the photogenerated electrons seems to open new electronic transfer paths, which lessen the production of $NO_2$ gas [37]. In order to obtain a better selective photocatalyst, the study of grafted $Fe^{3+}$ ions–$TiO_2$ samples was also carried out. The improved selectivity of this photocatalyst was attributed to the generation of Fe(IV) species on the surface of the catalyst, which exhibited enhanced reactivity toward $NO_2$ [38].

Lanthanides elements were also an object of study [39]. The incorporation of $La^{3+}$ to the anatase structure originates a narrower band gap, from 2.98 to 2.75 eV, enabling the visible light activity. Moreover, the $e^-/h^+$ recombination is lessened, and a complete selectivity for the NO → $NO_3^-$ PCO process is observed. The NO removal efficiency is increased by 32% [40]. In a similar way, cerium was incorporated into the $TiO_2$ lattice [41]. In a different study, the NO removal efficiency of $La^{3+}$, $Ba^{2+}$, and $Zr^{4+}$-doped titania was compared. The zirconium-containing samples exhibited outstanding 100% removal values (after 35 minutes of light irradiation), which were associated with the combined effects of increased surface area, optimal anatase/rutile ratio, smaller band gap, and increased production of ROS radical species [42]. The incorporation of noble metals, such as Pt and Au, also increases the visible light photo-response and $DeNO_x$ activity [43, 44]. However, Pt–$TiO_2$ does not always increase its UV light activity compared with the standard P25 [43, 45].

Another interesting approach makes use of quantum dots to functionalize the $TiO_2$ photocatalyst. $TiO_2$ P25 was functionalized with thioglycolic acid (TGA)-capped CdTe colloidal quantum dots (QDs), 3.5–4.0 nm in diameter [46]. There is a synergistic operation between TGA, CdTe, and $TiO_2$ components leading to the enhancement of the NO abatement. The TGA-CdTe QDs promote the photocatalytic conversion of the just formed $NO_2$, increasing the selectivity of the PCO process. This strategy is of interest as only minuscule concentrations of a target compound are required to optimize a $DeNO_x$ $TiO_2$ photocatalyst. Similarly, 5–8 nm Bi nanoparticles were deposited on $TiO_2$ P25 using a precipitation method [47]. The surface plasmon resonance (SPR) of the attached bismuth nanoparticles enhances the visible light and favors the electron/hole separation. Consequently, a better photocatalytic activity was measured. This is an interesting alternative solution to the use of the expensive noble metals to create advanced photocatalysts. In a similar way, a visible light-driven plasmonic photocatalyst Ag-$TiO_2$ with enhanced $DeNO_x$ activity was reported [48, 49]. Successful results were also obtained for Ag [50] and N-doped carbon [51] quantum dot titania composites.

### 7.2.3 Defects, Vacancies, and Crystal Facets in the $TiO_2$ Nanostructure

The introduction of defects in the $TiO_2$ lattice, such as O vacancy and $Ti^{3+}$, leads to the appearance of new intermediate energy levels in the band gap. This modulates the ability to absorb sunlight and the electronic pathways. In this sense, a high NO removal efficiency of the titania photocatalyst is obtained by introducing Ov [52]. The introduction of Ov in the $TiO_2$ structure is obtained not only by doping strategies (as previously mentioned) but also with the use of specific procedures of synthesis. Thus, $TiO_2$ hollow microspheres were prepared from the calcination of hydrogen titanate and urea. The introduction of Ov had several benefits such as the promotion of visible light response, the delay of $e^-/h^+$ recombination, a favorable adsorption, and activation of NO and $O_2$ on the surface. The enhanced photocatalytic activity associated with the existence of surface Ov was not observed in samples with bulk Ov [53]. Additionally, it is reported that the Ov site is a preferable location for the formation of $\cdot O_2^-$ radicals and, therefore, ideal in addressing the direct oxidation of NO to nitrate. Thus, without $NO_2$ release, titania turns into a highly selective photocatalyst [54].

The preparation of mesocrystals of $TiO_2$ as a $DeNO_x$ photocatalyst was studied by Chen and Yen et al. [55]. The anatase $TiO_2$ mesocrystals, obtained by simple hydrolysis reaction $TiCl_3$ and polyethylene glycol, exhibit a single-crystal-like structure constituted by 1.5–4.5 nm nanocrystal subunits. After vacuum annealing, the porous crystals with $Ti^{3+}$-induced sites in the anatase lattice are obtained. This type of structure allows a high performance to be obtained in the PCO of NO under visible light. Similar mesocrystals were obtained by Sánchez et al. using a hydrothermal approach with tetraisopropyl orthotitanate and triethanolamine as precursor materials [56]. This time, the outstanding photocatalytic activity of NO abatement was corroborated in long periods of light irradiation (five hours). Noticeably, apart from the ability of these systems to work with UV and visible light, the mesoporous network facilitates the quasi-complete NO photochemical oxidation, reaching $DeNO_x$ selectivity values as high as 89%.

A superior performance, compared with $TiO_2$ P25, was also observed with the use of urchin-like mesoporous titania hollow spheres (UMTHS). Thus, Wang et al. [57] used a hydrothermal method for the reaction of amorphous fluorinated $TiO_2$ solid spheres with polyvinylpyrrolidone. During the process, the spheres were selectively etched and hollowed by fluoride ions. Meanwhile, single-crystal [101] anatase nanothorns grew, radially distributed over the surface of the hollow spheres. The superior photocatalytic activity was associated with their large specific surface area (128.6 m$^2$ g$^{-1}$), the hierarchical architecture facilitating the mass transport, the improved light scattering among the well dispersed nanothorns, and the enhancement of charge carrier mobility in the 1D anatase single-crystal.

The preferable orientation in the $TiO_2$ crystal growth has also been reported as a parameter with influence on its photocatalytic properties. It is known that the (001) surface of titania is more reactive than the thermodynamically stable [101] surface [58]. In this sense, Trapalis et al. prepared fluoride-free $TiO_2$ anatase nanoplates with dominant {001} facets, with a specific surface area of 113.2 m$^2$g$^{-1}$, which doubled the NO efficiency abatement measured for $TiO_2$ P25 [59]. The key role of {001} facets in $DeNO_x$ reactions was also confirmed by De Abreu et al. on anatase nanoparticles endowed with a different crystal morphology [60].

Finally, with the aim of summarizing the aforementioned different strategies used to tailor the energy band gap of $TiO_2$, the modifications accounted for the schematic electronic band structure, together with the light absorption ability (UV or visible) and the new pathways for the charge carriers, are represented in Figure 7.3.

### 7.2.4 Composites/Substrates

Several studies have confirmed that the $DeNO_x$ activity of the $TiO_2$ photocatalyst can be improved by its immobilization onto suitable substrates. In this sense, Trapalis et al. [61] studied 1 : 1 $TiO_2$/mineral homogeneous dispersions in the following minerals: talc ($Mg_3Si_4(OH)_2$), hydrotalcite ($Mg_6Al_2CO_3(OH)_{16.4}H_2O$), and kunipia clay $F(Ca_{0.11}Na_{0.891}(Si_{7.63}Al_{0.37})(Al_{3.053}Mg_{0.65}Fe_{0.245}Ti_{0.015})O_{20}(OH)_4)$. In comparison with $TiO_2$, an enhanced $DeNO_x$ photocatalytic response was found in composites, which was ascribed to the appropriate dispersion of the titania

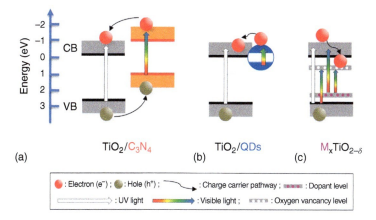

**Figure 7.3** Schematic electronic band structure for TiO$_2$ modified by coupling (a) with a semiconductor, (b) with quantum dots, or (c) with dopant elements. Source: L. Sánchez.

photoactive component and to the alkaline content of the substrates. The latter assists the complete oxidation of NO to NO$_3^-$ without the formation of NO$_2$. A similar strategy to enhance the activity of the photocatalyst, through a suitable dispersion of TiO$_2$ over different substrates (recycled clay brick sands, cement, glass, wood, and stainless steel) was reported [62, 63]. For this study they used mussel adhesive protein, which acts as a dispersing agent and enables a strong and stable adhesion of TiO$_2$ particles. In a separate work, hydroxyapatite was successfully used as the substrate [64]. Low amounts of released NO$_2$ occur when using this photocatalyst, this being associated with the stronger NO chemisorption originated by the OH groups present on the surface. In addition, a higher production of ·OH and ·O$_2^-$ radicals was detected when this composite was used, in comparison with TiO$_2$. This suggests the existence of alternative electron pathways favoring a higher mobility for the charge carriers.

With the aim of facilitating the adsorption of NOx gases, Ozensoy et al. prepared TiO$_2$/Al$_2$O$_3$ binary oxides [65, 66]. In this photocatalyst composite, alumina sites on the surface capture NO gas molecules, facilitating their oxidation to nitrites/nitrates. Thus, the release of the toxic NO$_2$ gas is significantly decreased, obtaining a more selective DeNO$_x$ TiO$_2$ photocatalyst. This photocatalytic system was later improved by CaO addition. The photoactivity of CaO/TiO$_2$/Al$_2$O$_3$ ternary oxides was favored, as the high content of CaO molecules promotes the strong adsorption of NO$_2$ gas [67]. In a similar way, the selectivity of titania was increased by using zeolite substrates [68, 69].

### 7.2.5 Titanium-Based Oxides

Titanium-based perovskite semiconductors are characterized by exhibiting chemical stability and excellent charge carrier mobility. These types of structures have been the object of study as photocatalysts for solar cells and electronic applications [70]. Because of their interesting characteristics, they have also been considered as DeNO$_x$

photocatalysts. Thus, Ag–SrTiO$_3$ nanocomposites exhibited an enhanced DeNO$_x$ performance, thanks to the plasmonic effect of Ag. The removal of NO is doubled under visible light for SrTiO$_3$ samples with only 5% of silver load. A decrease in the NO$_2$ emission is also observed, which is associated with the basic character of the Sr sites favoring the NOx storage. The improved DeNO$_x$ visible light ability of SrTiO$_3$ was also reported for Fe loading samples [71]. Interesting conclusions were obtained from the study of Cr$^{3+}$ doped and non-stoichiometric SrTiO$_3$, on which the photocatalytic performance for NO removal proved to be different depending on the light intensity [72]. The Cr dopant and the presence of oxygen vacancies gave rise to different energy band structures, favoring the NO removal photoactivity with low-intensity and high-intensity light, respectively. The creation of oxygen vacancies was also related to the enhanced DeNO$_x$ performance in the Fe$^{3+}$-doped SrTiO$_3$ photocatalyst [73]. Apart from an improved visible light activity for $\lambda > 420$ nm, the presence of Ov promoted the adsorption of oxygen, the formation of $\cdot O_2^-$ radicals, and a depressed charge recombination. In addition, Fe$^{3+}$ sites favor the adsorption and photoreaction of NO molecules.

The photocatalytic activity of SrTiO$_3$ perovskites was also modulated by the creation of interesting heterojunctions. Thus, LaTiO$_3$–SrTiO$_3$ composites exhibited enhanced visible light harvesting, enlarged surface area, and lower e$^-$/h$^+$ recombination [74]. Interestingly, the use of SrCO$_3$ to prepare SrTiO$_3$ photocatalysts not only improves the photocatalytic activity but also inhibits the poisoning effect of the active sites. At a low cost, the SrCO$_3$ acts as a noble metal accepting the photoinduced electrons coming from SrTiO$_3$ in its CB, improving the separation of charge carriers. On the other hand, the affinity of SrTiO$_3$ toward O$_2$ and NO adsorption, both forming nitrates under a light induced reaction, leads to a rapid poisoning of the active sites. This is suppressed by the presence of SrCO$_3$ acting as alternative sites for oxygen adsorption [75]. Finally, it is worth noting that Na$_{0.5}$Bi$_{0.5}$TiO$_3$ and Pb$_2$Bi$_4$Ti$_5$O$_{18}$ perovskites also exhibited superior activity on the photochemical NO abatement compared with P25 [76, 77].

## 7.3 Alternative Advanced Photocatalysts

In spite of the enhanced photocatalytic activity of TiO$_2$ under UV and visible light being reached following the strategies previously described, there is great interest in finding DeNO$_x$ photocatalysts with even better light harvesting ability, selectivity, and lower costs. In the following, the main compositions (Figure 7.4) and findings reported for transition metal oxides alternatives to TiO$_2$, are commented.

### 7.3.1 Bismuth Oxides

Bismuth-based oxides have been studied in depth as DeNO$_x$ photocatalysts because of their ability to work with only visible light. Most bismuth-based photocatalytic compounds can fall into one of three categories: binary oxides (including subcarbonates), multicomponent oxides, and oxyhalides.

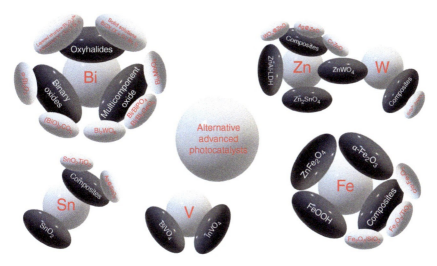

**Figure 7.4** Transition metal oxide DeNO$_x$-photocatalyst alternatives to TiO$_2$. Source: M. Cruz-Yusta.

In relation to binary oxides, the work of Ai et al. [78] reports the behavior of the commercial monoclinic α-Bi$_2$O$_3$ phase in contrast to that obtained by autoclave synthesis, finding a slightly higher photocatalytic efficiency of the sample prepared in the laboratory vs. the commercial one (35% vs. 24%). This behavior could be explained on the basis that laboratory synthesis allows a greater control of the particle size, morphology, and microstructure of the material. The morphology, the response to visible light, and the electronic configuration were key aspects that explain the differences.

To improve the DeNO$_x$ activity, Chen [79] and Li [80] research groups propose activating the compound with Bi metallic. Although they work from different approaches (the use of nanospheres or amorphous material), in both cases the Bi SPR effect (Figure 7.5a), an effect produced by the resonant oscillation of electrons in the CB, enables the improvement of the generation of charge carriers and the adsorption of visible light, leading to a greater efficiency in removing NO gas (Figure 7.5b).

As has been commented previously, Bi sub-carbonates can be considered here because they are a family of Aurivillius-type lamellar oxides whose structure consists of alternating sheets of Bi$_2$O$_2{}^{2+}$ y CO$_3{}^{2-}$ [81]. Dong et al. [82] studied pure (BiO)$_2$CO$_3$ finding that its photocatalytic behavior was very poor, with only 8% photocatalytic efficiency in visible light, mainly due to the rapid electron–hole recombination. Chen et al. [83], studying the same compound, found differences if the photocatalytic reaction was carried out on the facet {001} or the facet {110} of the structure. In this case, the best behavior was found on the facet {001}, due to a greater interaction between the reagents and the photocatalyst surface on the {001} facet.

In a similar way to TiO$_2$, the creation of Ov on the structure of the bismuth compounds is another way to improve the photocatalytic efficiency. Oxygen vacancies create intermediate levels in the energy band structure, resulting in a better use of

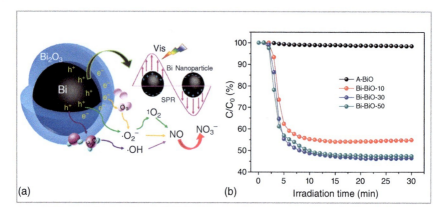

**Figure 7.5** (a) Scheme of the photocatalytic oxidation mechanism of NO with Bi@Bi$_2$O$_3$ and SPR effect. Source: Reprinted from Li et al. [79], with permission from ACS Publications. (b) Evolution of NO concentration under visible light irradiation. Source: Reprinted from Li et al. [80], with permission from Elsevier.

visible light and improving the effective separation of the charge carriers [84]. In any case, these works do not show an improvement in the photocatalytic efficiency of this compound, as in no case does it exceed 15% efficiency in the DeNO$_x$ process.

For better results, doping with nitrogen (N-(BiO)$_2$CO$_3$) is chosen. This doping action facilitates the formation of N—O surface bonds that causes an improvement in the adsorption of visible light and charge separation [85]. Moreover, Dong et al. [86] deposited platinum quantum dots on N-(BiO)$_2$CO$_3$. In this case, platinum has a double effect: a Schottky defect barrier that promotes the dispersion and separation of electrons and holes and a hierarchical structure that gives rise to the surface scattering and reflecting (SSR) effect. The SSR effect allows a better use of visible light. This effect was also claimed by Jin et al. [87] for (GO)/N-(BiO)$_2$CO$_3$ composites, for which the performance improves considerably reaching 62% efficacy in the photocatalytic process. Other authors have had similar results decorating N-(BiO)$_2$CO$_3$ with metals such as Bi [88, 89] or Fe [90]. In the first case, Bi acts as a plasmon, on the one hand improving the adsorption of light in the visible range and, on the other, generating H$_2$O$_2$ and capturing electrons from the defective states by reducing O$_2$ and inhibiting the formation of NO$_2$ (a highly selective process of 98%). In the second case, the deposition of Fe clusters on N-(BiO)$_2$CO$_3$ facilitates the separation of charge and the creation of ·OH radicals.

The category of Bi multicomponent oxides includes a series of oxysalts that can be considered as hybrid oxides between bismuth oxide and another metal oxide. One of these multicomponent oxides is the Bi$_2$WO$_6$ (BWO), with a perovskite-like lamellar structure with a layer of WO$_6$ (which is composed of a double [O]$^{2-}$ and a single [WO$_2$]$^{2+}$) and another [Bi$_2$O$_2$]$^{2+}$ [91]. In an initial study, Huo et al. [92] proposed a synthesis with a concentration gradient to introduce oxygen vacancies and studied their behavior. They found that oxygen vacancies allow the modification of the structure of bands and the surface chemical state, improving the efficiency in the separation of the charge carriers, reaching 47% efficiency in the photocatalytic process. Simultaneously, they carried out another study [93] in which they controlled the oxygen vacancies and intercalate carbonate in the structure, causing a delay in the recombination of the electron–hole pairs and improving the transformation of the charge carriers in active species. In this way, they reach 55% photocatalytic efficiency in the DeNO$_x$ process.

$Bi_2MoO_6$ is another attractive candidate as DeNO$_x$ photocatalyst. Its structure is similar to the previous one. In this case the material itself presents poor efficiency ($\approx$10%) in the DeNO$_x$ process. To improve these results, doping is studied with different materials such as bismuth [94], bromine [95], or carbonate [96]. In all three cases, the presence of a dopant element modulates the electronic state of $Bi_2MoO_6$, enhancing the charge separation and the formation of ROS species.

In general, these materials do not have a very high DeNO$_x$ process efficiency. To improve the efficiency, it is necessary to look for a modification of their structure, for example, by treating them with metallic bismuth; some examples are Bi/BiPO$_4$ [97], Bi/Bi$_2$O$_2$SiO$_3$ [98], and Bi/Bi$_2$GeO$_5$ systems [99].

In the category of bismuth oxyhalides, two different types of structures can be found: a layered structure of alternated [Bi$_2$O$_2$]$^{2+}$ layers intercalated with layers of halogen (anion X$^-$) or a solid solution consisting of a homogeneous crystalline structure formed by the solution of two or more crystalline phases under solid-state conditions. Zhang et al. [100] prepared BiOX with X = Cl, Br, and I in a 2D configuration. They observed that both the band gap and the thermal stability of the compound decreased when the size of the halogen increased. The compound with bromine showed the highest efficiency in the DeNO$_x$ process because it presented a greater specific surface and an appropriate band structure. However, the efficiency of the DeNO$_x$ process for these compounds is not very high ($\approx$25%). For the BiOCl, it was found that the creation of superficial oxygen vacancies on the facet {001} favors the formation of the $\cdot O_2^-$ radical, achieving a DeNO$_x$ process efficiency close to 50% [101]. To improve the efficiency of the photocatalytic reaction for this compound, the SPR effect is used again. In this sense, Zhang et al. [102] propose the preparation of self-assembly Ag/AgCl nanoparticles on the BiOCl. Also, Wang et al. [103] propose the deposition of Bi metallic, which induces an electromagnetic field between both layers, enhancing the separation of charges in photogenerated carriers. In both cases, the adsorption of light in the visible range was improved, but the DeNO$_x$ process efficiency levels of the untreated compound were not enhanced. On the other hand, the doping of BiOBr and BiOI with Iodine considerably improves the efficiency of the DeNO$_x$ process [104, 105]. In both cases, the iodine is located in the interlayer, forming channels between the two layers. Thus, the band gap is modulated introducing an intermediate level in the prohibited interval and improves the oxidation ability.

The formation of a solid solution allows a tailored band gap between 3.3 and 2.75 eV to be obtained by controlling the amount of halogen in solid solutions of BiOCl$_x$Br$_{1-x}$ with different morphologies [106, 107]. For these samples, the best performance in photocatalytic efficiency was around 60% for a value of $x = 0.5$. Another kind of solid solution is that constituted of compounds from different families. For example, Ou et al. [108] studied the (BiO)$_2$CO$_3$/BiOI solid solution. With this combination they were able to modify the band gap in the 1.89–2.68 eV range, being active in the visible light range. In these conditions, they reach up to 51% efficiency in the photocatalytic process.

Finally, the behavior of bismuth oxide-based materials used to form electronic heterojunctions was commented. The main type of heterojunction in bismuth-based compounds is the p–n junction. This promotes the effective electron-hole separation, leading to a greater efficiency of the DeNO$_x$ process [109]. The so-called Z-scheme heterojunction is typically used in different works, and those carried out by Li et al. [110], who studied the heterojunction formed by BiOBr/Bi$_{12}$O$_{17}$Br$_2$ systems, can be highlighted. They found that this heterojunction is responsible for lessening the e$^-$/h$^+$ recombination and allows greater efficiency in the DeNO$_x$ process (53%) when the ratio between the two components is 0.8. Similarly, high efficiencies of 57%, 61%, and 67% were obtained for (BiO)$_2$CO$_3$/MoS$_2$, (BiO)$_2$CO$_3$/BiO$_{2-x}$/graphene, and black-P/Bi$_2$WO$_6$ systems, respectively [111–113]. Another type of heterojunctions is made with g-C$_3$N$_4$. In this case, due to the way of preparing the sample, nanosheets are the most common morphology. The efficiencies in the DeNO$_x$ process for these materials vary depending on the system, but in general are slightly lower: 32%, 35%, and 51% for BiOBr/g-C$_3$N$_4$, Bi$_2$O$_2$CO$_3$/g-C$_3$N$_4$, and g-C$_3$N$_4$/Bi$_4$O$_5$I$_2$ systems, respectively [114–116].

The study of the Ag/AgCl-(BiO)$_2$CO$_3$ system was carried out by Cui et al. [117]. In this case, they reached efficiencies of 56% as a result of combining the SSR effect, which allows a better use of light with the SPR effect caused by silver. Using this SPR effect, but using metallic bismuth, values higher than 50% efficiency are achieved in the DeNO$_x$ process [118, 119].

The formation of heterojunctions between oxyhalides of the BiOX/BiOI type is another method used to improve the efficiency of the DeNO$_x$ process with these materials. Dong et al. studied the BiOI/BiOCl system finding a ≈55% efficiency in the DeNO$_x$ process when the relationship between the two compounds is 0.25/0.75; the coupling allows an improvement in the adsorption of light in the visible range [120]. Shi et al. studied the BiOBr/BiOI system, the ultrafine structure that increases the specific surface area and the creation of oxygen vacancies, which were responsible for a photocatalytic efficiency of 57% [121]. In the case of the BiOIO$_3$/BiOI system, the exposed facets of the BiOIO$_3$ ({010}) and the BIOI ({001}), together with the intimate contact between both structures, were the key factors in the improvement of the DeNO$_x$ process [122]. The use of heterojunctions allows reasonable efficiency values to be obtained, even when insulating materials such as the SrCO$_3$-BiOI system are used, where efficiency reaches 49% thanks to its core-shell structure [123].

### 7.3.2 Tin- and Zinc-Based Oxides

Another kind of advanced photocatalysts described in the literature are tin-based (for the visible range) or zinc-based photocatalysts (for the UV range). In relation to tin-based photocatalysts, two groups can be distinguished: one based on binary oxides, mainly SnO$_2$, and another based on multicomponent oxides, specifically on the Zn$_2$SnO$_4$ spinel-like structure. In reference to both groups, there are also works based on heterojunctions. In the so-called binary oxides, the work of Pham et al.

with $SnO_2$ nanoparticles [124] can be highlighted. They achieve an efficiency in the photocatalytic process of approximately 64%, but preparing the composite between Ag and $SnO_2$ nanoparticles (Ag@$SnO_2$), the efficiency of the photocatalytic process increases to 70% [125]. In both studies, the selectivity in respect to NO is quite high. The percentage of $NO_2$ formed is minimal, and they also show good retention of capacity after five cycles. Electron–hole recombination is the key to these good results. Finally, the same group studied the heterojunction $SnO_2$/$TiO_2$ nanotube. In this case, the results were slightly worse, reaching 59% efficiency [126].

In the so-called multicomponent oxides, Ai et al. [127] found that the $Zn_2SnO_4$ spinel with a band gap of 3.25 eV presents better results (70%) than pure $SnO_2$ (33%) or ZnO (42%). The better stability and photocatalytic activity were explained on the basis of a porous structure, which favors the diffusion of reaction intermediates and final products, and an electronic structure, which improves the charge carrier separation. Regarding the studies with heterojunctions, Li et al. studied the $SnO_2$/$Zn_2SnO_4$ system, with a Z-scheme on graphene, Figure 7.6a [128]. They found that the presence of graphene affected the formation of $SnO_2$ and Sn vacancies in $Zn_2SnO_4$, the formation of these vacancies being the key to better adsorption of visible light. In addition, the photoluminescence spectra of the sample in the visible region clearly indicated an efficient transportation and separation of carriers in samples containing graphene (Figure 7.6b). On the other hand, they studied the photocatalytic efficiency of NO removal, finding the best results after the addition of 3% of graphene where the photocatalytic efficiency reached 59% compared with 7.5% of the same compound without graphene, Figure 7.6c. The excellent photocatalytic activity was explained on the basis of the proposed Z-scheme heterojunction. Under illumination, the electrons in the CB of $SnO_2$ can migrate to the VB of $Zn_2SnO_4$, which can further be excited to the CB of $Zn_2SnO_4$. These are easily transferred to graphene, where superoxide is produced. On the other hand, the holes in the VB of $SnO_2$ promote the formation of hydroxyl radical. This effective separation of charge carriers facilitates the formation of ROS species improving the photocatalytic efficacy.

The other groups of interesting advanced photocatalysts in metal oxides are those based on ZnO. One of the first works reporting the photocatalytic efficiency in the $DeNO_x$ process with ZnO photocatalyst was developed by Long et al. [129]. They studied the effect of the particle morphology on the $DeNO_x$ efficiency, finding that the key factor in an improved efficiency was the surface area. The photocatalytic efficiency was approximately 30% with illumination in the UV light range and 12% under visible light. Later, Li et al. prepared a hierarchical structure of ZnO porous microspheres. The peculiar structure allows the efficiency of the $DeNO_x$ process to be improved to values of 50% and 16% under UV and visible light irradiation, respectively [130]. Luévano-Hipólito et al. extensively studied the preparation of ZnO particles using a sol–gel methodology [131]. They studied the effect of the precursor concentration (Zn and $NH_4OH$) present in the synthesis of ZnO and found a wide variability in the efficiency of the $DeNO_x$ process from 70% to 20%. As in the previous study, this variability is related to the surface area of the particles, and not to the morphology. In this sense, the concentration of a Zn precursor plays an

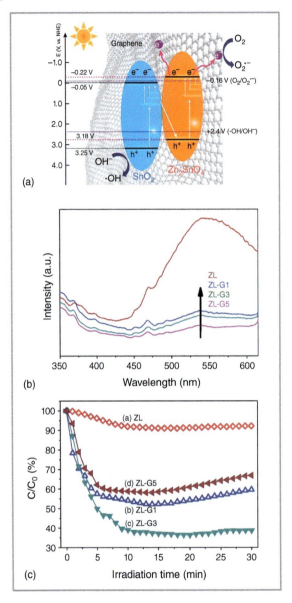

**Figure 7.6** (a) Scheme of Z-scheme heterojunction proposed for $SnO_2$–$Zn_2SnO_4$/graphene composite, (b) photoluminescence spectra of the photocatalysts, and (c) evolution of NO concentration under light irradiation Source: Reprinted from Li et al. [128], with permission from Elsevier.

important role in obtaining ZnO samples with a larger specific surface area. On the other hand, they also found a higher photocatalytic activity for the samples with preferential orientation in the plane (002).

In general, the efficiencies in the $DeNO_x$ photocatalytic process are much lower than in the other systems described and mainly work in the region of UV light; in the range of visible light, the ZnO-based system is barely around 15% efficient. To improve these efficiency values, different methodologies are proposed, such as the formation of heterojunctions or the surface deposition of different compounds. Sánchez et al. prepared a $ZnO@SiO_2$ composite using rice husk [132]. The spherical

ZnO particles grew scattered over the siliceous skeleton of the rice husk ashes (RHAs), allowing smaller ZnO particles to be obtained and giving rise to an increase in the efficiency of the photocatalytic process, reaching NO removal values of 70% and an outstanding selectivity of over 90%. The high selectivity values were attributed to the sensitivity of ZnO to absorb $NO_2$ molecules. In this case, they use solar light, which does not discriminate between the UV and visible region. This work exemplifies that the $DeNO_x$ activity of a ZnO photocatalyst can be superior to that exhibited for some titanium-based oxides.

Another way to approach the improvement in the efficiency of the photocatalytic process with these materials is the preparation of thin films. Liu et al. prepared ZnO sheets coated with Ag nanoparticles [133]. These Ag nanoparticles have a higher Fermi level than ZnO, so a Schottky defect barrier on the interface between the metal and the semiconductor occurs and promotes the separation of charge carriers, since Ag acts as an electron reservoir. The addition of Ag particles improves the efficiency of the photocatalytic process from 10% to 55%, maintaining this efficiency for five cycles. In the same way, Gasparotto et al. found similar results working with ZnO thin films covered with $WO_3$ nanoparticles [134]. The efficiency of the photocatalytic process was improved by 15% and the selectivity by 10%, in comparison with the pure ZnO film. A different way of addressing the problem is by doping. Son et al. doped the ZnO nanoparticles with Cr and found improvements in both the efficiency (17% vs. 24%) and selectivity (7.5% vs. 17.3%) of the photocatalytic process [135].

Using multicomponent oxides, Sánchez et al. reported a pioneer work using layered double hydroxides (LDH) based on Zn and Al as the $DeNO_x$ photocatalyst [136]. LDH has the general formula of $[M^{2+}_{1-x}M^{3+}_{x}(OH)_2]^{x+} A^{m-}{}_{x/m} \cdot nH_2O$, in which M includes a large variety of transition metal cations and A being a large variety of inorganic or organic anions. This versatile formula, along with the easy and low-cost preparation, make these materials suitable for photocatalytic applications. In the case of the $Zn_3Al\text{-}CO_3$ photocatalyst, NO conversion efficiency values of 50% and selectivity of 90% have been reported. These values increased to 55–60% and 92%, respectively, when $Al^{3+}$ is substituted in the LDH structure by $Cr^{3+}$ or $Fe^{3+}$ [137, 138]. In the case of LDHs containing iron(III), the metallic substitution induced changes in the structure, morphology, and optical properties of the LDHs, which favor the light harvesting of the visible light and mitigates the $e^-/h^+$ recombination. Substituted samples showed poorer crystallized particles (Figure 7.7a) giving rise to a larger specific surface area. Moreover, the presence of $Fe^{3+}$ introduces energy levels into the band gap of the ZnAl-LDH semiconductor favoring the visible light absorption in the doped samples, Figure 7.7b. The new electronic structure of the LDH samples promotes a new deactivation pathway for the photocharges and increases the ability to produce $\cdot O_2^-$ and $\cdot OH$ radicals, Figures 7.7c and 7.7d. The whole features promoted an increase of 4–11% in the photochemical NO abatement (Figure 7.7e). In these systems, the $DeNO_x$ efficiency was superior to that exhibited by $TiO_2$-P25 tested in similar experimental conditions, which was maintained for an extended period (six hours), to highlight the favorable reusability and outstanding selectivity shown by these photocatalysts.

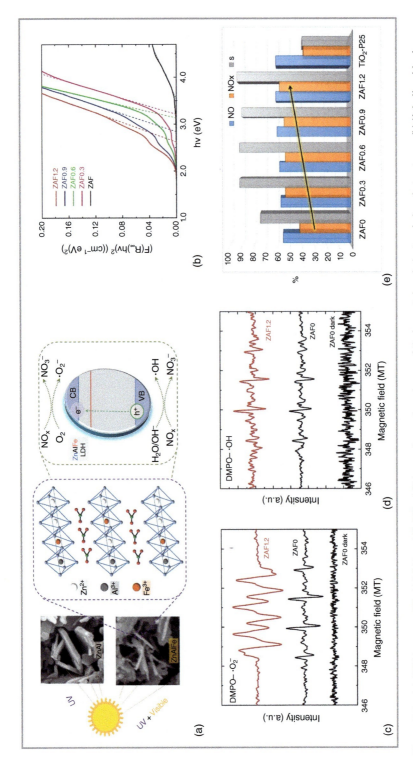

**Figure 7.7** (a) SEM images of ZnAl-LDH and ZnAlFe-LDH samples and a scheme of their crystal and electronic band structures. (b) Kubelka–Munk transformed reflectance spectra for LDH samples. DMPO spin-trapping EPR spectra of the ZAF0 and ZAF1.2 samples under UV–visible light irradiation for 15 minutes in (c) methanol solution for ·$O_2^-$ and (d) aqueous solution for ·OH. (e) NO conversion, NOX conversion, and selectivity values (%) for the LDH and $TiO_2$-P25 samples. Source: Reprinted from Pastor et al. [138], with permission from Elsevier.

With oxide mixtures, Pei et al. studied a TiO$_2$/ZnO mixture prepared by electrospinning followed by calcination [139]. They found that different ratios of the TiO$_2$ anatase and rutile phases are obtained depending on the calcination temperature. The optimal combination of these three semiconductors reduces the band gap energy compared to ZnO alone and improves the charge separation by reducing the electron-gap recombination. An increase in the calcination temperature results in a larger grain size, smaller specific surface area, and an inferior performance of the photocatalytic process. Finally, the formation of metal oxide frameworks (MOFs) could be an option in the improvement of the photocatalytic efficiency of the ZnO system. Even though these types of materials exhibit very large specific surface areas, one of their main drawbacks is the fast electron–hole recombination. Wei et al. prepared a composite between carbon quantum dots and a Zn zeolite, giving rise to the so-called CQDs/ZIF-8 (carbon quantum dots/zeolite imidazole frameworks) [140]. The efficiency of the photocatalytic process reached 4%, and after five cycles it lost less than 10%.

### 7.3.3 Transition Metal Oxides

Due to their intrinsic semiconductor-like behavior, many transition metal-based compounds have been studied as photocatalysts for the abatement of NOx gases. The first attempt in this field was proposed by the L. Sánchez research group reporting on the use of hematite α-Fe$_2$O$_3$, as the DeNO$_x$ photocatalyst. Hematite, the most stable iron oxide, exhibits n-type semiconducting properties with a band gap in the 1.9–2.2 eV. Therefore, α-Fe$_2$O$_3$ may act as a visible light photocatalyst for which interesting applications have been reported [141–143]. The first evidence of the DeNO$_x$ ability of iron oxides was found in the study of iron-containing industrial wastes. The appropriate thermal treatment of the wastes allowed compounds enriched in the hematite phase to be obtained, which resulted in significant photocatalytic activity when added as additives to DeNO$_x$ building materials [144, 145]. The first work reporting the ability of pure iron oxides toward the photochemical oxidation of NO gas molecules considered the preparation of nanostructured α-and β-Fe$_2$O$_3$ materials [146]. Thin film compounds (<840 nm in thickness) were prepared by using the CVD technique, the deposition temperature being a key parameter in determining the nano-organization of the systems. The metastable β-Fe$_2$O$_3$ phase consisted of interconnected pyramidal nanostructures (230 ± 30 nm particle size), while α-Fe$_2$O$_3$ crystallized as rounded structures similar in size. The different nano-morphology leads to significant differences in the roughness of both deposits, α-Fe$_2$O$_3$ being the roughest, and, therefore, possessing a larger effective surface area. In fact, the efficiency in the NO photodegradation was higher for the α-Fe$_2$O$_3$ film.

A deeper study on hematite DeNO$_x$ ability was carried out with nano-α-Fe$_2$O$_3$ powders (40–90 nm in size; 9.0 m$^2$ g$^{-1}$), Figure 7.8a [147]. Through the use of the IR spectroscopy, a DeNO$_x$ photochemical mechanism similar to that proposed for a TiO$_2$ photocatalyst was deduced. Thus, NOx gases were retained on the hematite surface as HNO$_2$/NO$_3^-$ under light irradiation. Of significant

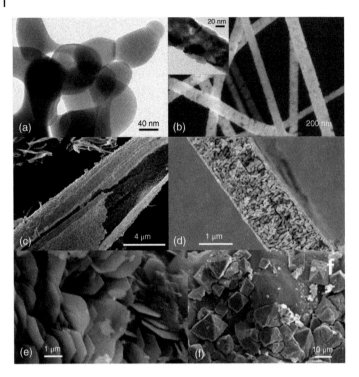

**Figure 7.8** Different particle morphology for α-Fe$_2$O$_3$ DeNO$_x$ photocatalysts: (a) nanopowders, (b) nanofibers, (c) porous microtubes, (d) flakes (cross-sectional view), (e) hexagonal plates, and (f) truncated pseudocubes crystallized over silica substrate. Source: L. Sánchez.

importance, the photocatalyst performed DeNO$_x$ action under both UV and visible light, but very low performance was observed. The poor photocatalytic behavior observed was linked to the inherent electronic characteristics of hematite. First, the inadequate positioning of the CB does not promote the formation of superoxide radicals. Moreover, the short lifetime of the excited states (c. 1 ps) and the short length of hole diffusion (c. 2–4 nm) facilitated a rapid recombination of the charge carriers. In this way, with the aim of enhancing the DeNO$_x$ performance exhibited by this photocatalyst, the authors proposed the preparation of carbon-hematite composites. Thus, by using resorcinol–formaldehyde as a template and carbon source, C/α-Fe$_2$O$_3$ samples were prepared with a carbon content of 0.9% and 3.5%. The composite samples exhibited a 17% increase in DeNO$_x$ efficiency, and the release of NO$_2$ gas was also mitigated, these favored findings being associated with the ability of carbon to adsorb NO$_2$ molecules and to lessen the hole/electron recombination. In fact, a similar strategy was used in the preparation of carbon quantum dots (CQDs)/FeOOH, nanocomposites that exhibited enhanced and high selectivity NO abatement efficiency (22%) compared with pure FeOOH [148]. Photo-electrochemical and EPR results demonstrate that the use of CQDs promotes a greater mobility of charge carriers and better generation of ROS radicals.

In the search of a higher efficiency of hematite toward the photochemical oxidation of NOx gases, new architectures and composites have been reported. Thus, α-Fe$_2$O$_3$ nanofibers prepared by the electrospinning technique exhibited 43% superior NO removal compared to hematite nanopowders [149]. These hollow fibers were constituted of a mosaic of flat nanocrystals with all their facets exposed to the contact with the reactant molecules, Figure 7.8b. The superior photoactivity of this hematite sample was attributed to the large surface-to-volume ratio aspect, resulting in a considerable specific surface area (22 m$^2$ g$^{-1}$). In a similar way, the same research group prepared porous architectures using polyvinyl pyrrolidone as the template [150]. Thus, highly porous interconnected nano/microtubes and flakes were obtained both with and without the use of the electrospinning technique, Figures 7.8c and 7.8d. Both types of samples showed a heterogeneous microstructure in which the micropores were dominant. These systems exhibited a good DeNO$_x$ performance, similar to that of TiO$_2$ P25, this high efficiency being reported for the first time in the hematite photocatalyst. Moreover, thanks to the facilitated NO$_2$ adsorption in the micropore network, the DeNO$_x$ selectivity measured was superior to that of TiO$_2$ P25. Additionally, RHA was used to prepare Fe$_2$O$_3$/SiO$_2$ composites [151]. By controlling the temperature calcination, hematite grows as irregular nanoparticles or microparticles onto the silica skeleton, which in prolonged calcination times are transformed into hexagonal plates and large truncated pseudocube crystals, Figures 7.8e and 7.8f.

The use of SiO$_2$ as a support allows an Fe$_2$O$_3$ photocatalyst with a specific surface of around 20 m$^2$ g$^{-1}$ to be obtained, for which the NO conversion value doubled that observed for hematite nanopowders [147]. Interestingly, the highest DeNO$_x$ selectivity values were associated with the presence of crystalline particles exhibiting {001} facets.

Another reported hematite composite was the Fe$_2$O$_3$/TiO$_2$ system. In this case, iron oxide thin films were deposited as interconnected leaflike nanostructures by means of the plasma-enhanced CVD technique [152]. The hematite films were subsequently functionalized with tailored TiO$_2$ amounts. The Fe$_2$O$_3$/TiO$_2$ nanocomposites allow the harvesting of the UV and visible solar light to be enhanced and facilitate the photogenerated charge transfer on the interface of the semiconductors. Thus, high selectivity and good values for the removal of NOx gases were found for the composites in contrast to the negligible DeNO$_x$ action observed for the bare Fe$_2$O$_3$ film. Based on this type of heterojunction, a visible light and persistent fluorescence-assisted photochemical NO removal was found for the CaAl$_2$O$_4$:(Eu,Nd)/(Ta,N)-co-doped TiO$_2$/Fe$_2$O$_3$ composite [153].

Zinc ferrites have also been studied for DeNO$_x$ applications. ZnFe$_2$O$_4$ possesses a band gap of 1.96 eV, which makes it attractive as a visible light photocatalyst, but, as in the case of the hematite photocatalyst, the rapid recombination of their photogenerated electron–hole pairs restricts the photocatalytic activity. Thus, ZnFe$_2$O$_4$ is not able to produce superoxide radicals (·O$_2^-$) for NO oxidation under visible light radiation. However, in the case of CDQs/ZnFe$_2$O$_4$ photocatalysts, the production of ROS species is greatly improved as the carbon dots trap the electrons present in the CB of ZnFe$_2$O$_4$, lessening the e$^-$/h$^+$ recombination [154], leading to a superior NO removal performance. Similar composites have been studied for zinc–cobalt spinels with outstanding results. Thus, photocatalytic NO conversion values greater than 80% were reported for rGO@ZnCo$_2$O$_4$ nanocomposites under visible and solar light irradiation [155]. On the other hand, with the same objective to enhance the separation efficiency of photogenerated electron–hole pairs, ZnFe$_2$O$_4$ was mixed with Bi$_2$O$_2$CO$_3$, a layered carbonate with a large band gap (~3.3 eV) [156]. The prepared Bi$_2$O$_2$CO$_3$/ZnFe$_2$O$_4$ composite resulted in a p-n heterojunction photocatalyst with enhanced photocatalytic removal of NO (35%) and low NO$_2$ selectivity under visible light irradiation. Finally, an interesting contribution of iron oxides to the NOx removal consists in the use of Fe$_3$O$_4$/SiO$_2$ composites to capture NIR light with the aim of enhancing the activity of the photocatalyst via a photothermal effect [157]. Thus, the NO removal capability of TiO$_2$ P25 was greatly increased by the preparation of TiO$_2$ P25/Fe$_3$O$_4$@SiO$_2$ composites.

Other interesting DeNO$_x$ photocatalyts are the W-based oxides. In a preliminary study, ZnWO$_4$ wolframite was prepared as porous photocatalysts by ultrasonic spray pyrolysis [158]. The synthesis temperature controls the preparation of the ZnWO$_4$ microstructures, leading to semiconductors of variable band gap (3.4–3.7 eV) and specific surface area (10–15 m$^2$ g$^{-1}$). In the best of the cases, a cyclic NO removal around 35% under solar light was reported. This DeNO$_x$ efficiency was nearly doubled in Bi/Bi$^{3+}$-ZnWO$_4$ composite [159], due to the new efficient electronic structure obtained as a result of the oxygen vacancies induced by the presence of Bi$^{3+}$ in the ZnWO$_4$ structure, together with the heterojunction performed with the loaded Bi nanoparticles. Moreover, the composite photocatalyst was also active under visible light.

In a different study, the scheelite type compounds ABO$_4$ (A = Ca, Pb; B = W, Mo) were investigated [160]. The photocatalytic efficiency found was different for each compound decreasing in the following order: CaMoO$_4$ > PbWO$_4$ > PbMoO$_4$ > CaWO$_4$. The highest NO removal value of 35% measured for the CaMoO$_4$ photocatalyst was associated with its favorable energy band positions, which promoted the formation of superoxide radicals.

Finally, the vanadium-based oxides have also been the object of study in this research field. The BiVO$_4$ crystallized in monoclinic scheelite structure, with a band gap of 2.2 eV, performs as a visible light photocatalyst. A hydrothermal method was used to prepare 3D BiVO$_4$ superstructures with different morphologies [161]. The NO photocatalytic removal efficiency, under visible light irradiation, was dependent on the type of morphology, 50% being the best value obtained for that sample with the larger surface area. In a similar way, InVO$_4$ (band gap = 2.3 eV) was studied as a visible light photocatalyst. This compound was used to prepare a CNQDs/GO-InVO$_4$ (CNQDs = graphitic C$_3$N$_4$ quantum dots, GO = graphene oxide) aerogel composite in the search of a favorable heterojunction between semiconductors [162]. In fact, a large amount of NO gas molecules (~65%) was removed. The favorable DeNO$_x$ response was attributed to the advantages of the prepared heterojunction: large visible light harvesting, good electron conductivity, and minimized e$^-$/h$^+$ recombination.

## 7.4 New Insights into the NOx Gases Photochemical Oxidation Mechanism

The photochemical oxidation of NOx gases begins once the ROS species are produced, when oxygen and water molecules adsorbed on the photocatalyst surface interact with the photogenerated electron and holes [163]. Basically, because of its strong oxidant power, superoxide radical assists the oxidation of NO to $NO_3^-$ [15, 164]. Simultaneously, the hydroxyl radical causes a sequential oxidative process: $NO \rightarrow NO_2^- \rightarrow NO_2 \rightarrow NO_3^-$ [7, 15]. For a desired high selective process [165], the nitrate ions are the major product deposited on the catalyst (which are easily removed with water during the photocatalyst regeneration). On the other hand, due to its high toxicity, the production of $NO_2$ gas intermediate is undesired [166].

The first studies on the $DeNO_x$ PCO process showed quite a complex mechanism. Thus, apart from ·OH and ·$O_2^-$ radicals, atomic oxygen (O), $O^-$ and ·$HO_2$ radicals are also involved in the photochemical process [167]. The physisorbed and chemisorbed nitrogen oxide molecules produce different intermediates. The NO adsorption, which is an important stage, leads to the formation of surface adsorbed $NO^-$, $N_2O_2^{2-}$, and $NO_3^-$ [168]. On the other hand, the physisorbed $NO_2$ concentration determines the pathways, leading to the $HNO_2$ and $HNO_3$ formation, through the $N_2O_4$ and $NO^+$ intermediates [166–172]. Moreover, the presence of water molecules plays an important role. Even though the increase of relative humidity values reduces the oxidation rate, as the $H_2O$ molecules mask the active sites on the photocatalyst surface [173, 174], their presence is crucial in the generation of hydroxyl radicals and in the adsorption of the NO molecules through interaction with their hydroxyl groups [175–177].

In the last few years, new light has been shed on the photochemical oxidation of NOx gases. Due to the use of in situ diffuse reflectance infrared Fourier transform spectroscopy (DRIFTS), it is possible to observe the evolution of related molecules or bonds accounted for on the catalyst surface during the $DeNO_x$ process [96, 178, 179]. The DRIFTS study is usually performed in two steps, the gas adsorption stage in dark conditions and the photochemical reaction under light irradiation. In the first step, once the catalyst is preconditioned under vacuum or argon atmosphere, the catalyst is subjected to a NO flow (and/or $NO_2$) in the presence of $O_2$, in order to ascertain the interactions of nitrogen oxide molecules on the catalyst surface before the beginning of the photochemical process. In this step, the physical adsorption of NO molecules occurs and, subsequently, their chemical adsorption and polymerization. Thus, NO, NOH, $N_2O$, $N_2O_3$, $N_2O_2$, $NO_2^+$, and $NO_2$ species are detected by DRIFTS [92, 99, 123, 180]. The $NO_2$ molecules – formed once NO reacts with $O_2$ – polymerize into $N_2O_4$, which undergoes disproportionation in the presence of water and forms nitrite, the latter inducing subsequent reactions with $NO_2$. On the other hand, when hydroxyl groups are present on the catalyst surface, the disproportionation of NO to $NO^-$ and $NO_2$ occurs [89, 93, 180]. Figure 7.9 summarizes the physicochemical pathways for NO on the catalyst surface in the absence of light irradiation.

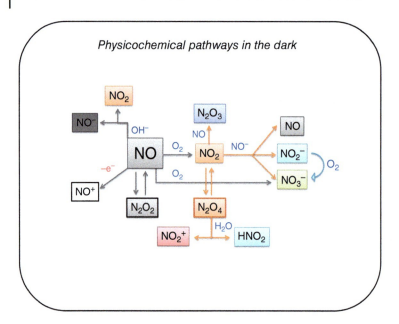

**Figure 7.9** The conversion pathways for the NO adsorbed on the catalyst surface in the absence of light. Source: L. Sánchez.

Once the catalyst is irradiated, the nitrogen species originated by the photochemical process are observed in the DRIFTS study. The nitrate and nitrite species are the major final products obtained in the oxidation process participated by ·OH and ·$O_2^-$ radicals [21, 69, 123, 181]. However, reaction intermediates such as $NO^+$, $N_2O_3$, $NO_2$, and $NO_2^+$ are also observed [123]. The presence of $N_2O$, together with $N_2O_3$, has also been observed, due to the coupling reaction between NO and $NO_2$, or the NO with N atoms from the catalyst framework [21]. Also, $N_2O_3$ could be formed as a result of the direct reaction of NO with oxidant species [83, 89]. Regarding the nitrate species, its appearance as monodentate nitrate is transformed with time into more stable bidentate and bridged linkages [182]. In a similar way, the nitrite ion – which occurs when $NO_2$ receives an electron – usually occurs in bridging form [102]. In some cases, it is observed that both NO and $NO_2$ molecules can obtain electrons from the catalyst to form $NO^-$ and $N_2O_2^{2-}$ [21]. Moreover, $NO^-$ also occurs from the disproportionation of NO adsorbed on the surface of the photocatalyst [85, 88]. It has also been reported for some catalysts that NO, once it makes contact with the surface, tends to lose electrons and form active $NO^+$ [178, 180, 181, 183]. Finally, Araña et al. [184] reported the formation of stable [$(NO_3^-)(H_2O)_n(NO_2)$] complexes, explaining a non-efficient photocatalytic conversion of the $NO_2$ gas. Moreover, the presence of this complex also inhibits the renoxification reaction of NO with $NO_3^-$ to form $NO_2$.

The multiple photochemical oxidation pathways for NOx molecules present on the catalyst surface under light irradiation are summarized in Figure 7.10.

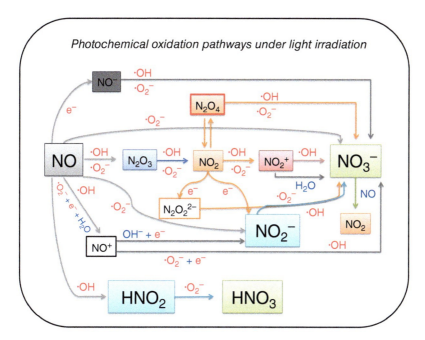

**Figure 7.10** The conversion pathways for the NOx molecules on the catalyst surface under light irradiation. Source: L. Sánchez.

## 7.5 Field Studies in Urban Areas

### 7.5.1 Photocatalytic Construction Materials

Nowadays, the incorporation of photoactive $TiO_2$ in several photocatalytic construction materials, such as self-cleaning windows [185], self-cleaning tiles [186], or photocatalytic coatings applied on building surfaces [187], is quite extended. A review on $TiO_2$-based photocatalytic cementitious composites and their application has recently been reported by Hamidi and Aslani [188]. $TiO_2$-based building materials are considered for quite a wide number of applications in urban environments, both for exterior construction and for interior furnishing. Most of these materials have been applied for improving indoor air quality, mainly for the removal of volatile pollutants in low concentrations [189].

In recent decades, the application of photocatalytic treatments, mainly containing $TiO_2$ nanoparticles, appears as one of the most promising alternatives for the NOx removal from the atmosphere in highly polluted areas. Indeed, the Environmental Industries Commission (EIC) in the United Kingdom proposes the application of titanium dioxide-based coatings as one of the options for tackling air pollution [190]. Two main patents owned by Mitsubishi Materials Corporation [191] and Italcementi S.p.A [192] have been produced for the application of $TiO_2$ in concrete paving blocks for the European market.

Some experience on real scale applying commercial $TiO_2$-white cement (BiancoTX Millennium) in different constructions is worth mentioning [193]: schools in Mortara (Italy, 1999), the Cité de la Musique in Chambéry (France, 2000), and the Dives in Misericordia Church in Rome (Italy, 2003). The main limitation in all these real construction applications is that the performance of these materials has been laboratory tested, but the efficiency of the mentioned constructions in reducing air pollution in field conditions is not currently being monitored.

One significant progress on the assessment of photocatalytic surfaces in the oxidation of NOx is the possibility of comparing the efficiencies of different photocatalytic coated surfaces using the protocolized methodology for measuring sample activity described in ISO standard (BS ISO 22197-1 2016) [194]. Several studies can be found in the literature giving values on the efficiency of a variety of photocatalytic construction materials determined using the ISO standard, mainly photocatalytic concrete [195, 196], but also photocatalytic paints [197, 198] and photocatalytic asphalts [199]. Recently, a modification of the ISO protocol for testing even low activity materials has been proposed by Mills et al. [200].

These results enable the characterization of the photocatalytic construction materials in laboratories, under controlled environmental conditions. ISO standard has been also extensively employed for comparing the performance of different nano-$TiO_2$ additions in a cementitious matrix to optimize the composition of the photocatalytic concrete [173], and for comparing the effectiveness of a cementitious concrete paving containing $TiO_2$-nanoparticles, with the effectiveness of a concrete coated with a spray of $TiO_2$ nanoparticles [201]. Furthermore, the ISO standard allows the comparative assessment of a photocatalytic pavement performance before and after exposure to real environments, in order to evaluate the ageing of the material and the loss of photocatalytic activity after working in field conditions for a specific period [196, 202].

Although most of the literature about photocatalytic construction materials is based on the use of $TiO_2$ nanoparticles, some recent studies have reported on the incorporation of different photocatalysts with improved performance in the UV–visible spectra, such as doped $TiO_2$ [203] or ZnO [132, 204]. Furthermore, photocatalytic concrete pavements with additional functionalities have been proposed by García-Calvo et al. [205].

### 7.5.2 Field Studies of NOx Abatement in Real Environments

Although photocatalytic blocks for pavements can be found on the market (Betonella®), photocatalytic construction materials are not extensively incorporated as a real alternative for NOx abatement in construction projects, and hardly any information on the performance of the photocatalytic can be found in the literature. This lack of information generates many uncertainties in assessing and modelling the effect of photocatalytic treatments in urban streets that must be solved before the use of photocatalytic construction materials becomes a reality for the atmospheric NOx removal in urban areas.

In the last two decades, several field studies dealing with the application of photocatalytic construction materials in urban areas have been reported. In the frame of PICADA project, Maggos' research group [206, 207] published some field studies on the application of photocatalytic construction materials for NOx abatement in urban areas. In Paris [206], the ceiling of a car park was coated with an acrylic $TiO_2$ photocatalytic paint, resulting in 19% reduction in NO and 20% reduction in $NO_2$ gas after five hours of exposure to UV lamps. In 2008, a study [207] was reported on a pilot site in a 20 m-long canyon street with panels covered with a $TiO_2$-containing mortar mixture on the street walls. Another artificial canyon street was created using panels without $TiO_2$, as a reference situation. An on-site monitoring system was installed for continuous NOx measurements. Lower NOx concentrations were registered in the $TiO_2$-containing canyon in comparison with the reference canyon (a decrease of up to 82% was recorded).

In London, in 2006, an eastern wall of the Sir John Cass School was painted with a photocatalytic paint for a trial field, and NOx levels were monitored before and after the coating application [208]. Although a 19% decrease in the NOx content was registered, it was not possible to attribute this reduction to the paint activity as a comparable NOx decrease during daylight and non-daylight hours. In 2016, Tremper and Green, also in London, reported a field trial [209], at the Artworks Elephant, to assess the photocatalytic activity of $DeNO_x$ paint by monitoring the NO, $NO_2$, and NOx concentrations before and after applying the paint. However, after monitoring for nine months before the paint application and 6 months afterwards, the authors concluded that no significant changes in NOx and $NO_2$ could be attributed to the photocatalytic paint.

In recent years, several in situ trial applications have been developed in Belgium aiming to demonstrate the effectiveness of photocatalytic construction materials in large-scale applications [210]. In Antwerp, 10 000 $m^2$ of a photocatalytic pavement was laid down on a parking lane. In this case, the durability of the pavement blocks was assessed by taking out blocks at different exposure periods and measuring their remaining photocatalytic activity at the laboratory [211]. In Wijnegen, an industrial region near Antwerp, a double-layered concrete with photocatalytic $TiO_2$-based material in the top layer, was used for the construction of a road [210, 211]. Trial sections of 30 m long and 3 m wide were constructed, and a complete testing program at the laboratory was defined for the preliminary evaluation of the significant parameters in order to optimize the photocatalytic performance of the construction material under field conditions. Between June 2011 and January 2013, a three-step field campaign was carried out in the Leopold II tunnel in Brussels within the Life+ project PhotoPAQ [212]. No significant reduction of NOx was obtained, despite the passivation of the photocatalytic material in the tunnel. Another interesting field trial was carried out in Rome during renovation work of the "Umberto I" tunnel [213], by painting the vault with two layers of a photocatalytic cement-based paint. A monitoring program was designed for measuring not only NOx concentrations, but also environmental variables such as humidity, temperature, wind parameters, and traffic flow. In this case, the effectiveness of the paint in reducing NOx levels was confirmed (a pollution level decrease, ranging from 51% to 64%, was reported).

Ballari and Brouwers reported a full-scale demonstration [214] in a street in Hengelo (The Netherlands). One part of the street (150 m long and the full width) was covered with an air-purifying pavement (DeNO$_x$ street), and the part of the street (100 m long) was covered with normal pavement (control street). The traffic intensity, NO and NO$_2$ concentrations, and the weather parameters were monitored. The NOx reduction attributed to the pavement was insignificant, so a further treatment with an additional TiO$_2$-based coating was applied. The effectiveness of this coating was also lost after some months due to weathering. The authors reached the tentative conclusion that a significant reduction of NOx concentrations can be expected, although a further comparable experiment in a new project was recommended. The performance of photocatalytic paving blocks has also been tested in a field study in an urban area in Copenhagen [214]. A central street was used as the testing area; on the southern end of the road, a 100 m-long section (and the full width) was covered with a photocatalytic paving. A 100 m-long control area at the northern end of the road was covered with ordinary concrete. NO levels were monitored before and after the application of the pavement, both near the photocatalytic region and near the control area. Although quite a notable reduction in NO content was observed, depending on the seasonal variations, a low selectivity of the photocatalytic materials was reported, and further research to overcome this limitation was recommended.

Recently, outdoor demonstrator platforms of different substrates containing selected photocatalytic materials have been tested as a pilot project in the frame of LIFE-PHOTOSCALING [215]. Two platforms, each consisting of three photoactive benches and one non-active bench, were built in two different zones in the Madrid region with different pollution conditions and seasonal humidity variations. Ten different photocatalytic TiO$_2$-based construction materials were tested through the in situ application of different commercial photocatalysts on both open asphalts and concrete tiles. Low cost AQmesh sensors were installed for monitoring NO$_2$ concentration [216], and the photocatalytic activity was evaluated in situ with PHOTONSITE [217], an in-house developed experimental device [218]. The authors concluded that the efficiency of the photocatalytic materials depended on NO, NO$_2$ ambient concentrations, but also on atmospheric variables such as irradiation and relative humidity.

## 7.6 Conclusions and Perspectives

Air pollution caused by NOx gases is a threat causing thousands of early deaths. Photocatalysis arises as a suitable methodology to remove these gases in urban environments, as demonstrated by the application of TiO$_2$-containing building materials. The irradiation of light over the photocatalyst surface promotes the excitation of electrons in the semiconductor VB to the CB, creating a pair of mobile charges e$^-$/h$^+$, which will initiate the redox processes. Thus, sequential oxidative processes take place, promoting the removal of the nitrogen gases (NO and NO$_2$) from air as they are transformed into nitrite/nitrate species. Even though it might seem to be a simple

## 7.6 Conclusions and Perspectives

oxidation process between nitrogen species, the whole mechanism is quite complex. In the absence of light, the NO gas molecules adsorbed onto the photocatalyst can be transformed into NOH, $N_2O$, $N_2O_3$, $N_2O_2$, $NO_2^+$, or $NO_2$ species, and during the photochemical process reaction, intermediates such as $NO^+$, $NO^-$, $N_2O$, $N_2O_2^{2-}$, $N_2O_3$, $NO_2$, and $NO_2^+$ are also occasionally observed.

Research in the field concentrates on overcoming the drawbacks of $TiO_2$ as a DeNO$_x$ photocatalyst and focuses the studies on the search of alternative highly efficient and selective, low cost, and sustainable photocatalysts.

Many studies report modifications on titania photocatalysts in an attempt to increase surface area, to perform visible light absorption ability, and to lessen the $e^-/h^+$ recombination, all with the aim of enhancing the efficiency and selectivity of the photochemical oxidation of NOx gases in air. In fact, the enhanced mobility of the charge carriers and the promotion of photocatalytic activity under visible light is performed by tailoring the energy band gap of the $TiO_2$ semiconductor and its edge potentials, for which the creation of semiconductor heterojunctions and the use of dopants are successful strategies. Thus, the DeNO$_x$ activity of $TiO_2$ under visible light is clearly enhanced when g-$C_3N_4$, graphene, graphene oxide, ZnO, $Bi_2O_3$, $WO_3$, and $MoO_3$ are used to form electronic heterojunctions. On the other hand, B, C, N, Fe(III), La(III), Ce, and W were incorporated into the anatase structure, creating oxygen vacancies in some cases. The presence of a dopant usually decreases the band gap, due to the appearance of intermediate energy levels between VB and CB, leading to an enhancement of the visible light absorption and charge mobility, respectively. Similar effects are obtained when nanoparticles of Pt, Au, Ag, Bi, and Carbon or CdTe quantum dots are deposited onto $TiO_2$, some of them promoting an SPR.

Moreover, the DeNO$_x$ photoactivity of $TiO_2$ is also promoted by the introduction of modifications in the crystalline structure. Defects in the lattice structure, such as O vacancy and $Ti^{3+}$, give a better light harvesting and a slower $e^-/h^+$ recombination. The creation of meso-structures favors the selectivity of the photochemical process, while a favored growth of the crystals with dominant {001} facets increases the reactivity.

The selectivity of the DeNO$_x$ process aims to complete the NO → $NO_2^-$ → $NO_2$ → $NO_3^-$ sequential photooxidative process in the best way, avoiding the emissions of the highly toxic $NO_2$ gas. As the efficiency of the photocatalytic activity is enhanced by the aforementioned strategies, the final oxidation of $NO_2$ toward nitrate is also improved. Moreover, the release of $NO_2$ is also mitigated when the adsorption of these molecules is favored by the existence of a mesopore structure or components with a basic character, being retained at the surface until their oxidation.

On the other hand, interesting results are obtained by preparing composites of $TiO_2$ with talc, hydrotalcite, clays, hydroxyapatite, or $Al_2O_3$, apart from the use of titanium-based perovskites.

A vast number of semiconductor materials have been studied as DeNO$_x$ photocatalyst alternatives to $TiO_2$. In all cases, similar strategies to those commented for titania are used in order to enhance their photoactivity. An interesting family of compounds is that constituted by bismuth-based oxides. In particular, these compounds exhibit good ability to perform the DeNO$_x$ action by using only visible light. Among them, Bi$_2$O$_3$, (BiO$_2$)CO$_3$, Bi$_2$WO$_6$, Bi$_2$MoO$_6$, and BiOX type semiconductors are being studied. However, most of them usually exhibit very poor photocatalytic efficiency. In order to achieve acceptable DeNO$_x$ efficiencies (around 50%), actions such as the use of N as the dopant, the deposition of small clusters of Bi, Fe, or Pt, the creation of Ov, the preparation of solid solutions or the formation of heterojunctions are necessary.

Another group of interest is that formed of Sn- and Zn-based oxides. With the appropriate design of the SnO$_2$ or ZnO photocatalysts, similar or higher efficiencies (around 70%) than those usually reported for TiO$_2$-based compounds can be reached under solar light irradiation. Thus, the DeNO$_x$ behavior of Ag@SnO$_2$, Zn$_2$SnO$_4$, ZnO@SiO$_2$, and ZnAl-LDHs deserves to be highlighted. Of significant importance, the zinc-based compounds exhibit outstanding selectivity values ($S > 90\%$) because of the sensitivity of Zn to adsorb NO$_2$.

Transition metal oxides are highly sustainable photocatalysts for DeNO$_x$ actions as some of them combine interesting features such as being earth abundant, low cost, and chemically stable materials, which are safe for human health. Interesting results have been obtained for α-Fe$_2$O$_3$, FeO(OH), ZnFe$_2$O$_4$, ZnCo$_2$O$_4$, ZnWO$_4$, or BiVO$_4$ or InVO$_4$. Usually, they exhibit low band gap energy values, which are able to absorb the visible light directly. They perform a good light harvesting, but the inadequate position of the energy band edges, or the very rapid e$^-$/h$^+$ recombination, leads to a low DeNO$_x$ performance. However, with an adequate strategy to prepare the photocatalyst, these semiconductors could reach even better efficiencies than that of TiO$_2$-P25, as is the case of the highly porous interconnected hematite nano/microtubes.

In spite of the vast number of studies that have reported on the enhancement of titania-based photocatalysts and alternative compounds, many of them with brilliant results concerning the removal of nitrogen oxide gases from air and a high selectivity in the DeNO$_x$ process, field studies are still limited. It should be highlighted that all the limited experiments existing on field trials to study the performance of photocatalytic construction materials in real urban environments is focused on the application of TiO$_2$ as photocatalysts. Nonconclusive studies confirming the efficiency of photocatalytic construction materials in NOx abatement, when applied in real urban environments, have been found in the literature. Several factors affecting the performance of photocatalytic materials in field conditions make it quite difficult to give quantitative values of efficiency. The weathering of photocatalytic construction materials and low selectivity have been reported as limiting factors that should be studied further. Furthermore, the incorporation of new photocatalysts capable of improving the performance of photocatalytic construction materials in real environmental irradiation conditions appears as a hot topic for advancing in the incorporation of these materials as a real alternative for NOx abatement, contributing to the air purification of highly polluted urban areas.

This contribution highlights the efficacy of innovative photocatalysts to be used in DeNO$_x$ environmental actions. Even though successful results are observed in many studies, some considerations would be useful in order to advance in this field of research. As the final application of these materials is on the surfaces of urban architecture, low costs and nontoxicity are mandatory characteristics. Currently, a DeNO$_x$ photocatalyst is usually employed in cement-based materials and paints, products that certainly need a large amount of additives to be effective in the removal of NOx gases. Another alternative would be the use of thin films of photocatalytic DeNO$_x$ materials that could be easily deposited onto glass or metallic substrates, which are plentiful in urban architecture, although reported results in the literature on this type of materials are scarce. In fact, the preparation of transition metal oxide thin films is well known, and with tiny amounts of photocatalyst covering the surfaces, it is quite probable that high efficiencies in DeNO$_x$ action could be obtained. Concerning the associated human health risks, and in spite of the good results in the laboratory as DeNO$_x$ photocatalysts, the use of nanometric titania or Bi-based compounds must be avoided. To advance in the study of safer and low cost transition metal oxides, those based on iron or cobalt would be recommendable. Otherwise, the strategies to enhance the electronic properties of the semiconductor must be as simple as possible. One successful action is the use of quantum dots in order to improve the electronic efficiency. Even though the preparation of QDs is not an easy task, the miniscule amount necessary to highly improve the activity of a semiconductor makes investigating new, low cost, and sustainable methodologies to prepare QDs/photocatalysts an interesting prospect. Finally, the preparation of 2D photocatalysts is barely studied in this field. The unique properties of 2D materials, such as an extra-large specific surface area, the high charge carrier mobility, and the simple surface functionalization, make them a desired DeNO$_x$ photocatalyst. In summary, there are still new perspectives open to investigation in the preparation of advanced, low cost, and sustainable photocatalytic materials, which allow real use in urban architecture with the aim of effectively cleaning the atmosphere under real-world conditions.

## References

1 Agency, E. E. (2017). European Union emission inventory report 1990–2015 under the UNECE Convention on Long-range Transboundary Air Pollution (LRTAP); https://www.eea.europa.eu/publications/annual-eu-emissions-inventory-report.
2 Chen, B., Hong, C., and Kan, H. (2004). Exposures and health outcomes from outdoor air pollutants in China. *Toxicology* 198 (1): 291–300. http://www.sciencedirect.com/science/article/pii/S0300483X04001040.
3 Agency, E. E. (2015). Air quality in Europe — 2015 report. EEA Report No 5/2015. https://www.eea.europa.eu/publications/air-quality-in-europe-2015.
4 Analysis II for AS. 5,000 deaths annually from dieselgate in Europe. (2017).
5 Balbuena, J., Cruz-Yusta, M., Sánchez, L. et al. *J. Nanosci. Nanotechnol.* 15: 6373–6385. https://www.ingentaconnect.com/content/asp/jnn/2015/00000015/00000009/art00008.

**6** Ângelo, J., Andrade, L., Madeira, L.M., and Mendes, A. (2013). An overview of photocatalysis phenomena applied to NOx abatement. *J. Environ. Manage.* 129: 522–539. https://doi.org/10.1016/j.jenvman.2013.08.006.

**7** Devahasdin, S., Fan, C., Li, K., and Chen, D.H. (2003). $TiO_2$ photocatalytic oxidation of nitric oxide: transient behavior and reaction kinetics. *J. Photochem. Photobiol., A* 156 (1): 161–170. http://www.sciencedirect.com/science/article/pii/S1010603003000054.

**8** Poon, C.S. and Cheung, E. (2007). NO removal efficiency of photocatalytic paving blocks prepared with recycled materials. *Constr. Build Mater.* 21 (8): 1746–1753. http://www.sciencedirect.com/science/article/pii/S0950061806001929.

**9** Folli, A., Campbell, S.B., Anderson, J.A., and Macphee, D.E. (2011). Role of $TiO_2$ surface hydration on NO oxidation photo-activity. *J. Photochem. Photobiol. A Chem.* 220 (2): 85–93. http://www.sciencedirect.com/science/article/pii/S1010603011001225.

**10** Ballari, M.M., Hunger, M., Hüsken, G., and Brouwers, H.J.H. (2010). NOx photocatalytic degradation employing concrete pavement containing titanium dioxide. *Appl. Catal., B* 95 (3): 245–254. http://www.sciencedirect.com/science/article/pii/S0926337310000081.

**11** de Melo, J.V.S. and Trichês, G. (2012). Evaluation of the influence of environmental conditions on the efficiency of photocatalytic coatings in the degradation of nitrogen oxides (NOx). *Build Environ.* 49: 117–123. http://www.sciencedirect.com/science/article/pii/S0360132311002952.

**12** Dillert, R., Stötzner, J., Engel, A., and Bahnemann, D.W. (2012). Influence of inlet concentration and light intensity on the photocatalytic oxidation of nitrogen(II) oxide at the surface of Aeroxide® $TiO_2$ P25. *J. Hazard. Mater.* 211–212: 240–246. http://www.sciencedirect.com/science/article/pii/S0304389411014075.

**13** Sugrañez, R., Álvarez, J.I., Cruz-Yusta, M. et al. (2013). Enhanced photocatalytic degradation of NOx gases by regulating the microstructure of mortar cement modified with titanium dioxide. *Build. Environ.* [Internet] 69: 55–63. http://www.sciencedirect.com/science/article/pii/S0360132313002114.

**14** Folli, A., Pochard, I., Nonat, A. et al. (2010). Engineering photocatalytic cements: understanding $TiO_2$ surface chemistry to control and modulate photocatalytic performances. *J. Am. Ceram. Soc.* 93 (10): 3360–3369.

**15** Dalton, J.S., Janes, P., Jones, N. et al. (2001). Photocatalytic oxidation of NOx gases using $TiO_2$: a spectroscopic approach. *Acta Univ. Carolinae, Geol.* 45 (1): 8.

**16** Agency, E.C. (2017). Annex 2 Response to comments document (RCOM) to the Opinion proposing harmonised classification and labelling at EU level of Titanium dioxide. Comm Risk Assesment, RAC. https://echa.europa.eu/es/-/titanium-dioxide-proposed-to-be-classified-as-suspected-of-causing-cancer-when-inhaled..

**17** Giannakopoulou, T., Papailias, I., Todorova, N. et al. (2017). Tailoring the energy band gap and edges' potentials of g-$C_3N_4$/$TiO_2$ composite photocatalysts for NOx removal. *Chem. Eng. J.* 310: 571–580. http://www.sciencedirect.com/science/article/pii/S138589471501774X.

18 Papailias, I., Todorova, N., Giannakopoulou, T. et al. (2017). Photocatalytic activity of modified g-$C_3N_4$/$TiO_2$ nanocomposites for NOx removal. *Catal. Today* 280: 37–44. http://www.sciencedirect.com/science/article/pii/S0920586116304333.

19 Song, X., Hu, Y., Zheng, M., and Wei, C. (2016). Solvent-free in situ synthesis of g-$C_3N_4$/{001}$TiO_2$ composite with enhanced UV- and visible-light photocatalytic activity for NO oxidation. *Appl. Catal., B* 182: 587–597. http://www.sciencedirect.com/science/article/pii/S0926337315301892.

20 Ma, J., Wang, C., and He, H. (2016). Enhanced photocatalytic oxidation of NO over g-$C_3N_4$-$TiO_2$ under UV and visible light. *Appl. Catal., B* 184: 28–34. http://www.sciencedirect.com/science/article/pii/S0926337315302496.

21 Jiang, G., Cao, J., Chen, M. et al. (2018). Photocatalytic NO oxidation on N-doped $TiO_2$/g-$C_3N_4$ heterojunction: enhanced efficiency, mechanism and reaction pathway. *Appl. Surf. Sci.* 458: 77–85. http://www.sciencedirect.com/science/article/pii/S016943321831986X.

22 Yang, Y., Li, Y., Wang, J. et al. (2017). Graphene-$TiO_2$ mesoporous spheres assembled by anatase and rutile nanowires for efficient NO photooxidation. *J. Alloys Compd.* 699: 47–56. http://www.sciencedirect.com/science/article/pii/S092583881634124X.

23 Yang, W., Li, C., Wang, L. et al. (2015). Solvothermal fabrication of activated semi-coke supported $TiO_2$-rGO nanocomposite photocatalysts and application for NO removal under visible light. *Appl. Surf. Sci.* 353: 307–316. http://www.sciencedirect.com/science/article/pii/S0169433215009903.

24 Trapalis, A., Todorova, N., Giannakopoulou, T. et al. (2016). $TiO_2$/graphene composite photocatalysts for NOx removal: a comparison of surfactant-stabilized graphene and reduced graphene oxide. *Appl. Catal., B* 180: 637–647. http://www.sciencedirect.com/science/article/pii/S0926337315300291.

25 Pei, C.C., Kin Shing Lo, K., and Leung, W.W.-F. (2017). Titanium-zinc-bismuth oxides-graphene composite nanofibers as high-performance photocatalyst for gas purification. *Sep .Purif. Technol.* 184: 205–212. http://www.sciencedirect.com/science/article/pii/S1383586617303088.

26 Luévano-Hipólito, E., Martínez-de la Cruz, A., López-Cuellar, E. et al. (2014). Synthesis, characterization and photocatalytic activity of $WO_3$/$TiO_2$ for NO removal under UV and visible light irradiation. *Mater. Chem. Phys.* 148 (1): 208–213. http://www.sciencedirect.com/science/article/pii/S0254058414004726.

27 Balayeva, N.O., Fleisch, M., and Bahnemann, D.W. (2018). Surface-grafted $WO_3$/$TiO_2$ photocatalysts: enhanced visible-light activity towards indoor air purification. *Catal. Today* 313: 63–71. http://www.sciencedirect.com/science/article/pii/S0920586117308271.

28 Wang, L., Zhang, L., Jiang, Y., and Li, P. (2020). Morphology control of molybdenum titanium oxide and its enhanced NO removal performance. *Catal. Lett.* 150 (6): 1707–1713. https://doi.org/10.1007/s10562-019-03076-z.

**29** Wu, Z., Sheng, Z., Liu, Y. et al. (2011). Deactivation mechanism of PtOx/TiO2 photocatalyst towards the oxidation of NO in gas phase. *J. Hazard. Mater.* 185 (2): 1053–1058. http://www.sciencedirect.com/science/article/pii/S0304389410013014.

**30** He, D., Li, Y., Inshu, W. et al. (2017). Carbon wrapped and doped $TiO_2$ mesoporous nanostructure with efficient visible-light photocatalysis for NO removal. *Appl. Surf. Sci.* 391: 318–325. http://www.sciencedirect.com/science/article/pii/S0169433216314155.

**31** Sitthisang, S., Komarneni, S., Tantirungrotechai, J. et al. (2012). Microwave-hydrothermal synthesis of extremely high specific surface area anatase for decomposing NOx. *Ceram. Int.* 38 (8): 6099–6105. http://www.sciencedirect.com/science/article/pii/S0272884212003641.

**32** Amadelli, R., Samiolo, L., Borsa, M. et al. (2013). $N-TiO_2$ Photocatalysts highly active under visible irradiation for NOX abatement and 2-propanol oxidation. *Catal. Today* 206: 19–25. http://www.sciencedirect.com/science/article/pii/S0920586111008078.

**33** Ai, Z., Zhu, L., Lee, S., and Zhang, L. (2011). NO treated $TiO_2$ as an efficient visible light photocatalyst for NO removal. *J. Hazard. Mater.* 192 (1): 361–367. http://www.sciencedirect.com/science/article/pii/S0304389411006728.

**34** Cha, J.-A., An, S.-H., Jang, H.-D. et al. (2012). Synthesis and photocatalytic activity of N-doped $TiO_2/ZrO_2$ visible-light photocatalysts. *Adv. Powder Technol.* 23 (6): 717–723. http://www.sciencedirect.com/science/article/pii/S092188311100152X.

**35** Li, H., Yin, S., and Sato, T. (2011). Novel luminescent photocatalytic deNOx activity of $CaAl_2O_4$: (Eu,Nd)/$TiO_2$–xNy composite. *Appl. Catal., B* 106 (3): 586–591. http://www.sciencedirect.com/science/article/pii/S0926337311002955.

**36** Ding, X., Song, X., Li, P. et al. (2011). Efficient visible light driven photocatalytic removal of NO with aerosol flow synthesized B, N-codoped $TiO_2$ hollow spheres. *J. Hazard. Mater.* 190 (1): 604–612. http://www.sciencedirect.com/science/article/pii/S0304389411004158.

**37** Folli, A., Bloh, J.Z., Armstrong, K. et al. (2018). Improving the selectivity of photocatalytic NOx abatement through improved $O_2$ reduction pathways using Ti0.909W0.091O2Nx semiconductor nanoparticles: from characterization to photocatalytic performance. *ACS Catal.* 8 (8): 6927–6938. https://doi.org/10.1021/acscatal.8b00521.

**38** Patzsch, J., Spencer, J.N., Folli, A., and Bloh, J.Z. (2018). Grafted iron(iii) ions significantly enhance $NO_2$ oxidation rate and selectivity of $TiO_2$ for photocatalytic NOx abatement. *RSC Adv.* 8 (49): 27674–27685. http://dx.doi.org/10.1039/C8RA05017A.

**39** Ho, C.-C., Kang, F., Chang, G.-M. et al. (2019). Application of recycled lanthanum-doped $TiO_2$ immobilized on commercial air filter for visible-light photocatalytic degradation of acetone and NO. *Appl. Surf. Sci.* 465: 31–40. http://www.sciencedirect.com/science/article/pii/S0169433218325522.

**40** Huang, Y., Cao, J.-J., Kang, F. et al. (2017). High selectivity of visible-light-driven La-doped $TiO_2$ photocatalysts for NO removal. *Aerosol. Air Qual. Res.* 17 (10): 2555–2565. http://dx.doi.org/10.4209/aaqr.2017.08.0282.

**41** Cao, X., Yang, X., Li, H. et al. (2017). Investigation of Ce-$TiO_2$ photocatalyst and its application in asphalt- based specimens for NO degradation. *Constr. Build Mater.* 148: 824–832. http://www.sciencedirect.com/science/article/pii/S0950061817309844.

**42** Silvestri, S., Szpoganicz, B., Schultz, J. et al. (2016). Doped and undoped anatase-based plates obtained from paper templates for photocatalytic oxidation of NOX. *Ceram. Int.* 42 (10): 12074–12083. http://www.sciencedirect.com/science/article/pii/S0272884216305478.

**43** Hernández Rodríguez, M.J., Pulido Melián, E., García Santiago, D. et al. (2017). NO photooxidation with $TiO_2$ photocatalysts modified with gold and platinum. *Appl. Catal., B* 205: 148–157. http://www.sciencedirect.com/science/article/pii/S0926337316309377.

**44** Song, S., Sheng, Z., Liu, Y. et al. (2012). Influences of pH value in deposition-precipitation synthesis process on Pt-doped $TiO_2$ catalysts for photocatalytic oxidation of NO. *J. Environ. Sci.* 24 (8): 1519–1524. http://www.sciencedirect.com/science/article/pii/S1001074211609807.

**45** Hu, Y., Song, X., Jiang, S., and Wei, C. (2015). Enhanced photocatalytic activity of Pt-doped $TiO_2$ for NOx oxidation both under UV and visible light irradiation: a synergistic effect of lattice Pt4+ and surface PtO. *Chem. Eng. J.* 274: 102–112. http://www.sciencedirect.com/science/article/pii/S1385894715004726.

**46** Balci Leinen, M., Dede, D., Khan, M.U. et al. (2019). CdTe quantum dot-functionalized P25 titania composite with enhanced photocatalytic $NO_2$ storage selectivity under UV and Vis irradiation. *ACS Appl. Mater. Interfaces* 11 (1): 865–879. https://doi.org/10.1021/acsami.8b18036.

**47** Zhao, Z., Zhang, W., Lv, X. et al. (2016). Noble metal-free Bi nanoparticles supported on $TiO_2$ with plasmon-enhanced visible light photocatalytic air purification. *Environ. Sci. Nano* 3 (6): 1306–1317. http://dx.doi.org/10.1039/C6EN00341A.

**48** Duan, Y., Zhang, M., Wang, L. et al. (2017). Plasmonic Ag-$TiO_{2-x}$ nanocomposites for the photocatalytic removal of NO under visible light with high selectivity: the role of oxygen vacancies. *Appl. Catal., B* 204: 67–77. http://www.sciencedirect.com/science/article/pii/S0926337316308827.

**49** Cerrato, G., Galli, F., Boffito, D.C. et al. (2019). Correlation preparation parameters/activity for micro$TiO_2$ decorated with SilverNPs for NOx photodegradation under LED light. *Appl. Catal., B* 253: 218–225. http://www.sciencedirect.com/science/article/pii/S0926337319303789.

**50** Duan, Y., Luo, J., Zhou, S. et al. (2018). $TiO_2$-supported Ag nanoclusters with enhanced visible light activity for the photocatalytic removal of NO. *Appl. Catal., B* 234: 206–212. http://www.sciencedirect.com/science/article/pii/S0926337318303709.

**51** Martins, N.C.T., Ângelo, J., Girão, A.V. et al. (2016). N-doped carbon quantum dots/TiO$_2$ composite with improved photocatalytic activity. *Appl. Catal., B* 193: 67–74. http://www.sciencedirect.com/science/article/pii/S092633731630282X.

**52** Hu, Z., Li, K., Wu, X. et al. (2019). Dramatic promotion of visible-light photoreactivity of TiO$_2$ hollow microspheres towards NO oxidation by introduction of oxygen vacancy. *Appl. Catal., B* 256: 117860. http://www.sciencedirect.com/science/article/pii/S092633731930606X.

**53** Shen, X., Dong, G., Wang, L. et al. (2019). Enhancing photocatalytic activity of NO removal through an in situ control of oxygen vacancies in growth of TiO$_2$. *Adv. Mater. Interfaces* 6 (19): 1–10.

**54** Shang, H., Li, M., Li, H. et al. (2019). Oxygen vacancies promoted the selective photocatalytic removal of NO with blue TiO$_2$ via simultaneous molecular oxygen activation and photogenerated hole annihilation. *Environ. Sci. Technol.* 53 (11): 6444–6453. https://doi.org/10.1021/acs.est.8b07322.

**55** Tan, B., Zhang, X., Li, Y. et al. (2017). Anatase TiO$_2$ mesocrystals: green synthesis, in situ conversion to porous single crystals, and self-doping Ti$^{3+}$ for enhanced visible light driven photocatalytic removal of NO. *Chem. - A Eur. J.* 23 (23): 5478–5487.

**56** Balbuena, J., Calatayud, J.M., Cruz-Yusta, M. et al. (2018). Mesocrystalline anatase nanoparticles synthesized using a simple hydrothermal approach with enhanced light harvesting for gas-phase reaction. *Dalton Trans.* 47 (18): 6590–6597. http://dx.doi.org/10.1039/C8DT00721G.

**57** Pan, J.H., Wang, X.Z., Huang, Q. et al. (2014). Large-scale synthesis of urchin-like mesoporous TiO$_2$ hollow spheres by targeted etching and their photoelectrochemical properties. *Adv. Funct. Mater.* 24 (1): 95–104.

**58** Zhang, D., Li, G., Wang, H. et al. (2010). Biocompatible anatase single-crystal photocatalysts with tunable percentage of reactive facets. *Cryst. Growth Des.* 10 (3): 1130–1137.

**59** Sofianou, M.V., Trapalis, C., Psycharis, V. et al. (2012). Study of TiO$_2$ anatase nano and microstructures with dominant {001} facets for NO oxidation. *Environ. Sci. Pollut. Res.* 19 (9): 3719–3726.

**60** de Abreu, M.A.S., Morgado, E., Jardim, P.M., and Marinkovic, B.A. (2012). The effect of anatase crystal morphology on the photocatalytic conversion of NO by TiO$_2$-based nanomaterials. *Cent. Eur. J. Chem.* 10 (4): 1183–1198. https://doi.org/10.2478/s11532-012-0040-3.

**61** Todorova, N., Giannakopoulou, T., Karapati, S. et al. (2014). Composite TiO$_2$/clays materials for photocatalytic NOx oxidation. *Appl. Surf. Sci.* 319: 113–120. http://www.sciencedirect.com/science/article/pii/S0169433214015451.

**62** Chen, X.-F., Lin, S., and Kou, S.-C. (2018). Effect of composite photo-catalysts prepared with recycled clay brick sands and nano-TiO2 on methyl orange and NOx removal. *Constr. Build Mater.* 171: 152–160. http://www.sciencedirect.com/science/article/pii/S0950061818305890.

**63** Kim, J.H., Han, J.H., Jung, Y.C., and Kim, Y.A. (2019). Mussel adhesive protein-coated titanium oxide nanoparticles for effective NO removal from versatile substrates. *Chem. Eng. J.* 378: 122164. http://www.sciencedirect.com/science/article/pii/S138589471931558X.

64 Yao, J., Zhang, Y., Wang, Y. et al. (2017). Enhanced photocatalytic removal of NO over titania/hydroxyapatite ($TiO_2$/HAp) composites with improved adsorption and charge mobility ability. *RSC Adv.* 7 (40): 24683–24689. http://dx.doi.org/10.1039/C7RA02157G.

65 Polat, M., Soylu, A.M., Erdogan, D.A. et al. (2015). Influence of the sol–gel preparation method on the photocatalytic NO oxidation performance of $TiO_2$/$Al_2O_3$ binary oxides. *Catal. Today* 241: 25–32. http://www.sciencedirect.com/science/article/pii/S0920586114003071.

66 Soylu, A.M., Polat, M., Erdogan, D.A. et al. (2014). $TiO_2$–$Al_2O_3$ binary mixed oxide surfaces for photocatalytic NOx abatement. *Appl. Surf. Sci.* 318: 142–149. http://www.sciencedirect.com/science/article/pii/S0169433214003584.

67 Çağlayan, M., Irfan, M., Ercan, K.E. et al. (2020). Enhancement of photocatalytic NOx abatement on titania via additional metal oxide NOx-storage domains: interplay between surface acidity, specific surface area, and humidity. *Appl. Catal., B* 263: 118227. http://www.sciencedirect.com/science/article/pii/S0926337319309749.

68 Mendoza, J.A., Lee, D.H., and Kang, J.H. (2016). Photocatalytic removal of NOx using $TiO_2$-coated zeolite. *Environ. Eng. Res.* 21 (3): 291–296.

69 Guo, G., Hu, Y., Jiang, S., and Wei, C. (2012). Photocatalytic oxidation of NOx over $TiO_2$/HZSM-5 catalysts in the presence of water vapor: effect of hydrophobicity of zeolites. *J. Hazard. Mater.* 223–224 (x): 39–45. http://dx.doi.org/10.1016/j.jhazmat.2012.04.043.

70 Grabowska, E. (2016). Selected perovskite oxides: characterization, preparation and photocatalytic properties-A review. *Appl. Catal., B* 186: 97–126. http://dx.doi.org/10.1016/j.apcatb.2015.12.035.

71 Li, H., Yin, S., Wang, Y., and Sato, T. (2013). Microwave-assisted hydrothermal synthesis of $Fe_2O_3$-Sensitized $SrTiO_3$ and its luminescent photocatalytic de NOx activity with $CaAl_2O_4$:(Eu, Nd) assistance. *J. Am. Ceram. Soc.* 96 (4): 1258–1262.

72 Li, H., Yin, S., Wang, Y. et al. (2013). Roles of $Cr^{3+}$ doping and oxygen vacancies in $SrTiO_3$ photocatalysts with high visible light activity for NO removal. *J. Catal.* 297: 65–69. http://dx.doi.org/10.1016/j.jcat.2012.09.019.

73 Zhang, D., Guo, Y., and Zhao, Z. (2018). Porous defect-modified graphitic carbon nitride via a facile one-step approach with significantly enhanced photocatalytic hydrogen evolution under visible light irradiation. *Appl. Catal., B* 226 (October 2017): 1–9. https://doi.org/10.1016/j.apcatb.2017.12.044.

74 Zhang, Q., Huang, Y., Peng, S. et al. (2017). Perovskite $LaFeO_3$-$SrTiO_3$ composite for synergistically enhanced NO removal under visible light excitation. *Appl. Catal., B* 204: 346–357. http://dx.doi.org/10.1016/j.apcatb.2016.11.052.

75 Jin, S., Dong, G., Luo, J. et al. (2018). Improved photocatalytic NO removal activity of $SrTiO_3$ by using $SrCO_3$ as a new co-catalyst. *Appl. Catal., B* 227 (January): 24–34.

**76** Ai, Z., Lu, G., and Lee, S. (2014). Efficient photocatalytic removal of nitric oxide with hydrothermal synthesized $Na_{0.5}Bi_{0.5}TiO_3$ nanotubes. *J. Alloys Compd.* 613: 260–266. http://dx.doi.org/10.1016/j.jallcom.2014.06.039.

**77** Hailili, R., Dong, G., Ma, Y. et al. (2017). Layered perovskite $Pb_2Bi_4Ti_5O_{18}$ for excellent visible light-driven photocatalytic NO removal. *Ind. Eng. Chem. Res.* 56 (11): 2908–2916.

**78** Ai, Z., Huang, Y., Lee, S., and Zhang, L. (2011). Monoclinic α-$Bi_2O_3$ photocatalyst for efficient removal of gaseous NO and HCHO under visible light irradiation. *J. Alloys Compd.* 509 (5): 2044–2049. http://dx.doi.org/10.1016/j.jallcom.2010.10.132.

**79** Chen, M., Li, Y., Wang, Z. et al. (2017). Controllable synthesis of core-shell Bi@amorphous $Bi_2O_3$ nanospheres with tunable optical and photocatalytic activity for NO removal. *Ind. Eng. Chem. Res.* 56 (37): 10251–10258.

**80** Li, X., Sun, Y., Xiong, T. et al. (2017). Activation of amorphous bismuth oxide via plasmonic Bi metal for efficient visible-light photocatalysis. *J. Catal.* 352: 102–112. http://dx.doi.org/10.1016/j.jcat.2017.04.025.

**81** Liu, Y., Wang, Z., Huang, B. et al. (2010). Preparation, electronic structure, and photocatalytic properties of $Bi_2O_2CO_3$ nanosheet. *Appl. Surf. Sci.* 257 (1): 172–175. http://dx.doi.org/10.1016/j.apsusc.2010.06.058.

**82** Dong, F., Ho, W.-K., Lee, S.C. et al. (2011). Template-free fabrication and growth mechanism of uniform $(BiO)_2CO_3$ hierarchical hollow microspheres with outstanding photocatalytic activities under both UV and visible light irradiation. *J. Mater. Chem.* 21 (33): 12428–12436. http://dx.doi.org/10.1039/C1JM11840D.

**83** Chen, P., Sun, Y., Liu, H. et al. (2019). Facet-dependent photocatalytic NO conversion pathways predetermined by adsorption activation patterns. *Nanoscale* 11 (5): 2366–2373. http://dx.doi.org/10.1039/C8NR09147A.

**84** Yu, S., Zhang, Y., Dong, F. et al. (2018). Readily achieving concentration-tunable oxygen vacancies in $Bi_2O_2CO_3$: triple-functional role for efficient visible-light photocatalytic redox performance. *Appl. Catal., B* 226 (July 2017): 441–450. https://doi.org/10.1016/j.apcatb.2017.12.074.

**85** Zhou, Y., Zhao, Z., Wang, F. et al. (2016). Facile synthesis of surface N-doped $Bi_2O_2CO_3$: origin of visible light photocatalytic activity and in situ DRIFTS studies. *J. Hazard. Mater.* 307: 163–172. http://dx.doi.org/10.1016/j.jhazmat.2015.12.072.

**86** Dong, X., Zhang, W., Cui, W. et al. (2017). Pt quantum dots deposited on N-doped $(BiO)_2CO_3$: enhanced visible light photocatalytic NO removal and reaction pathway. *Catal. Sci. Technol.* 7 (6): 1324–1332. http://dx.doi.org/10.1039/C6CY02444K.

**87** Jin, R., Jiang, X., Zhou, Y., and Zhao, J. (2016). Microspheres of graphene oxide coupled to N-doped $Bi_2O_2CO_3$ for visible light photocatalysis. *Cuihua Xuebao/Chinese J. Catal.* 37 (5): 760–768. http://dx.doi.org/10.1016/S1872-2067(15)61079-8.

**88** Lu, Y., Huang, Y., Zhang, Y. et al. (2019). Effects of $H_2O_2$ generation over visible light-responsive Bi/$Bi_2O_2$–$xCO_3$ nanosheets on their photocatalytic NOx removal performance. *Chem. Eng. J.* 363 (January): 374–382. https://doi.org/10.1016/j.cej.2019.01.172.

**89** Chen, P., Liu, H., Sun, Y. et al. (2020). Bi metal prevents the deactivation of oxygen vacancies in $Bi_2O_2CO_3$ for stable and efficient photocatalytic NO abatement. *Appl. Catal., B* 264 (August 2019): 118545. https://doi.org/10.1016/j.apcatb.2019.118545.

**90** Feng, X., Zhang, W., Sun, Y. et al. (2017). Fe(iii) cluster-grafted $(BiO)_2CO_3$ superstructures: in situ DRIFTS investigation on IFCT-enhanced visible light photocatalytic NO oxidation. *Environ. Sci. Nano* 4 (3): 604–612. http://dx.doi.org/10.1039/C6EN00637J.

**91** Wu, L., Bi, J., Li, Z. et al. (2008). Rapid preparation of $Bi_2WO_6$ photocatalyst with nanosheet morphology via microwave-assisted solvothermal synthesis. *Catal. Today* 131 (1–4): 15–20.

**92** Huo, W.C., Dong, X., Li, J.Y. et al. (2019). Synthesis of $Bi_2WO_6$ with gradient oxygen vacancies for highly photocatalytic NO oxidation and mechanism study. *Chem. Eng. J.* 361 (December 2018): 129–138. https://doi.org/10.1016/j.cej.2018.12.071.

**93** Huo, W., Xu, W., Cao, T. et al. (2019). Carbonate-intercalated defective bismuth tungstate for efficiently photocatalytic NO removal and promotion mechanism study. *Appl. Catal., B* 254 (April): 206–213. https://doi.org/10.1016/j.apcatb.2019.04.099.

**94** Ding, X., Ho, W., Shang, J., and Zhang, L. (2016). Self doping promoted photocatalytic removal of no under visible light with $Bi_2MoO_6$: indispensable role of superoxide ions. *Appl. Catal., B* 182 (3): 316–325. http://dx.doi.org/10.1016/j.apcatb.2015.09.046.

**95** Wang, S., Ding, X., Yang, N. et al. (2020). Insight into the effect of bromine on facet-dependent surface oxygen vacancies construction and stabilization of $Bi_2MoO6$ for efficient photocatalytic NO removal. *Appl. Catal., B* 265 (September 2019): 118585. https://doi.org/10.1016/j.apcatb.2019.118585.

**96** Huo, W., Xu, W., Cao, T. et al. (2019). Carbonate doped $Bi_2MoO_6$ hierarchical nanostructure with enhanced transformation of active radicals for efficient photocatalytic removal of NO. *J. Colloid. Interface Sci.* 557 (September): 816–824. https://doi.org/10.1016/j.jcis.2019.09.089.

**97** Li, J., Zhang, W., Ran, M. et al. (2019). Synergistic integration of Bi metal and phosphate defects on hexagonal and monoclinic $BiPO_4$: enhanced photocatalysis and reaction mechanism. *Appl. Catal., B* 243 (October 2018): 313–321. https://doi.org/10.1016/j.apcatb.2018.10.055.

**98** Li, X., Zhang, W., Li, J. et al. (2019). Transformation pathway and toxic intermediates inhibition of photocatalytic NO removal on designed Bi metal@defective $Bi_2O_2SiO_3$. *Appl. Catal., B* 241 (July 2018): 187–195. https://doi.org/10.1016/j.apcatb.2018.09.032.

**99** Li, X., Zhang, W., Cui, W. et al. (2019). Reactant activation and photocatalysis mechanisms on Bi-metal@$Bi_2GeO_5$ with oxygen vacancies: a combined experimental and theoretical investigation. *Chem. Eng. J.* 370 (April): 1366–1375. https://doi.org/10.1016/j.cej.2019.04.003.

**100** Zhang, W., Zhang, Q., and Dong, F. (2013). Visible-light photocatalytic removal of NO in air over BiOX (X = Cl, Br, I) single-crystal nanoplates prepared at room temperature. *Ind. Eng. Chem. Res.* 52 (20): 6740–6746.

**101** Li, H., Shang, H., Cao, X. et al. (2018). Oxygen vacancies mediated complete visible light NO oxidation via side-on bridging superoxide radicals. *Environ. Sci. Technol.* 52 (15): 8659–8665.

**102** Zhang, W., Dong, X., Liang, Y. et al. (2018). Ag/AgCl nanoparticles assembled on BiOCl/$Bi_{12}O_{17}Cl_2$ nanosheets: enhanced plasmonic visible light photocatalysis and in situ DRIFTS investigation. *Appl. Surf. Sci.* 455 (December 2017): 236–243. https://doi.org/10.1016/j.apsusc.2018.05.171.

**103** Wang, H., Zhang, W., Li, X. et al. (2018). Highly enhanced visible light photocatalysis and in situ FT-IR studies on Bi metal@defective BiOCl hierarchical microspheres. *Appl. Catal., B* 225 (November 2017): 218–227. https://doi.org/10.1016/j.apcatb.2017.11.079.

**104** Li, R., Liu, J., Zhang, X. et al. (2018). Iodide-modified $Bi_4O_5Br_2$ photocatalyst with tunable conduction band position for efficient visible-light decontamination of pollutants. *Chem. Eng. J.* 339 (January): 42–50. https://doi.org/10.1016/j.cej.2018.01.109.

**105** Sun, Y., Xiong, T., Dong, F. et al. (2016). Interlayer-I-doped BiOIO3 nanoplates with an optimized electronic structure for efficient visible light photocatalysis. *Chem. Commun.* 52 (53): 8243–8246. http://dx.doi.org/10.1039/C6CC03630A.

**106** Bai, Y., Yang, P., Wang, P. et al. (2018). Solid phase fabrication of Bismuth-rich $Bi_3O_4Cl_xBr_{1-x}$ solid solution for enhanced photocatalytic NO removal under visible light. *J. Taiwan Inst. Chem. Eng.* 82: 273–280. https://doi.org/10.1016/j.jtice.2017.10.021.

**107** Wu, T., Li, X., Zhang, D. et al. (2016). Efficient visible light photocatalytic oxidation of NO with hierarchical nanostructured 3D flower-like $BiOCl_xBr_{1-x}$ solid solutions. *J. Alloys Compd.* 671: 318–327.

**108** Ou, M., Dong, F., Zhang, W., and Wu, Z. (2014). Efficient visible light photocatalytic oxidation of NO in air with band-gap tailored $(BiO)_2CO_3$-BiOI solid solutions. *Chem. Eng. J.* 255: 650–658. http://dx.doi.org/10.1016/j.cej.2014.06.086.

**109** He, R., Cao, S., Zhou, P., and Yu, J. (2014). Recent advances in visible light Bi-based photocatalysts. *Cuihua Xuebao/Chinese J. Catal.* 35 (7): 989–1007. http://dx.doi.org/10.1016/S1872-2067(14)60075-9.

**110** Li, R., Xie, F., Liu, J. et al. (2019). Room-temperature hydrolysis fabrication of BiOBr/$Bi_{12}O_{17}Br_2$ Z-Scheme photocatalyst with enhanced resorcinol degradation and NO removal activity. *Chemosphere [Internet]* 235: 767–775. https://doi.org/10.1016/j.chemosphere.2019.06.231.

**111** Xiong, T., Wen, M., Dong, F. et al. (2016). Three dimensional Z-scheme $(BiO)_2CO_3/MoS_2$ with enhanced visible light photocatalytic NO removal. *Appl. Catal., B* 199: 87–95. http://dx.doi.org/10.1016/j.apcatb.2016.06.032.

**112** Jia, Y., Li, S., Gao, J. et al. (2019). Highly efficient $(BiO)_2CO_3$-$BiO_{2-x}$-graphene photocatalysts: Z-Scheme photocatalytic mechanism for their enhanced photocatalytic removal of NO. *Appl. Catal., B* 240 (July 2018): 241–252.

**113** Hu, J., Chen, D., Mo, Z. et al. (2019). Z-scheme 2D/2D heterojunction of black phosphorus/monolayer $Bi_2WO_6$ nanosheets with enhanced photocatalytic activities. *Angew. Chem. Int. Ed.* 58 (7): 2073–2077.

**114** Sun, Y., Zhang, W., Xiong, T. et al. (2014). Growth of BiOBr nanosheets on $C_3N_4$ nanosheets to construct two-dimensional nanojunctions with enhanced photoreactivity for NO removal. *J. Colloid Interface Sci.* 418: 317–323. http://dx.doi.org/10.1016/j.jcis.2013.12.037.

**115** Wang, Z., Huang, Y., Ho, W. et al. (2016). Fabrication of $Bi_2O_2CO_3$/g-$C_3N_4$ heterojunctions for efficiently photocatalytic NO in air removal: in-situ self-sacrificial synthesis, characterizations and mechanistic study. *Appl. Catal., B* 199 (x): 123–133. http://dx.doi.org/10.1016/j.apcatb.2016.06.027.

**116** Tian, N., Zhang, Y., Liu, C. et al. (2016). g-$C_3N_4$/$Bi_4O_5I_2$ 2D–2D heterojunctional nanosheets with enhanced visible-light photocatalytic activity. *RSC Adv.* 6 (13): 10895–10903. http://dx.doi.org/10.1039/C5RA24672E.

**117** Cui, W., Li, X., Gao, C. et al. (2017). Ternary Ag/AgCl-$(BiO)_2CO_3$ composites as high-performance visible-light plasmonic photocatalysts. *Catal. Today* 284: 67–76. http://dx.doi.org/10.1016/j.cattod.2016.10.020.

**118** He, W., Sun, Y., Jiang, G. et al. (2018). Activation of amorphous $Bi_2WO_6$ with synchronous Bi metal and $Bi_2O_3$ coupling: photocatalysis mechanism and reaction pathway. *Appl. Catal., B* 232 (January): 340–347. https://doi.org/10.1016/j.apcatb.2018.03.047.

**119** He, W., Sun, Y., Jiang, G. et al. (2018). Defective $Bi_4MoO_9$/Bi metal core/shell heterostructure: enhanced visible light photocatalysis and reaction mechanism. *Appl. Catal., B* 239 (May): 619–627. https://doi.org/10.1016/j.apcatb.2018.08.064.

**120** Dong, F., Sun, Y., Fu, M. et al. (2012). Room temperature synthesis and highly enhanced visible light photocatalytic activity of porous BiOI/BiOCl composites nanoplates microflowers. *J. Hazard. Mater.* 219–220: 34–26. http://dx.doi.org/10.1016/j.jhazmat.2012.03.015.

**121** Shi, X., Wang, P., Li, W. et al. (2019). Change in photocatalytic NO removal mechanisms of ultrathin BiOBr/BiOI via $NO_3^-$ adsorption. *Appl. Catal., B* 243 (October 2018): 322–329.

**122** Dong, F., Xiong, T., Sun, Y. et al. (2015). Controlling interfacial contact and exposed facets for enhancing photocatalysis via 2D–2D heterostructures. *Chem. Commun.* 51 (39): 8249–8252. http://dx.doi.org/10.1039/C5CC01993A.

**123** Wang, H., Sun, Y., Jiang, G. et al. (2018). Unraveling the Mechanisms of Visible Light Photocatalytic NO Purification on Earth-Abundant Insulator-Based Core-Shell Heterojunctions. *Environ. Sci. Technol.* 52 (3): 1479–1487.

**124** Huy, T.H., Phat, B.D., Thi, C.M., and Van Viet, P. High photocatalytic removal of NO gas over $SnO_2$ nanoparticles under solar light. *Environ. Chem. Lett.* 2019, 17 (1): 527–531. https://doi.org/10.1007/s10311-018-0801-0.

**125** Bui, D.P., Nguyen, M.T., Tran, H.H. et al. (2020). Green synthesis of Ag@$SnO_2$ nanocomposites for enhancing photocatalysis of nitrogen monoxide removal under solar light irradiation. *Catal. Commun.* 136 (September 2019): 105902. https://doi.org/10.1016/j.catcom.2019.105902.

**126** Huy, T.H., Bui, D.P., Kang, F. et al. (2019). $SnO_2$/$TiO_2$ nanotube heterojunction: the first investigation of NO degradation by visible light-driven photocatalysis. *Chemosphere* 215: 323–332.

**127** Ai, Z., Lee, S., Huang, Y. et al. (2010). Photocatalytic removal of NO and HCHO over nanocrystalline $Zn_2SnO_4$ microcubes for indoor air purification. *J. Hazard. Mater.* 179 (1–3): 141–150. http://dx.doi.org/10.1016/j.jhazmat.2010.02.071.

**128** Li, Y., Wu, X., Ho, W. et al. (2018). Graphene-induced formation of visible-light-responsive $SnO_2$-$Zn_2SnO_4$ Z-scheme photocatalyst with surface vacancy for the enhanced photoreactivity towards NO and acetone oxidation. *Chem. Eng. J.* 336 (November 2017): 200–210. https://doi.org/10.1016/j.cej.2017.11.045.

**129** Long, T., Dong, X., Liu, X. et al. (2010). Synthesis of ZnO crystals with unique morphologies by a low-temperature solvothermal process and their photocatalytic deNOx properties. *Res. Chem. Intermed.* 36 (1): 61–67.

**130** Le, T.H., Truong, Q.D., Kimura, T. et al. (2012). Synthesis of hierarchical porous ZnO microspheres and its photocatalytic DeNOx activity. *Ceram. Int.* 38 (6): 5053–5059. http://dx.doi.org/10.1016/j.ceramint.2012.03.007.

**131** Luévano-Hipólito, E., Martínez-de la Cruz, A., and López Cuéllar, E. (2017). Performance of ZnO synthesized by sol-gel as photocatalyst in the photooxidation reaction of NO. *Environ. Sci. Pollut. Res.* 24 (7): 6361–6371.

**132** Pastor, A., Balbuena, J., Cruz-Yusta, M. et al. (2019). ZnO on rice husk: a sustainable photocatalyst for urban air purification. *Chem. Eng. J.* 368 (March): 659–667.

**133** Liu, H., Liu, H., Yang, J. et al. (2019). Microwave-assisted one-pot synthesis of Ag decorated flower-like ZnO composites photocatalysts for dye degradation and NO removal. *Ceram. Int.* 45 (16): 20133–20140. https://doi.org/10.1016/j.ceramint.2019.06.279.

**134** Gasparotto, A., Carraro, G., Maccato, C. et al. (2018). WO3-decorated ZnO nanostructures for light-activated applications. *CrystEngComm* 20 (9): 1282–1290. http://dx.doi.org/10.1039/C7CE02148H.

**135** Nguyen, S.N., Truong, T.K., You, S.J. et al. (2019). Investigation on photocatalytic removal of NO under visible light over Cr-doped ZnO nanoparticles. *ACS Omega* 4 (7): 12853–12859.

**136** Rodriguez-Rivas, F., Pastor, A., Barriga, C. et al. (2018). Zn-Al layered double hydroxides as efficient photocatalysts for NOx abatement. *Chem. Eng. J.* 346 (December 2017): 151–158. https://doi.org/10.1016/j.cej.2018.04.022.

**137** Rodriguez-Rivas, F., Pastor, A., de Miguel, G. et al. (2020). $Cr^{3+}$ substituted Zn-Al layered double hydroxides as UV–Vis light photocatalysts for NO gas removal from the urban environment. *Sci. Total Environ.* 706: 136009. https://doi.org/10.1016/j.scitotenv.2019.136009.

**138** Pastor, A., Rodriguez-Rivas, F., de Miguel, G. et al. (2020). Effects of $Fe^{3+}$ substitution on Zn-Al layered double hydroxides for enhanced NO photochemical abatement. *Chem. Eng. J.* 387 (September 2019): 124110. https://doi.org/10.1016/j.cej.2020.124110.

**139** Pei, C.C. and Leung, W.W.F. (2013). Enhanced photocatalytic activity of electrospun $TiO_2$/ZnO nanofibers with optimal anatase/rutile ratio. *Catal. Commun.* 37: 100–104. http://dx.doi.org/10.1016/j.catcom.2013.03.029.

**140** Wei, X., Wang, Y., Huang, Y., and Fan, C. (2019). Composite ZIF-8 with CQDs for boosting visible-light-driven photocatalytic removal of NO. *J. Alloys Compd.* 802: 467–476. https://doi.org/10.1016/j.jallcom.2019.06.086.

**141** Xie, H., Li, Y., Jin, S. et al. (2010). Facile fabrication of 3D-ordered macroporous nanocrystalline iron oxide films with highly efficient visible light induced photocatalytic activity. *J. Phys. Chem. C* 114 (21): 9706–9712.

**142** Sun, W., Meng, Q., Jing, L. et al. (2013). Facile synthesis of surface-modified nanosized α-$Fe_2O_3$ as efficient visible photocatalysts and mechanism insight. *J. Phys. Chem. C* 117 (3): 1358–1365.

**143** Barreca, D., Carraro, G., Gasparotto, A. et al. (2013). Surface functionalization of nanostructured $Fe_2O_3$ polymorphs: from design to light-activated applications. *ACS Appl. Mater. Interfaces* 5 (15): 7130–7138.

**144** Sugrañez, R., Cruz-Yusta, M., Mármol, I. et al. (2013). Preparation of sustainable photocatalytic materials through the valorization of industrial wastes. *ChemSusChem* 6 (12): 2340–2347.

**145** Balbuena, J., Sánchez, L., and Cruz-Yusta, M. (2019). Use of steel industry wastes for the preparation of self-cleaning mortars. *Materials (Basel)* 12 (4): 621.

**146** Carraro, G., Sugrañez, R., Maccato, C. et al. (2014). Nanostructured iron(III) oxides: from design to gas- and liquid-phase photo-catalytic applications. *Thin Solid Films* 564: 121–127. http://dx.doi.org/10.1016/j.tsf.2014.05.048.

**147** Sugrañez, R., Balbuena, J., Cruz-Yusta, M. et al. (2015). Efficient behaviour of hematite towards the photocatalytic degradation of NOx gases. *Appl. Catal., B* 165 (X): 529–536.

**148** Huang, Y., Gao, Y., Zhang, Q. et al. (2018). Biocompatible FeOOH-Carbon quantum dots nanocomposites for gaseous NOx removal under visible light: improved charge separation and High selectivity. *J. Hazard. Mater.* 354 (February): 54–62. https://doi.org/10.1016/j.jhazmat.2018.04.071.

**149** Balbuena, J., Cruz-Yusta, M., Cuevas, A.L. et al. (2016). Enhanced activity of α-$Fe_2O_3$ for photocatalytic NO removal. *RSC Adv.* 6 (95): 92917–92922. http://dx.doi.org/10.1039/C6RA19167C.

**150** Balbuena, J., Cruz-Yusta, M., Cuevas, A.L. et al. (2019). Hematite porous architectures as enhanced air purification photocatalyst. *J. Alloys Compd.* 797 (X): 166–173.

**151** Balbuena, J., Cruz-Yusta, M., Pastor, A., and Sánchez, L. (2018). α-$Fe_2O_3$/$SiO_2$ composites for the enhanced photocatalytic NO oxidation. *J. Alloys Compd.* 735 (2): 1553–1561.

**152** Balbuena, J., Carraro, G., Cruz, M. et al. (2016). Advances in photocatalytic NOx abatement through the use of $Fe_2O_3$/$TiO_2$ nanocomposites. *RSC Adv.* 6 (78): 74878–74885. http://dx.doi.org/10.1039/C6RA15958C.

**153** Li, H., Yin, S., Wang, Y., and Sato, T. (2013). Efficient persistent photocatalytic decomposition of nitrogen monoxide over a fluorescence-assisted $CaAl_2O_4$:(Eu, Nd)/(Ta, N)-codoped $TiO_2$/$Fe_2O_3$. *Appl. Catal., B* 132–133: 487–492. http://dx.doi.org/10.1016/j.apcatb.2012.12.026.

**154** Huang, Y., Liang, Y., Rao, Y. et al. (2017). Environment-friendly carbon quantum dots/$ZnFe_2O_4$ photocatalysts: characterization, biocompatibility, and mechanisms for NO removal. *Environ. Sci. Technol.* 51 (5): 2924–2933.

**155** Xiao, S., Pan, D., Liang, R. et al. (2018). Bimetal MOF derived mesocrystal $ZnCo_2O_4$ on rGO with High performance in visible-light photocatalytic NO oxidization. *Appl. Catal., B* 236 (May): 304–313. https://doi.org/10.1016/j.apcatb.2018.05.033.

**156** Huang, Y., Zhu, D., Zhang, Q. et al. (2018). Synthesis of a $Bi_2O_2CO_3$/$ZnFe_2O_4$ heterojunction with enhanced photocatalytic activity for visible light irradiation-induced NO removal. *Appl. Catal., B* 234 (February): 70–78. https://doi.org/10.1016/j.apcatb.2018.04.039.

**157** Hu, J., Wang, H., Dong, F., and Wu, Z. (2017). A new strategy for utilization of NIR from solar energy—Promotion effect generated from photothermal effect of $Fe_3O_4$@$SiO_2$ for photocatalytic oxidation of NO. *Appl. Catal., B* 204: 584–592. http://dx.doi.org/10.1016/j.apcatb.2016.12.009.

**158** Huang, Y., Gao, Y., Zhang, Q. et al. (2016). Hierarchical porous $ZnWO_4$ microspheres synthesized by ultrasonic spray pyrolysis: characterization, mechanistic and photocatalytic NOx removal studies. *Appl. Catal., A* 515 (x): 170–178. http://dx.doi.org/10.1016/j.apcata.2016.02.007.

**159** Li, S., Chang, L., Peng, J. et al. (2020). Bi0 nanoparticle loaded on $Bi^{3+}$-doped $ZnWO_4$ nanorods with oxygen vacancies for enhanced photocatalytic NO removal. *J. Alloys Compd.* 818: 152837. https://doi.org/10.1016/j.jallcom.2019.152837.

**160** Luévano-Hipólito, E. (2017). Martínez-de la Cruz A. Photooxidation of NOx using scheelite-type ABO4 (A = Ca, Pb; B = W, Mo) phases as catalysts. *Adv. Powder Technol.* 28 (6): 1511–1518.

**161** Ai, Z. and Lee, S. (2013). Morphology-dependent photocatalytic removal of NO by hierarchical BiVO 4 microboats and microspheres under visible light. *Appl. Surf. Sci.* 280: 354–359. http://dx.doi.org/10.1016/j.apsusc.2013.04.160.

**162** Hu, J., Chen, D., Li, N. et al. (2018, 52). Fabrication of graphitic-$C_3N_4$ quantum dots/graphene-InVO4 aerogel hybrids with enhanced photocatalytic NO removal under visible-light irradiation. *Appl. Catal., B* 236 (May): 45. https://doi.org/10.1016/j.apcatb.2018.04.080.

**163** Linsebigler, A.L., Lu, G., and Yates, J.T. (1995). Photocatalysis on $TiO_2$ surfaces: principles, mechanisms, and selected results. *Chem. Rev.* 95 (3): 735–758.

**164** Hashimoto, K., Wasada, K., Toukai, N. et al. (2000). Photocatalytic oxidation of nitrogen monoxide over titanium(IV) oxide nanocrystals large size areas. *J. Photochem. Photobiol., A* 136 (1–2): 103–109.

**165** Bloh, J.Z., Folli, A., and Macphee, D.E. (2014). Photocatalytic NOx abatement: why the selectivity matters. *RSC Adv.* 4 (86): 45726–45734. http://dx.doi.org/10.1039/C4RA07916G.

**166** Peters, R.W. (1991). Dangerous properties of industrial materials, 7th edn. (a three-volume set), N. Irving Sax and Richard J. Lewis, Jr., Van Nostrand Reinhold, New York, NY, (1989). *Environ. Prog.* 10 (3): A7–A8. https://doi.org/10.1002/ep.670100308.

**167** Kaneko, M. and Okura, I. (2002). *Photocatalysis: Science and Technology*, vol. 360, 143–155. Berlin, Heidelberg: Springer. (Biological and Medical Physics, Biomedical Engineering). https://books.google.es/books?id=Ttwh4HtVFngC.

**168** Hadjiivanov, K. and Knözinger, H. (2000). Species formed after NO adsorption and $NO+O_2$ co-adsorption on $TiO_2$: an FTIR spectroscopic study. *Phys. Chem. Chem. Phys.* 2 (12): 2803–2806. http://dx.doi.org/10.1039/B002065F.

**169** Bedjanian, Y. and El Zein, A. (2012). Interaction of $NO_2$ with $TiO_2$ surface under UV irradiation: products study. *J. Phys. Chem. A* 116 (7): 1758–1764.

**170** Goodman, A.L., Underwood, G.M., and Grassian, V.H. (1999). Heterogeneous reaction of NC2: characterization of gas-phase and adsorbed products from the reaction, $2NO_2(g) + H_2O(a) \rightarrow HONO(g) + HNO_3(a)$ on hydrated silica particles. *J. Phys. Chem. A* 103 (36): 7217–7223.

**171** Ramazan, K.A., Syomin, D., and Finlayson-Pitts, B.J. (2004). The photochemical production of HONO during the heterogeneous hydrolysis of NO2. *Phys. Chem. Chem. Phys.* 6 (14): 3836–3843. http://dx.doi.org/10.1039/B402195A.

**172** Finlayson-Pitts, B.J. (2009). Reactions at surfaces in the atmosphere: integration of experiments and theory as necessary (but not necessarily sufficient) for predicting the physical chemistry of aerosols. *Phys. Chem. Chem. Phys.* 11 (36): 7760–7779. http://dx.doi.org/10.1039/B906540G.

**173** Hüsken, G., Hunger, M., and Brouwers, H.J.H. (2009). Experimental study of photocatalytic concrete products for air purification. *Build Environ.* 44 (12): 2463–2474. http://dx.doi.org/10.1016/j.buildenv.2009.04.010.

**174** Gustafsson, R.J., Orlov, A., Griffiths, P.T. et al. (2006). Reduction of $NO_2$ to nitrous acid on illuminated titanium dioxide aerosol surfaces: implications for photocatalysis and atmospheric chemistry. *Chem. Commun.* 37: 3936–3938. http://dx.doi.org/10.1039/B609005B.

**175** Yu, Q.L. and Brouwers, H.J.H. (2009). Indoor air purification using heterogeneous photocatalytic oxidation. Part I: experimental study. *Appl. Catal., B* 92 (3–4): 454–461.

**176** Rosseler, O., Sleiman, M., Montesinos, V.N. et al. (2013). Chemistry of NOx on $TiO_2$ surfaces studied by ambient pressure XPS: products, effect of UV irradiation, water, and coadsorbe $K^+$. *J. Phys. Chem. Lett.* 4 (3): 536–541.

**177** Li, S.C., Jacobson, P., Zhao, S.L. et al. (2012). Trapping nitric oxide by surface hydroxyls on rutile $TiO_2$ (110). *J. Phys. Chem. C* 116 (2): 1887–1891.

**178** Cui, W., Li, J., Dong, F. et al. (2017). Highly efficient performance and conversion pathway of photocatalytic NO oxidation on SrO-clusters@amorphous carbon nitride. *Environ. Sci. Technol.* 51 (18): 10682–10690.

**179** Wu, J.C.S. and Cheng, Y.T. (2006). In situ FTIR study of photocatalytic NO reaction on photocatalysts under UV irradiation. *J. Catal.* 237 (2): 393–404.

**180** Liao, J., Cui, W., Li, J. et al. (2020). Nitrogen defect structure and NO+ intermediate promoted photocatalytic NO removal on $H_2$ treated g-$C_3N_4$. *Chem. Eng. J.* 379 (May 2019): 122282. https://doi.org/10.1016/j.cej.2019.122282.

**181** Cui, W., Li, J., Cen, W. et al. (2017). Steering the interlayer energy barrier and charge flow via bioriented transportation channels in g-$C_3N_4$: enhanced photocatalysis and reaction mechanism. *J. Catal.* 352: 351–360. http://dx.doi.org/10.1016/j.jcat.2017.05.017.

**182** Wu, Q., Yang, C.C., and Van De Krol, R. (2014). A dopant-mediated recombination mechanism in Fe-doped $TiO_2$ nanoparticles for the photocatalytic decomposition of nitric oxide. *Catal. Today* 225: 96–101. http://dx.doi.org/10.1016/j.cattod.2013.09.026.

**183** Cui, W., Chen, L., Li, J. et al. (2019). Ba-vacancy induces semiconductor-like photocatalysis on insulator $BaSO_4$. *Appl. Catal., B* 253 (April): 293–299. https://doi.org/10.1016/j.apcatb.2019.04.070.

**184** Araña, J., Garzón Sousa, D., González Díaz, O. et al. (2019). Effect of $NO_2$ and $NO_3^-$/$HNO_3$ adsorption on no photocatalytic conversion. *Appl. Catal., B* 244 (2): 660–670.

**185** Kumar Babu, B., Ghosh, S., and Chakrabortty, S. (2020). Recent developments in smart window engineering: from antibacterial activity to self-cleaning behavior [Internet]. In: *Energy Saving Coating Materials*, 227–263. Elsevier. http://dx.doi.org/10.1016/B978-0-12-822103-7/00010-8.

**186** da Silva, A.L., Dondi, M., Raimondo, M., and Hotza, D. (2018). Photocatalytic ceramic tiles: challenges and technological solutions. *J. Eur. Ceram. Soc.* 38 (4): 1002–1017. https://doi.org/10.1016/j.jeurceramsoc.2017.11.039.

**187** Padmanabhan, N. and John, H. (2020). Titanium dioxide based self-cleaning smart surfaces: a short review. *J. Environ. Chem. Eng.* 8 (5): 104211. https://doi.org/10.1016/j.jece.2020.104211.

**188** Hamidi, F. and Aslani, F. (2019). $Tio_2$-based photocatalytic cementitious composites: materials, properties, influential parameters, and assessment techniques. *Nanomaterials* 9 (10): 1444.

**189** Guo, S., Wu, Z., and Zhao, W. (2009). $TiO_2$-based building materials: above and beyond traditional applications. *Chin. Sci. Bull.* 54 (7): 1137–1142.

**190** Matthew, F.C.F. (2015). *A Clear Choice for the UK: Technology Options for Tackling Air Pollution*, 20. The Environmental Industries Commision. http://eic-uk.co.uk/wp-content/uploads/2018/03/EIC-Air-quality-report-2015-digital.pdf.

**191** Yoshihiko, M., Hideo, T., Hiroshi, O., and Murata, K. (2003). *NOX-Cleaning Paving Block*. EP 0 786 283 B1, 20. European Patent Office. https://patentimages.storage.googleapis.com/c9/9d/f3/07ec7dba0b561f/EP0786283B1.pdf.

192 Luigi. C. and Pepe C. (1997). Paving Tile Comprising an Hydraulic Binder and Photocatalyst Particles. EP 1 600 430 A1, . p. 21. https://patentimages.storage.googleapis.com/a0/76/35/41ae4be6d197b0/EP1600430A1.pdf.

193 Cassar, L. (2004). Photocatalysis of cementitious materials: clean buildings and clean air. *MRS Bull.* 29 (5): 328–331. https://www.cambridge.org/core/article/photocatalysis-of-cementitious-materials-clean-buildings-and-clean-air/9BB0557274134CE41738C6EA1D341D53.

194 ISO. (2016). ISO 22197-1:2016 Fine ceramics (advanced ceramics, advanced technical ceramics) - Test method for air-purification performance of semiconducting photocatalytic materials - Part 1: Removal of Nitric oxide. https://www.iso.org/standard/65416.html.

195 Ballari, M.M., Yu, Q.L., and Brouwers, H.J.H. (2011). Experimental study of the NO and $NO_2$ degradation by photocatalytically active concrete. *Catal. Today* 161 (1): 175–180.

196 Zouzelka, R. and Rathousky, J. (2017). Photocatalytic abatement of NOx pollutants in the air using commercial functional coating with porous morphology. *Appl. Catal., B* 217: 466–476. http://dx.doi.org/10.1016/j.apcatb.2017.06.009.

197 Águia, C., Ângelo, J., Madeira, L.M., and Mendes, A. (2011). Photo-oxidation of NO using an exterior paint - Screening of various commercial titania in powder pressed and paint films. *J Environ Manage [Internet].* 92 (7): 1724–1732. http://dx.doi.org/10.1016/j.jenvman.2011.02.010.

198 Ângelo, J., Andrade, L., and Mendes, A. (2014). Highly active photocatalytic paint for NOx abatement under real-outdoor conditions. *Appl. Catal., A* 484 (x): 17–25. http://dx.doi.org/10.1016/j.apcata.2014.07.005.

199 Wang, D., Leng, Z., Hüben, M. et al. (2016). *Constr. Build Mater.* 107: 44–51. http://dx.doi.org/10.1016/j.conbuildmat.2015.12.164.

200 Mills, A., Andrews, R., Han, R. et al. (2020). Supersensitive test of photocatalytic activity based on ISO 22197-1:2016 for the removal of NO. *J. Photochem. Photobiol. A Chem.* 400 (April): 112734. https://doi.org/10.1016/j.jphotochem.2020.112734.

201 Guo, M.Z., Maury-Ramirez, A., and Poon, C.S. (2015). Photocatalytic activities of titanium dioxide incorporated architectural mortars: effects of weathering and activation light. *Build Environ.* 94: 395–402. http://dx.doi.org/10.1016/j.buildenv.2015.08.027.

202 De Melo, J.V.S., Trichês, G., Gleize, P.J.P., and Villena, J. (2012). Development and evaluation of the efficiency of photocatalytic pavement blocks in the laboratory and after one year in the field. *Constr. Build. Mater.* 37: 310–319.

203 Pérez-Nicolás, M., Navarro-Blasco, I., Fernández, J.M., and Alvarez, J.I. (2017). Atmospheric NOx removal: study of cement mortars with iron- and vanadium-doped $TiO_2$ as visible light–sensitive photocatalysts. *Constr. Build. Mater.* 149: 257–271.

204 Bica, B.O. and de Melo, J.V.S. (2020). Concrete blocks nano-modified with zinc oxide (ZnO) for photocatalytic paving: performance comparison with titanium dioxide ($TiO_2$). *Constr. Build Mater.* 252: 119120. https://doi.org/10.1016/j.conbuildmat.2020.119120.

**205** García Calvo, J.L., Carballosa, P., Castillo, A. et al. (2019). Expansive concretes with photocatalytic activity for pavements: enhanced performance and modifications of the expansive hydrates composition. *Constr. Build. Mater.* 218: 394–403.

**206** Maggos, T., Bartzis, J.G., Liakou, M., and Gobin, C. (2007). Photocatalytic degradation of NOx gases using $TiO_2$-containing paint: a real scale study. *J. Hazard. Mater.* 146 (3): 668–673.

**207** Maggos, T., Plassais, A., Bartzis, J.G. et al. (2008). Photocatalytic degradation of NOx in a pilot street canyon configuration using $TiO_2$-mortar panels. *Environ. Monit. Assess.* 136 (1–3): 35–44.

**208** Barratt, B. (2007). *Statistical Analysis of Monitoring Results from the City of London's NOX-Reducing Paint Study*, 1–19. Sir John Cass.

**209** Tremper Anja, Green D. (2016). Artworks D-NOx Paint trial Report 2-24.

**210** Boonen, E. and Beeldens, A. (2014). Recent photocatalytic applications for air purification in Belgium. *Coatings* 4 (3): 553–573.

**211** Boonen, E. and Beeldens, A. (2013). Photocatalytic roads: from lab tests to real scale applications. *Eur. Transp. Res. Rev.* 5 (2): 79–89.

**212** Gallus, M., Akylas, V., Barmpas, F. et al. (2015). Photocatalytic de-pollution in the Leopold II tunnel in Brussels: NOx abatement results. *Build. Environ.* 84 (2): 125–133.

**213** Guerrini, G.L. (2012). Photocatalytic performances in a city tunnel in Rome: NOx monitoring results. *Constr. Build Mater.* 27 (1): 165–175. http://dx.doi.org/10.1016/j.conbuildmat.2011.07.065.

**214** Folli, A., Strøm, M., Madsen, T.P. et al. (2015). Field study of air purifying paving elements containing $TiO_2$. *Atmos. Environ.* 107 (2): 44–51. http://dx.doi.org/10.1016/j.atmosenv.2015.02.025.

**215** Cordero, J.M., Hingorani, R., Jimenez-Relinque, E. et al. (2020 Jun). NO(x) removal efficiency of urban photocatalytic pavements at pilot scale. *Sci. Total Environ.* 719: 137459.

**216** Cordero, J.M., Borge, R., and Narros, A. (2018). Using statistical methods to carry out in field calibrations of low cost air quality sensors. *Sensors Actuators, B* 267 (2): 245–254. https://doi.org/10.1016/j.snb.2018.04.021.

**217** Nevshupa, R., Jimenez-Relinque, E., Grande, M. et al. (2020 Oct). Assessment of urban air pollution related to potential nanoparticle emission from photocatalytic pavements. *J. Environ. Manage.* 272: 111059.

**218** Marta Castellote AP. (2017). Device for determining photocatalytic properties of materials [Internet]. WO 2017/085342 A1. p. 19, filed 26 May 2017 and issued 14 November 2016. https://digital.csic.es/bitstream/10261/176033/1/WO2017085342A1.pdf.

# 8

# Synthesis and Characterization of Oxide Photocatalysts for $CO_2$ Reduction

*Fernando Fresno[1] and Patricia García-Muñoz[2]*

[1] Instituto de Catálisis y Petroleoquímica (CSIC), C/ Marie Curie 2, Cantoblanco, Madrid 28049, Spain
[2] Escuela Técnica Superior de Ingenieros Industriales, Universidad Politécnica de Madrid, Department of Industrial Chemical & Environmental Engineering, C/ José Gutiérrez Abascal 2, Madrid 28006, Spain

## 8.1 Introduction

Following the directives of the Paris Agreement, adopted in December 2015 under the UN Framework Convention on Climate Change and ratified in the following Conferences of the Parties, the developed countries should lead the way to reach global peaking of greenhouse gas emissions, with the aim of holding the increase in the worldwide average temperature below 1.5 °C above preindustrial levels. The actions to be taken in this frame include the reduction of anthropogenic greenhouse gas emissions by 2030 to 40 Gt/y [1]. Although enhancing the efficiency of the current energy system is an undoubtedly needed measure to reduce these emissions, the previously mentioned aim seems difficult to achieve if the primary energy demand, which has doubled over the last 40 years and is expected to double again by 2050, has to be met by increasing fossil fuel combustion. Therefore, it is crucial to slowly escape from the fossil fuel-based energy system by increasing the weight of renewable sources in the energy mix, according not only to the future scarcity of fossil sources but also to the current need of fighting global warming. Among all renewable sources, solar energy is by far the largest exploitable one, delivering more energy on the Earth surface per hour than the total energy consumed by human activities in one year. As it is known, however, this energy source presents as main drawbacks its dilute nature and its inherent discontinuity, which make it necessary not only to convert it but also to store it into manageable and transportable forms of energy. Photosynthetic organisms have performed this complex task for millions of years, and mimicking this ability in an artificial way has become one of the most elusive challenges to be faced. Mechanisms are mostly known, but the efficiency, selectivity, and stability issues that can lead to a practical device have not been addressed yet [2]. Nevertheless, significant progress has been made in the last years and the path forward is promising [3]. Artificial photosynthesis (AP) processes should encompass both the transformation of $CO_2$ into chemically and energetically useful molecules and the splitting of water into hydrogen and oxygen, which is indeed the core of

*Tailored Functional Oxide Nanomaterials: From Design to Multi-Purpose Applications*, First Edition.
Edited by Chiara Maccato and Davide Barreca.
© 2022 WILEY-VCH GmbH. Published 2022 by WILEY-VCH GmbH.

natural photosynthesis [4]. The so-obtained *solar fuels* would provide not only a direct way of storing solar energy into chemical vectors but also the reutilization of $CO_2$ emissions to complement carbon capture and storage (CCS) technologies [5]. In recent years, research on AP technologies has been increasingly encouraged in countries such as the United States, China, or Japan. For instance, in 2010 the US Department of Energy (DOE) established the Joint Center for AP (JCAP), which has recently begun a five-year renewal phase with a $75 M investment focused on solar carbon dioxide reduction. Accordingly, the European Union (EU) is promoting international agreements and programs to control climate change and support low-carbon energy technologies.

Among the different pathways to solar fuels via AP, photochemical – photocatalytic – approaches involve a direct solar-to-chemical energy conversion and thus exhibit relatively high theoretical efficiencies. This fact, together with environmental and social aspects, makes them especially promising [6], although the practical efficiencies reported to date are still low and their technological development is still far from implementation. The central role in the overall efficiency of photocatalytic processes is played by the semiconductor photocatalyst, which is able to generate and manage electron–hole pairs upon irradiation with light of photon energy equal to or greater than its band gap. In order to efficiently promote photocatalytic reactions, semiconductors must have the appropriate thermodynamic (band positions), kinetic (surface chemistry), and stability (lack of deactivation or photocorrosion) characteristics. In this respect, a number of studies related to the development of efficient photocatalyst materials have been reported [7–9]. One of the key challenges is to explore suitable photocatalytic materials that meet the following criteria: i) broad range of solar spectrum absorption; ii) high photochemical stability; iii) efficient use of photogenerated electrons and holes; iv) suitable band edge positions; v) low overpotential for the pursued reactions; vi) and low cost. Typically, it is difficult for a single semiconductor to achieve all of the previously mentioned requirements. Early studies focused on $TiO_2$ [10, 11], which, in spite of its relatively high activity, has the main drawback of absorbing only in the ultraviolet (UV) region, which accounts for c. 4% of the incoming solar energy on the Earth's surface, and thus misses a large portion of the available solar spectrum. Therefore, an overarching objective in this field is to expand the action spectrum of photocatalysts, for which a plethora of materials have been studied; most of them are based not only on metal oxides or chalcogenides [12] but also with other compositions, with $C_3N_4$ as an archetypical example [13]. However, the reported solar-to-fuels energy conversion efficiencies in photocatalytic and photoelectrochemical (PEC) systems are still low, especially when water is used as the only hole scavenger, i.e. without the use of any sacrificial reagent. In the case of water splitting, the maximum solar-to-hydrogen (STH) efficiency achieved so far is 16.2% [14], and in the case of $CO_2$ reduction systems, a 10% of solar-to-fuels energy conversion efficiency has been accomplished [15].

This chapter aims at providing an insightful overview of the development of nanosized photocatalysts for $CO_2$ reduction, particularly those based on oxides, with special emphasis on the employed synthetic methods and the characteristics of the materials attained by them. Since an exhaustive collection of assayed materials would probably be impractical and little informative, we have tried to highlight the most relevant cases and the knowledge that can be generated from them.

## 8.2 Fundamentals of Heterogeneous Photocatalysis

The International Union of Pure and Applied Chemistry (IUPAC) defines the term *photocatalysis* as the "increase of the rate of a chemical reaction or its initiation under the action of ultraviolet, visible or infrared radiation in the presence of a substance — the photocatalyst — that absorbs light and is involved in the chemical transformation of the reaction partners." Accordingly, a *photocatalyst* is defined as a "substance able to produce, by absorption of ultraviolet, visible, or infrared radiation, chemical transformations of the reaction partners, repeatedly coming with them into intermediate chemical interactions and regenerating its chemical composition after each cycle of such interactions" [16]. In this way, heterogeneous photocatalysis can be defined as a chemical process in which a solid (generally a broadband semiconductor), after the absorption of light of the appropriate wavelength, catalyzes a reaction in the liquid or gas phase. The starting point of current research in heterogeneous photocatalysis was marked by the discovery by Fujishima and Honda, in 1972, of the photocatalytic dissociation of water on $TiO_2$ electrodes [10]. Since then, numerous investigations have been carried out aimed at both understanding the fundamental processes of photocatalysis and improving the photocatalytic activity of the used materials, as well as the search for new applications.

The irradiation of a semiconductor material with light of photon energy greater than or equal to its band gap causes the excitation of electrons ($e^-$) from the valence band (VB) to the conduction band (CB), leaving empty states or holes ($h^+$) in the former that behave as positively charged particles. Once the excitation has occurred, the electron–hole pairs generated can follow different paths. A diagram of the different de-excitation processes is shown in Figure 8.1.

The half-life of the charged species formed after excitation is of the order of nanoseconds [17], sufficient for them to migrate to the surface of the solid (Figure 8.1, paths 1 and 2) and produce electron exchange reactions with adsorbed species. Therefore, the semiconductor can donate electrons to an acceptor species, reducing it (path 5), while holes can combine with electrons from a donor species, so that it converts into its oxidized form (path 6). For this to occur, as in any other oxidation–reduction reaction, the relative positions of the reduction potentials of the oxidizing and reducing species must be adequate. In the case of photogenerated electrons and holes, their reduction potentials are determined by the energy of the VBs and CBs. In turn, the energy difference between these bands (i.e. the *band gap* of the semiconductor photocatalyst) determines the wavelength necessary for light activation. Charge separation is facilitated in high dielectric constant semiconductors because the carriers are shielded by the network. On the other hand, the ability of charge carriers to move toward the surface depends on structural factors (crystallinity, defects, etc.) and electronic properties (orbitals involved) of

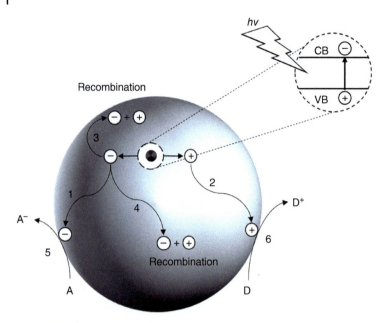

**Figure 8.1** Schematic diagram showing the charge carrier dynamics upon irradiation of a photocatalyst: (1) electron transport to the surface, (2) hole transport to the surface, (3) surface recombination, (4) bulk recombination, (5) electron transfer to an acceptor molecule, and (6) hole trapping by a donor molecule. Source: Adapted with permission from Linsebigler et al. [17]. Copyright 1995, American Chemical Society.

the semiconductor. As in any heterogeneous catalytic process, the adsorption of reactants and desorption of products depend on the textural properties (surface area and porosity) and surface chemistry of the semiconductor. Finally, as stated previously, the energy levels (and thus the reduction potentials) of the conduction and VBs determine the reduction and oxidation half-reactions that can occur, and this will be particularly important in photocatalytic $CO_2$ reduction, as we will see later. The parameters that determine the reaction rate in heterogeneous photocatalysis (mass of catalyst, concentration of reactants, temperature, pressure, etc.) are similar to those of any heterogeneous catalytic reaction except that the wavelength of irradiation and the irradiance are also involved and the effect of temperature is generally limited to its influence on the adsorption and desorption of reactants and products [18].

In competition with photocatalytic reactions, holes and electrons can also recombine, releasing energy in the form of radiation or heat, which can occur both on the surface (Figure 8.1, path 3) and in the bulk of the semiconductor particle (Figure 8.1, path 4). The *quantum yield* (QY) of a photocatalytic process is defined as the ratio between the number of reacted molecules and the number of photons absorbed by the catalyst. In the hypothetical case that there were no recombination, the QY would take the ideal value of 1. However, given the difficulty of experimentally determining the actual number of photons absorbed by the photocatalyst, due to the additional scattering and reflection phenomena, the term *apparent quantum yield*

(AQY) or *photonic yield* (PY) was introduced, in which the number of *absorbed* photons is substituted by that of *incident* photons. Both QY and AQY are measured under *monochromatic* light. In the case, closer to the usual experimental conditions, that *polychromatic* light is employed, these terms turn into *quantum efficiency* (QE) and *photonic efficiency* (PE) for absorbed and incident photons, respectively [16]. PE is probably the efficiency measurement that can be adapted to most photocatalytic materials and experimental conditions [19] and is defined by Eq. (8.1):

$$\Im = \frac{dN/dt}{\int_{\lambda_1}^{\lambda_3} q_{p,\lambda}^0 dt} \tag{8.1}$$

where $\Im$ is the PE in the wavelength interval between $\lambda_1$ and $\lambda_2$, $q_{p,\lambda}^0$ is the incident spectral photon flux, and $N$ is the number of entities (e.g. molecules or radicals) consumed or produced by the reaction.

## 8.3 Applications of Heterogeneous Photocatalysis

Due to the great variety of chemical reactions that can be favored by heterogeneous photocatalysis, it has a wide range of applications. Among them, those related to the elimination of pollutants are in a higher state of technological development than energy or synthetic applications.

Starting with environmental applications, those for self-cleaning and antifogging surfaces are related to the formation of oxygen vacancies, which make the $TiO_2$ surface superhydrophilic, so that the contact angle of the water is practically zero, and it forms a film instead of droplets. This effect is accompanied by the oxidation of organic contaminants on the surface [20]. In this way, a surface coated with titanium oxide (essentially, although other semiconductors also present this characteristic) accumulates less dirt, which has been used in already commercial materials, mainly glass but also others like polyvinyl chloride (PVC) and other plastics. In addition, the oxidative power of the holes generated in $TiO_2$ enables the application of photocatalytic films to be extended to the disinfection of surfaces by inactivation of bacteria and viruses [21, 22].

The elimination of pollutants in air and water is enabled by their catalytic oxidation by means of oxygen, generally through the formation of highly oxidizing radicals such as hydroxyl, which is why photocatalysis is included in the so-called *advanced oxidation processes* (AOPs) [23]. It allows the degradation of a great variety of compounds and complex mixtures, precisely because of the nonspecificity of these radicals, under mild conditions of pressure and temperature, and using air as an oxidant and ideally sunlight as an energy source. The most common target pollutants in air are different volatile organic compounds and nitrogen oxides [24], while in water some of the most frequent are pesticides and emerging pollutants, the latter of particular importance due to the low concentration in which they are found [23]. Regarding the elimination of pollutants in air, projects to incorporate $TiO_2$ into architectural elements for urban decontamination are common, and there are applications, for example, in sound barriers on highways or on facades.

For indoor air, there are filters that incorporate a photocatalyst ($TiO_2$) and UV lamps, which allow treating the low concentrations, although higher than outside, of pollutants in these environments and which incorporate the possibility of recirculation. Although its scientific relevance is equally high, the removal of contaminants from water is less commercially developed. There are demonstration plants, for example, in wastewater treatment plants, in which the elimination of pollutants of emerging concern has been successfully tested [25].

Perhaps the least explored application of photocatalysis is organic synthesis, most commonly partial oxidation reactions, precisely because of the difficulty of controlling the selectivity of photocatalytic oxidation, particularly if water is used as a solvent, due to the formation of the hydroxyl radicals mentioned earlier [26]. However, improvements in selectivity toward certain products have been described by modifying both the characteristics of the catalyst and the reaction conditions. For example, in a first approach, the use of organic solvents instead of water limits the formation of hydroxyl radicals and therefore modifies the selectivity of the reaction. On the other hand, different organic solvents can give rise to different selectivities because they modify the adsorption on the catalyst, as, for example, in the oxidation of cyclohexane on titanium dioxide [27]. Another approach is that of the reactor design, as in the intensification of the process through the use of microfluidics, so that the reaction yield can be increased, as, for example, in a Diels–Alder cycloaddition reaction in which 95% conversion was obtained [28]. Regarding the characteristics of the catalyst, in the case of $TiO_2$, examples of control of different selectivities have been published using its different crystalline phases or modifying its surface with fluoride to reduce hydroxylation, as e.g. in the conversion of phenol into catechol [29]. On the other hand, the synthesis of $TiO_2$ with control of exposed facets has also led to a greater control of selectivity in certain reactions, as in the case of the reduction of nitrobenzene to aniline [30]. These changes in the selectivity of oxidation reactions with different parameters have also been described for other materials in addition to $TiO_2$, as in the case of the oxidation of sulfoxides to sulfones on bismuth vanadate [31] or, more recently, the bifunctionalization of aromatics under visible light on graphitic carbon nitride [32].

Currently, more studies are focused on the energy applications of photocatalysis, the so-called AP, which encompass the production of hydrogen by decomposition of water or photoreforming and the reduction of $CO_2$, both aimed at obtaining *solar* fuels and chemicals. The development of these products is among the so-called "grails" of modern chemistry [4], and it would respond to some of its most important challenges, according to leading researchers [33].

The photocatalytic decomposition of water uses photogenerated electrons and holes to obtain hydrogen and oxygen, respectively. There are a series of semiconductors thermodynamically capable of carrying out this reaction, which involves the direct conversion of solar energy into chemical energy. It has been estimated that this technology could cover a third of the world's energy demand using 1% of the desert areas of the planet [34]. Taking as a reference the electrolysis of water using a photovoltaic cell, to be competitive this reaction should reach efficiencies from solar to hydrogen of around 14%, for which it has been estimated that the necessary QY is 30% at 600 nm [35]. The maximum efficiency from solar to hydrogen achieved so far is slightly over 1%, with an apparent QY of 30–420 nm [36], which is still a long way from the stated goal. In the water decomposition reaction, the rate-limiting half-reaction is oxidation, which requires the transfer of four electrons and four protons per produced oxygen molecule. To circumvent this limitation, an oxidizing sacrificial agent can be used instead of water, thus promoting hydrogen production and avoiding hydrogen–oxygen separation. The problem is that reagents are consumed and waste is generated. Therefore, for this procedure to make sense from an environmental point of view, the use of biomass derivatives as sacrificial agents has been proposed, thus closing a virtually carbon neutral cycle. Some of these derivatives may be the by-product glycerol from the production of biodiesel, lignocellulosic residues from the paper industry, or carbohydrates left over from agriculture [37]. A recent outstanding work in this regard is the direct photoreforming of lignocellulose using CdS/CdO photocatalysts [38]. However, this process presents the problem of the complexity of the oxidation of this substrate, which gives rise to a series of by-products that are difficult to oxidize further and which even make it difficult for the catalyst to absorb light. On the other hand, precisely this complexity can be taken advantage of through the concept of *photobiorefinery* [39], a process by which biomass is used in photocatalytic reactions not only as a sacrificial reaction for the production of hydrogen but also to obtain organic molecules of interest, such as, for example, diesel fuel precursors obtained with high selectivity on a ruthenium-doped indium zinc thiospinel, as recently published [40].

Finally, photocatalytic $CO_2$ reduction represents, as outlined in the introduction, a highly appealing approach to the use of both renewable energy and virtually inexhaustible feedstock, since it offers the possibility of storing solar energy and using $CO_2$ in the same process, offering an alternative to fossil fuels both as energy source and as raw material.

## 8.4 Photocatalytic $CO_2$ Reduction: State of the Art and Main Current Issues

The photocatalytic reduction of $CO_2$ has grown considerably in scientific relevance in recent years [2], although it is still far from the rest of the applications. For this reason, this technology is considered relevant in the medium-long term for the direct conversion of solar energy into chemical energy. Figure 8.2 compares the state of the art of three ways of $CO_2$ conversion (photo-, thermo-, and electro-catalytic) based on different figures of merit. As can be seen, the photocatalytic pathway stands out in aspects related to the potential of the process, such as theoretical efficiency and scalability, and with recent scientific activity, although it lags behind in aspects like selectivity control and level of knowledge and development [41]. Indeed, in spite of the efforts devoted to developing photocatalytic $CO_2$ reduction in the last 40 years, and the growing number of works related to it, there are still many unknowns about different aspects of the reaction. Some of these unsolved and key aspects of the

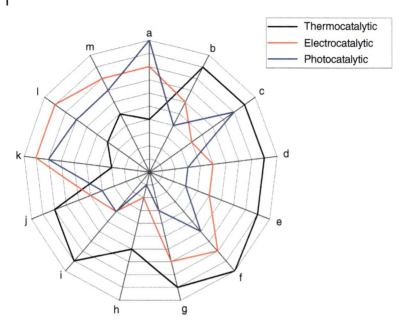

**Figure 8.2** Web graph comparing different catalytic pathways for $CO_2$ conversion. Values increasing outward indicate the state of development according to (a) ideal predicted energy efficiency of these technologies, (b) potential breakdown of commercial barriers, (c) relative scalability, (d) compared activity of state-of-the-art catalysts, (e) attained shifts to longer-chain hydrocarbons, (f) selectivity to methane, (g) selectivity to olefins, (h) selectivity to long-chain hydrocarbons, (i) currently reported catalyst stability, (j) current mechanism knowledge, (k) recent catalyst development, (l) emergence of high-impact breakthroughs, and (m) catalyst synthesis easiness. Source: Adapted with permission from Mota and Kim [41]. Copyright 2019, Royal Society of Chemistry.

process are the activation of the $CO_2$ molecule, the $H_2O$ oxidation half-reaction, the competition of water reduction, the deactivation of catalysts, and the origin of carbon products.

Regarding $CO_2$ activation, Figure 8.3 shows the main reduction potentials involved in $CO_2$ transformation compared to the valence and CB positions of several semiconductors employed as photocatalysts for this reaction. As seen in Figure 8.3, the standard potential for the one-electron reduction of $CO_2$ to form the $CO_2^{-}$ radical anion, often regarded as the first step in the photoreduction process, would be largely unfavored with most of the semiconductors, including the usual $TiO_2$. Therefore, this activation must occur either by a proton-assisted multiple reaction transfer to yield protonated products with lower reduction potentials such as methane, methanol, or formic acid [42] or by the transfer of one electron to a $CO_2$ molecule that exists in a previously activated, more reactive from. Some works have pointed out that, prior to electron transfer, $CO_2$ adsorbs on the surface of the photocatalyst in a partially charged, activated $CO_2^*$ species with a bent rather than linear structure, to which the transfer of an electron is thermodynamically favored [43, 44]. Alternatively, activation paths by two-electron transfer have been proposed

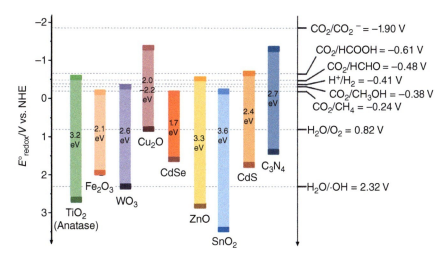

**Figure 8.3** Conduction and valence band potentials and band gap energies of several semiconductors relative to the redox potentials at pH 7 involved in $CO_2$ reduction. Source: Reproduced with permission from Fresno et al. [2]. Copyright 2018, American Chemical Society.

to be more thermodynamically favorable than one-electron processes on the basis of theoretical calculations [45].

As indicated previously, the oxidation of water is a kinetically hindered process. However, contrarily to the case of water splitting, the use of sacrificial agents to reduce $CO_2$ is rare, although some works report the use of sulfites, amines, or alcohols [46]. Despite this, many studies do not include $O_2$ detection data, or indicate its absence, or the amount is not stoichiometric, and the presence or absence of oxygen among the reaction products does not apparently depend on the catalyst, reaction phase, or reactor configuration [2]. Different factors may contribute to this, such as the experimental difficulty of oxygen quantification [47], the reoxidation of the reaction products or the oxidation of organic contamination on the surface of the catalyst by the formed oxygen or water oxidation intermediates [48], incomplete water oxidation [49], or the role of carbonates as actual hole acceptors [50].

The role of water in $CO_2$ photoreduction, on the other hand, is not only to act as hole acceptor and proton donor. In fact, water is also a competitor of $CO_2$ for photoexcited electrons, leading to the formation of hydrogen at the expense of $CO_2$ reduction. Indeed, the reduction of protons to form $H_2$ is considerably more favorable than $CO_2$ conversion from both the thermodynamic and the kinetic viewpoints [51], and it occurs in most of the reported photocatalytic $CO_2$ experiments [2]. Even if hydrogen is a desirable product, a high shift of this competition toward the formation of hydrogen would question the practical utility of the photocatalytic $CO_2 + H_2O$ reaction against a two-step process in which hydrogen is produced by water splitting and then used for the more favorable reaction of $CO_2$ hydrogenation, either in a pure thermo-catalytic reaction or by the increasingly relevant photothermal $CO_2$ hydrogenation process [52].

One of the major issues that still need to be understood and then avoided is catalyst deactivation. In $CO_2$ reduction, most of the reported photocatalytic systems have shown to be active only for a few hours [53]. The main proposed reasons that have been proposed to account for catalyst deactivation are poisoning of the catalyst by active site blocking by intermediates (e.g. carbonates or coke-like deposits), slow kinetics for the desorption of products or intermediates, agglomeration or sintering of cocatalysts, and depletion of active sites due to changes in surface chemistry during the reaction [2].

Last but not least, given the low conversion levels that have been reported to date, it is becoming increasingly important for academic works to prove the origin of the detected carbon products at the reactor outlet. These products could arise not only from the conversion or $CO_2$ (or carbonate/bicarbonate) but also from reactions of organic surface contamination by means of photoreforming or Boudouard-like reactions. Actually, these product sources have been shown in some cases to contribute to the main reaction products in the same order of magnitude as the actual $CO_2$ conversion [54]. Exhaustive surface cleaning protocols and the use of ultra-pure reaction conditions help elucidating the contribution of true reactants and spurious carbon sources [48]. In any case, direct proof of $CO_2$ conversion by means of $^{13}C$ isotopic labeling followed by mass spectrometry or infrared spectroscopy analysis becomes key to ascertain the reaction mechanisms [19].

### 8.4.1 $TiO_2$-Based Photocatalysts for $CO_2$ Reduction

As previously indicated, $TiO_2$ is the most widely used photocatalyst in all photocatalytic applications, and $CO_2$ reduction does not make an exception. Actually, $TiO_2$ is the studied or used catalyst in approximately half of the publications on $CO_2$ reduction to date [2, 55].

Titanium is the ninth most abundant element on Earth (it constitutes 0.63% of the Earth's crust), and there are abundant mineral deposits that contain it. For this reason, titanium dioxide is a low-cost material, which is mainly used, in addition to its catalytic and photocatalytic applications, as a white pigment in paints, plastics, and paper. Its low toxicity makes it also suitable for use in products for human consumption, and thus it is used as an excipient in medicines, as a component of high protection factor sunscreens, in toothpaste, and even as a food colorant [56]. From the ecological point of view, its negative environmental effects are only appreciable in the case of massive spills, although in the form of nanoparticles it can hold effects on human health [57]. Under the conditions of a photocatalytic reaction, in addition to its high activity compared to other semiconductors, it is stable against cathodic photocorrosion, and, although thermodynamically it is not against the anodic one, it is already strong since the oxidation reaction of water is more favorable energetically [58]. On the other hand, it is possible to prepare it in a great variety of forms, from single crystals to nanoparticles, which makes it possible to modulate its properties according to technological needs. For these reasons, extensive research has been devoted to optimizing its photocatalytic activity, as we will see in the following section.

TiO$_2$ exists mainly in three crystalline forms: anatase, rutile, and brookite, although the third of them is rare and is not generally used in photocatalysis. Of these three polymorphs, rutile is the thermodynamically stable one. Another structure called TiO$_2$ (B), similar to anatase, but with limited technological importance, has been described. In addition, two high-pressure forms, TiO$_2$ (II) and TiO$_2$ (H), have also been synthesized [56]. TiO$_2$ is an oxide with a strong ionic character, so it can be considered as being formed by Ti$^{4+}$ and O$^{2-}$ ions. In anatase and rutile, the Ti$^{4+}$ ions are surrounded, in their first coordination sphere, by six O$^{2-}$ ions, in distorted octahedral coordination. In anatase, Ti–Ti distances are longer than in rutile, while Ti–O are shorter. Both crystal structures differ in the arrangement of the "TiO$_6$" octahedra.

Regarding its electronic characteristics, TiO$_2$ is an n-type semiconductor due to a small amount of oxygen vacancies that are compensated by the presence of Ti$^{3+}$ centers [56]. In a simplified way, it can be considered that the VB is formed mainly by the overlap of the occupied 2p oxygen orbitals and the CB by the empty 3d orbitals with t$_{2g}$ symmetry of the Ti$^{4+}$. The band gaps of the different polymorphs are different, namely, 3.2 eV for anatase and brookite and 3.0 eV for rutile as a general rule, although size effects and other factors can noticeably affect the electronic properties [59]. Of the two forms of TiO$_2$ commonly used in heterogeneous photocatalysis, anatase and rutile, the first has generally shown greater activity [17, 60, 61], which has been related to a higher concentration of surface hydroxyls, as well as its different band structure with respect to rutile. This increased activity has also been related to the formation of peroxo species on the surface of anatase [62].

In the particular case of CO$_2$ reduction, TiO$_2$ was logically the catalyst employed in early studies. In 1979, Inoue et al. reported for the first time the photocatalytic and photoelectrochemical (PEC) reduction of CO$_2$ in aqueous suspensions of semiconductor powders and on semiconductor electrodes, respectively, to produce formic acid, formaldehyde, and methanol, with the highest QYs obtained with TiO$_2$ and SiC, in the latter case due to the high-energy CB of silicon carbide [11], although the wide band gap of SiC has later made it more interesting as a support rather than a photocatalyst itself [63]. After that seminal work, significant effort has been devoted to the investigation and further improvement of the efficiency of photocatalytic reduction of CO$_2$ on titania [55, 64–66].

As in the rest of applications of photocatalysis, anatase is generally found to be more active than the other TiO$_2$ polymorphs for CO$_2$ reduction [64]. However, the simultaneous presence of two phases can improve the photocatalytic activity with respect to a single anatase phase due to the formation of a heterojunction. Actually, the most commonly used commercial TiO$_2$ photocatalyst, including for CO$_2$ reduction [55], Evonik P25, is formed by a mixture of anatase and rutile, and this phase composition has been invoked for a long time already as one of the reasons behind its high activity [67]. Furthermore, synthetic approaches have been developed in order to obtain highly active CO$_2$ reduction photocatalysts by means of phase composition control. In a simple approximation, the anatase–rutile ratio can be controlled by varying the annealing temperature. For example, Tan et al. examined TiO$_2$ samples obtained by a sol–gel procedure followed by thermal treatment at temperatures from 200 to 800 °C, by which they obtained $TiO_2$ samples with different phase compositions between pure anatase and pure rutile, respectively [68]. The photocatalytic activity for the conversion of $CO_2$ into $CH_4$ showed a dependence of the phase composition in a volcano-type curve with a maximum at 17.5 wt% rutile (Figure 8.4), although other temperature-dependence factors like crystallinity and surface area should be also taken into account. The authors ascribed the results to a predominant role of phase composition, by which a charge transfer from rutile to anatase in a type-II heterojunction (as will be seen in the corresponding Section 8.5 of this chapter) occurs. This phase combination has been also exploited in hierarchical $TiO_2$ nanofibers obtained using by a combination of the sol–gel method and the electrospinning technique [69], in which the phase composition depended on the annealing conditions, with a static Ar atmosphere leading to a higher rutile amount than an argon flow. As demonstrated by electrochemical impedance spectroscopy and photoluminescence measurements, lower electron–hole recombination rates were observed in biphasic samples, which led to higher $CO_2$ photoreduction activities and higher selectivities toward highly electron-demanding products, such as methanol and methane. Anatase and brookite can also lead to high activities, as shown by Reli et al., who observed a better performance in such phase combination than in the anatase–rutile mixture of P25 for the liquid-phase conversion of $CO_2$ into methane, which they ascribed to a combination of factors including the higher activity *per se* of brookite compared with rutile, a large surface area and the electron transfer from the CB of brookite to that of anatase [70].

Control of the morphology of $TiO_2$ can be achieved by means of different synthetic approaches. A plethora of synthesis methods, including sol–gel, microemulsion, hydro- or solvo-thermal, CVD, microwave, etc., have indeed been used to produce different $TiO_2$ nanostructures [71]. Some of these forms have been tested for photocatalytic $CO_2$ reduction by different authors. For example, the previously mentioned electrospun $TiO_2$ nanofibers [69], composed of hierarchically arranged $TiO_2$ nanoparticles interconnected with large interfaces, gave rise to higher conversions than single nanoparticles obtained by the sol–gel method, as a result of a faster charge transport along the nanofiber grain boundaries as shown by photoluminescence and electrochemical impedance spectroscopy measurements. Other 1D $TiO_2$ nanostructures such as nanorods [72] and nanotubes [73] have shown interesting performances for $CO_2$ reduction. In the latter case, $TiO_2$ nanotube arrays, which can be obtained in a highly controlled manner and find a plethora of technological applications [74], have shown a high activity in photocatalytic reduction of $CO_2$ to alcohol, mainly methanol and ethanol, for which the good charge separation across the nanotube growing direction plays an essential role [73]. Similar structures like $TiO_2$ nanocolumns obtained by glancing-angle deposition with sputtering have shown improved properties for other light-activated applications such as photovoltaics [75] and self-cleaning surfaces [76]. In turn, 2D nanostructures like nanosheets have been shown by some authors to be more active than 1D nanotubes or 3D nanoparticles in the reduction of $CO_2$ [77], which is closely related to the preferentially exposed $TiO_2$ facets. Indeed, selectively exposed facets represent another strategy for finely tuning the photocatalytic properties of titanium dioxide. In the

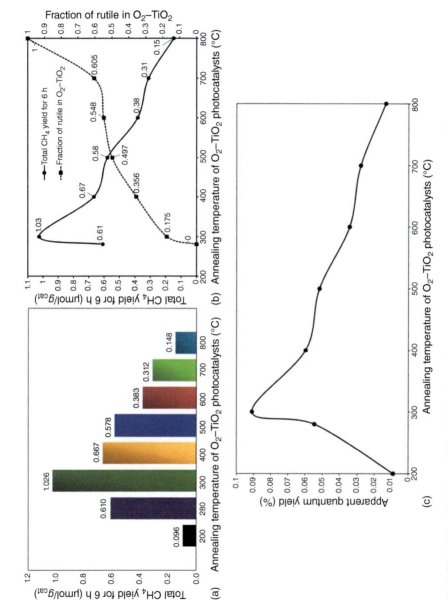

**Figure 8.4** (a) Total $CH_4$ yield attained over $TiO_2$ calcined in $O_2$ at different temperatures, (b) dependence of A–R fraction and photoactivity on the annealing temperature, and (c) apparent quantum yields vs. annealing temperature. Source: Reproduced with permission from Tan et al. [68]. Copyright 2016, Elsevier.

particular case of $CO_2$ reduction, different facets have shown different adsorption, activation, and reactivity toward $CO_2$ and $H_2O$ molecules, both theoretically and experimentally. For example, it has been theoretically shown that the most stable and predominantly exposed (101) facet of anatase has a fundamental role in the adsorption of $CO_2$ and therefore its further reduction by CB electrons [78]. However, it has been also shown that the (001) facets are more reactive in photocatalytic reduction of $CO_2$ due to a higher density of under-coordinated Ti atoms and oxygen-related active sites [79]. Hence, a plausible strategy to enhance activity consists of controlling the exposure of both facets in order to optimize the relations stability–adsorption–reactivity. This can be performed by finely controlling the synthesis conditions as shown by Yu et al., who obtained an optimized production of methane with a 55 : 45 combination of (001) and (101) exposed facets with the formation of a kind of "surface heterojunction" [80].

Metal doping has been a "classical" strategy followed in photocatalysis research to enhance photocatalytic activity [81]. Cation doping provides an interesting synthetic flexibility given that it can modify the structural, electronic, and surface characteristics of the host solid, depending on the characteristics of the guest cation. For example, Ce doping has been shown to change the redox potentials of CB electrons and VB holes [82, 83] and to reduce their recombination [84], with an improved $CO_2$ reduction activity as result. Reduced recombination, with the concomitant improvement in electrons and holes availability and therefore in photoconductivity, has also been reported for Cu-doped titania for $CO_2$ photoreduction [85], and a similar result was reported for Mo-doped $TiO_2$ nanotubes [86]. In $Bi^{3+}$-doped titania, improved charge separation is coupled with higher capacity for $CO_2$ adsorption with respect to pure $TiO_2$ [87], while cobalt doping modifies the surface chemistry as well by provoking oxygen vacancy formation, which results in different product distribution [88]. $In^{3+}$ is also an interesting dopant for $TiO_2$ since it can decrease the electron–hole recombination rate by the induced formation of electronic states related to oxygen vacancies [89]. Tahir and Amin have reported in several papers the activity of In-doped titania for this reaction [90, 91].

Nonmetal doping has been also used to extend $TiO_2$ response toward the visible light and to increase its $CO_2$ conversion activity [92]. For example, N-doped $TiO_2$ materials have been widely used as photocatalysts. Liu and coworkers synthesized nanosheets of N-doped anatase $TiO_2$ using a hydrothermal method that starts from titanium nitride (TiN). The doped catalysts showed considerably higher activity than pure $TiO_2$ for the conversion of $CO_2$ into $CH_4$ [93]. Similarly, Zhang et al. obtained nanocolumns of $TiO_2$ doped with nitrogen by means of a hydrothermal method that used $N_2H_4$ or $NH_3$ as source of nitrogen and indicated that using different nitrogen sources influenced the selectivity of the photoreduction of $CO_2$ [94]. Thus, the main product was $CH_4$ and CO when hydrazine and ammonia were used, respectively, likely due to hydrazine-induced N–N species. Additionally, ordered phosphorus-doped $TiO_2$ nanotubes have been prepared by pyrolytic phosphating and electrochemical oxidation [95]. These structures showed an improvement in charge transfer and light harvesting. P atoms substituted for Ti, incorporating into the crystal lattice and forming Ti—O—P bonds, which decreased electron–hole recombination. In addition, the P-doped material had a narrower band gap, and, as a consequence, a high production of methanol was obtained under visible light, using water as reductant. Sulfur-doped $TiO_2$ anatase has also shown good properties. For example, it can be synthesized by a sonochemical method, by which S atoms are incorporated predominantly in the +4 oxidation state into the crystal lattice [96]. According to theoretical calculations, sulfur doping can induce the formation of intra-band gap states near the VB and improve charge transfer and photocatalytic activity under visible light. Thus, a relatively high production of $CH_4$ and methanol has been achieved, although using triethanolamine as a sacrificial agent [96]. Nanosheets of $TiO_2$ doped with iodine have also been reported by a two-step hydrothermal treatment [97]. After doping, the creation of I—O—Ti and I—O—I bonds was observed, which led to an extension of the photocatalytic response into the visible spectrum. In the gas phase, $CH_4$ and CO were reported as main products.

## 8.4.2 Other Oxide Photocatalysts

Zinc oxide is probably the most studied oxide photocatalyst after $TiO_2$ [98]. ZnO is a wide band gap n-type semiconductor ($E_g = 3.2$ eV). The bottom of the CB and the top of the VB are primarily formed by the Zn 4s orbitals and the oxygen 2p orbitals, respectively. ZnO crystallizes in three possible polymorphs: hexagonal wurtzite, cubic zinc blende, and cubic rock salt, among which the former is the most commonly reported and the most relevant one in photocatalysis. The applications of ZnO in environmental photocatalysis have been extensively studied and reviewed, being anodic photocorrosion and solubility in strong acids and alkalis its main limitations [35]. In the case of $CO_2$ reduction, $Zn_5(OH)_6(CO_3)_2$ has been observed during the reaction, suggesting that ZnO is not very stable as photocatalyst [99]. Indeed, undoped, single-phase ZnO photocatalysts have been rarely reported. Actually, ZnO-based catalysts have exhibited lower $CO_2$ conversions than $TiO_2$-based ones, with CO and $H_2$ as the main reaction products [100]. One of the advantages that ZnO presents is its synthetic versatility, which allows to obtain it in a myriad of nanoshapes that have permitted to establish some relations between structure/morphology and photocatalytic activity [101]. As an example of this versatility, Li et al. synthesized two different types of porous ZnO nanocatalysts by a hydrothermal method, with preferentially exposed (110) and (001) facets depending on the starting precursors. It was found that the porous ZnO catalyst with the exposed (110) facets exhibited a higher photocatalytic activity than the one with the exposed (001) facets, which was mainly ascribed to a synergy between high charge separation efficiency, good $CO_2$ adsorption ability, and effective activation of $CO_2$ [102]. Chemical modifications of ZnO have also been reported to improve $CO_2$ photoreduction ability. For example, it has been found that synthetically induced defects can serve as separation centers to facilitate the separation of photogenerated electrons and holes, as well as the adsorption and activation of $CO_2$ on the ZnO surface [99]. Peroxidation of the ZnO surface has also led to positive results. Guo et al. demonstrated the successful synthesis of

$ZnO_2$-promoted ZnO nanorods, obtained from ZnO by $H_2O_2$ treatment. This surface modification promoted the adsorption of $CO_2$ forming bidentate-$CO_3^{2-}$ and $HCO_3^{-}$. Bidentate-$CO_3^{2-}$ was proposed as the active intermediate of $CH_4$ formation, while $HCO_3^{-}$ species may be more active for $CH_3OH$ production [103]. By contrast, nitrogen doping has shown no significant effect on the $CO_2$ photoreduction activity of zinc oxide [104].

Copper oxides can be particularly relevant in $CO_2$ photoreduction since they tend to promote the formation of C—C bonds and therefore can lead to the obtainment of ethylene or other $C_{2+}$ products [105]. In addition, both cupric (CuO) and cuprous oxide ($Cu_2O$) are direct p-type semiconductors, and their narrow band gaps (1.2 and 2.2 eV, respectively) allow them to absorb a significant fraction of the solar spectrum. However, due to stability issues and fast electron–hole recombination, their relevance is greater in combination with other materials forming heterojunctions (see Section 8.5) than by themselves [106]. Nevertheless, recent studies have demonstrated that these drawbacks may be partially overcome by controlling the morphology at the nanoscale. For example, nano-foam-shaped $Cu_2O$ showed photocatalytic activity to convert $CO_2$ into acetaldehyde and methane [107]. Furthermore, the influence of the exposed facets in $Cu_2O$ on activity and on selectivity against the competing water reduction was demonstrated by Handoko and Tang [108], who showed that $Cu_2O$ particles with exposed (100) facets were more selective toward $CO_2$ reduction while the $Cu_2O$ (111) facets predominantly favored $H_2$ production. The authors ascribed this to the different adsorption of $CO_2$ onto different facets. Further insight was given by Yang et al. [109], who demonstrated that the surface coverage of hydroxyl groups in (111) facets played an essential role on this selectivity.

$CeO_2$ has received attention as a photocatalyst thanks to its interesting properties: stability under illumination and strong absorption of both UV and visible light [35]. Another known advantage of cerium oxide, namely, its oxygen storage and release capacity in the $Ce^{4+}/Ce^{3+}$ redox cycle, has been taken advantage of for $CO_2$ photoreduction. In this regard, Zhu et al. have studied the synergy between oxygen defects and hydroxyl groups in this reaction over $CeO_2$ (110) and $CeO_2$ (100) surfaces [110], proving, by means of experimental results and density functional theory (DFT) calculations, that the synergistic interactions of Lewis acidity and basicity, i.e. oxygen defects and hydroxyl groups, enhance the $CO_2$ photoreduction activity in $CeO_2$ (110) compared with $CeO_2$ (100), which is ascribed to the preferential generation of carboxylate species and $CO_2^{\cdot-}$ radicals in the former instead of carbonate in the latter. In a different approach, Wang et al. stabilized oxygen vacancies in $CeO_2$ by means of Cu doping [111]. According to analyses by Raman and X-ray photoelectron spectroscopies, they evidenced that Cu introduction benefits the chemical stabilization of O vacancies in $CeO_2$ during photocatalytic $CO_2$ reduction, which in turn promotes the electron–hole separation and transfer, prolonging the charge carriers lifetime, and eventually provides more active sites for adsorbed $CO_2$, which results in the enhanced activity and stability of $Cu-CeO_{2-x}$. A reaction mechanism was proposed on the basis of in situ FTIR spectra. Similarly, iron [112] and chromium [113] doped cerium oxides, obtained by a nanocasting method, have been reported

## 8.4 Photocatalytic $CO_2$ Reduction: State of the Art and Main Current Issues

to show extended spectral response and enhanced $CO_2$ photoreduction activity with respect to pure ceria.

Interesting semiconductor photocatalysts for $CO_2$ reaction are ternary oxides with perovskite structure, like alkaline niobates and tantalates, which possess a CB structure with sufficiently high energy to transfer electrons to the $CO_2$ molecule and, at the same time, a VB with low enough energy to promote water oxidation. Some works have revealed the potential of these structures for $CO_2$ photocatalytic reduction. Both $NaNbO_3$ and $NaTaO_3$ possess interesting intrinsic activities for the photocatalytic reduction of $CO_2$ under UV irradiation, and, interestingly, they drive the competition between $CO_2$ and $H_2O$ for CB electrons toward the reduction of carbon dioxide [114]. Ye's group have extensively studied both Nb- and Ta-based mixed oxides for $CO_2$ photoreduction. In the case of tantalum, a comparison of lithium, sodium, and potassium tantalates shows that the activity for the production of methane from $CO_2$ and $H_2O$ grows in the series K < Na < Li, which is the same order as the band gap and the energy of the CB, suggesting that the reduction power of the CB electrons is the determining factor for activity [115]. Nanostructured $NaNbO_3$ has shown superior performance than bulk one. Thus, $NaNbO_3$ nanowires exhibited a much higher photocatalytic activity for $CH_4$ production than samples prepared by a standard solid state reaction method, possibly due to crystallinity, surface-to-volume ratio, and anisotropic aspect [116]. As for the different crystalline phases of $NaNbO_3$, samples with cubic perovskite structure synthesized at low temperature by an organic ligand induced process have shown higher activity than high-temperature, orthorhombically distorted ones [117]. Nevertheless, these metallates present the limitation of very wide band gaps and therefore very limited light absorption within the solar spectrum. Introducing cations with filled outer shells instead of alkali or alkaline earth metals rises the VB position, leading to lower band gaps while keeping the CB reduction potential. With this approach, $CO_2$ reduction activity under simulated solar light has been demonstrated in $InTaO_4$, which absorbs wavelengths below 460 nm [118], and in $InNbO_4$, with an absorption onset at 475 nm [119]. Other visible light-absorbing metallates that have received attention for $CO_2$ photo-conversion are $Bi_2WO_6$ and $BiVO_4$. The former has been synthesized in a variety of nanoshapes that have shown interesting implications of morphology in photocatalytic activity. For instance, hydrothermally synthesized hierarchical microspheres formed by $Bi_2WO_6$ nanoplatelets have shown activity under visible light (420 nm < $\lambda$ < 620 nm) to obtain mainly CO from a dilute $CO_2$ stream and water [120], which authors ascribed to enhanced crystallinity, reduced recombination rate, and strengthened $CO_2$ adsorption sites. Further enhancement of the activity of this kind of structures has been later achieved by employing as catalyst support an activated carbon with high $CO_2$ adsorption capacity [121]. $Bi_2WO_6$ nanosheets with predominantly exposed (001) facets have been obtained by Kong et al. by means of a topotactic transformation of BiOBr through a hydrothermal anion exchange reaction with $Na_2WO_4$, during which 2D nanostructure and dominantly exposed facets were preserved [122]. These structures showed a ninefold improvement of $CH_4$ production with respect to thicker $Bi_2WO_6$ nanoplates synthesized in a one-step hydrothermal process, which

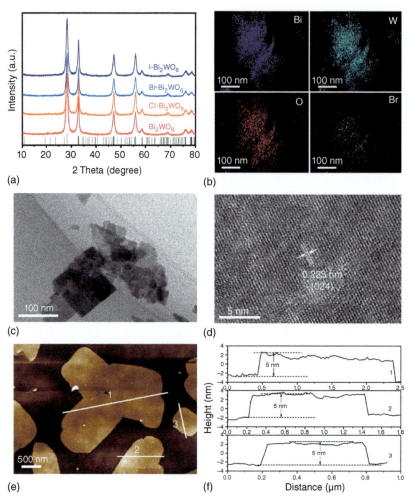

**Figure 8.5** (a) XRD patterns of X-$Bi_2WO_6$ (X = Cl, Br, and I), (b) EDX elemental maps, (c) TEM image and (d) high-resolution TEM image of Br-$Bi_2WO_6$, (e) AFM image of Br-$Bi_2WO_6$, and (f) the corresponding thickness profiles. Source: Reproduced from Liu et al. [124]. Copyright, Wiley-VCH.

the authors attributed to several synergistic factors such as facet reactivity, increased specific surface area, and simultaneous band gap narrowing and CB upshifting. Hydrophobicity was introduced by Liu et al. as an additional factor leading to improved photo-methanation of $CO_2$ on atomically thin $Bi_2WO_6$ nanosheets under simulated solar radiation [123]. Surface halogenation of this kind of 2D $Bi_2WO_6$ nanostructures (Figure 8.5) has been reported by Liu et al. to accelerate photoexcited charge separation and transport, optimize the process of adsorption and activation of $CO_2$ molecules, and facilitate the $CO_2$ conversion [124]. DFT calculations showed that the surface inclusion of halogen atoms on $Bi_2WO_6$ can lead to a decreased formation energy of the COOH reaction intermediate.

Known as a good water oxidation catalyst [3], $BiVO_4$ is most commonly reported in $CO_2$ photoreduction as a component of heterojunction systems (see Section 8.5). However, some works have reported the activity of this mixed oxide by itself. For instance, Mao et al. reported the preparation of lamellar $BiVO_4$ through a surfactant-assisted hydrothermal process [125]. The obtained monoclinic $BiVO_4$ was used as photocatalyst for the reduction of $CO_2$ and exhibited a selective methanol production under visible light irradiation. This kind of $BiVO_4$ structures have also been reported to induce the concurrent formation of methanol and acetone in the liquid-phase photoreduction of $NaHCO_3$ aqueous solutions [126]. Furthermore, single-unit-cell $BiVO_4$ layers with induced vanadium vacancies have been shown to be highly efficient and stable for the photo-production of methanol from $CO_2$, owing to increased photoabsorption and superior electronic conductivity, as revealed by DFT calculations, which results in higher surface photovoltage intensity and hence higher carriers separation efficiency, confirmed by increased carriers lifetime revealed by time-resolved fluorescence emission decay measurements [127].

Adding to visible-light photocatalytic activity the practical possibility of easy catalyst separation in aqueous systems, magnetic iron mixed oxides such as ferrites were early reported as $CO_2$ photoreduction catalysts, although they have received less attention than other oxides [128]. Methanol formation has been reported for $CaFe_2O_4$ [129] and with complex ferrites containing magnesium, strontium, barium, lead, and bismuth [130], while CB electrons in photo-activated $ZnFe_2O_4$ have been utilized to reduce aqueous bicarbonate into mainly acetaldehyde and ethanol while using triethanolamine as a sacrificial agent to scavenge photoproduced holes [131].

## 8.5 Oxide-Based Heterojunctions and Z-Scheme Photocatalytic Systems

Provided that the relative positions of their valence and CBs are adequate and the contact between them is intimate enough, the combination of two semiconductors with different band structures can lead to a band alignment and a physical separation of photogenerated electrons and holes by means of charge carrier transfer across the interface. That is, a so-called *heterojunction* can be formed. This phenomenon has been largely explored in photocatalysis research for all kinds of applications [132]. Depending on the relative band positions between both components, different types of heterojunctions can be formed, as illustrated in Figure 8.6.

Nevertheless, for the particular case of energy applications of photocatalysis, where solar-to-chemical energy storage is intended, these systems present the drawback that the redox potentials of electrons and holes are somewhat sacrificed for the sake of charge separation, since electrons migrate to a lower energy band and holes to a higher energy band, losing part of their reductive and oxidative power, respectively. In order to overcome this limitation, the so-called Z-scheme photocatalytic system was proposed. This name derives from that given to the movement of electrons that occurs in natural photosynthesis, in which the oxidation of

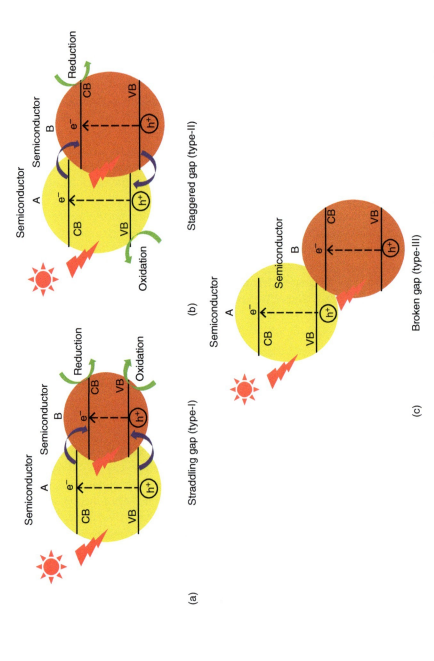

**Figure 8.6** Schematic depiction of the different types of band alignment and charge carrier separation in conventional heterojunction photocatalysts: (a) type-I, (b) type-II, and (c) type-III heterojunctions. Source: Reproduced with permission from Low et al. [132]. Copyright 2017, Wiley-VCH.

## 8.5 Oxide-Based Heterojunctions and Z-Scheme Photocatalytic Systems | 297

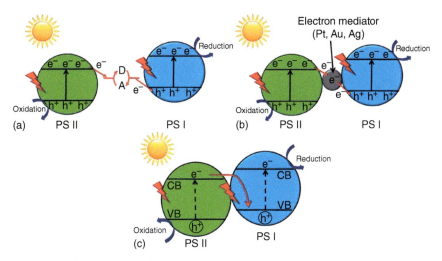

**Figure 8.7** Three types of Z-scheme photocatalytic systems: (a) solution electron mediator, (b) all-solid Z-scheme system, and (c) direct Z-scheme. Source: Reproduced with permission from Low et al. [132]. Copyright 2017, Wiley-VCH.

water occurs in photosystem 2, and electrons arrive through the transport chain to photosystem 1, where NADP$^+$ is reduced to NADPH, which in turn passes into the Calvin cycle [133]. As seen in Figure 8.7, even if the band structure is similar to the type II depicted in Figure 8.6, the charge transfer occurs in a different way. Thus, the electrons in the CB of with lower energy recombine with the VB holes of higher energy, while maintaining the positions of the higher-energy electrons and lower-energy holes. This maintains the strongest redox power of electrons and holes, improving photocatalytic performance in comparison to conventional heterojunctions [134].

Again, three types of Z-scheme systems appear, this time depending on how electron–hole recombination between one semiconductor and the other occurs. Initially proposed systems made use of a redox mediator (e.g. $IO_3^-/I^-$, $Fe^{3+}/Fe^{2+}$) in solution, with no physical contact between both semiconductors (Figure 8.7a), by imitation of the previously mentioned electron transport chain of natural photosynthesis. However, even if a very efficient charge separation can be attained with this system, it can only be used in liquid-phase reactions, limiting its applications, and it implies the use of potentially costly and harmful chemicals that can even eventually suffer from back reactions. As a consequence, later developments introduced the concept of all-solid Z-scheme systems, in which the electron mediator is a conducting solid rather than a redox pair in solution [135], as depicted in Figure 8.7b. Electron mediators can be metal nanoparticles [135] or other conducting materials such as graphene or reduced graphene oxide (RGO) [136].

As mentioned earlier, many different semiconductor couples have been utilized to form heterojunctions in general photocatalysis applications [137]. In the particular case of $CO_2$ photoreduction, several oxide/oxide and oxide/non-oxide systems have been reported. As mentioned in Section 8.5, copper oxides can be particularly relevant in $CO_2$ photoreduction, especially when forming heterojunction systems. Thus, $Cu_2O$ and $TiO_2$ have such relative band positions that they can form a direct Z-scheme for charge separation in which reduction takes place at the highly reducing CB of the copper oxide while oxidation occurs via the highly oxidative holes in the VB of titania [138]. The coupling of $Cu_xO$ with $TiO_2$ has been shown not only to improve charge separation, but also to induce modifications in morphology, surface chemistry, and surface area that tend to facilitate $CO_2$ adsorption. For instance, CuO hollow nanocubes, obtained via a multi-template synthesis starting from $Cu_3N$ and coupled to $TiO_{2-x}N_x$, have led to improved methane production from $CO_2$ and water [139]. Interestingly, $Cu_2O$ can modify the surface properties of $Pt/TiO_2$ catalysts in such a way that the reaction between $CO_2$ and $H_2O$ can be directed selectively toward the conversion of $CO_2$ into $CH_4$ while suppressing the parallel formation of hydrogen from water even in the presence of a paradigmatic $H_2$ evolution catalyst like Pt [140]. Copper oxides also form an interesting binary system with $BiVO_4$ in which the good water oxidation properties of the latter are combined with the reduction power of the former. This has been exploited for the formation of an all-solid-state Z-scheme heterojunction with a conductive carbon layer as electron mediator [141]. $BiVO_4$, as mentioned in the Section 8.5, is actually an interesting heterojunction component due to this water oxidation ability, to which its absorption in the visible range can be added [142]. In addition to copper oxides, its combination with other photocatalysts to obtain $CO_2$ reduction Z-scheme systems like $CuGaS_2/RGO/BiVO_4$ [143], CdS/BiVO4 [144], $SrTiO_3/BiVO_4$ [145], and $C_4N_3/BiVO_4$ [146] has been reported. Ferrites constitute another class of mixed oxides that have been studied in the form of heterojunctions for AP. Different ferrites like $Fe_3O_4$, $CaFe_2O_4$, $MgFe_2O_4$, $CuFe_2O_4$, $NiFe_2O_4$, $ZnFe_2O_4$, or $BiFeO_3$ have been studied for this purpose, forming heterostructures between them or in combination with simpler oxides like $TiO_2$ or ZnO, in most of the cases with improved activity with respect to the single photocatalysts due to reduced recombination rates, as usually shown by steady-state or time-resolved photoluminescence spectroscopy [128]. In addition to electronic modifications, and particularly for liquid-phase reactions, ferrite include the possibility of easy catalyst recovery due to their magnetic properties.

Graphitic carbon nitride, $g-C_3N_4$, has gained much interest in recent years as a $CO_2$ reduction photocatalyst [147], and, due to the high reduction potential of its CB and its absorption in the visible region, it is an interesting component of heterojunctions in combination with oxide photocatalysts. Coupled with $TiO_2$, $g-C_3N_4$ forms a type-II heterojunction that has been exploited by several groups to improve charge separation in $CO_2$ photoreduction [148]. For instance, Dehkordi et al. obtained hierarchical $g-C_3N_4/TiO_2$ hollow spheres with high visible light activity for the reduction of $CO_2$ into mainly methanol [149]. Photoluminescence measurements together with UV–vis spectra and PEC experiments have shown improved charge separation in the $C_3N_4/TiO_2$ systems by means of the formation of a type-II heterojunction [150]. This type of heterojunctions has been also reported for $g-C_3N_4$ coupled to other metal oxides for application to photocatalytic $CO_2$ reduction. For example, a type-II heterojunction forms between ZnO and graphitic carbon nitride, which not only minimizes the recombination of photogenerated electrons and holes, but also improves the light absorption range and the adsorption of $CO_2$ on ZnO, giving rise to improved photocatalytic activity for the conversion of $CO_2$ into CO and $CH_4$ [151]. Coupled to metal oxides of different electronic structure, $C_3N_4$ can lead to the formation of a direct Z-scheme heterojunction system. Examples with binary oxides with high oxidative power of VB holes such as $Fe_2O_3$ [152], $WO_3$ [153], $CeO_2$, and $V_2O_5$ [154] have been reported, as well as with ternary oxides like $BiVO_4$ [155], $FeWO_4$ [156], or $La_2Ti_2O_7$ [157], in all cases with improved catalytic performance and/or extended activity spectrum with respect to the single components.

### 8.5.1 Cocatalysts for $CO_2$ Reduction: Metal-Oxide Synergies

Cocatalysts have a fundamental role in $CO_2$ photoreduction, since they catalyze surface reactions, favor charge separation and transfer, improve the stability of the catalyst, inhibit reverse reactions, and, in addition, modify the selectivity of the reaction [158]. Indeed, the versatility that selectivity control provides to the process can be one of the most attractive characteristics of photocatalysis compared to other $CO_2$ conversion pathways. In this respect, the addition of metal nanoparticle as cocatalysts on the semiconductor surface has been proven to modify selectivity, which could be later tuned by modifying the chemistry, size, or shape of the nanoparticles [159]. Thus, some metals, like platinum, gold, silver, or copper, to mention only the most relevant ones, have been described as selectivity modifiers with respect to the naked semiconductor and usually lead to the formation of carbon products in a highly reduced state like methanol [160–162] or methane [100, 163, 164]. This has been traced back not only to improved electron transfer [106, 165], but also to modifications in the chemistry of the surface active sites [166, 167].

Among the metals mentioned previously, silver is considered as an interesting case due to its relatively low price and its photophysical and chemical properties, which may alter not only the surface characteristics of the photocatalytic material, by playing the roles of catalytic active site and electron withdrawal and transfer center [165, 168], but also by providing localized surface plasmon resonance (LSPR) properties that can positively influence the light harvesting ability of the system [100, 161]. Regarding electronic modifications and their effect on catalytic activity and selectivity, in a study that combines *operando* spectroscopic characterization with theoretical calculations, an electronic mechanism for the preferential formation of methane on $Ag/TiO_2$, vs. carbon monoxide on $TiO_2$, has been proposed. This mechanism is based on the transfer of charge from the semiconductor to the metal and on the formation of surface electronic states [165]. These intra-gap electronic states would also be responsible for the activity observed under visible light, which, differently to the case of UV irradiation, led to the preferential formation of methanol. This difference in selectivity between UV and visible irradiation has been later observed on $Pt/C_3N_4$ catalysts too and has been related to the possibility that excitation with higher photon energy may favor the transfer of a higher number of electrons simultaneously, driving the reaction toward the highest reduced product, $CH_4$, under UV light [147]. Some authors have even pointed at the possibility of tuning the selectivity between carbon monoxide and methane by modifying the silver loading, as it has been reported in the case of brookite nanocubes with exposed (210) and (001) facets [168]. Selectivity enhancement toward hydrocarbons vs. CO by using silver nanoparticles as cocatalyst has also been reported in ZnO photocatalysts [100]. The selectivity toward methane when using silver nanoparticles as cocatalysts has been also related, in addition to electronic and charge transfer effects, to the modification of the surface chemistry and adsorption properties of the semiconductor, in this case sodium niobate. Thus, Ag/NaNbO$_3$ catalyst produces mainly methanol, while NaNbO$_3$ gives rise to CO as the major product [169]. Adsorbed CO was observed by in situ DRIFTS on Ag/NaNbO$_3$, but not on bare NaNbO$_3$, revealing that this product shows a stronger interaction with the surface of the catalyst in the presence of the silver cocatalyst. This interaction was proposed to induce further reduction of CO and shift the selectivity of the reaction toward more reduced compounds (Figure 8.8).

Gold nanoparticles have also shown attractive properties as cocatalysts for photocatalytic CO$_2$ reduction. On the one hand, the Fermi level of gold is close to the CB energy of TiO$_2$, which favors the role of the former as electron sink and transfer center, thus promoting the formation of highly reduced products; on the other hand, the LSPR band of gold nanoparticles lies in the visible light range, rendering it possible to employ this effect to extend the photocatalytic activity to visible wavelengths [170]. LSPR excitation of Au nanoparticles on TiO$_2$ has thus been studied by several groups. Under UV irradiation, the large improvements in activity with respect to bare TiO$_2$ have been assigned to different possible mechanisms. On the one hand, improved charge carrier separation due to electron transfer from TiO$_2$ to metal NPs has been identified by both in situ LSPR spectroscopy measurements and laser-based microsecond–second transient absorption decay recordings [163]. This electron transfer gives rise to improved activity not only in terms of CO$_2$ conversion, but also of enhanced selectivity toward highly reduced products such as methane [163]. In contrast, under visible light, rather an opposite mechanism seems to operate, where electrons arising from the plasmonic excitation of Au nanoparticles may be transferred to TiO$_2$ to initiate the photocatalytic reactions [171]. In any case, the activities found for Au/TiO$_2$ photocatalysts under visible light are considerably lower than those under UV light [163], and the best photocatalytic results, for both CO$_2$ reduction and other reactions, are found when the plasmon resonance wavelength of the metal nanoparticles matches the band gap energy of the semiconductor support [172, 173], which would be related to a plasmon-induced improvement of light absorption rather to charge transfer from the metal to the semiconductor.

Known as a very active metal in catalysis, platinum has been extensively studied as cocatalyst for photocatalytic CO$_2$ reduction, particularly, but not only, supported on titania [159]. Among noble metals, platinum displays the highest work function and thus the lowest Fermi level and the largest capacity for electron extraction. Generally, the increased activity obtained with Pt/TiO$_2$ systems compared to unmodified titania is traced back to improved electron lifetime due to this ability of Pt nanoparticles to withdraw the photoexcited electrons from the TiO$_2$ CB. This has been actually proven with photoluminescence and transient absorption

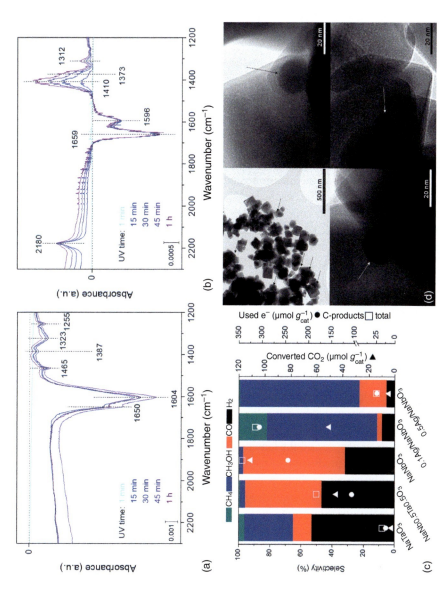

**Figure 8.8** (a)–(d) Difference in situ DRIFT spectra of $NaNbO_3$ and $0.1Ag/NaNbO_3$ catalysts in a $CO_2$ and $H_2O$ atmosphere under different UV irradiation times. Product selectivities (colored bars), $CO_2$ conversions (triangle symbols), and electron use (circle symbols: in C-products; open square symbols: in C-products + $H_2$) in photocatalytic reactions with different Nb and Ta-based catalysts with or without silver cocatalyst. TEM images of a $0.1Ag/NaNbO_3$ catalyst. Source: Reproduced with permission from Fresno et al. [169]. Copyright 2021, Elsevier.

spectroscopies [174–177]. The most prominent catalytic feature of platinum in photocatalytic $CO_2$ conversion is its selectivity toward methane against other products like CO or methanol. However, compared with other metals for which the mentioned electron withdrawal has been proved, platinum has shown an exceptional selectivity to methane [159, 177]. In this regard, additional aspects to those based on purely electronic modifications have been proposed to play a key role, such as surface chemistry and adsorption properties [177, 178]. Similarly to the case of silver mentioned earlier, a study combining infrared spectroscopy after the photocatalytic reaction and in situ high-pressure X-ray photoelectron spectroscopy proposed a mechanism to explain the outstanding methane selectivity of platinum cocatalysts, based on the known strong chemisorption of carbon monoxide on Pt, which indeed leads usually to its deactivation in catalytic reactions. This chemisorption would favor further reduction of CO until producing methane, in contrast to the case of bare titania catalysts in which this initially formed CO becomes the major product [166]. However, even if Pt is the most studied cocatalyst for photocatalytic $CO_2$ reduction and shows the mentioned interesting characteristics, it presents the drawback of being as well the best hydrogen evolution catalyst, which means that it uses a considerable fraction of the withdrawn electrons to reduce $H^+$ to $H_2$ instead of reducing $CO_2$ in the $CO_2 + H_2O$ reaction. Several strategies, like controlling the size, location, and facets of Pt nanoparticles, or loading them with a $CuO_x$ layer, have been developed in order to circumvent this drawback and finely tune the selectivity in $CO_2$ conversion [158].

As mentioned in the preceding sections, copper is a very interesting metal for photocatalytic $CO_2$ conversion, and this involves its use not only as the sole photocatalyst or as a component of heterojunctions, but also as cocatalyst for other semiconductors. Thus, it has been loaded in the form of nanoparticles on different semiconducting oxides like $TiO_2$, ZnO, $ZrO_2$, titanates, or niobates, again with $TiO_2$ as the most studied support [159]. Apart from the high natural abundance of copper compared with noble metals, its most prominent property for $CO_2$ reduction is the selectivity control than can be exerted by controlling the chemistry of the nanoparticles. Thus, $CuO_x$ cocatalysts tend to favor the formation of CO or methanol depending on the oxidation state of copper and the loading amount [104, 179, 180]. In turn, metallic copper can favor the production of methane or even the formation of longer-chain hydrocarbons by carbon–carbon coupling. Thus, Cook et al. reported PEC $CH_4$ formation on Cu–SiC electrodes immersed in a $KHCO_3$ electrolyte saturated with $CO_2$ under UV irradiation [181]. Methane evolution became larger with more acidic electrolyte, and in the same conditions, ethylene and ethane were also observed though with considerably lower formation rates. In purely photocatalytic conditions, Adachi et al. reported the generation of methane and ethylene over Cu–$TiO_2$ suspended in $CO_2$-saturated water under irradiation with a Xe lamp [182], while Park et al. explored a ternary photocatalyst composed of metallic copper and cadmium sulfide quantum dots supported on sodium titanate nanotubes for selectively producing C1–C3 hydrocarbons from carbon dioxide and water [183]. In order to avoid copper oxidation and keep the mentioned selectivity to hydrocarbons, Cu nanoparticles have been alloyed with palladium, the activity and selectivity depending on the relative content of the two metals [184]. Additionally, coating copper nanoparticles with black titania has also been attempted [185].

Other metals in addition to the ones highlighted previously deserve to be mentioned as $CO_2$ reduction cocatalysts. Palladium has been studied in a number of works, with selectivity toward carbon monoxide, methane, or even higher hydrocarbons depending on the type of photocatalytic system and the reaction conditions [159]. Ruthenium is also an interesting metal for this application. For instance, in the specific $Ru/TiO_2$ systems, $RuO_x$ particles are generally found as good oxidation catalysts, while rather metallic species tend to promote reduction reactions [186, 187], due to their different band alignment with respect to $TiO_2$ [188].

## 8.6 Conclusions and Future Perspectives

In conclusion, photocatalytic $CO_2$ reduction is a promising technology in the medium–long term that nevertheless will need a breakthrough in order to be fully developed. In our opinion, this breakthrough will be brought about by the development of highly efficient, stable, and selective photocatalysts, as well as by the full understanding the mechanisms underpinning the reaction pathways and the main bottlenecks that still prevent the process from being totally understood and developed. The last matter has necessarily to be faced by integrated in situ and *operando* time resolved spectroscopies in combination with theoretical studies. Even if other types of semiconductors, like $C_3N_4$ or sulfides, have attracted a great deal of attention in the last years, oxide-based photocatalysts appear as a safe bet due to their synthetic versatility and their long-term stability. Decoration with suitable cocatalysts allows to tune the selectivity of the reaction, which is one of the characteristics that can convert photocatalytic $CO_2$ conversion into an attractive and competitive process for the sustainable production of energy vectors and chemicals. Given the thermodynamic restrictions that govern water oxidation and $CO_2$ reduction, the possibility of combining two different semiconductors, each specialized in one of the half-reactions, will make Z-scheme photocatalytic systems a very promising approach to reach the target efficiency. Finally, for $CO_2$ photoreduction to be fully developed, it is imperative that the scientific community agrees on a series of operational criteria that allow a better sharing of results and facilitate future decision making.

## References

1 Rogelj, J., Den Elzen, M., Höhne, N. et al. (2016). Paris agreement climate proposals need a boost to keep warming well below 2 °C. *Nature* 534 (7609): 631–639.
2 Fresno, F., Villar-García, I.J., Collado, L. et al. (2018). Mechanistic view of the main current issues in photocatalytic $CO_2$ reduction. *J. Phys. Chem. Lett.* 9 (24): 7192–7204.

3 House, R.L., Iha, N.Y.M., Coppo, R.L. et al. (2015). Artificial photosynthesis: where are we now? Where can we go? *J. Photochem. Photobiol., C* 25: 32–45.

4 Nocera, D.G. (2017). Solar fuels and solar chemicals industry. *Acc. Chem. Res.* 50 (3): 616–619.

5 Cuéllar-Franca, R.M. and Azapagic, A. (2015). Carbon capture, storage and utilisation technologies: a critical analysis and comparison of their life cycle environmental impacts. *J. CO2 Util.* 9: 82–102.

6 Dufour, J., Serrano, D.P., Gálvez, J.L. et al. (2012). Life cycle assessment of alternatives for hydrogen production from renewable and fossil sources. *Int. J. Hydrogen Energy* 37 (2): 1173–1183.

7 Xie, S., Zhang, Q., Liu, G., and Wang, Y. (2016). Photocatalytic and photoelectrocatalytic reduction of $CO_2$ using heterogeneous catalysts with controlled nanostructures. *Chem. Commun.* 52: 35–59.

8 Kim, W., Edri, E., and Frei, H. (2016). Hierarchical inorganic assemblies for artificial photosynthesis. *Acc. Chem. Res.* 49 (9): 1634–1645.

9 Bai, S., Yin, W., Wang, L. et al. (2016). Surface and interface design in cocatalysts for photocatalytic water splitting and $CO_2$ reduction. *RSC Adv.* 6 (62): 57446–57463.

10 Fujishima, A. and Honda, K. (1972). Electrochemical photolysis of water at a semiconductor electrode. *Nature* 238 (5358): 37–38.

11 Inoue, T., Fujishima, A., Konishi, S., and Honda, K. (1979). Photoelectrocatalytic reduction of carbon dioxide in aqueous suspensions of semiconductor powders. *Nature* 277 (5698): 637–638.

12 Roger, I., Shipman, M.A., and Symes, M.D. (2017). Earth-abundant catalysts for electrochemical and photoelectrochemical water splitting. *Nat. Rev. Chem.* 1 (1): 3.

13 Fu, J., Yu, J., Jiang, C., and Cheng, B. (2018). g-$C_3N_4$-Based heterostructured photocatalysts. *Adv. Energy Mater.* 8 (3): 1–31.

14 Young, J.L., Steiner, M.A., Döscher, H. et al. (2017). Direct solar-to-hydrogen conversion via inverted metamorphic multi-junction semiconductor architectures. *Nat. Energy* 2 (4): 17028.

15 Zhou, X., Liu, R., Sun, K. et al. (2016). Solar-driven reduction of 1 atm of $CO_2$ to formate at 10% energy-conversion efficiency by use of a $TiO_2$-protected III–V tandem photoanode in conjunction with a bipolar membrane and a Pd/C cathode. *ACS Energy Lett.* 1 (4): 764–770.

16 Braslavsky, S.E., Braun, A.M., Cassano, A.E. et al. (2011). Glossary of terms used in photocatalysis and radiation catalysis (IUPAC recommendations 2011). *Pure Appl. Chem.* 83 (4): 931–1014.

17 Linsebigler, A.L., Lu, G., and Yates, J.T. (1995). Photocatalysis on $TiO_2$ surfaces: principles, mechanisms, and selected results. *Chem. Rev.* 95 (3): 735–758.

18 Herrmann, J.M. (2010). Photocatalysis fundamentals revisited to avoid several misconceptions. *Appl. Catal., B* 99 (3–4): 461–468.

19 Melchionna, M. and Fornasiero, P. (2020). Updates on the roadmap for photocatalysis. *ACS Catal.* 10 (10): 5493–5501.

20 Zhang, L., Dillert, R., Bahnemann, D., and Vormoor, M. (2012). Photo-induced hydrophilicity and self-cleaning: models and reality. *Energy Environ. Sci.* 5 (6): 7491.

21 Rodríguez-González, V., Obregón, S., Patrón-Soberano, O.A. et al. (2020). An approach to the photocatalytic mechanism in the $TiO_2$-nanomaterials microorganism interface for the control of infectious processes. *Appl. Catal., B* 270: 118853.

22 Zhang, C., Li, Y., Shuai, D. et al. (2019). Progress and challenges in photocatalytic disinfection of waterborne viruses: a review to fill current knowledge gaps. *Chem. Eng. J.* 355: 399–415.

23 Xu, X., Pliego, G., Zazo, J.A. et al. (2017). An overview on the application of advanced oxidation processes for the removal of naphthenic acids from water. *Crit. Rev. Env. Sci. Technol.* 47: 1–34.

24 Boyjoo, Y., Sun, H., Liu, J. et al. (2017). A review on photocatalysis for air treatment: from catalyst development to reactor design. *Chem. Eng. J.* 310: 537–559.

25 Rizzo, L., Malato, S., Antakyali, D. et al. (2019). Consolidated vs new advanced treatment methods for the removal of contaminants of emerging concern from urban wastewater. *Sci. Total Environ.* 655: 986–1008.

26 Parrino, F., Bellardita, M., García-López, E.I. et al. (2018). Heterogeneous photocatalysis for selective formation of high-value-added molecules: some chemical and engineering aspects. *ACS Catal.* 8 (12): 11191–11225.

27 Almquist, C.B. and Biswas, P. (2001). The photo-oxidation of cyclohexane on titanium dioxide: an investigation of competitive adsorption and its effects on product formation and selectivity. *Appl. Catal., A* 214 (2): 259–271.

28 Baghbanzadeh, M., Glasnov, T.N., and Kappe, C.O. (2013). Continuous-flow production of photocatalytically active titanium dioxide nanocrystals and its application to the photocatalytic addition of N,N-dimethylaniline to N-methylmaleimide. *J. Flow Chem.* 3 (4): 109–113.

29 Lv, K. and Lu, C.S. (2008). Different effects of fluoride surface modification on the photocatalytic oxidation of phenol in anatase and rutile $TiO_2$ suspensions. *Chem. Eng. Technol.* 31 (9): 1272–1276.

30 Liu, L., Gu, X., Ji, Z. et al. (2013). Anion-assisted synthesis of $TiO_2$ nanocrystals with tunable crystal forms and crystal facets and their photocatalytic redox activities in organic reactions. *J. Phys. Chem. C* 117 (36): 18578–18587.

31 Zhang, B., Li, J., Zhang, B. et al. (2015). Selective oxidation of sulfides on $Pt/BiVO_4$ photocatalyst under visible light irradiation using water as the oxygen source and dioxygen as the electron acceptor. *J. Catal.* 332: 95–100.

32 Ghosh, I., Khamrai, J., Savateev, A. et al. (2019). Organic semiconductor photocatalyst can bifunctionalize arenes and heteroarenes. *Science* 365 (6451): 360–366.

33 Stanchak, J. (2016). *10 Big Ideas No Scientist Can Afford to Ignore*. ACS Axial.

34 Maeda, K. and Domen, K. (2010). Photocatalytic water splitting: recent progress and future challenges. *J. Phys. Chem. Lett.* 1 (18): 2655–2661.

**35** Hernández-Alonso, M.D., Fresno, F., Suárez, S., and Coronado, J.M. (2009). Development of alternative photocatalysts to $TiO_2$: challenges and opportunities. *Energy Environ. Sci.* 2 (12): 1231.

**36** Wang, Q., Hisatomi, T., Jia, Q. et al. (2016). Scalable water splitting on particulate photocatalyst sheets with a solar-to-hydrogen energy conversion efficiency exceeding 1%. *Nat. Mater.* 15 (6): 611–615.

**37** Granone, L.I., Sieland, F., Zheng, N. et al. (2018). Photocatalytic conversion of biomass into valuable products: a meaningful approach? *Green Chem.* 20: 1169–1192.

**38** Wakerley, D.W., Kuehnel, M.F., Orchard, K.L. et al. (2017). Solar-driven reforming of lignocellulose to $H_2$ with a $CdS/CdO_x$ photocatalyst. *Nat. Energy* 2 (4): 17021.

**39** Butburee, T., Chakthranont, P., Phawa, C., and Faungnawakij, K. (2020). Beyond artificial photosynthesis: prospects on photobiorefinery. *ChemCatChem* 12: 1–19.

**40** Luo, N., Montini, T., Zhang, J. et al. (2019). Visible-light-driven coproduction of diesel precursors and hydrogen from lignocellulose-derived methylfurans. *Nat. Energy* 4 (7): 575–584.

**41** Mota, F.M. and Kim, D.H. (2019). From $CO_2$ methanation to ambitious long-chain hydrocarbons: alternative fuels paving the path to sustainability. *Chem. Soc. Rev.* 48 (1): 205–259.

**42** Habisreutinger, S.N., Schmidt-Mende, L., and Stolarczyk, J.K. (2013). Photocatalytic reduction of $CO_2$ on $TiO_2$ and other semiconductors. *Angew. Chem. Int. Ed.* 52 (29): 7372–7408.

**43** Lee, J., Sorescu, D.C., and Deng, X. (2011). Electron-induced dissociation of $CO_2$ on $TiO_2(110)$. *J. Am. Chem. Soc.* 133 (110): 10066–10069.

**44** Calaza, F., Stiehler, C., Fujimori, Y. et al. (2015). Carbon dioxide activation and reaction induced by electron transfer at an oxide–metal interface. *Angew. Chem. Int. Ed.* 54 (42): 12484–12487.

**45** Ji, Y. and Luo, Y. (2016). Theoretical study on the mechanism of photoreduction of $CO_2$ to $CH_4$ on the anatase $TiO_2(101)$ surface. *ACS Catal.* 6 (3): 2018–2025.

**46** Tu, W., Zhou, Y., and Zou, Z. (2014). Photocatalytic conversion of $CO_2$ into renewable hydrocarbon fuels: state-of-the-art accomplishment, challenges, and prospects. *Adv. Mater.* 26 (27): 4607–4626.

**47** Yang, C.C., Yu, Y.H., Van Der Linden, B. et al. (2010). Artificial photosynthesis over crystalline $TiO_2$-based catalysts: fact or fiction? *J. Am. Chem. Soc.* 132 (24): 8398–8406.

**48** Dilla, M., Schlögl, R., and Strunk, J. (2017). Photocatalytic $CO_2$ reduction under continuous flow high-purity conditions: quantitative evaluation of $CH_4$ formation in the steady-state. *ChemCatChem* 9 (4): 696–704.

**49** Fresno, F., Reñones, P., Alfonso, E. et al. (2018). Influence of surface density on the $CO_2$ photoreduction activity of a DC magnetron sputtered $TiO_2$ catalyst. *Appl. Catal., B* 224: 912–918.

50 Dimitrijevic, N.M., Shkrob, I.A., Gosztola, D.J. et al. (2011). Role of water and carbonates in photocatalytic transformation of $CO_2$ to $CH_4$ on titania. *J. Am. Chem. Soc.* 133 (11): 3964–3971.

51 Stolarczyk, J.K., Bhattacharyya, S., Polavarapu, L., and Feldmann, J. (2018). Challenges and prospects in solar water splitting and $CO_2$ reduction with inorganic and hybrid nanostructures. *ACS Catal.* 8: 3602–3635.

52 Iglesias-Juez, A. and Coronado, J.M. (2018). Light and heat joining forces: methanol from photothermal $CO_2$ hydrogenation. *Chem* 4 (7): 1490–1491.

53 Thompson, W.A., Sanchez Fernandez, E., and Maroto-Valer, M.M. (2020). Probability Langmuir-Hinshelwood based $CO_2$ photoreduction kinetic models. *Chem. Eng. J.* 384: 123356.

54 Grigioni, I., Dozzi, M.V., Bernareggi, M. et al. (2017). Photocatalytic $CO_2$ reduction vs. $H_2$ production: the effects of surface carbon-containing impurities on the performance of $TiO_2$-based photocatalysts. *Catal. Today* 281: 214–220.

55 Moustakas, N.G. and Strunk, J. (2018). Photocatalytic $CO_2$ reduction on $TiO_2$-based materials under controlled reaction conditions: systematic insights from a literature study. *Chem. Eur. J.* 24 (49): 12739–12746.

56 Carp, O., Huisman, C.L., and Reller, A. (2004). Photoinduced reactivity of titanium dioxide. *Prog. Solid State Chem.* 32 (1–2): 33–177.

57 Landsiedel, R., Ma-Hock, L., Kroll, A. et al. (2010). Testing metal-oxide nanomaterials for human safety. *Adv. Mater.* 22 (24): 2601–2627.

58 Fox, M.A. and Dulay, M.T. (1993). Heterogeneous photocatalysis. *Chem. Rev.* 93 (1): 341–357.

59 Kubacka, A., Fernández-García, M., and Colón, G. (2012). Advanced nanoarchitectures for solar photocatalytic applications. *Chem. Rev.* 112 (3): 1555–1614.

60 Mills, A., Davies, R.H., and Worsley, D. (1993). Water purification by semiconductor photocatalysis. *Chem. Soc. Rev.* 22 (6): 417–425.

61 Hoffmann, M.R., Martin, S.T., Choi, W., and Bahnemann, D.W. (1995). Environmental applications of semiconductor photocatalysis. *Chem. Rev.* 95 (1): 69–96.

62 Li, Y.F., Aschauer, U., Chen, J., and Selloni, A. (2014). Adsorption and reactions of $O_2$ on anatase $TiO_2$. *Acc. Chem. Res.* 47 (11): 3361–3368.

63 García-Muñoz, P., Fresno, F., Lefevre, C. et al. (2020). Ti-Modified $LaFeO_3$/β-SiC alveolar foams as immobilized dual catalysts with combined photofenton and photocatalytic activity. *ACS Appl. Mater. Interfaces* 12 (51): 57025–57037.

64 Dhakshinamoorthy, A., Navalon, S., Corma, A., and Garcia, H. (2012). Photocatalytic $CO_2$ reduction by $TiO_2$ and related titanium containing solids. *Energy Environ. Sci.* 5 (11): 9217.

65 Nguyen, T.P., Nguyen, D.L.T., Nguyen, V.-H. et al. (2020). Recent advances in $TiO_2$-based photocatalysts for reduction of $CO_2$ to fuels. *Nanomaterials* 10: 337.

66 Santalucia, R., Mino, L., Cesano, F. et al. (2020). Surface processes in photocatalytic reduction of $CO_2$ on $TiO_2$-based materials. *J. Photocatal.* 2 (1): 10–24.

67 Ohno, T., Sarukawa, K., Tokieda, K., and Matsumura, M. (2001). Morphology of a $TiO_2$ photocatalyst (Degussa, P-25) consisting of anatase and rutile crystalline phases. *J. Catal.* 203 (1): 82–86.

**68** Tan, L.L., Ong, W.J., Chai, S.P., and Mohamed, A.R. (2016). Visible-light-activated oxygen-rich $TiO_2$ as next generation photocatalyst: importance of annealing temperature on the photoactivity toward reduction of carbon dioxide. *Chem. Eng. J.* 283: 1254–1263.

**69** Reñones, P., Moya, A., Fresno, F. et al. (2016). Hierarchical $TiO_2$ nanofibres as photocatalyst for $CO_2$ reduction: influence of morphology and phase composition on catalytic activity. *J. CO2 Util.* 15: 24–31.

**70** Reli, M., Kobielusz, M., Matějová, L. et al. (2017). $TiO_2$ Processed by pressurized hot solvents as a novel photocatalyst for photocatalytic reduction of carbon dioxide. *Appl. Surf. Sci.* 391: 282–287.

**71** Chen, X. and Mao, S.S. (2007). Titanium dioxide nanomaterials: synthesis, properties, modifications and applications. *Chem. Rev.* 107 (7): 2891–2959.

**72** Wang, P.Q., Bai, Y., Liu, J.Y. et al. (2012). One-pot synthesis of rutile $TiO_2$ nanoparticle modified anatase $TiO_2$ nanorods toward enhanced photocatalytic reduction of $CO_2$ into hydrocarbon fuels. *Catal. Commun.* 29: 185–188.

**73** Ping, G., Wang, C., Chen, D. et al. (2013). Fabrication of self-organized $TiO_2$ nanotube arrays for photocatalytic reduction of $CO_2$. *J. Solid State Electrochem.* 17 (9): 2503–2510.

**74** Grimes, C.A. and Mor, G.K. (2009). *$TiO_2$ Nanotube Arrays*. Boston, MA: Springer.

**75** Hu, Z., García-Martín, J.M., Li, Y. et al. (2020). $TiO_2$ Nanocolumn arrays for more efficient and stable perovskite solar cells. *ACS Appl. Mater. Interfaces* 12 (5): 5979–5989.

**76** Fresno, F., González, M.U., Martínez, L. et al. (2021). Photo-induced self-cleaning and wettability in $TiO_2$ nanocolumn arrays obtained by glancing-angle deposition with sputtering. *Adv. Sustain. Syst.*: 2100071.

**77** He, Z., Wen, L., Wang, D. et al. (2014). Photocatalytic reduction of $CO_2$ in aqueous solution on surface-fluorinated anatase $TiO_2$ nanosheets with exposed {001} facets. *Energy Fuels* 28 (6): 3982–3993.

**78** He, H., Zapol, P., and Curtiss, L.A. (2012). Computational screening of dopants for photocatalytic two-electron reduction of $CO_2$ on anatase (101) surfaces. *Energy Environ. Sci.* 5 (3): 6196–6205.

**79** Han, X., Kuang, Q., Jin, M. et al. (2009). Synthesis of titania nanosheets with a high percentage of exposed (001) facets and related photocatalytic properties. *J. Am. Chem. Soc.* 131 (9): 3152–3153.

**80** Yu, J., Low, J., Xiao, W. et al. (2014). Enhanced photocatalytic $CO_2$-reduction activity of anatase $TiO_2$ by coexposed {001} and {101} facets. *J. Am. Chem. Soc.* 136 (25): 8839–8842.

**81** Fresno, F., Portela, R., Suárez, S., and Coronado, J.M. (2014). Photocatalytic materials: recent achievements and near future trends. *J. Mater. Chem. A* 2 (9): 2863–2884.

**82** Matějová, L., Kočí, K., Reli, M. et al. (2014). Preparation, characterization and photocatalytic properties of cerium doped $TiO_2$: on the effect of Ce loading on the photocatalytic reduction of carbon dioxide. *Appl. Catal., B* 152–153: 172–183.

83 Kočí, K., Matějová, L., Ambrožová, N. et al. (2016). Optimization of cerium doping of $TiO_2$ for photocatalytic reduction of $CO_2$ and photocatalytic decomposition of $N_2O$. *J. Sol-Gel Sci. Technol.* 78 (3): 550–558.

84 Xiong, Z., Zhao, Y., Zhang, J., and Zheng, C. (2015). Efficient photocatalytic reduction of $CO_2$ into liquid products over cerium doped titania nanoparticles synthesized by a sol–gel auto-ignited method. *Fuel Process. Technol.* 135: 6–13.

85 She, H., Zhao, Z., Bai, W. et al. (2020). Enhanced performance of photocatalytic $CO_2$ reduction via synergistic effect between chitosan and $Cu:TiO_2$. *Mater. Res. Bull.* 124: 110758.

86 Nguyen, N.H., Wu, H.Y., and Bai, H. (2015). Photocatalytic reduction of $NO_2$ and $CO_2$ using molybdenum-doped titania nanotubes. *Chem. Eng. J.* 269 (2): 60–66.

87 Lee, J.H., Lee, H., and Kang, M. (2016). Remarkable photoconversion of carbon dioxide into methane using Bi-doped $TiO_2$ nanoparticles prepared by a conventional sol–gel method. *Mater. Lett.* 178: 316–319.

88 Wang, T., Meng, X., Liu, G. et al. (2015). In situ synthesis of ordered mesoporous Co-doped $TiO_2$ and its enhanced photocatalytic activity and selectivity for the reduction of $CO_2$. *J. Mater. Chem. A* 3 (18): 9491–9501.

89 Kumaravel, V., Rhatigan, S., Mathew, S. et al. (2019). Indium-doped $TiO_2$ photocatalysts with high-temperature anatase stability. *J. Phys. Chem. C* 123 (34): 21083–21096.

90 Tahir, M. and Amin, N.S. (2013). Photocatalytic $CO_2$ reduction and kinetic study over $In/TiO_2$ nanoparticles supported microchannel monolith photoreactor. *Appl. Catal., A* 467: 483–496.

91 Tahir, M. and Amin, N.A.S. (2015). Indium-doped $TiO_2$ nanoparticles for photocatalytic $CO_2$ reduction with $H_2O$ vapors to $CH_4$. *Appl. Catal., B* 162: 98–109.

92 Chen, X. and Jin, F. (2019). Photocatalytic reduction of carbon dioxide by titanium oxide-based semiconductors to produce fuels. *Front. Energy* 13 (2): 207–220.

93 Akple, M.S., Low, J., Qin, Z. et al. (2015). Nitrogen-doped $TiO_2$ microsheets with enhanced visible light photocatalytic activity for $CO_2$ reduction. *Chin. J. Catal.* 36 (12): 2127–2134.

94 Zhang, Z., Huang, Z., Cheng, X. et al. (2015). Product selectivity of visible-light photocatalytic reduction of carbon dioxide using titanium dioxide doped by different nitrogen-sources. *Appl. Surf. Sci.* 355: 45–51.

95 Wang, K., Yu, J., Liu, L. et al. (2016). Hierarchical P-doped $TiO_2$ nanotubes array@Ti plate: towards advanced $CO_2$ photocatalytic reduction catalysts. *Ceram. Int.* 42 (14): 16405–16411.

96 Olowoyo, J.O., Kumar, M., Jain, S.L. et al. (2018). Reinforced photocatalytic reduction of $CO_2$ to fuel by efficient $S-TiO_2$: significance of sulfur doping. *Int. J. Hydrogen Energy* 43 (37): 17682–17695.

97 He, Z., Yu, Y., Wang, D. et al. (2016). Photocatalytic reduction of carbon dioxide using iodine-doped titanium dioxide with high exposed {001} facets under visible light. *RSC Adv.* 6 (28): 23134–23140.

98 García-Rodríguez, S. (2013). Alternative metal oxide photocatalysts. In: *Design of Advanced Photocatalytic Materials for Energy and Environmental Applications* (eds. J.M. Coronado, F. Fresno, M.D. Hernández-Alonso and R. Portela), 103–122. London: Springer London.

99 Li, P., Zhu, S., Hu, H. et al. (2019). Influence of defects in porous ZnO nanoplates on $CO_2$ photoreduction. *Catal. Today* 335: 300–305.

100 Collado, L., Jana, P., Sierra, B. et al. (2013). Enhancement of hydrocarbon production via artificial photosynthesis due to synergetic effect of Ag supported on $TIO_2$ and ZnO semiconductors. *Chem. Eng. J.* 224 (1): 128–135.

101 Kumar, S.G. and Rao, K.S.R.K. (2015). Zinc oxide based photocatalysis: tailoring surface-bulk structure and related interfacial charge carrier dynamics for better environmental applications. *RSC Adv.* 5 (5): 3306–3351.

102 Li, P., Hu, H., Luo, G. et al. (2020). Crystal facet-dependent $CO_2$ photoreduction over porous ZnO nanocatalysts. *ACS Appl. Mater. Interfaces* 12 (50): 56039–56048.

103 Guo, Q., Zhang, Q., Wang, H., and Zhao, Z. (2018). $ZnO_2$-promoted ZnO as an efficient photocatalyst for the photoreduction of carbon dioxide in the presence of water. *Catal. Commun.* 103: 24–28.

104 Núñez, J., De La Peña O'Shea, V.A., Jana, P. et al. (2013). Effect of copper on the performance of ZnO and $ZnO_{1-x}N_x$ oxides as $CO_2$ photoreduction catalysts. *Catal. Today* 209: 21–27.

105 Peterson, A.A., Abild-Pedersen, F., Studt, F. et al. (2010). How copper catalyzes the electroreduction of carbon dioxide into hydrocarbon fuels. *Energy Environ. Sci.* 3 (9): 1311–1315.

106 Christoforidis, K.C. and Fornasiero, P. (2019). Photocatalysis for hydrogen production and $CO_2$ reduction: the case of copper-catalysts. *ChemCatChem* 11 (1): 368–382.

107 Ovcharov, M.L., Mishura, A.M., Shcherban, N.D. et al. (2016). Photocatalytic reduction of $CO_2$ using nanostructured $Cu_2O$ with foam-like structure. *Sol. Energy* 139: 452–457.

108 Handoko, A.D. and Tang, J. (2013). Controllable proton and $CO_2$ photoreduction over $Cu_2O$ with various morphologies. *Int. J. Hydrogen Energy* 38 (29): 13017–13022.

109 Yang, P., Zhao, Z.-J., Chang, X. et al. (2018). The functionality of surface hydroxy groups on the selectivity and activity of carbon dioxide reduction over cuprous oxide in aqueous solutions. *Angew. Chem. Int. Ed.* 57 (26): 7724–7728.

110 Zhu, C., Wei, X., Li, W. et al. (2020). Crystal-plane effects of $CeO_2\{110\}$ and $CeO_2\{100\}$ on photocatalytic $CO_2$ reduction: synergistic interactions of oxygen defects and hydroxyl groups. *ACS Sustainable Chem. Eng.* 8 (38): 14397–14406.

111 Wang, M., Shen, M., Jin, X. et al. (2019). Oxygen vacancy generation and stabilization in $CeO_{2-x}$ by Cu introduction with improved $CO_2$ photocatalytic reduction activity. *ACS Catal.* 9 (5): 4573–4581.

112 Wang, Y., Wang, F., Chen, Y. et al. (2014). Enhanced photocatalytic performance of ordered mesoporous Fe-doped $CeO_2$ catalysts for the reduction of $CO_2$ with $H_2O$ under simulated solar irradiation. *Appl. Catal., B* 147: 602–609.

113 Wang, Y., Bai, X., Wang, F. et al. (2019). Nanocasting synthesis of chromium doped mesoporous $CeO_2$ with enhanced visible-light photocatalytic $CO_2$ reduction performance. *J. Hazard. Mater.* 372: 69–76.

114 Fresno, F., Jana, P., Reñones, P. et al. (2017). $CO_2$ reduction over $NaNbO_3$ and $NaTaO_3$ perovskite photocatalysts. *Photochem. Photobiol. Sci.* 16 (1): 17–23.

115 Zhou, H., Li, P., Guo, J. et al. (2015). Artificial photosynthesis on tree trunk derived alkaline tantalates with hierarchical anatomy: towards $CO_2$ photo-fixation into CO and $CH_4$. *Nanoscale* 7 (1): 113–120.

116 Shi, H., Wang, T., Chen, J. et al. (2011). Photoreduction of carbon dioxide over $NaNbO_3$ nanostructured photocatalysts. *Catal. Lett.* 141 (4): 525–530.

117 Li, P., Ouyang, S., Zhang, Y. et al. (2013). Surface-coordination-induced selective synthesis of cubic and orthorhombic $NaNbO_3$ and their photocatalytic properties. *J. Mater. Chem. A* 1 (4): 1185–1191.

118 Liou, P.-Y., Chen, S.-C., Wu, J.C.S. et al. (2011). Photocatalytic $CO_2$ reduction using an internally illuminated monolith photoreactor. *Energy Environ. Sci.* 4 (4): 1487.

119 Lee, D.S., Chen, H.J., and Chen, Y.W. (2012). Photocatalytic reduction of carbon dioxide with water using $InNbO_4$ catalyst with NiO and $Co_3O_4$ cocatalysts. *J. Phys. Chem. Solids* 73 (5): 661–669.

120 Sun, Z., Yang, Z., Liu, H. et al. (2014). Visible-light $CO_2$ photocatalytic reduction performance of ball-flower-like $Bi_2WO_6$ synthesized without organic precursor: effect of post-calcination and water vapor. *Appl. Surf. Sci.* 315 (1): 360–367.

121 Ren, G., Zhang, X., Zhang, C. et al. (2020). Synergetic effect of $Bi_2WO_6$ micro-spheres and activated carbon mm-spheres for enhancing photoreduction activity of $CO_2$ to CO. *Mater. Lett.* 264: 127201.

122 Kong, X.Y., Tong, T., Ng, B.J. et al. (2020). Topotactic transformation of bismuth oxybromide into bismuth tungstate: bandgap modulation of single-crystalline {001}-faceted nanosheets for enhanced photocatalytic $CO_2$ reduction. *ACS Appl. Mater. Interfaces* 12 (24): 26991–27000.

123 Liu, Y., Shen, D., Zhang, Q. et al. (2021). Enhanced photocatalytic $CO_2$ reduction in $H_2O$ vapor by atomically thin $Bi_2WO_6$ nanosheets with hydrophobic and nonpolar surface. *Appl. Catal., B* 283: 119630.

124 Liu, Z., Jiang, W., Liu, Z. et al. (2020). Optimizing the carbon dioxide reduction pathway through surface modification by halogenation. *ChemSusChem* 13 (21): 5638–5646.

125 Mao, J., Peng, T., Zhang, X. et al. (2012). Selective methanol production from photocatalytic reduction of $CO_2$ on $BiVO_4$ under visible light irradiation. *Catal. Commun.* 28: 38–41.

126 Corradini, P.G., De Brito, J.F., Boldrin Zanoni, M.V., and Mascaro, L.H. (2020). Artificial photosynthesis for alcohol and 3-C compound formation using $BiVO_4$-lamelar catalyst. *J. CO2 Util.* 36: 187–195.

127 Gao, S., Gu, B., Jiao, X. et al. (2017). Highly efficient and exceptionally durable $CO_2$ photoreduction to methanol over freestanding defective single-unit-cell bismuth vanadate layers. *J. Am. Chem. Soc.* 139 (9): 3438–3445.

**128** Garcia-Muñoz, P., Fresno, F., de la Peña O'Shea, V.A., and Keller, N. (2020). Ferrite materials for photoassisted environmental and solar fuels applications. *Top. Curr. Chem.* 378 (1) 6, 56 pages.

**129** Matsumoto, Y., Obata, M., and Hombo, J. (1994). Photocatalytic reduction of carbon dioxide on p-type $CaFe_2O_4$ powder. *J. Phys. Chem.* 98: 2950–2951.

**130** Matsumoto, Y. (1996). Energy positions of oxide semiconductors and photocatalysis with iron complex oxides. *J. Solid State Chem.* 126 (2): 227–234.

**131** Xiao, J., Yang, W., Gao, S. et al. (2018). Fabrication of ultrafine $ZnFe_2O_4$ nanoparticles for efficient photocatalytic reduction $CO_2$ under visible light illumination. *J. Mater. Sci. Technol.* 34 (12): 2331–2336.

**132** Low, J., Yu, J., Jaroniec, M. et al. (2017). Heterojunction photocatalysts. *Adv. Mater.* 29 (20): 1601694.

**133** Wen, F. and Li, C. (2013). Hybrid artificial photosynthetic systems comprising semiconductors as light harvesters and biomimetic complexes as molecular cocatalysts. *Acc. Chem. Res.* 46 (11): 2355–2364.

**134** Huang, D., Chen, S., Zeng, G. et al. (2019). Artificial Z-scheme photocatalytic system: what have been done and where to go? *Coord. Chem. Rev.* 385: 44–80.

**135** Tada, H., Mitsui, T., Kiyonaga, T. et al. (2006). All-solid-state Z-scheme in CdS–Au–$TiO_2$ three-component nanojunction system. *Nat. Mater.* 5 (10): 782–786.

**136** Wei, L., Yu, C., Zhang, Q. et al. (2018). $TiO_2$-based heterojunction photocatalysts for photocatalytic reduction of $CO_2$ into solar fuels. *J. Mater. Chem. A* 6: 22411–22436.

**137** Fresno, F. (2013). Heterojunctions: joining different semiconductors. *Green Energy Technol.* 71: 311–327.

**138** Aguirre, M.E., Zhou, R., Eugene, A.J. et al. (2017). $Cu_2O/TiO_2$ heterostructures for $CO_2$ reduction through a direct Z-scheme: protecting $Cu_2O$ from photocorrosion. *Appl. Catal., B* 217: 485–493.

**139** In, S.I., Vaughn, D.D., and Schaak, R.E. (2012). Hybrid $CuO-TiO_{2-x}N_x$ hollow nanocubes for photocatalytic conversion of $CO_2$ into methane under solar irradiation. *Angew. Chem. Int. Ed.* 51 (16): 3915–3918.

**140** Xiong, Z., Lei, Z., Kuang, C.C. et al. (2017). Selective photocatalytic reduction of $CO_2$ into $CH_4$ over Pt-$Cu_2O$ $TiO_2$ nanocrystals: the interaction between Pt and $Cu_2O$ cocatalysts. *Appl. Catal., B* 202: 695–703.

**141** Kim, C., Cho, K.M., Al-Saggaf, A. et al. (2018). Z-scheme photocatalytic $CO_2$ conversion on three-dimensional $BiVO_4$/carbon-coated $Cu_2O$ nanowire arrays under visible light. *ACS Catal.* 8 (5): 4170–4177.

**142** Longo, G., Fresno, F., Gross, S., and Štangar, U.L. (2014). Synthesis of $BiVO_4/TiO_2$ composites and evaluation of their photocatalytic activity under indoor illumination. *Environ. Sci. Pollut. Res.* 21 (19): 11189–11197.

**143** Iwase, A., Yoshino, S., Takayama, T. et al. (2016). Water splitting and $CO_2$ reduction under visible light irradiation using Z-scheme systems consisting of metal sulfides, $CoO_x$-loaded $BiVO_4$, and a reduced graphene oxide electron mediator. *J. Am. Chem. Soc.* 138 (32): 10260–10264.

**144** Wei, Z.H., Wang, Y.F., Li, Y.Y. et al. (2018). Enhanced photocatalytic $CO_2$ reduction activity of Z-scheme CdS/$BiVO_4$ nanocomposite with thinner $BiVO_4$ nanosheets. *J. CO2 Util.* 28: 15–25.

**145** Yoshino, S., Sato, K., Yamaguchi, Y. et al. (2020). Z-schematic $CO_2$ reduction to CO through interparticle electron transfer between $SrTiO_3$:RH of a reducing photocatalyst and $BiVO_4$ of a water oxidation photocatalyst under visible light. *ACS Appl. Energy Mater.* 3 (10): 10001–10007.

**146** Wu, J., Xiong, L., Hu, Y. et al. (2021). Organic half-metal derived erythroid-like $BiVO_4$/hm-$C_4N_3$ Z-scheme photocatalyst: reduction sites upgrading and rate-determining step modulation for overall $CO_2$ and $H_2O$ conversion. *Appl. Catal., B* 295: 120277.

**147** Tasbihi, M., Fresno, F., Álvarez-Prada, I. et al. (2021). A molecular approach to the synthesis of platinum-decorated mesoporous graphitic carbon nitride as selective $CO_2$ reduction photocatalyst. *J. CO2 Util.* 50: 101574.

**148** Que, M., Cai, W., Chen, J. et al. (2021). Recent advances in g-$C_3N_4$ composites within four types of heterojunctions for photocatalytic $CO_2$ reduction. *Nanoscale* 13 (14): 6692–6712.

**149** Dehkordi, A.B., Ziarati, A., Ghasemi, J.B., and Badiei, A. (2020). Preparation of hierarchical g-$C_3N_4$@$TiO_2$ hollow spheres for enhanced visible-light induced catalytic $CO_2$ reduction. *Sol. Energy* 205: 465–473.

**150** Wang, H., Li, H., Chen, Z. et al. (2020). $TiO_2$ modified g-$C_3N_4$ with enhanced photocatalytic $CO_2$ reduction performance. *Solid State Sci.* 100: 106099.

**151** Guo, Q., Fu, L., Yan, T. et al. (2020). Improved photocatalytic activity of porous ZnO nanosheets by thermal deposition graphene-like g-$C_3N_4$ for $CO_2$ reduction with $H_2O$ vapor. *Appl. Surf. Sci.* 509: 144773.

**152** Jiang, Z., Wan, W., Li, H. et al. (2018). A hierarchical Z-scheme α-$Fe_2O_3$/g-$C_3N_4$ hybrid for enhanced photocatalytic $CO_2$ reduction. *Adv. Mater.* 30 (10): 1706108.

**153** Li, X., Song, X., Ma, C. et al. (2020). Direct Z-scheme $WO_3$/graphitic carbon nitride nanocomposites for the photoreduction of $CO_2$. *ACS Appl. Nano Mater.* 3 (2): 1298–1306.

**154** Kumar, A., Sharma, S.K., Sharma, G. et al. (2020). $CeO_2$/g-$C_3N_4$/$V_2O_5$ ternary nano hetero-structures decorated with CQDs for enhanced photo-reduction capabilities under different light sources: dual Z-scheme mechanism. *J. Alloys Compd.* 838: 155692.

**155** Lu, M., Li, Q., Zhang, C. et al. (2020). Remarkable photocatalytic activity enhancement of $CO_2$ conversion over 2D/2D g-$C_3N_4$/$BiVO_4$ Z-scheme heterojunction promoted by efficient interfacial charge transfer. *Carbon N. Y.* 160: 342–352.

**156** Bhosale, R., Jain, S., Vinod, C.P. et al. (2019). Direct Z-scheme g-$C_3N_4$/$FeWO_4$ nanocomposite for enhanced and selective photocatalytic $CO_2$ reduction under visible light. *ACS Appl. Mater. Interfaces* 11 (6): 6174–6183.

**157** Wang, K., Jiang, L., Wu, X., and Zhang, G. (2020). Vacancy mediated Z-scheme charge transfer in a 2D/2D $La_2Ti_2O_7$/g-$C_3N_4$ nanojunction as a bifunctional photocatalyst for solar-to-energy conversion. *J. Mater. Chem. A* 8 (26): 13241–13247.

**158** Li, X., Yu, J., Jaroniec, M., and Chen, X. (2019). Cocatalysts for selective photoreduction of $CO_2$ into solar fuels. *Chem. Rev.* 119 (6): 3962–4179.

**159** Ran, J., Jaroniec, M., and Qiao, S. (2018). Cocatalysts in semiconductor-based photocatalytic $CO_2$ reduction: achievements, challenges, and opportunities. *Adv. Mater.* 30: 1704649.

**160** Ohno, T., Higo, T., Murakami, N. et al. (2014). Photocatalytic reduction of $CO_2$ over exposed-crystal-face-controlled $TiO_2$ nanorod having a brookite phase with co-catalyst loading. *Appl. Catal., B* 152–153: 309–316.

**161** Pathak, P., Meziani, M.J., Castillo, L., and Sun, Y.P. (2005). Metal-coated nanoscale $TiO_2$ catalysts for enhanced $CO_2$ photoreduction. *Green Chem.* 7 (9): 667–670.

**162** Liu, J.-Y., Garg, B., and Ling, Y.-C. (2011). CuxAgyInzZnkSm solid solutions customized with $RuO_2$ or $Rh_{1.32}Cr_{0.66}O_3$ co-catalyst display visible light-driven catalytic activity for $CO_2$ reduction to $CH_3OH$. *Green Chem.* 13 (8): 2029–2031.

**163** Collado, L., Reynal, A., Coronado, J.M. et al. (2015). Effect of Au surface plasmon nanoparticles on the selective $CO_2$ photoreduction to $CH_4$. *Appl. Catal., B* 178: 177–185.

**164** Wang, Y., Zhao, J., Li, Y., and Wang, C. (2018). Selective photocatalytic $CO_2$ reduction to $CH_4$ over $Pt/In_2O_3$: significant role of hydrogen adatom. *Appl. Catal., B* 226: 544–553.

**165** Collado, L., Reynal, A., Fresno, F. et al. (2018). Unravelling the effect of charge dynamics at the plasmonic metal/semiconductor interface for $CO_2$ photoreduction. *Nat. Commun.* 9 (1): 4986.

**166** Tasbihi, M., Fresno, F., Simon, U. et al. (2018). On the selectivity of $CO_2$ photoreduction towards $CH_4$ using $Pt/TiO_2$ catalysts supported on mesoporous silica. *Appl. Catal., B* 239: 68–76.

**167** Lu, L., Wang, S., Zhou, C. et al. (2018). Surface chemistry imposes selective reduction of $CO_2$ to CO over $Ta_3N_5/LaTiO_2N$ photocatalyst. *J. Mater. Chem. A* 6 (30): 14838–14846.

**168** Li, K., Peng, T., Ying, Z. et al. (2016). Ag-loading on brookite $TiO_2$ quasi nanocubes with exposed {210} and {001} facets: activity and selectivity of $CO_2$ photoreduction to $CO/CH_4$. *Appl. Catal., B* 180: 130–138.

**169** Fresno, F., Galdón, S., Barawi, M. et al. (2021). Selectivity in UV photocatalytic $CO_2$ conversion over bare and silver-decorated niobium-tantalum perovskites. *Catal. Today* 361: 85–93.

**170** Linic, S., Christopher, P., and Ingram, D.B. (2011). Plasmonic-metal nanostructures for efficient conversion of solar to chemical energy. *Nat. Mater.* 10 (12): 911–921.

**171** Yu, Y., Wen, W., Qian, X.-Y. et al. (2017). UV and visible light photocatalytic activity of $Au/TiO_2$ nanoforests with Anatase/Rutile phase junctions and controlled Au locations. *Sci. Rep.* 7: 1–13.

**172** Sellappan, R., Nielsen, M.G., González-Posada, F. et al. (2013). Effects of plasmon excitation on photocatalytic activity of $Ag/TiO_2$ and $Au/TiO_2$ nanocomposites. *J. Catal.* 307: 214–221.

173 Dilla, M., Pougin, A., and Strunk, J. (2017). Evaluation of the plasmonic effect of Au and Ag on Ti-based photocatalysts in the reduction of $CO_2$ to $CH_4$. *J. Energy Chem.* 26 (2): 277–283.

174 Wang, W., An, W., Ramalingam, B. et al. (2012). Size and structure matter: enhanced $CO_2$ photoreduction efficiency by size-resolved ultra fine Pt nanoparticles on $TiO_2$ single crystals. *J. Am. Chem. Soc.* 134: 11276–11281.

175 Mao, J., Ye, L., Li, K. et al. (2014). Pt-loading reverses the photocatalytic activity order of anatase $TiO_2$ {001} and {010} facets for photoreduction of $CO_2$ to $CH_4$. *Appl. Catal., B* 144: 855–862.

176 Wang, Y., Lai, Q., Zhang, F. et al. (2014). High efficiency photocatalytic conversion of $CO_2$ with $H_2O$ over $Pt/TiO_2$ nanoparticles. *RSC Adv.* 4 (84): 44442–44451.

177 Xie, S., Wang, Y., Zhang, Q. et al. (2014). MgO- and Pt-promoted $TiO_2$ as an efficient photocatalyst for the preferential reduction of carbon dioxide in the presence of water. *ACS Catal.* 4 (10): 3644–3653.

178 Feng, X., Sloppy, J.D., LaTempa, T.J. et al. (2011). Synthesis and deposition of ultrafine Pt nanoparticles within high aspect ratio $TiO_2$ nanotube arrays: application to the photocatalytic reduction of carbon dioxide. *J. Mater. Chem.* 21: 13429.

179 Júnior, M.A.M., Morais, A., and Nogueira, A.F. (2016). Boosting the solar-light-driven methanol production through $CO_2$ photoreduction by loading $Cu_2O$ on $TiO_2$-pillared $K_2Ti_4O_9$. *Microporous Mesoporous Mater.* 234: 1–11.

180 Fang, B., Xing, Y., Bonakdarpour, A. et al. (2015). Photodriven reduction of $CO_2$ to $CH_4$. *ACS Sustainable Chem. Eng.* 3 (10): 2381–2388.

181 Cook, R.L., MacDuff, R.C., and Sammells, A.F. (1988). Photoelectrochemical carbon dioxide reduction to hydrocarbons at ambient temperature and pressure. *J. Electrochem. Soc.* 135 (12): 3069–3070.

182 Adachi, K., Ohta, K., and Mizuno, T. (1994). Photocatalytic reduction of carbon dioxide to hydrocarbon using copper-loaded titanium dioxide. *Sol. Energy* 53 (2): 187–190.

183 Park, H., Ou, H.-H., Colussi, A.J., and Hoffmann, M.R. (2015). Artificial photosynthesis of $C_1$–$C_3$ hydrocarbons from water and $CO_2$ on titanate nanotubes decorated with nanoparticle elemental copper and CdS quantum dots. *J. Phys. Chem. A* 119 (19): 4658–4666.

184 Long, R., Li, Y., Liu, Y. et al. (2017). Isolation of Cu atoms in Pd lattice: forming highly selective sites for photocatalytic conversion of $CO_2$ to $CH_4$. *J. Am. Chem. Soc.* 139 (12): 4486–4492.

185 Zhao, J., Li, Y., Zhu, Y. et al. (2016). Enhanced $CO_2$ photoreduction activity of black $TiO_2$–coated Cu nanoparticles under visible light irradiation: role of metallic Cu. *Appl. Catal., A* 510: 34–41.

186 Maeda, K., Abe, R., and Domen, K. (2011). Role and function of ruthenium species as promoters with TaON-based photocatalysts for oxygen evolution in two-step water splitting under visible light. *J. Phys. Chem. C* 115 (7): 3057–3064.

**187** Vignolo-González, H.A., Laha, S., Jiménez-Solano, A. et al. (2020). Toward standardized photocatalytic oxygen evolution rates using $RuO_2@TiO_2$ as a benchmark. *Matter* 3 (2): 464–486.

**188** Ismail, A.A., Bahnemann, D.W., and Al-Sayari, S.A. (2012). Synthesis and photocatalytic properties of nanocrystalline Au, Pd and Pt photodeposited onto mesoporous $RuO_2$-$TiO_2$ nanocomposites. *Appl. Catal., A* 431–432: 62–68.

# 9

# Functionalized Titania Coatings for Photocatalytic Air and Water Cleaning

*Ksenija Maver[1], Andraž Šuligoj[1,2], Urška Lavrenčič Štangar[2], and Nataša Novak Tušar[1,3]*

[1] *National Institute of Chemistry, Department of Inorganic Chemistry and Technology, Hajdrihova 19, Ljubljana SI-1000, Slovenia*
[2] *University of Ljubljana, Faculty of Chemistry and Chemical Technology, Večna pot 113, Ljubljana SI-1000, Slovenia*
[3] *University of Nova Gorica, Graduate School, Vipavska 13, Nova Gorica SI-5000, Slovenia*

## 9.1 Introduction

### 9.1.1 Titania as a Photocatalyst for Air and Water Cleaning

Our modern society intensively consumes natural resources, but a crucial element for sustainability is the balance between availability and consumption of resources [1]. Science and technology are the cornerstones to finding a greener way to use energy and materials more efficiently. Photocatalysis is an advanced oxidation process (AOP) that uses a combination of photocatalyst and light irradiation working under ambient conditions with the possibility of employing sunlight. For this reason, photocatalysis is considered as a sustainable process with a wide range of applications, especially in the areas of pollutant removal from air and water [2, 3]. The main function of AOP is to generate highly reactive free hydroxyl radicals that can completely decompose water-soluble and volatile organic pollutants or convert them into more biodegradable compounds as a pretreatment step.

Among metal oxides, titanium dioxide ($TiO_2$) is the most widely used photocatalyst. Due to its particular physiochemical properties, biocompatibility, and availability at low cost, this photocatalyst is used in numerous applications [4–6]: water purification, air pollutant degradation, pesticide residue removal, self-cleaning coatings, dye-sensitized solar cells, water splitting, etc. It occurs naturally in a few crystallographic phases [7], such as tetragonal anatase and rutile or orthorhombic brookite. Among the phases, anatase is considered to be the most photocatalytically active [8].

The mechanism of photocatalysis in $TiO_2$ is explained from the concepts of semiconductor theory [9]. In semiconductors, there is a range of energies occupied by electrons (called a valence band, VB) and a range of unoccupied electronic states (called a conduction band, CB). The energy difference between those bands is called a band gap ($E_g$) of the semiconductors, and if the incident photon's energy is greater

*Tailored Functional Oxide Nanomaterials: From Design to Multi-Purpose Applications*, First Edition.
Edited by Chiara Maccato and Davide Barreca.
© 2022 WILEY-VCH GmbH. Published 2022 by WILEY-VCH GmbH.

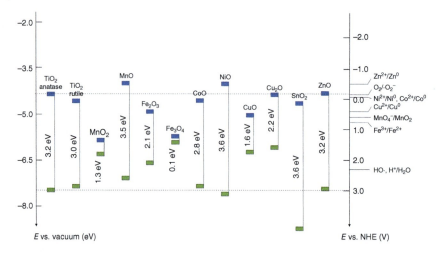

**Figure 9.1** Schematic representation of the band positions of various bulk metal oxides. The potentials (in V vs. NHE) for single-electron transfer of the selected transition metal ions in water are plotted on the right. The CB and VB of a selected semiconductor determine the edges of the energy potential levels within which surface molecules can be photocatalyzed or metal oxides/clusters can be used to reduce the recombination rate of charge carriers. Source: Adapted from Šuligoj et al. [10] and [11].

than or equal to the band gap energy, the process of photoexcitation happens. After absorbing the photon energy, the electron ($e^-$) is excited and ejected from the VB into the CB, leaving behind a hole ($h^+$) in the VB. As a result, electron–hole pairs ($e^-$–$h^+$) are formed. Also, the opposite reaction happens, so that electrons and holes annihilate with each other in the process called recombination, leaving behind heat or emitting light. However, the photogenerated electrons and holes can also successfully migrate to the solid surface without recombination. In this case, they are available for the redox reactions at the surface: the reducing power of electrons and highly oxidizing holes can participate in the reaction with the adsorbed molecules.

From a photocatalytic point of view, $TiO_2$ has a particular electronic structure among the semiconductors. The relative position of the CB and VB is such (Figure 9.1) that it simultaneously allows the reduction of oxygen (via photogenerated electrons) and the oxidation of hydrogen (via photogenerated holes) [12]. In this way, the reactive oxygen species (ROS) are formed, and when $TiO_2$ is used to degrade organic molecules, these can be completely mineralized to form carbon dioxide ($CO_2$) and water ($H_2O$) as final products.

However, the two main drawbacks of $TiO_2$ lead scientists to search for a way to improve the efficiency of the photocatalyst:

(i) The gap value (about 3.2 eV for anatase) classifies $TiO_2$ among semiconductors with a wide band gap width. Since $TiO_2$ is active in the UV region of sunlight, which is only 5% of the spectrum coming from the sun, it is important to find modifications that reduce the band gap of $TiO_2$.
(ii) The recombination rate of photogenerated electrons and holes in $TiO_2$ is very fast (10 ns timescale) compared to the processes of charge transfer to the

adsorbed molecules on the surface (hundreds of nanoseconds to milliseconds) [13]. Prolonging the lifetime of the photogenerated electron and holes by their physical separation is one way to improve the efficiency of the photocatalyst.

## 9.1.2 Titania Functionalization

Metal doping and combination of metal oxides have proven to be a successful strategy to obtain photocatalysts with an improved optical response toward visible light and to hinder $e^--h^+$ recombination [7, 14]. Figure 9.1 shows the relative position of the band gap energies of some metal oxides. The reduction potentials of the selected metal ions in water are plotted on the right. The positions of the energy levels of a semiconductor determine the chemical potentials of the electrons and holes generated after light absorption. These states set the thermodynamic limits of the processes that can be photocatalyzed: the corresponding potential level of the acceptor molecules must be below the semiconductor CB potential, and the potential level of the donor must be above the VB position of the semiconductor in order to donate an electron to the hole at the surface of the semiconductor. In addition, the energy levels of CB and VB in a semiconductor determine the required energy edges at which photogenerated charge transfer from the semiconductor to the metal oxide or cluster is possible, altering the charge transfer properties at the surface and reducing the recombination rate of the charge carriers. When the metal centers at the surface of the semiconductor introduce additional energy levels within the energy gap, the shift of light absorption to smaller wavelengths is possible.

Metal dopants can occupy substitutional or interstitial sites at the titania lattice, creating intra-band gap states that can lead to a redshift, which means a photocatalytic response toward visible light. Depending on the valence state of the dopant, p-type (cations with a lower valence state than $Ti^{4+}$) or n-type doping (cations with a higher valence state than $Ti^{4+}$) can be achieved. When larger metal ions are used or the metal doping concentration is increased, segregated metal oxide clusters form on the surface. In this case, the presence of defects associated with the formation of oxygen vacancies can contribute to the absorption of visible light.

Another approach is the modification of the surface of $TiO_2$ to boost the performance by so-called grafting. It acts via the loading of small metal clusters on the surface of the original photocatalyst. In this case, the metal clusters are good receivers of electrons, which are then passed on to target molecules: they usually reduce oxygen molecules ($O_2$) to form ROS or are used in direct reduction reactions with organic molecules. This kind of modification is not introducing any mid-band states, rather the oxidation potential of the photogenerated holes stays unchanged, and only the electrons are affected by the presence of surface clusters [15].

Several transition metals were used in this approach, such as Fe [16], Cu [17], Zn [18], Ni [19], and others. However, there has been some discrepancy among the various reports regarding the transition metal grafting of titania for visible light applications. For example, Tobaldi et al. showed that Zn clusters gave better photocatalytic performance than Cu, while for antibacterial applications, the situation was reversed. On the other hand, introduction of Ni quantum dots on the surface of $TiO_2$ was shown to improve the photocatalytic activity, which is also an example discussed in Section 9.2.3.

Another approach, which is also an effective way to improve the overall efficiency of the photocatalyst, is to ensure a high specific surface area and adsorption capacity of the pollutant to be treated. For this purpose, functionalization of $TiO_2$ with porous supports is an attractive option [20]. Additionally, it was shown experimentally that in aerated conditions, which are present in porous solid–solid interface, the presence of $O_2$ is of paramount importance for successful electron transfer from the oxo-clusters to $O_2$. Porous silicates such as microporous zeolites and mesoporous silicates are superior supports for the incorporation of photocatalyst nanoparticles (NPs) because they are chemically inert, have a large surface area, are transparent to UV radiation, have great physical stability, and have a hydrophobic character [21–23]. Mesoporous silica ($SiO_2$) also known as SBA-15 (Santa Barbara Amorphous) [24], with highly ordered hexagonal straight pore arrangement, thick pore walls, and high surface area, has many advantages compared with other porous silica supports. It can be prepared over a wide range of pore sizes (5–15 nm) and pore wall thicknesses (3–6 nm) at low temperatures (35–100 °C) and has excellent adsorption properties. As already mentioned, in addition to the adjustable silica pore system, strategies for functionalization of $TiO_2$ with porous silica, e.g. selective immobilization of $TiO_2$ NPs within the channels of the porous support, are crucial (see Section 9.2.1).

### 9.1.3 Fabrication of Titania-Based Coatings

The powder form of photocatalysts is common in many photocatalytic applications. However, working with suspensions (e.g. in water purification) is not always technologically practicable due to the high cost of the filtration process to recover or remove the photocatalysts from the liquid. Moreover, a special reactor design is required to ensure proper illumination and fluidization of the photocatalytic particle suspension, because the availability of photons in a liquid is limited by the absorption of light and scattering by the particles [12]. Immobilization of the photocatalyst on a carrier as a coating enables to overcome some of these drawbacks [25].

The application of the photocatalyst as a coating opens a variety of possibilities for the treatment of pollutants in aqueous or gaseous phase. Current approaches of immobilization of the photocatalyst in the form of a coating focus on improving the methods of fixing $TiO_2$ on the carrier (appropriate adherence and chemically inert material), increasing the adsorption capacity of the photocatalyst (high specific surface area of the support), or increasing the illuminating area (photoreactor design).

Focusing on photocatalytically active coatings instead of powders has several advantages in air treatment. Although the physical drawbacks of catalyst fouling, aggregation, and homogeneous light distribution are not as severe as in the liquid phase due to the low mechanical stress on the catalyst, there are still circumstances in which catalyst immobilization offers several advantages. First, the design of the reactor can be much broader, as the catalyst can be supported on materials of many shapes, offering a variety of more rational designs. Second, there is no need for complete homogenization of the powder catalyst with the air stream, which opens the applicability of such reactors and increases the modes of operation, e.g. one could use different air velocities at different times, depending on the need and area of application. Finally, the applicability aspect is again improved, as the installation of such air purifiers is easier and safer.

Carriers of a photocatalyst include glass slides, aluminum plates, textiles, polymers, etc. The choice of the carrier is an important issue and is always tailored to the application [5]. The first key characteristic of a suitable carrier is that it provides good mechanical attachment and support for the semiconductor. Second, it is important that when the photocatalyst is in contact with the carrier, it does not affect the photocatalytic performance of the semiconductor. In addition, the carrier must be able to withstand the erosion processes during photocatalysis. Besides, a technical solution for an efficient light irradiation must also be considered.

While it is possible to immobilize the nanopowders to the substrate directly [26], in the past, researchers have used inorganic binders, surfactants, or sol–gel-derived nanostructures to encapsulate the NPs near to the carrier. Alternatively, in the case of titania, the nanostructure can be grown directly from the titanium surface [27]. In any case, the effects of structure, mesoporosity, pore connectivity, NP distribution, and others are important for understanding the best usage of binders/growing nanostructures in immobilized photocatalytic materials. When two or more oxide phases are combined, the final result may be different from a simple mixture of the two. Therefore, it is important to study the effect of incorporating NPs into another inorganic phase that serves as a support, such as silica, alumina, ceria, or others.

There are several methods to deposit $TiO_2$ as a coating on the material surface (carriers), such as dip-coating, spin-coating, spraying, chemical vapor deposition (CVD), and others [28]. The dip-coating sol–gel method and thermal treatment method are commonly recognized as convenient and practical methods for immobilizing the photocatalytic materials to the carriers due to their simplicity, low operating cost, and tunable output of the final properties of the materials. However, due to the high consumption of the sols, alternatives have been used, such as brush deposition and, in case of flat surfaces, the doctor blade method. Ceramics, glass, metals, and other materials have mostly been employed for the manufacture of carriers. Soda–lime glass represents a commonly used $TiO_2$ film carrier, although it contains sodium, which has a detrimental effect on photoactivity if the deposited film is treated at higher temperatures. The negative effect of sodium could be eliminated by introducing an amorphous $SiO_2$ barrier layer between $TiO_2$ film and carrier or by using mixed $TiO_2$–$SiO_2$ films. The addition of $SiO_2$ to $TiO_2$ also promotes the well-known synergetic effect of both phases: the formation of Si–O–Ti cross-linking bonds and oxygen vacancies in titania.

### 9.1.4 Characterization of Titania-Based Materials

Basic structural characterization of titania-based materials (in the form of powder or coating) is usually performed using X-ray diffraction (XRD), scanning electron microscopy (SEM), transmission electron microscopy (TEM), scanning transmission electron microscopy (STEM), elemental analysis (energy dispersion analysis by X-ray, EDAX), thermogravimetric analysis (TG/DTG, DSC), $N_2$ physisorption, UV–Vis spectroscopy (UV–VIS), and atomic force microscopy (AFM). XRD provides the identification of the material crystal structure, phase purity, and information about the degree of crystallinity. The size and morphology of the crystals are evaluated by SEM and TEM. High-temperature powder XRD and TG/DTG yield information on structure stability. With $N_2$ adsorption isotherms, we analyze the porous structure of the material. The optical properties of the material in the form of coating are investigated by UV–VIS, while their surface morphology by AFM.

Suitable spectroscopic techniques are used for the study of the local environment of titanium: UV–VIS spectroscopy, infrared (IR) spectroscopy, X-ray photoelectron spectroscopy (XPS), X-ray absorption spectroscopy (XAS) methods (extended X-ray absorption fine structure, EXAFS, and X-ray absorption near edge structure, XANES), and electron paramagnetic resonance (EPR). Combined use of XAS and XPS is an excellent characterization tool to elucidate the local environment of a titanium atom.

XAS became available with the development of synchrotron radiation sources. Powerful experimental methods of XAS (XANES and EXAFS) for the investigation of atomic and molecular structures of materials enable the identification of the local structure around the selected type of atoms in the sample. The analysis can be applied to crystalline, nanostructural, and amorphous materials, liquids, and molecular gases. In XANES analysis, the valence state of the selected type of the atom (e.g. titanium atom) in the sample and its unoccupied orbitals can be deduced from the information hidden in the shape and energy shift of the X-ray absorption edge. In EXAFS analysis, the number and species of neighbor atoms, their distance from the selected atom, and the thermal or structural disorder of their positions can be determined.

XPS spectroscopy is a surface-sensitive method with analytical depth of up to 10 nm. This method is based on irradiation of sample surface with soft monochromatic X-ray beam in ultrahigh vacuum and subsequent energy analysis of photo-emitted core-level electrons from the surface atoms. The energy distribution of emitted electrons presents the XPS spectrum, from which one can deduce the surface composition, and in addition, from the chemical shift of the XPS spectra, one can conclude about chemical bonds, i.e. valence states of selected atoms (e.g. titanium atoms). In combination with Ar-ion sputtering, it is possible to obtain information on depth distribution of elements up to a few hundreds of nanometer in depth.

EPR is a technique for studying bulk and surface defects in titania and radical intermediates at the surface [29]. By determining the g-value (i.e. a dimensionless quantity that determines the magnetic and angular moment of a particle), it is a very sensitive tool to obtain information about the symmetry of the paramagnetic center (i.e. $Ti^{3+}$ or oxygen vacancies), and in the case of the presence of nuclei with nonzero nuclear spin interacting with the unpaired electron, it allows a detailed mapping of the electron spin density via the so-called hyperfine interaction. The technique is particularly useful in the study of metal oxide photocatalysts due to the low frequency of the electromagnetic radiation required for electron spin resonance (in the range of microwaves), which allows the recording of EPR spectra either in the dark or under irradiation with higher frequency light (UV, visible, infrared) without interference [30].

## 9.2 Case Studies

### 9.2.1 SiO$_2$-Supported TiO$_2$ for Removal of Volatile Organic Pollutants from Indoor Air Under UV Light

To study the influence of the mesoporous SiO$_2$ structure upon TiO$_2$ insertion, we synthesized organically free 10 nm large pure anatase titania nanoparticles (further denoted as TiO$_2$ NP) and incorporated them in the ordered well-known mesoporous silica called SBA-15 (further denoted as mesoporous SiO$_2$) [28], obtaining mixed oxide mesoporous TiO$_2$–SiO$_2$ (further denoted as mesoporous TiO$_2$–SiO$_2$). The Ti : Si molar ratio was set to 1 : 1, which was shown to offer the greatest benefits when combining TiO$_2$ with SiO$_2$ [31]. Such mixed oxide materials were later immobilized onto glass slides and thermally treated at 150 °C for better adherence. The photocatalytic efficiency was tested in the gas phase for two of the major indoor air pollutants, toluene and formaldehyde.

We were interested in the location and aggregation of the TiO$_2$ NP in the whole mixed oxide structure. From Figure 9.2, it is seen that the loading of titania NPs into mesoporous SiO$_2$ support led to a decreased specific surface area, pore volume, and multimodal porosity. It can be concluded that TiO$_2$ NPs have been dispersed inside of the support, which was confirmed also by TEM (Figure 9.3a,b). However, TiO$_2$ NPs were also seen outside the pores, a fact which was additionally confirmed by IR spectroscopy, where not much Ti–Si bond vibration was detected at ~970 cm$^{-1}$, which would appear if most particles were present inside the pores. This, together with the decrease of specific surface area (from 855 m$^2$ g$^{-1}$ for mesoporous SiO$_2$ to 460 m$^2$g$^{-1}$ for mesoporous TiO$_2$–SiO$_2$), showed that the pore structure of mesoporous SiO$_2$ was partially destroyed, and TiO$_2$ NPs were present across the material rather homogeneously. The latter is one of the main goals of inserting photocatalytic NPs into any matrix. Their homogeneous distribution and prevention of aggregation are beneficial in terms of effective light absorption and increased accessibility of active sites.

Related to the performance in photocatalysis, the encapsulation of TiO$_2$ NP in mesoporous SiO$_2$ improved the crystallinity of titania. We characterized this by a lower number of structural defects, which can act as recombination sites for the generated charge carriers, hence decreasing their lifetime and lowering the rate of e$^-$–h$^+$ couple generation per incident photon (quantum efficiency). As it is known that toluene is adsorbed on the surface Ti–OH as well as Si–OH groups via OH–$\pi$ electron-type interaction [32], the fast degradation in the case of mesoporous TiO$_2$–SiO$_2$ sample is a strong indication that toluene molecules adsorb on surface titanol groups and not only silanol ones, as that would result in high disappearance of toluene but not complete decomposition to CO$_2$ as is the case here. On the other

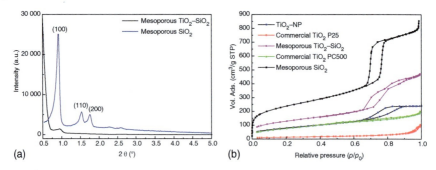

**Figure 9.2** Physicochemical characterization of the samples: anatase titania nanoparticles (TiO$_2$ NP) are 10 nm in size, in the mesoporous TiO$_2$–SiO$_2$ mixed oxide sample, the molar ratio of Ti : Si is 1 : 1, and commercial titania P25 and PC500 are added as a reference. Low-angle XRD patterns (a) and N$_2$-physisorption isotherms (b) are shown. Source: Šuligoj et al. [28]/with permission of Elsevier Science & Technology Journals.

hand, this could also mean that toluene adsorbs primarily on Si–OH groups close to active sites, i.e. TiO$_2$, since titania NPs were located inside as well as outside the mesoporous SiO$_2$ pores.

The turnover frequencies (TOF) of the catalysts are shown in Figure 9.3c. The TOF is a criterion for catalyst activity on the molecular scale: number of moles of reactant converted to product per mole of catalytic active sites. The addition of mesoporous SiO$_2$ improved the activity of TiO$_2$ NP and even surpassed the comparable standard commercial TiO$_2$ materials (Degussa P25 from Evonik and Millennium PC500 from Millennium Inorganic Chemicals) in the case of toluene. In the case of formaldehyde, other researchers have shown [33] that TiO$_2$ photocatalysts were able to decompose formaldehyde to a certain degree, whereas the 10 wt.% addition of other metal oxides, i.e. CrO$_2$, WO$_3$, MnO$_3$, and ZnO, generally decreased the decomposition efficiency, except for SiO$_2$ where the efficiency was increased up to 94% [34], which is in accordance with our results. We explained the phenomenon by considering the band gap ($E_g$) values of the catalysts. Although commercial PC500 exhibited high $E_g$, total decomposition of formaldehyde was not reached, hence this confirms the claim that the high $E_g$ value in this sample is partly a consequence of the presence of the amorphous phase of titania. On the other hand, the decomposition efficiency of TiO$_2$ NP was improved from 87.9% to 91.3% when grafted to mesoporous SiO$_2$, which is a consequence of improved crystallinity and good access of O$_2$ and the model molecule to catalytic sites, as discussed earlier.

Therefore, the improved distribution of TiO$_2$ NP incorporated in/on mesoporous SiO$_2$ affected the accessibility of the active sites to the pollutant, thus improving the degradation kinetics. On the other hand, better crystallinity of TiO$_2$ NP in/on mesoporous SiO$_2$ prolonged the lifetime of charge carriers, thus improving the efficiency of the reaction. Thus, titania in colloidal form, with the size of NPs small enough to be encapsulated into mesopores of silica, is appropriate for the impregnation of such SiO$_2$. Accordingly, the mesoporous TiO$_2$–SiO$_2$ sample has proven to be the most suitable for toluene as well as formaldehyde decomposition, and the addition of

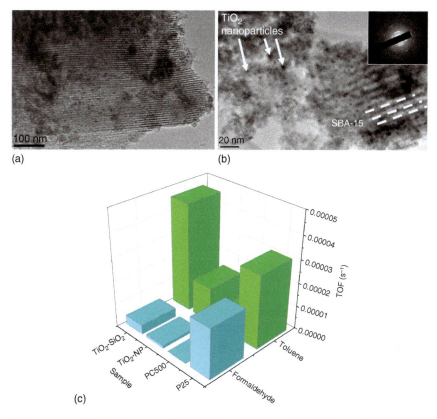

**Figure 9.3** TEM images and catalytic results. In (a), pristine mesoporous $SiO_2$ is shown with the ordered hexagonal pore channels, and in (b), partially destructed mesoporous $SiO_2$ structure can be seen upon incorporating the $TiO_2$ NP, confirmed by the inserted diffractogram. $TiO_2$ NP are visible as dark spots due to higher atomic number of Ti vs. Si. In (c), the TOF values of formaldehyde and toluene in the presence of different catalysts are shown. Source: Modified from Šuligoj et al. [28]/with permission of Elsevier Science & Technology Journals.

mesoporous silica is a purposeful way of improving the properties of photocatalytic materials.

### 9.2.2 Sn-Functionalized $TiO_2$ as a Photocatalytic Thin Coating for Removal of Organic Pollutants from Water Under UV Light

Since one of the disadvantages of $TiO_2$ photocatalyst is a high electron–hole recombination rate, coupling with other semiconductors is one of the strategies to improve its efficiency [35]. This method is only possible if the direct bonding (coupling) of the two semiconductors is efficient [36]. The first choice among the possible semiconductors is the coupling of two crystal phases of titania, i.e. anatase and rutile. The most commercially known mixed-phase titania is Degussa P25, which is a mixture of anatase and rutile (80%/20%) and shows greatly improved photoactivity compared to the pure anatase phase [37]. Due to its superior photoactivity, it is used in a variety

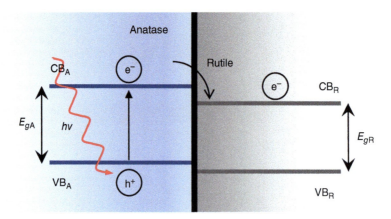

**Figure 9.4** Relative band position of anatase and rutile crystalline phase. Source: Maver [11].

of applications and as a standard photocatalyst for evaluating other photocatalytic materials in powder form.

To explain an improved photocatalytic activity of the anatase/rutile mixture system, the energy diagram of the relative band gap energy positions is crucial to describe the mechanism of the photocatalytic process. The band gaps of rutile and anatase $TiO_2$ polymorphs are 3.0 and 3.2 eV, respectively, but are slightly shifted relative to each other [38]. Figure 9.4 shows a possible relative position of the band gaps.

As long as the relative energies of the bands and the contact between the components are sufficient, the junction between the two semiconductors can lead to a physical separation of the photogenerated charge carriers. In this case, electrons migrate from the higher energy CB to the lower energy one, while holes migrate from the lower VB to the higher energy one. In this way, the physical separation of charge carriers leads to their prolonged lifetime, making them available for redox reactions at the surface and consequently increasing the photoefficiency of the material.

Among the crystalline $TiO_2$ forms, rutile is the stable phase. Anatase is thermodynamically metastable and transforms to rutile at temperatures above 550 °C. Some cations promote anatase-to-rutile transformation at much lower temperatures than thermodynamically possible in undoped $TiO_2$. They act as nucleation sites for the anatase-to-rutile conversion [39]. In our case [11], we used Sn as a promoter to induce the formation of the rutile phase around them. This is possible because the ionic radii of Sn(IV) and Ti(IV) are similar (0.083 nm and 0.0745 nm for Sn(IV) and Ti(IV), respectively, an increase of about 10%). More importantly, both have the same valence state, and $Sn^{4+}$ can easily replace Ti(IV) in the crystal lattice site.

The Sn-modified $TiO_2$ photocatalysts were synthesized by a low-temperature sol–gel method based on organic titanium and tin precursors with different molar concentrations of Sn to Ti (symbolized as $Sn_xTi_{1-x}O_2$, where $x = 0.1$–20 mol%) [35]. The precursors underwent sol–gel processes together. After a prolonged time of refluxing, a stable yellowish-white sol was obtained. The sol–gels were deposited on the glass substrates (Figure 9.5) by a dip-coating technique to obtain

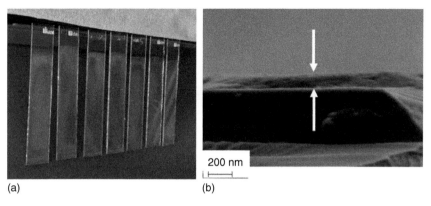

**Figure 9.5** (a) Sn-modified photocatalysts with increasing Sn concentration (from left to right for 0–20 mol% of Sn) deposited on glass substrate by a dip-coating technique. (b) SEM image of Sn-modified (0.1 mol% Sn) $TiO_2$ thin films dried at 150 °C with the estimated thicknesses of 120 nm. Source: Maver[11].

the photocatalysts in the form of a thin film (thickness of ≈120 nm, SEM image in Figure 9.5b). The films were dried and then heated at 150 °C. The films have good adhesion to the substrate and can be reused in several cycles of photocatalytic experiments.

The photocatalytic activity of the films was determined by measuring the degradation of an azo-dye (Plasmocorinth B, PB) as a function of UV irradiation time (Figure 9.6a). Only one layer of the film was sufficient to show adequate photocatalyst performance. The kinetics of the degradation curves shows an exponential decay of the PB dye over the UV irradiation time. The most photocatalytically active films are the Sn-modified $TiO_2$ photocatalysts with low Sn cation concentrations (0.1–1 mol%). These samples completely degrade the dye earlier than 100 minutes and show an increase in photocatalytic activity of up to 40% compared with unmodified $TiO_2$ (Figure 9.6b). When the Sn cation concentrations increase above 1 mol%, the photocatalytic activity becomes lower than the activity of unmodified $TiO_2$ and decreases significantly with increasing Sn concentration (10 and 20 mol%).

To explain the mechanism of improved photoactivity, the crystalline phase composition of the samples was determined by XRD. The results (see Table 9.1) show that there is no difference in crystalline composition in the unmodified $TiO_2$ and in the samples with low Sn concentration (0.1–1 mol%). These samples consist of about 70% anatase and 30% brookite (ineffective for photocatalytic reactions). When the Sn concentration increases above 1 mol%, the rutile phase increases significantly and reaches 83% in the samples with 20 mol% Sn. Thus, at higher Sn concentrations, the reduced sample photocatalytic activity can be explained by the excessive amount of rutile, which is a photocatalytically less active phase compared to anatase.

Other factors determining the photocatalytic properties were also investigated (the results presented in Table 9.1): crystallite size (determined by XRD), specific surface area (determined from $N_2$-physisorption isotherms), and morphology (observed by

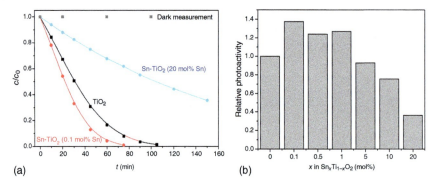

**Figure 9.6** (a) PB dye concentrations as a function of UV illumination time for unmodified $TiO_2$ and Sn-modified $TiO_2$ catalysts with 0.1 and 20 mol% Sn – only the most and least photocatalytically active Sn-modified catalysts shown, respectively. (b) Histogram of relative photocatalytic activities of unmodified and Sn-modified $TiO_2$ samples dried at 150 °C. $x$ denotes mol% of Sn in Sn-modified $TiO_2$ photocatalysts, where $x = 0$ means unmodified $TiO_2$. Source: Modified from Maver et al. [35].

TEM). All the samples with the Sn concentration below 1 mol% are composed of nanocrystals less than 6 nm in size, have a high surface area ($\approx 200\,m^2/g$), and are composed of quasi-round NPs. The obtained results, which showed no significant differences between the samples in the concentration range below 1 mol%, could not explain the improved photocatalytic performance.

To unambiguously explain the structural composition around the Sn cations, the structural features below 1 nm had to be identified. For this purpose, the Sn K-edge EXAFS analysis (Figure 9.7) was used to determine the local structure around the Sn centers and their incorporation site on the titania NPs.

The EXAFS results of the Sn-modified photocatalysts show that the Sn centers act as nucleation sites of the anatase-to-rutile conversion at temperatures far below the thermodynamic conversion in the undoped $TiO_2$. The Sn cations are already incorporated at the Ti sites of rutile $TiO_2$ during the synthesis process of sol formation. The rutile $TiO_2$ phase in the form of small NPs forms around the Sn cations in the liquid Sn–$TiO_2$ sol, and then the same local Sn structure is preserved in the catalysts after drying at 150 °C, regardless of the Sn concentration. This is supplementary information to the XRD and TEM results, which, due to the detection limit of the methods, could not confirm the presence of the rutile $TiO_2$ phase in the form of small NPs in the dried Sn-modified photocatalysts with low Sn concentrations (0.1–1 mol%) showing the highest photocatalytic activity.

We can conclude that in the low Sn concentration range (0.1–1 mol%), an optimal relative ratio between the anatase and rutile $TiO_2$ phases is achieved, with an adequate contact between them that effectively enables the separation of the photogenerated electrons and holes (as shown in Figure 9.4). The physical separation of the charge carriers prolongs their lifetime, which consequently translates into enhanced photocatalytic activity.

**Table 9.1** Structural properties of TiO$_2$ and Sn-modified TiO$_2$ photocatalysts: crystalline composition and crystalline size (determined from XRD) and specific surface area (obtained from nitrogen adsorption isotherms). Uncertainty of the last digit is given in parentheses.

| Sample | Crystalline composition | Crystallite size (nm) | Specific surface area (m$^2$/g) |
|---|---|---|---|
| TiO$_2$ | 68% anatase<br>32% brookite<br>rutile not detected | 4–6 | 208(1) |
| 0.1 mol% Sn mod. TiO$_2$ | 67% anatase<br>33% brookite<br>rutile not detected | 4–6 | 202(1) |
| 0.5 mol% Sn mod. TiO$_2$ | 69% anatase<br>31% brookite<br>rutile not detected | 4–6 | 184(1) |
| 1 mol% Sn mod. TiO$_2$ | 69% anatase<br>33% brookite<br>rutile not detected | 4–6 | 194(1) |
| 5 mol% Sn mod. TiO$_2$ | 68% anatase<br>32% brookite<br>1% rutile | 4–6  10(1) | 188(1) |
| 10 mol% Sn mod. TiO$_2$ | 53% anatase<br>39% brookite<br>8% rutile | 4–6  8(1) | 192(1) |
| 20 mol% Sn mod. TiO$_2$ | 0% anatase<br>17% brookite<br>83% rutile | 5(1) 4(1) | 171(2) |

### 9.2.3 SiO$_2$-Supported TiO$_2$ Functionalized with Transition Metals for Removal of Organic Pollutants from Water Under Visible Light

For the purpose of increasing the visible light absorption properties of titania, we used a selective choice of transition metals loaded as nano-oxo-clusters on the surface of pure anatase titania and studied their influence on both sides of redox reactions – oxidation and reduction [10]. While most research papers are focused on the oxidation side of the reaction, this does not suffice to completely describe the photocatalytic system, especially when it is modified by the introduction of another phase. We used the reaction of oxidation of terephthalic acid by holes and the reaction of reduction of resazurin to resorufin by employing photogenerated electrons. Both reactions were conducted in a solid–solid interface using the same modified TiO$_2$, which was entrapped in the SiO$_2$ mesh for mechanical integrity (Figure 9.8).

The photocatalytic results showed an optimal concentration of metal grafting to be about 1% (not presented here). The reasons for the optimal concentration of clusters present on the surface can be explained as follows. Under suboptimal loading, there is a lack of surface species, and all photogenerated electrons reach the surface Ni

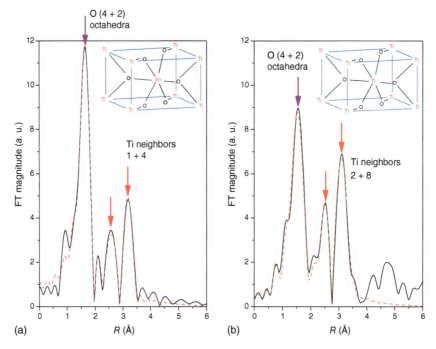

**Figure 9.7** Fourier transform magnitude of the $k^3$-weighted Sn K-edge EXAFS spectrum measured on Sn-modified TiO$_2$ (10 mol% Sn) photocatalyst dried at 150 °C (a). The spectrum of $k^3$-weighted Ti K-edge EXAFS spectrum of crystalline rutile TiO$_2$ reference is added for comparison (b). The FT-spectra are calculated in the $k$-range of 4.5–14.5 Å$^{-1}$ (black solid line, experiment; red dashed line, EXAFS model calculated in the R-range of 1.2–3.8 Å). The corresponding crystal structures of Sn cation incorporated at Ti site of rutile TiO$_2$ and the reference rutile TiO$_2$ crystal cell are inserted. Source: Modified from Maver [11].

oxo-clusters that mediate the reduction of $O_2$ to $O_2\cdot^-$. It has been shown experimentally and theoretically that the presence of excess NiO on the surface of titania leads to a rise in the level of VB (TiO$_2$), thus reducing the oxidation strength of holes [40].

Furthermore, this was true for both oxidation and reduction reactions. The oxidation of terephthalic acid (Figure 9.9a) showed the beneficial role of Ni and Zn, while the reduction of Resazurin ink (Figure 9.9b) showed a significant improvement. Such a result was reported before by the Tada group [40], where only NiO surface clusters (but not, for instance, FeO$_x$ clusters) were shown to be beneficial for both oxidation and reduction photocatalytic reactions. We will discuss this in the later paragraphs. Such corroboration gave us a hint that the nature of nanoclusters is at least partially oxidic.

To further elucidate the nature of such clusters on the surface of TiO$_2$, additional characterization was done. The spatial distribution of co-catalysts on the surface of TiO$_2$ is crucial for a successful mechanism; they must be well separated, since aggregation leads to poor performance and decreases available active sites. The absence of diffraction peaks in XRD analysis confirmed the small nature of grafted species on the surface, while the anatase TiO$_2$ structure was affected (Figure 9.10g).

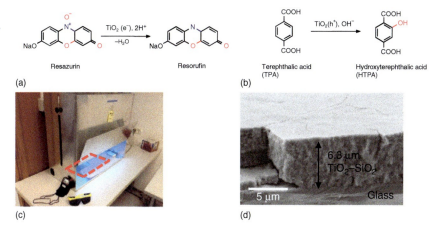

**Figure 9.8** Experimental model compounds used in the study (a–b), experimental setup (c), and the SEM crosscut, showing the thickness of the layers on the glass carrier (d). Source: Šuligoj et al. [10]/with permission of Royal Society of Chemistry.

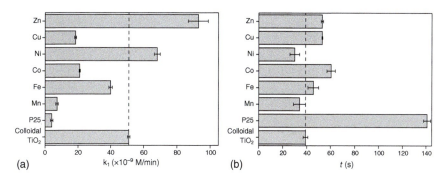

**Figure 9.9** Catalytic results of solar light oxidation of terephthalic acid (a) and reduction of Resazurin ink (b). Note that in reduction reaction the lower values of bleaching time mean higher reducing capacity of the samples. Source: Modified from Šuligoj et al. [10].

A superb distribution of Ni clusters is seen by TEM in Figure 9.10f – Ni is evenly deposited on the surface of $TiO_2$ and in the interparticle pores also. The micrographs (Figure 9.10a–e) also confirm the nature of particles on the surface, i.e. nanoclusters. The size distribution of oxo-clusters on the surface of titania is homogeneous and monomodal in size. The electron energy loss spectra (EELS) taken from the center of the $TiO_2$ particles (not shown here) demonstrated the typical $Ti^{4+}$ octahedral nature of Ti, while the ones taken closer to the outside of titania missed the $t_{2g}$ and $e_g$ splitting – some $Ti^{3+}$ is present on the surface; this effect was also confirmed by the EPR technique as is discussed in the next paragraph.

EPR spectra recorded at 20 K showed the formation of oxygen radicals ($O_2^{-}\cdot$) and $Ni^{2+}$ centers upon visible light irradiation, depicted as the difference in Figure 9.11. A broad signal at the low magnetic field was ascribed to very small Ni NPs, with an effective g-value of 2.28 [41]. These can be observed since Ni particles with a size significantly less than 50 nm (indicated also with TEM) may have a non-stoichiometric

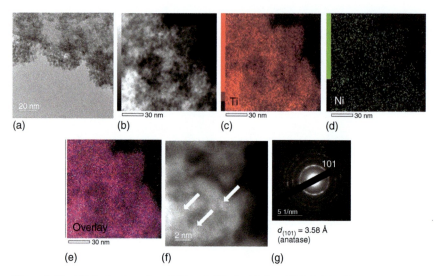

**Figure 9.10** TEM micrographs of TiO$_2$ modified with 1 wt% Ni: bright-field micrographs (a), EDAX – Ti and Ni elemental mapping (b–e), high-angle annular dark-field (HAADF) STEM (f), and electron diffraction on the selected area (g). Source: Šuligoj et al. [10]/with permission of Royal Society of Chemistry.

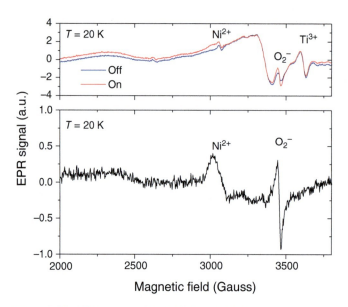

**Figure 9.11** EPR spectra of 1% Ni-TiO$_2$ at 20 K; blue and red colors in the upper plot represent spectra with the light on and off, respectively; the lower plot denotes the difference between the two. Source: Based on Šuligoj et al. [10].

nature, resulting in a significant portion of $Ni^{3+}$ and thus exhibit paramagnetism [42]. The presence of disorder in such clusters causes a reduction in size-dependent exchange interaction between two neighboring $Ni^{2+}$ ions mediated through oxygen ions and induces anisotropy of $Ni^{3+}$. Upon irradiation, the presence of $e^-$ from $TiO_2$ increased the interactions in the surrounding nickel, thus increasing the $Ni^{2+}$ signal, together with $O_2^{\cdot-}$ ($g = 2.002$). $Ti^{3+}$ centers were not seen, since these are already present in some proportion in the parent (non-irradiated) materials, as indicated by the EELS (not shown) from TEM. Thus, we confirmed the nature of Ni clusters as oxo-clusters in a size smaller than 2 nm.

To better determine the fine structure of the clusters, XAS was employed. From the XANES study (Figure 9.12a), it can be seen that the Ni-modified $TiO_2$ sample exhibited the same Ni K-edge energy position and similar edge profile as those of the reference $Ni^{2+}$ compounds Ni hyaluronate complex [43] and NiO [44], where $Ni^{2+}$ is octahedrally coordinated to six oxygens in the first coordination shell, indicating that Ni cations in the Ni-modified $TiO_2$ sample were also in a divalent state, octahedrally coordinated to six oxygen atoms in its first coordination sphere.

In Fourier transform Ni K-edge EXAFS spectrum of the Ni-modified $TiO_2$ sample, contributions of Ni neighbor shells up to about 3 Å were resolved (Figure 9.12b). Ni cations were found to be coordinated to six oxygen atoms in the first coordination shell at 2.04 Å and, on average, to one Ti and one Ni neighbor at larger distances. The Ni–O–Ti and Ni–O–Ni bridges detected in EXAFS analysis suggest that $Ni^{2+}$ cations were partially attached on the surface of the photocatalytically active $TiO_2$ NPs and partially agglomerated in NiO NPs. This fact was confirmed by measuring solid-state absorption of Ni-modified catalyst with the aid of diffuse reflectance spectroscopy (DRS) (Figure 9.12c). The fact that the redshift appeared non-monotonically confirmed this notion. In other words, an increase in concentration of Ni did not necessarily lead to a shift in absorption to longer wavelengths. The latter is an indication of successful doping of material, which did not occur in this case, hence the confirmation that the clusters were present on the surface of $TiO_2$ NPs.

In conclusion, the main reasons for improvements in activities of the samples grafted with Ni, Zn, and, in some manner, Mn are discussed in terms of the reduction potentials of metals present as nanoclusters on the surface of $TiO_2$ (Figure 9.1). The small size of metal nanoclusters allows acceptance/release of electrons from $TiO_2$ VB, and they behave mainly as metal ions in water; this is governed mainly via the position of the reduction potential of such species as close as possible to the redox potential of $O_2/O_2^{\cdot-}$ ($-0.33$ V vs. normal hydrogen electrode [NHE]). If this condition is fulfilled, the contribution of such species is still limited by the position of the VB of their corresponding oxides, since these are also present on the surface of $TiO_2$. We proved this by monitoring the formation of $\cdot OH$, where the oxides of deposited metals whose bands were bracketed by $TiO_2$ bands resulted in decreased solar light activity, probably due to higher recombination rate of electrons and holes, for which such clusters offered additional trapping sites.

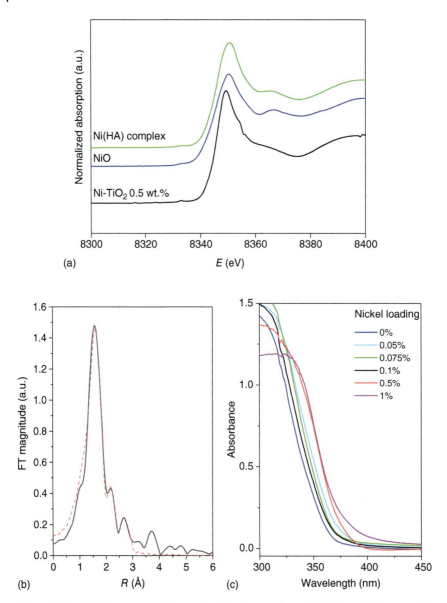

**Figure 9.12** XANES spectra of Ni-modified titania together with NiO and Ni-humic acid complex (a) and $k^3$-weighted Ni K-edge EXAFS spectrum of 0.5% Ni-TiO$_2$ (b): black solid line, experiment; red dashed line, EXAFS model. (c) DRS UV–Vis spectra of Ni-modified samples with different Ni loadings. Source: Based on Šuligoj et al. [10]. Reproduced with permission of Royal Society of Chemistry.

## 9.3 Conclusion and Further Outlook

Metal doping and combination of metal oxides have proven to be a successful strategy to obtain $TiO_2$ photocatalysts with improved optical response toward visible light and to hinder electron–hole recombination. Immobilization of such photocatalysts as a coating on the appropriately chosen carrier makes them even more applicable for water and air treatment.

It is projected that the number of commercial and institutional buildings will triple by 2050 compared with 2010, thus further design and development of coatings for photocatalytic outdoor air cleaning and self-cleaning surfaces is needed. Additionally, industrial activities generate ever-increasing amounts of wastewater, causing serious threat to the living environment and human health. Unregulated trace organic contaminants of emerging concerns (CECs), including pharmaceuticals, illicit drugs and personal care products, originate mainly from municipal wastewater effluents.

CECs cannot be efficiently removed by conventional municipal wastewater treatment using already established technologies based on biological, physical, or chemical methods. In the last decade, various systems of AOPs [25] have been intensively investigated as supplementary chemical methods, because they can completely degrade organic pollutants to inorganic compounds such as $H_2O$, $CO_2$, and inorganic salts. However, it is important to estimate the intermediate degradation products of CEC and their possible toxicity. Besides, it is also necessary to evaluate the efficiency of the prepared photocatalysts in the real wastewater conditions, where different pH and salts drastically reduce the effectivity of the AOPs.

Photocatalysts with the ability to harvest a broader spectrum of the sunlight, together with the possibility to be immobilized to appropriate carriers on larger scales, represent one of the most promising opportunities of AOP. Technical development and financial investment in such units are opening a great challenge for industry to build a greener future.

## References

1 Rockström, J., Steffen, W., Noone, K. et al. (2009). A safe operating space for humanity. *Nature* 461: 472–475. https://doi.org/10.1038/461472a.
2 Paumo, H.K., Dalhatou, S., Katata-Seru, L.M. et al. (2021). $TiO_2$ assisted photocatalysts for degradation of emerging organic pollutants in water and wastewater. *J. Mol. Liq.* 331: 115458. https://doi.org/10.1016/j.molliq.2021.115458.
3 Manisalidis, I., Stavropoulou, E., Stavropoulos, A., and Bezirtzoglou, E. (2020). Environmental and health impacts of air pollution: a review. *Front. Public Health* 8: 14. https://doi.org/10.3389/fpubh.2020.00014.
4 Janczarek, M., Klapiszewski, Ł., Jędrzejczak, P. et al. (2021). Progress of functionalized $TiO_2$-based nanomaterials in the construction industry: a comprehensive review. *Chem. Eng. J.*: 132062. https://doi.org/10.1016/j.cej.2021.132062.

5 Coronado, J., Fresno, F., Hernández-Alonso, M. et al. (2013). *Design of Advanced Photocatalytic Materials for Energy and Environmental Applications*. London: Springer https://doi.org/10.1007/978-1-4471-5061-9.

6 Fresno, F., Portela, R., Suárez, S., and Coronado, J.M. (2014). Photocatalytic materials: recent achievements and near future trends. *J. Mater. Chem. A* 2: 2863–2884. https://doi.org/10.1039/C3TA13793G.

7 Rahimi, N., Pax, R.A., and Gray, E.M.A. (2016). Review of functional titanium oxides. I: $TiO_2$ and its modifications. *Prog. Solid State Chem.* 44: 86–105. https://doi.org/10.1016/j.progsolidstchem.2016.07.002.

8 Luttrell, T., Halpegamage, S., Tao, J. et al. (2015). Why is anatase a better photocatalyst than rutile? - model studies on epitaxial $TiO_2$ films. *Sci. Rep.* 4: 1–8. https://doi.org/10.1038/srep04043.

9 Nosaka, Y. and Nosaka, A. (2016). *Introduction to Photocatalysis*. The Royal Society of Chemistry.

10 Šuligoj, A., Arčon, I., Mazaj, M. et al. (2018). Surface modified titanium dioxide using transition metals: nickel as a winning transition metal for solar light photocatalysis. *J. Mater. Chem. A* 6: 9882–9892. https://doi.org/10.1039/C7TA07176K.

11 Maver, K. (2021). Sn-modified $TiO_2$ thin film photocatalysts prepared by low-temperature sol-gel processing. Doctoral dissertation. University of Nova Gorica (Slovenia). http://repozitorij.ung.si/IzpisGradiva.php?id=6553.

12 Ola, O. and Maroto-Valer, M.M. (2015). Review of material design and reactor engineering on $TiO_2$ photocatalysis for $CO_2$ reduction. *J. Photochem. Photobiol., C* 24: 16–42. https://doi.org/10.1016/j.jphotochemrev.2015.06.001.

13 Qian, R., Zong, H., Schneider, J. et al. (2019). Charge carrier trapping, recombination and transfer during $TiO_2$ photocatalysis: an overview. *Catal. Today* 335: 78–90. https://doi.org/10.1016/j.cattod.2018.10.053.

14 Khaki, M.R.D., Shafeeyan, M.S., Raman, A.A.A., and Daud, W.M.A.W. (2017). Application of doped photocatalysts for organic pollutant degradation – a review. *J. Environ. Manage.* 198: 78–94. https://doi.org/10.1016/j.jenvman.2017.04.099.

15 Liu, M., Qiu, X., Miyauchi, M., and Hashimoto, K. (2011). Cu(II) oxide amorphous nanoclusters grafted $Ti^{3+}$ self-doped $TiO_2$: an efficient visible light photocatalyst. *Chem. Mater.* 23: 5282–5286. https://doi.org/10.1021/cm203025b.

16 Patzsch, J., Spencer, J.N., Folli, A., and Bloh, J.Z. (2018). Grafted iron(III) ions significantly enhance $NO_2$ oxidation rate and selectivity of $TiO_2$ for photocatalytic $NO_x$ abatement. *RSC Adv.* 8: 27674–27685. https://doi.org/10.1039/C8RA05017A.

17 Čižmar, T., Štangar, U.L., and Arčon, I. (2017). Correlations between photocatalytic activity and chemical structure of cu-modified $TiO_2$–$SiO_2$ nanoparticle composites. *Catal. Today* 287: 155–160. https://doi.org/10.1016/j.cattod.2016.11.039.

18 Tobaldi, D., Pullar, R., Škapin, A. et al. (2014). Visible light activated photocatalytic behaviour of rare earth modified commercial $TiO_2$. *Mater. Res. Bull.* 50: 183–190. https://doi.org/10.1016/j.materresbull.2013.10.033.

**19** Fan, L., Long, J., Gu, Q. et al. (2014). Single-site nickel-grafted anatase $TiO_2$ for hydrogen production: toward understanding the nature of visible-light photocatalysis. *J. Catal.* 320: 147–159. https://doi.org/10.1016/j.jcat.2014.09.020.

**20** MiarAlipour, S., Friedmann, D., Scott, J., and Amal, R. (2018). $TiO_2$/porous adsorbents: recent advances and novel applications. *J. Hazard. Mater.* 341: 404–423. https://doi.org/10.1016/j.jhazmat.2017.07.070.

**21** Qian, X., Fuku, K., Kuwahara, Y. et al. (2014). Design and functionalization of photocatalytic systems within mesoporous silica. *ChemSusChem* 7: 1528–1536. https://doi.org/10.1002/cssc.201400111.

**22** Kuwahara, Y. and Yamashita, H. (2011). Efficient photocatalytic degradation of organics diluted in water and air using $TiO_2$ designed with zeolites and mesoporous silica materials. *J. Mater. Chem.* 21: 2407–2416. https://doi.org/10.1039/C0JM02741C.

**23** López-Muñoz, M.-J., van Grieken, R., Aguado, J., and Marugán, J. (2005). Role of the support on the activity of silica-supported $TiO_2$ photocatalysts: structure of the $TiO_2$/SBA-15 photocatalysts. *Catal. Today* 101: 307–314. https://doi.org/10.1016/j.cattod.2005.03.017.

**24** Zhao, D., Feng, J., Huo, Q. et al. (1998). Triblock copolymer syntheses of mesoporous silica with periodic 50 to 300 angstrom pores. *Science* 279: 548–552. https://doi.org/10.1126/science.279.5350.548.

**25** Murgolo, S., De Ceglie, C., Di Iaconi, C., and Mascolo, G. (2021). Novel $TiO_2$-based catalysts employed in photocatalysis and photoelectrocatalysis for effective degradation of pharmaceuticals (PhACs) in water: a short review. *Curr. Opin. Green Sustain. Chem.* 30: 100473. https://doi.org/10.1016/j.cogsc.2021.100473.

**26** Tasbihi, M., Călin, I., Šuligoj, A. et al. (2017). Photocatalytic degradation of gaseous toluene by using $TiO_2$ nanoparticles immobilized on fiberglass cloth. *J. Photochem. Photobiol., A* 336: 89–97. https://doi.org/10.1016/j.jphotochem.2016.12.025.

**27** Krivec, M., Žagar, K., Suhadolnik, L. et al. (2013). Highly efficient $TiO_2$-based microreactor for photocatalytic applications. *ACS Appl. Mater. Interfaces* 5: 9088–9094. https://doi.org/10.1021/am402389t.

**28** Šuligoj, A., Štangar, U.L., Ristić, A. et al. (2016). $TiO_2$–$SiO_2$ films from organic-free colloidal $TiO_2$ anatase nanoparticles as photocatalyst for removal of volatile organic compounds from indoor air. *Appl. Catal., B* 184: 119–131. https://doi.org/10.1016/j.apcatb.2015.11.007.

**29** Chiesa, M., Paganini, M.C., Livraghi, S., and Giamello, E. (2013). Charge trapping in $TiO_2$ polymorphs as seen by electron paramagnetic resonance spectroscopy. *Phys. Chem. Chem. Phys.* 15: 9435–9447. https://doi.org/10.1039/C3CP50658D.

**30** Žerjav, G., Teržan, J., Djinović, P. et al. (2021). $TiO_2$-$\beta$-$Bi_2O_3$ junction as a leverage for the visible-light activity of $TiO_2$ based catalyst used for environmental applications. *Catal. Today* 361: 165–175. https://doi.org/10.1016/j.cattod.2020.03.053.

**31** Tasbihi, M., Štangar, U.L., Škapin, A.S. et al. (2010). Titania-containing mesoporous silica powders: structural properties and photocatalytic activity

**32** Nagao, M. and Suda, Y. (1989). Adsorption of benzene, toluene, and chlorobenzene on titanium dioxide. *Langmuir* 5: 42–47. https://doi.org/10.1021/la00085a009.

**33** Zhang, Y., Xiong, G., Yao, N. et al. (2001). Preparation of titania-based catalysts for formaldehyde photocatalytic oxidation from $TiCl_4$ by the sol–gel method. *Catal. Today* 68: 89–95. https://doi.org/10.1016/S0920-5861.

**34** Liao, Y., Xie, C., Liu, Y. et al. (2012). Review paper. *Ceram. Int.* 38: 4437–4444. https://doi.org/10.1016/j.ceramint.2012.03.016.

**35** Maver, K., Arčon, I., Fanetti, M. et al. (2021). Improved photocatalytic activity of anatase-rutile nanocomposites induced by low-temperature sol-gel Sn-modification of $TiO_2$. *Catal. Today* 361: 124–129. https://doi.org/10.1016/j.cattod.2020.01.045.

**36** Maver, K., Arčon, I., Fanetti, M. et al. (2021). Improved photocatalytic activity of $SnO_2$-$TiO_2$ nanocomposite thin films prepared by low-temperature sol–gel method. *Catal. Today* https://doi.org/10.1016/j.cattod.2021.06.018.

**37** Hurum, D.C., Agrios, A.G., Gray, K.A. et al. (2003). Explaining the enhanced photocatalytic activity of Degussa P25 mixed-phase $TiO_2$ using EPR. *J. Phys. Chem. B* 107: 4545–4549. https://doi.org/10.1021/jp0273934.

**38** Maheu, C., Cardenas, L., Puzenat, E. et al. (2018). UPS and UV spectroscopies combined to position the energy levels of $TiO_2$ anatase and rutile nanopowders. *Phys. Chem. Chem. Phys.* 20: 25629–25637. https://doi.org/10.1039/c8cp04614j.

**39** Hanaor, D.A.H. and Sorrell, C.C. (2011). Review of the anatase to rutile phase transformation. *J. Mater. Sci.* 46: 855–874. https://doi.org/10.1007/s10853-010-5113-0.

**40** Jin, Q., Ikeda, T., Fujishima, M., and Tada, H. (2011). Nickel(II) oxide surface-modified titanium(IV) dioxide as a visible-light-active photocatalyst. *Chem. Commun.* 47: 8814–8816. https://doi.org/10.1039/C1CC13096J.

**41** Alonso, F., Riente, P., Sirvent, J.A., and Yus, M. (2010). Nickel nanoparticles in hydrogen-transfer reductions: characterisation and nature of the catalyst. *Appl. Catal., A* 378: 42–51. https://doi.org/10.1016/j.apcata.2010.01.044.

**42** Yi, J.B., Ding, J., Feng, Y.P. et al. (2007). Size-dependent magnetism and spin-glass behavior of amorphous NiO bulk, clusters, and nanocrystals: experiments and first-principles calculations. *Phys. Rev. B: Condens. Matter.* 76: 2–6. https://doi.org/10.1103/PhysRevB.76.224402.

**43** Tratar Pirc, E., Arcon, I., Kodre, A., and Bukovec, P. (2004). Metal-ion environment in solid Mn(II), co(II) and Ni(II) hyaluronates. *Carbohydr. Res.* 339: 2549–2554. https://doi.org/10.1016/j.carres.2004.07.025.

**44** Marinšek, M., Gomilšek, J.P., Arčon, I. et al. (2007). Structure development of NiO-YSZ oxide mixtures in simulated citrate-nitrate combustion synthesis. *J. Am. Ceram. Soc.* 90: 3274–3281. https://doi.org/10.1111/j.1551-2916.2007.01924.x.

# 10

## Metal Oxides for Photoelectrochemical Fuel Production

*Gian Andrea Rizzi and Leonardo Girardi*

*Università degli Studi di Padova, Dipartimento di Scienze Chimiche, Via Marzolo 1, Padova, 35131, Italy*

## 10.1 Introduction to Photoelectrochemical Cells

Nowadays, the importance and the interest in the field of solar fuels have increased due to the necessity of carbon-free and renewable energy sources.

Various compounds have been studied as solar fuels like methanol, methane, and hydrogen. One of the most promising is hydrogen, which can be stored and used as a fuel or can be used to produce important chemicals like $NH_3$ (fundamental to produce fertilizer) or other important substances that can, as well, be employed as fuels for transport purposes. The steps depicted in the following text are part of the so-called hydrogen economy (Figure 10.1), a theoretical economy system based on carbon-free energy sources, like hydrogen and renewable energy sources.

The most common way to produce $H_2$ on a large scale is by steam reforming [1], partial oxidation, and coal gasification [2]. These processes are high-temperature processes (400–1100 °C) and involve the use of fossil fuels (oil and coal) that lead to the production of a large number of greenhouse gases (for 1 ton of $H_2$, 9–12 tons of $CO_2$ are produced). It is obvious that these processes are not sustainable and, therefore, it is important to find "greener" alternatives. There is no doubt that one of these alternatives is the "water splitting" (WS) process:

$$2H_2O_{(g)} \rightarrow 2H_{2(g)} \uparrow + O_{2(g)} \uparrow \qquad \Delta G = 228.7 \text{ kJ/mol} \qquad (10.1)$$

This simple reaction is very promising because the reagent is water (more than 70% of the Earth is covered by it), and the products are simply hydrogen and oxygen that are important reactants for a lot of industrial reactions. This reaction can be performed by thermal decomposition [3, 4], by radiolysis [5], and by photoelectrochemical (PEC) reactions [6, 7]. This last option has been widely studied, and it is already used in laboratory-scale implants. The PEC WS must be optimized to have a sufficiently low cost per kg of $H_2$ and be economically appealing. The hydrogen production from fossil fuels costs 1–1.8 $/kg, while the production by renewable

---

*Tailored Functional Oxide Nanomaterials: From Design to Multi-Purpose Applications*, First Edition.
Edited by Chiara Maccato and Davide Barreca.
© 2022 WILEY-VCH GmbH. Published 2022 by WILEY-VCH GmbH.

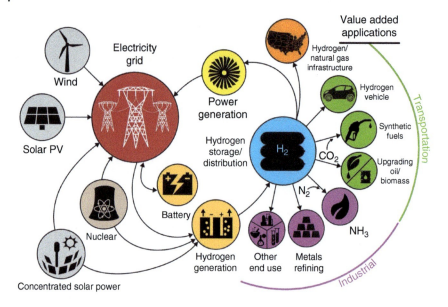

**Figure 10.1** Scheme of the production and the uses of hydrogen in the hydrogen economy. Source: H2@Scale program, U.S. Department of Energy. https://www.energy.gov/eere/fuelcells/h2scale.

sources has a cost in the range of 2.50–6.80 $/kg (https://www.rechargenews.com/transition/a-wake-up-call-on-green-hydrogen-the-amount-of-wind-and-solar-needed-is-immense/2-1-776481).

$$2H_2O_{(l)} \rightarrow 2H_{2(g)} \uparrow + O_{2(g)} \uparrow \qquad \Delta E = 1.23\,V \tag{10.2}$$

To perform this reaction, it is necessary to apply an electric potential between two electrodes immersed in water (usually an electrolytic solution). An electrode can be defined as a material that electrically connects an electric conductor (metal, semiconductor) with an ionic one (electrolytic solution) and allows to have a flow of electrons from the solution or to the solution upon application of a potential.

An electrochemical reaction involves a charge transfer between the electrode and species in solution. The transfer can be an electron that goes into the solution (reduction) or from the solution into the electrode (oxidation). These reactions will take place only if there is a good overlap between the energy levels of the electrode (Fermi level) and the ones of the substance (highest occupied molecular orbital [HOMO]–lowest unoccupied molecular orbital [LUMO]) in the solution. The applied potential can shift the energy levels of the electrode until a good match (overlap) between these energy levels is reached.

In principle, it is necessary to apply a potential higher than 1.23 V to start the WS process, when nonideal electrodes and solutions are considered. In addition, the kinetics of the mechanism introduces some overpotential ($\eta$). The overall reaction can be divided into two half-reactions: on the electrode at the higher potential, the

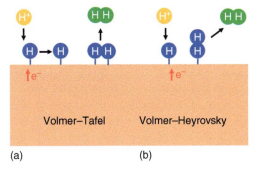

**Figure 10.2** Schematic representation of the HER mechanisms. Adsorption and reduction processes of a proton on a surface: the adsorbed hydrogen atom migrates on different surface sites until another proton is found in close proximity. In (a) the atoms react to form an adsorbed hydrogen molecule that can desorb into the solution (Volmer–Tafel mechanism), while in (b) a proton impinges on top of an adsorbed hydrogen atom, and the two ions receive an electron from the substrate to form an adsorbed hydrogen molecule (Volmer–Heyrovsky mechanism). Source: Based on Zhou et al. [8].

oxidation of water oxygen evolution reaction (OER) will take place:

$$6H_2O \rightarrow O_2 + 4H_3O^+ + 4e^- \qquad E^\circ_a = 0.817 \text{ V} \qquad (10.3)$$

while at the electrode at the lower potential, the reduction of water occurs (hydrogen evolution reaction [HER]):

$$4H_2O + 4e^- \rightarrow 2H_2 + 4OH^- \qquad E^\circ_c = -0.413 \text{ V} \qquad (10.4)$$

where $E^\circ_c$ and $E^\circ_a$ are the equilibrium half-cell potentials and the pH = 7 in standard conditions (25 °C, 1 atm). As expected, the water molecule is very stable, and to split it, a high amount of energy is necessary.

The half-reactions are pH dependent due to the presence of $H_3O^+$ and $OH^-$ species. In the case of an acidic medium (pH = 0), the reactions at the electrodes are

$$4H_3O^+ + 4e^- \rightarrow 2H_2 + 4H_2O \qquad E^\circ_c = 0 \text{ V} \qquad (10.5)$$

$$2H_2O \rightarrow O_2 + 4H_3O^+ + 4e^- \qquad E^\circ_a = 1.23 \text{ V} \qquad (10.6)$$

The HER mechanism in acidic solutions (Figure 10.2) involves the surface adsorption of hydrogen ions [8] (Volmer adsorption) with a first reduction step, and, after that, two different paths can be followed: the adsorbed hydrogen atom can react with another one to produce the $H_2$ molecule (Volmer–Tafel mechanism), or an $H^+$ ion from the solution reacts with an adsorbed hydrogen atom (Volmer–Heyrovsky mechanism) to produce an $H_2$ molecule.

Instead, when the ambient is alkaline (pH ≫ 7) the reaction at the cathode is less favorable, while the reaction at the anode is enhanced:

$$4H_2O + 4e^- \rightarrow H_2 + 4OH^- \qquad E^\circ_c = -0.826 \text{ V} \qquad (10.7)$$

$$4OH^- \rightarrow O_2 + 2H_2O + 4e^- \qquad E^\circ_a = 0.404 \text{ V} \qquad (10.8)$$

It is important to note that this last anodic half-reaction is the bottleneck of the WS process due to the multi-electron charge transfer mechanism.

The OER pathway involves the adsorption of water on a catalytic site, $\Sigma$, and the oxidation to $\Sigma$-OH [9]. The reaction proceeds with further oxidation to $\Sigma$-O, and this last oxidation can be achieved through two different pathways that depend mainly on the system (Eqs. (10.10)–(10.12)).

$$\Sigma + H_2O \rightarrow \Sigma - OH + H^+ + e^- \tag{10.9}$$

$$\Sigma - OH \rightarrow \Sigma - O + H^+ + e^- \tag{10.10}$$

$$\Sigma - OH + \Sigma - OH \rightarrow \Sigma - O + \Sigma + H_2O \tag{10.11}$$

$$2\Sigma - O \rightarrow O_2 + 2\Sigma \tag{10.12}$$

In the case of alkaline media, the active site will adsorb the hydroxide ion that will oxidize to form $\Sigma$-OH, which will be further deprotonated and oxidized to form $\Sigma$-O. If 2 $\Sigma$-O species are close enough, they can react to produce $O_2$.

$$\Sigma + OH^- \rightarrow \Sigma - OH + e^- \tag{10.13}$$

$$\Sigma - OH + OH^- \rightarrow \Sigma - O^- + H_2O \tag{10.14}$$

$$\Sigma - O^- \rightarrow \Sigma - O + e^- \tag{10.15}$$

$$2\Sigma - O \rightarrow O_2 + 2\Sigma \tag{10.16}$$

Sometimes it happens that the $\Sigma$-OH bond will break to form a hydrogen peroxide molecule [9] that will eventually decompose into $O_2$.

$$\Sigma - OH + OH^- \rightarrow \Sigma + H_2O_2 + e^- \tag{10.17}$$

This oxidation process involves different steps characterized by energy levels that depend largely on the chemistry and morphology of the substrate. It is observed that the electrocatalytic behavior depends inversely on the bond energy of $OH^-$ with the substrate so that the higher this energy is, the lower the catalytic performances of the material. At the same time, the substrate —OH bond should be strong enough to allow the reaction to occur. This behavior is well summarized and depicted by the "volcano plot" that correlates the catalytic performances with the energetics of this reaction step (Figure 10.3). The HER has, similarly, a critical step defining the overall properties of the material, which is the hydrogen adsorption energy.

The maximum of the volcano plot corresponds to the materials with the best electrocatalytic performance. In this way, it is clear what materials are more suitable for that particular reaction: for HER (Figure 10.3a) Pt is the best, while for OER (Figure 10.3b), $IrO_x/SrIrO_x$ has the best catalytic performance. Unfortunately, these materials are all very expensive and, besides, are not abundant on the Earth [12] (about $10^{-6}$% w/w for both Pt and Ir). This means that the use of these materials will imply a very high cost of the electrodes, thus limiting the diffusion of this type

## 10.1 Introduction to Photoelectrochemical Cells

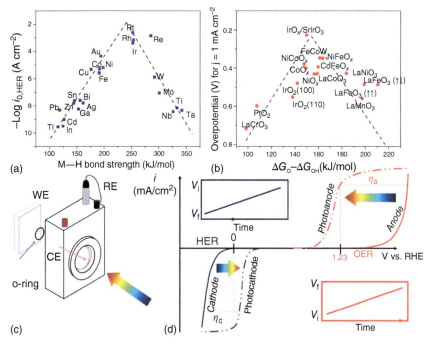

**Figure 10.3** (a) Volcano plots for HER from [10], and (b) for OER from [11]. (c) PEC cell scheme where the three electrodes are indicated as working electrode (WE), reference electrode (RE), and counter electrode (CE). (d) Example of a linear sweep voltammetry (LSV) measurement for HER and OER with the potential varied between $V_i$ and $V_f$ as function of time represented in the blue and red boxes, respectively. Source: (a) Modified from Conway [10], (b) Modified from She et al. [11].

of technology. For this reason, there is a lot of research in the development of good electrocatalyst materials composed of Earth-abundant elements. The materials are tested by an electrochemical setup, using a three-electrode cell (Figure 10.3c) coupled with a potentiostat. The cell contains the testing material, called working electrode (WE), an electrode with a stable potential that is used as a reference electrode (RE), and a noble metal or a stable electron conductor (Pt, Au, or a graphite rod) to close the electrochemical circuit (counter electrode [CE]). In this system the WE is polarized with respect to the RE so that it is possible to define a common potential scale according to the equation $V_{(RHE)} = V_{app} + V_{RE} + 0.0592\,\text{pH}$, where RHE stands for reversible hydrogen electrode. To evaluate the material (WE) (photo)-electrocatalytic properties, techniques like linear sweep voltammetry (LSV) and cyclic voltammetry (CV) that consist in applying a linear crescent in time potential ($V(t) = V_i + vt$) are commonly used. The potential is raised from a value where the reaction is not happening to a condition in which the reaction occurs (CV is a LSV with a return-step); see Figure 10.3d. Usually, precious information about the kinetics of the process can be obtained from LSV experiments [13] using the Tafel relation [14] $\eta = \pm b\,\text{Log}(i/i_0)$, where $i_0$ is the exchange currents, which represent the current for the process at equilibrium without overpotential ($\eta = 0$).

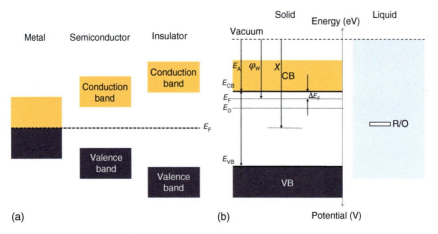

**Figure 10.4** (a) Schematic representation of the electronic structure of a metal, semiconductor, and insulator. $E_F$ indicates the Fermi energy. In semiconductors and insulators, the valence band ($E_{VB}$) and conduction band ($E_{CB}$) edges are separated by an energy gap ($E_G$). (b) Schematic representation of the electronic structure of an n-type semiconductor: $E_D$ indicates the energy level of donor species, $\chi$ is the absolute electronegativity of the semiconductor, $E_A$ is the electron affinity, and $\varphi_W$ indicates its work function.

This value allows to evaluate the electrocatalytic properties of a material (e.g. for HER, $-\text{Log}(i_0)$ is 3.1 for Pt and 12.3 for Hg). The Tafel slope, $b$, is a value that depends on the slowest mechanism process [14] (Figure 10.4).

## 10.1.1 The Photoelectrochemistry Approach

Until now we have considered electrocatalytic materials that work in dark conditions and require an external bias that could be in principle generated from a photovoltaic cell, converting the solar energy into the electrical power needed by the electrolyzer. This is an indirect approach to solar energy storage because it needs two energy conversion steps (radiative to electric and electric to chemical). If at least one of the electrodes of the electrolyzer is a semiconductor, it is possible to use UV–visible radiation to generate the minority carriers (holes or electrons) that can migrate toward the surface to participate in one half-reaction of the WS process.

The semiconductor electrode has the role of light absorption, and, in the case of a photoanode (the electrode that works for the OER half-reaction under illumination), the absorption process will lead to the generation of a hole in the valence band (VB) and the promotion of an electron in the conduction band (CB). These charges must remain separate and migrate, one, through the back contact, to the CE, and the other to the semiconductor surface, where it will be injected in the solution (electron for HER). In addition, it is required that the potential of the electron in the CB is more negative than the $H^+$ reduction potential, while the hole's potential in the VB must be more positive than the reduction potential of $O_2$. The VB and CB of an n-type semiconductor (Figure 10.5a), at the interface with an electrolyte, after the

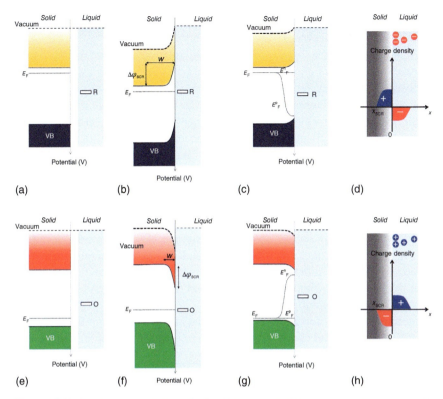

**Figure 10.5** Band scheme of the solid–liquid junction (SLJ) in the case of a photoanode (a)–(d) and of a photocathode (e)–(h). (a), (e) Semiconductor and solution band scheme before the contact. (b), (f) Junction formation between the semiconductor and the solution, and (c), (g) illumination of the semiconductor generating holes in the valence band, causing a reduced band bending. The Fermi level of the semiconductor is indicated by $E_F$, while the Fermi levels for holes and electrons are indicated by $E_F^p$ and $E_F^n$, respectively. $W$ is the width of the space charge region (or depletion zone) and $\Delta\varphi_{SCR}$ is the potential formed as the result of charge accumulation. (d), (h) Representation of the charge density accumulated on each part of junction causing the band bending for n and p semiconductors, respectively.

respective Fermi energies have aligned, undergo a band bending toward more negative potentials due to the difference in the position of Fermi levels (see Figure 10.5b). This band bending is the reason why an n-type semiconductor can only work as a photoanode (holes will be injected in solution). On the contrary, in the case of a p-type semiconductor, a band bending toward more positive potentials will occur, at the interface with the electrolyte. A p-type semiconductor will work as a photocathode. The thickness through which the band bending occurs is defined as the depletion region (or space charge region [SCR]). The SCR thickness ($W$) depends on the material, the pH, and the electrolyte:

$$W = \sqrt{\frac{2\varepsilon \Delta\varphi_{SCR}}{q_e N_D}} \tag{10.18}$$

where $\varepsilon$ is the dielectric constant of the material, $q_e$ is the charge of the electron, $N_D$ is the concentration of donor impurities, and $\Delta\varphi_{SCR}$ is the potential caused by the accumulation of charges on the interface.

With illumination (see Figure 10.5c), there will be a generation of minority carriers (holes in the case of photoanodes), and to describe the system a single value of Fermi energy of the semiconductor ($E_F$) is not enough. For this reason, it is important to introduce the quasi-Fermi level for each carrier type as depicted in Figure 10.5c: $E_F^p$ for holes and $E_F^n$ for electrons. The difference between quasi-Fermi level of the holes $E_F^p$ (electrons $E_F^n$) and the potential value for the OER (or HER) corresponds to the photovoltage produced by a photoanode (photocathode) material, under illumination.

In some conditions, it is possible to fine-tune the material properties, developing a composite material that, under illumination, allows the generation of the electrical potential required by the WS reaction. In this case, no bias is applied, and it is possible to directly convert the energy of the solar light into chemical species (solar fuels).

These systems are composed usually of a semiconductor or, more frequently, a junction between two different semiconductors ($TiO_2$, $BiVO_4$, ZnO, $WO_3$, $Fe_2O_3$, $Cu_2O$, etc.), decorated with some electrocatalysts (Pt, $IrO_x$, $Co_3O_4$, etc.) to improve the charge transfer in solution.

Nevertheless, one of the main problems is the charge separation. It is necessary that the photogenerated carriers do not undergo a recombination process and, at the same time, they must possess good mobility.

The energy conversion process is limited by several factors, such as charge separation efficiency, carrier mobility, the efficiency of charge injection in the solution, etc. The active material is, in the case of simple photocatalysis, usually in powder form and dispersed in the reaction solution (usually an alkaline media) to maximize the exposed area. In this case, the produced gases are not separated, so an additional effort is needed to separate the products. This means also that cross-reactions are always possible (oxygen reduction and hydrogen oxidation). Another approach consists of the separation of the two active materials while keeping an electric contact between them [15]. The two active materials are p-type (photocathode) and n-type (photoanode) semiconductors. A WS process that does not require the application of an external bias can be achieved by the fabrication of a device where the anodic and cathodic compartments are separated by a membrane.

The sum of the photovoltages developed by the photoanode and photocathode should, of course, be higher than 1.23 V. The illumination of photoanode and photocathode can be done in transmission mode [16, 17] (Figure 10.6a), but it requires a transparent and conductive supporting material (fluorine-doped tin oxide [FTO] or indium tin oxide (ITO)) and, of course, a partial absorption of the light occurring so that the anode and cathode are not equally illuminated (this depends on the light absorption characteristics of the material). A second possibility is to use photoanodes and photocathode in parallel mode [18] (Figure 10.6b), and, in this case, the light intensity is the same for the two electrodes. It is also possible to create a tandem device where one side performs the OER and the other one the HER (Figure 10.6c). Such a device can be obtained by growing the photocatalysts

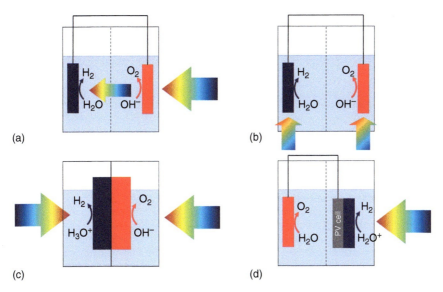

**Figure 10.6** Possible geometries of illumination and electrical contact for bias-free photoelectrochemical cells: (a) illumination in transmission mode, (b) in parallel mode with an external electrical connection, (c) scheme of a tandem device, and (d) a photovoltaic (PV) cell integrated with a photocathode for the HER. Source: (a) Based on Fan et al. [16], Song et al. [17], (b) Based on Wang et al. [18].

on both sides of the same supporting material, usually metal or semiconductor with a good conductivity [19, 20].

Another approach consists of the coupling of an electrocatalyst with a photovoltaic cell that generates the required potential to trigger both WS semi-reactions (Figure 10.6d). It is possible to use one photovoltaic (PV) cell for each electrode to increase the performance.

## 10.2 Metal Oxides Photoelectrode Candidate Materials

Fujishima was the first to use a semiconductor to perform this type of reaction, using $TiO_2$ ($E_g$ = 3.1 eV) [21]. After his pioneering work, many studies were performed on $TiO_2$ to improve its catalytic and solar light absorption properties. Titanium oxide has many positive properties like chemical stability, nontoxicity, and a high abundance. The main drawbacks of $TiO_2$ are a wide band gap that does not match with the solar spectrum (Figure 10.7), the kinetics of charge transfer in solution, a high recombination rate [23], and the low mobility of photogenerated carriers. To overcome these problems the first approach was to create heterojunctions with other semiconductors with lower band gaps [24, 25] or plasmonic nanoparticles [26], with the idea of enhancing the generation of carriers available for the electrochemical reaction. Other researchers also tried to coat the material with some cocatalysts to increase the reaction rate due to the presence of more active catalytic sites for

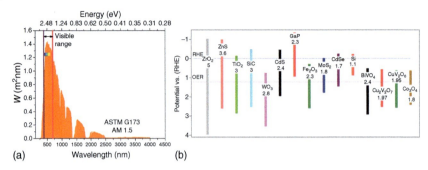

**Figure 10.7** Solar Spectra at 37° of latitude with air mass coefficient (AM) 1.5 (a) [22], and band edges position for several semiconductors (b). Below the chemical formula, the value of the band gap of the material is indicated. The approximate band edges position can be estimated from the absolute electronegativity value $\chi$ of a semiconductor. $\chi$ can be calculated as the geometrical average using the absolute Pearson's electronegativity values of each element according to Eq. (10.19) where $a$ and $b$ are the stoichiometric coefficients of elements $i$ and $j$ in the semiconductor. Since the electronegativity value is assumed to correspond to the center of the band gap (vacuum energy scale) and the electron affinity, $E_A$(vac), to the edge of the CB, the VB position is defined by Eq. (10.20). Source: (a) Based on Würfel and Würfel [22].

HER/OER [25, 27].

$$\chi_{sc} = \left( \chi_i^a \cdot \chi_j^b \right)^{(1/(a+b))} \tag{10.19}$$

$$E_{VB} = -\chi_{sc} - \frac{1}{2} E_g \tag{10.20}$$

Over the years, other materials were studied, in particular oxides (binary and ternary) [28, 29], as well as nitrides (e.g. $C_3N_4$ [30]), oxynitrides (e.g. TaON [31]), and, of course, pure elements (e.g. Si, Ge)[32], III–V semiconductors (e.g. GaAs, InP) [33, 34], II–VI semiconductors (e.g. CdS, CdSe) [35], sulfides and selenides (e.g. PbSe, PbS) [36], and even chalcogenides (e.g. $MoS_2$, $MoSe_2$, etc.) [37–39].

All these materials present good properties, but other characteristics are not suitable for large-scale applications. The ideal material for PEC WS should possess all the following properties:

- Low production cost and easy reparability of the raw materials.
- Stability under PEC working conditions.
- Suitable band gap to match the solar spectrum.
- Good charge injection in solution and charge separation (good efficiency).
- Band edge position close to the specific oxidation/reduction reaction potential.

There are two main ways to the research of the best PEC material: the first is a combinatory approach that consists in the preparation of a library of new phases [40, 41] that might in principle be suitable to be used as photoanodes or photocathodes, while the second one consists in the modification of an already rather good material trying to overcome its drawbacks. The modification involves doping, band-structure engineering, creating a junction with other materials, or the nanostructuring of the material to reduce the path of the photogenerated carriers to reach the surface.

Nowadays, a large number of semiconductor materials are known, but just a few of them are well-suitable for the PEC process.

### 10.2.1 Photoanodes

The number of photoanodes candidate materials is high, especially in the group of transition metal oxides (TMO). These oxides are usually n-type semiconductors, due to the presence of oxygen vacancies in the lattice, a common type of defect for this class of materials, which causes a partial reduction in the metal ions (decrease of the oxidation state) to balance the charge. The semi-reaction occurring on the surface of these materials is the OER that implies the $O_2$ evolution. Another possible reaction that can be exploited for the $H_2$ production on the CE is the oxidation of organic substances until a complete mineralization to $CO_2$ is eventually achieved. In this last case a photoanode can be used for the very important purpose of water remediation [42]. Oxidation of common hole scavengers like $H_2O_2$, $SO_3^{2-}$, and $NO_2^-$ is always possible and easy.

Important binary oxides are $TiO_2$, $Fe_2O_3$, and $WO_3$ [26, 43–45], while one of the most studied ternary oxide is $BiVO_4$ [28, 46], although other oxides like $FeVO_4$ and many copper vanadates (CVO) are also studied. Other possible interesting candidates to work as photoanodes are transition metal dicalcogenide (TMDC) ($MoS_2$, $WS_2$) [47–50], oxynitrides (TaON) [31], nitrides (g-$C_3N_4$) [30], and perovskites $ABO_3$ (e.g. $SrTiO_3$) [51, 52].

Among typical semiconductor materials, silicon is the best-known and studied also in the PEC WS field. It has a band gap of 1.1 eV, and it is characterized by the extrinsic nature of the carriers that limit its electronic conduction. This drawback can be easily overcome by the doping procedure (P or B) that greatly increases its conduction properties. The use of silicon wafers for the PEC WS is problematic due to the corrosion process that occurs in strongly alkaline conditions (the native $SiO_2$ layer will dissolve leading to continuous oxidation of Si by $H_2O$), which makes it impossible to use bare silicon wafers as photoanodes. A lot of work was done in studying and optimizing the deposition on the $SiO_2$/Si surface of protective oxide layers that catalyze the reaction for the oxidation of water [53] and, at the same time, prevent the corrosion of silicon. These protective layers are usually TMO [54–56].

### 10.2.2 Photocathodes

Photocathodes are p-type semiconductors and present a band bending with the liquid interface that allows to the electrons an "easy flow" toward the solution. The main semi-reaction occurring on a photocathode surface is the HER, but other reduction reactions can also occur, such as the $CO_2$ reduction. The $CO_2$ reduction usually occurs together with the HER since the reduction potentials of both products are similar. It is important to point out that a good catalyst for

HER is not suitable for $CO_2$ due to the selectivity toward the hydrogen evolution, but in principle a photocathode could work for both reactions. The electron's flow to the semiconductor surface can be helpful also in terms of the stability of the material. In fact, some semiconductors cannot be used for the OER due to the severe surface corrosion process (silicon, GaP, etc.), while they can be used for the HER (with opportune doping), thanks to the parallel reduction process given by the electrons that prevent the oxidation mechanism [57] in harsh environments. Some of these materials are, as well, TMO, and the most studied one is $Cu_2O$ [58, 59], which possesses good properties despite the tendency to photodegradation and the stability only at neutral and alkaline pHs that limits the kinetics of the HER process ($Cu_2O$ disproportionate to $Cu^{2+}$ and Cu has a pH value below 4). Other p-type oxide are NiO, $Co_3O_4$, the ternary oxides of copper, $CuBi_2O_4$ [60], iron, $CuFeO_2$ [61], $LaFeO_3$ [62], and $CaFe_2O_4$ [63]. Other important materials are of course the ones that dominate the photovoltaics sector like p-type Si, III–V semiconductors as p-InP [33], p-$GaInP_2$, II–VI semiconductors like CdTe [35], and $CdIn_{1-z}Ga_zSe_2$ (CIGS) [64]. The last one is studied a lot due to the possibility to tune its band gap (and the positions of the band edges) by changing the stoichiometry. This behavior is also typical of copper chalcopyrite $CuIn_xGa_yS_2$ [65]. The surface of these materials is usually decorated by islands of cocatalyst metals like Pt, Rh, Ru, or Re to boost the electrocatalyst properties [66, 67].

## 10.3 Tailoring Surface Catalytic Sites and Catalyst Use

Large band gap metal oxides are characterized by energy band positions suitable for WS but lack significant visible light harvesting. Many attempts have been made to modify their band gap, thus increasing their visible light absorption [68]. The creation of new states inside the band gap by forming oxygen vacancies or doping with non-metals or transition metals allows having smaller energy transitions, thus narrowing the optical band gap. For instance, the annealing in the air of $TiO_2$ modifies the number of oxygen vacancies [69], while the nitrogen of hydrogen doping, as performed by Wang et al., results in the production of a yellowish material that is characterized by visible light absorption [70, 71]. Nevertheless, in the case of hydrogen doping, there is not a general agreement to attribute the performance enhancement to the reduced band gap. Instead, the enhancement of the PEC performances is attributed to an increase of the majority carries concentration (n-type semiconductor behavior) [72]. Similar studies on band gap modification by doping have been done also on $WO_3$ and ZnO [73–76]. Unfortunately, these treatments do not allow more than a few percentage efficiency increase, since creating states inside the band gap also enhances the number of detrimental recombination paths for electron–hole pairs. More recently interest has been focused on plasmonic nanocrystals to enhance light absorption in the visible range. For instance, $TiO_2$ nanotubes decorated by glutathione-capped $Au_x$ nanoparticles present light absorption below 525 nm [77, 78]. Zhang et al. also proved that 20 nm Au particles not only increase the light absorption in the visible range but also enhance the UV conversion efficiency.

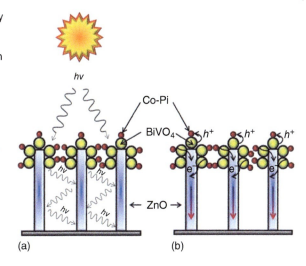

**Figure 10.8** Design strategy of a BiVO$_4$/ZnO heterojunction, involving (a) increased light absorption and charge generation in both BiVO$_4$ and ZnO in conjunction with light-trapping effect of the nanorods and (b) electron injection into ZnO nanorods followed by prompt electron transport along ZnO nanorods and simultaneous hole transfer to Co–Pi for efficient water oxidation. Source: Moniz et al. [93]. Licensed under CC BY-4.0.

Surface plasmon resonance of the Au nanocrystals is supposed to inject hot electrons to the TiO$_2$ CB with an increase from 1.22 to 2.25 mA cm$^{-2}$ at 1.23 V vs. RHE [79].

The optimization of material morphology by controlling their growth processes at the nanoscale opens a completely new and profitable field. Nanoscale structuration increases the surface area in contact with the electrolyte, and kinetic overpotentials are reduced by lower current densities. Also, light absorption and charge transport are enhanced because the light penetration depth and photogenerated charges diffusion length are decoupled. Thin films with a flat morphology require thicknesses suitable to absorb a significant amount of the incident light, while photogenerated charges need diffusion lengths short enough to reach the electrocatalytic surface and the back contact. In the presence of nano-architectures, the absorption distance for photons is decoupled from the diffusion length so that higher absorptions are possible, and, at the same time, the diffusion length is shorter resulting in a reduced current loss [80]. Nanostructured surfaces are also characterized by large areas in contact with the electrolyte, reducing local current densities and thus OER and HER overpotentials. TiO$_2$ nanorods and nanotubes [72, 81–84], WO$_3$ nanowires and nanoflakes [85, 86], CuV$_2$O$_6$ nano-stripes [87], BiVO$_4$ [88], Fe$_2$O$_3$ [89–91], and Cu$_2$O [59] have been obtained as nano-architectures with enhanced PEC performances. For instance, the Cu–Ti–O nanotube array, obtained by anodization, was characterized by a high surface area and high light absorption. The light absorption could be optimized by changing the Cu–Ti–O ratio, obtaining photocurrents from 0.035 to 0.065 mA cm$^{-2}$ in 1 : 1 methanol/H$_2$O electrolyte [92]. A good example of nanostructuration with light-trapping effect is the one involving BiVO$_4$/ZnO nanorod photoanodes exhibiting high photocurrent under visible light (c. 2 mA cm$^{-2}$ at 1.23 V vs. RHE). The introduction of a Co–Pi oxygen evolution catalyst further improved the photocurrent to 3 mA cm$^{-2}$ with an incident photon-to-current efficiency (IPCE) of 47% at 410 nm [93]. It is worth noting that in this case the efficiency was further improved due to the light-trapping effect of vertically aligned ZnO nanorods (see Figure 10.8).

Another example can be that of $Cu_2O$ nanowires protected by $ZnO/TiO_2/RuO$ that showed increased light absorption and charge separation because of their nanostructuration (5 mA cm$^{-2}$ in the case of the flat device vs. 8 mA cm$^{-2}$ in the case of the nanostructured system) [59, 94]. The combination of substrate nanostructuration and nanoscale material deposition results in improved light absorption and photogenerated charge creation. The deposition of Mo-doped $BiVO_4$ on nanocone-shaped substrates with Fe/NiOOH cocatalyst showed a photocurrent density up to 6 mA cm$^{-2}$ at 1.23 V vs. RHE [95]. The minimization of superficial electronic states that may act as trap states for the electron–hole recombination is also an interesting way to optimize the performances of a photoactive material. This goal can be achieved by incorporating other metals on the surface to fill the trap states and at the same time working as OER catalysts. In the case of $Fe_2O_3$ photoanodes, the surface recombination can be partly suppressed by decoration with $CoO_x$ [96] nanoparticles and by surface passivation with $Ga_2O_3$ [97] or $TiO_2$ [98]. In the case of $TiO_2$-based photoanodes, the surface recombination can be lowered by removing detrimental chlorine atoms with a 250 °C annealing treatment [81] or by increasing the surface hydroxyl groups that can act as hole trap sites [83]. The deposition of ultrathin films (ideally atomically thick) is also a very effective strategy to suppress detrimental surface states and recombination paths like in the case of $Co_3O_4/TiO_2$ nanotubes array coated by atomic layer deposition (ALD)-grown $HfO_2$ ultrathin films (Figure 10.9) [99].

Optimal post-fabrication temperature, time, and annealing atmosphere control may also play a crucial role for many metal oxides like α-$Fe_2O_3$ where morphology, Sn doping, and oxygen vacancies significantly improve the PEC performances when the annealing is carried out in a low oxygen partial pressure atmosphere [44]. A further example is the annealing temperature and time in the case of $Co_3O_4$ thin films grown of a $SiO_2/Si(100)$ wafer by magnetron sputtering. The formation of $Co_3O_4$ nanopetals and the interface formed with the $SiO_2$ native layer strongly depend on the annealing conditions. A lower annealing temperature and a longer annealing time favor the formation of $Co_3O_4$ nanopetals and, at the same time, the obtainment of a graded interface where CoO smoothly turns into $Co_3O_4$ [42, 56]. The presence of CoO in contact with $SiO_2$ favors an easier injection of holes from Si into the $CoO_x$ overlayer [56].

Finally, photoelectrodes like $Fe_2O_3$, $Cu_2O$, or $TiO_2$ have surface states that allow to directly perform the desired reaction, but the use of a cocatalyst can strongly enhance their PEC performances, thanks to favorable synergistic effects with the cocatalyst particles. Usually, these particles are Pt, $MoS_2$, or $RuO_x$ for the HER [59, 100, 101], while for OER Ni/FeOOH, Co–Pi (amorphous cobalt phosphate) or $IrO_x$ are often used [102–106]. $Cu_2O$ photocathodes are often decorated not only with Pt nanoparticles or $RuO_x$ but also with the much less expensive $MoS_2$ cocatalyst. For instance, amorphous $MoS_2$ on $Cu_2O$ led a 5.7 mA cm$^{-2}$ photocurrent density at 0 V vs. RHE at pH = 1 [101]. Different from HER, OER is a complex four-electron process, often characterized by slow kinetics. For this reason, the addition of a cocatalyst can often improve the performances of $TiO_2$ and $Fe_2O_3$ photoanodes [107]. For instance, the addition of Co–Pi to α-$Fe_2O_3$ photoanodes resulted in a more than 15% increase in

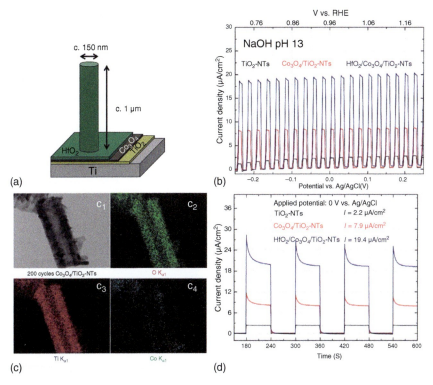

**Figure 10.9** TiO$_2$ nanotubes modified by atomic layer deposition of Co$_3$O$_4$ and by a further deposition of HfO$_2$ are schematized in (a). Transmission electron microscopy (TEM) images with composition analysis are shown in (c). The linear sweep voltammetry and chronoamperometry are shown in (b) and (d), respectively, both with chopped illumination. These measurements coupled with IMPS measurements (see Section 10.6) confirm that the overlayers increase the charge transfer efficiency and the absorption properties of the material. The HfO$_2$ layer further improves the properties, thanks to its passivating effect that limits the recombination processes. Source: (a) Modified from Huai et al. [99], (b), (d) Reproduced with permission of Huai et al. [99], (c) Huai et al. [99]/with permission of Elsevier.

photocurrent density and over 200 mV cathodic shift [108]. The electrodeposition of Co–Pi allowed also the improvement of BiVO$_4$ photoanodes up to 1.7 mA cm$^{-2}$ [109, 110]. Incorporation of FeOOH on BiVO$_4$ electrodes, as well, led to 2.0 mA cm$^{-2}$ photocurrent density in 0.1 M KH$_2$PO$_4$ + 0.1 M Na$_2$SO$_3$ electrolyte solution.

## 10.4 Metal Oxide Heterostructures

The formation of heterojunctions, which is the electrical connection between two materials, is a very common way to improve the light absorption properties, the efficiency of migration, and the separation of charges.

When two materials are in electrical contact, the difference of the Fermi levels acts as a driving force for the carrier's migration: electrons tend to move where $E_F$ is lower, while holes will move where it is higher (vs. the vacuum scale). This migration creates a charged interface between the two materials, and the result is the formation of an electric field that, as anticipated before, enhances the charges separation and migration after the generation process induced by illumination. This description is qualitative and does not require the use of equations, although for a complete and rigorous description of the formation of heterojunctions, reading of a book of solid-state physics or physics of semiconductors [111–113] is highly recommended.

When two semiconductors are in contact, the junction that can be obtained depends on the position of the band edges. In particular, if the VB and the CB of one material are both at higher energy with respect to the corresponding edges of the other semiconductor ($VB_1 > VB_2$, $CB_1 > CB_2$), the junction is defined as type I [110]. Instead, if the edges of one material are between the edges of the other ($VB_1 > VB_2$, $CB_1 < CB_2$), the heterojunction is known as type II. If one of the two materials is a metal, a Schottky junction is formed. In this case, the formation of a potential barrier (related to the position of the Fermi levels of the materials) creates an "obstacle" to the migration of electrons. This behavior is normally used in a diode to create the typical asymmetrical response as a function of the applied potential. The same type of junction is used to describe the dynamic of the solid–liquid interface (SLJ).

Another type of junction classification relies on the nature of the majority carriers of the two semiconductors. When the two materials have the same carrier's type, the junction is defined as n–n or p–p and usually leads to a moderate improvement of the PEC properties with respect to the bare material. In the case of different carrier types, there is the formation of a p–n junction [114, 115], characterized by the formation of an intense electric field (built-in potential). Until now, we are considering junction where the carriers are generated and migrate toward the interface with the electrolyte, thanks to the application of an external bias or the buildup of an internal electric field, but it is also possible that the generated carriers in one of the semiconductors recombine with carriers of opposite charges in other material. This condition is known as the Z-scheme, which can be defined as "direct" when the neutralization occurs directly between carriers in VB of one material (holes) and carriers in the CB of the other material (electrons) [116, 117]. The Z-scheme is defined as "redox-mediated" if the presence of a redox couple mediates the process [118, 119]. The mechanism through which these heterojunctions work is explained pictorially in Figure 10.10.

## 10.5 Metal Oxides as a Protective Anti-corrosion Layer in Photoelectrodes

PEC devices should ideally use photo absorbers characterized by a very long activity to be cost-effective. The material in contact with the electrolyte must be thermodynamically stable and possibly corrosion-resistant or, at least, characterized by very low corrosion kinetics. Typical corrosion mechanisms can be the material dissolution in the electrolyte, the formation of a passivating or insulating layer on the surface, and a chemical modification of the photo absorber. Pourbaix diagrams are,

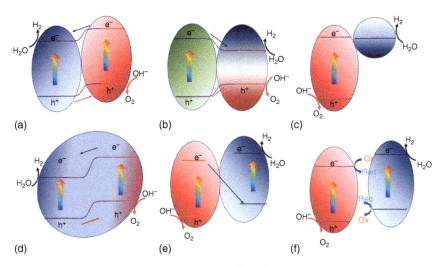

**Figure 10.10** Schematic representation of possible heterostructures between semiconductors: (a) junction type I; (b) type II; (c) Schottky junction with a metal; (d) p–n junction; (e) Z-scheme; and (f) redox-mediated Z-scheme.

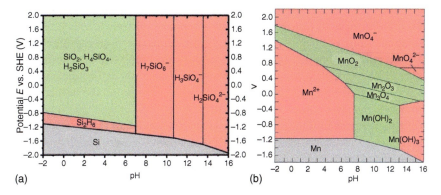

**Figure 10.11** Pourbaix diagram for silicon (a) and (b) manganese. Source: Weinrich et al. [120]. Licensed under CC BY-4.0, (b) Tem5psu Mn pourbaix diagram [121]. In these diagrams the regions in green represent the condition of passivation of the material with formation of solid species, the red ones represent the conditions of formation of water-soluble species (corrosion), and the gray areas represent the range of pHs and potentials that allow the presence of silicon and manganese.

of course, very important to predict, with some approximation, the behavior of a material as a function of the pH of the electrolyte, the potential applied, and the eventual presence of coordinating ions. Typical Pourbaix diagrams, calculated for Si and Mn are reported in Figure 10.11.

An example of a photoactive material, which in KOH electrolyte is very quickly corroded, is Si. The native $SiO_2$ layer (about 2 nm thick) that is always present on the Si surface reacts in an alkaline solution forming water-soluble silicates ($SiO_{2(s)} + 4$ $KOH_{(aq)} \rightarrow K_4SiO_{4(aq)} + 2H_2O_{(l)}$). In this way, the Si surface, directly in contact with the electrolyte, reacts with $H_2O$ ($Si_{(s)} + 2H_2O_{(l)} \rightarrow SiO_{2(l)} + H_{2(g)}$), and a new $SiO_2$ layer is formed. The progressive dissolution of $SiO_2$ and oxidation of Si by water leads to a very quick corrosion process. There are many ways to protect a semiconductor surface from corrosion, very well described in the review by Ros et al. [122]. The first method consists of the deposition of a thin metallic layer (thin enough to be transparent) that should work as a catalyst and/or forming a Schottky junction [123–127]. A second way consists in the deposition of a thin oxide layer, thin enough to be tunneled and resistant to harsh environment [128, 129]. A third method consists in the deposition of a thicker oxide layer, transparent to the radiation used to excite the semiconductor, and forming a heterojunction with the semiconductor [130]. The first protective layers, used for short band gap semiconductors, were metallic layers obtained by thermal or electron beam evaporation of noble metals like platinum, palladium, silver, ruthenium, or rhodium [123–125, 127]. Si, n-GaP, p-WSe, and p-InP were often protected in this way, although not significant stabilities were achieved [131, 132]. Interesting results were achieved by using protective metallic layers, but the full potential of the photo absorbers in long-term stabilities cannot be exploited because of significant light absorption in the metallic layer and significant recombination in the Schottky junction due to Fermi level pinning and extra states at the interface [133]. Insulating metal oxide thin films have been studied to prevent corrosion of photoelectrodes used in the WS reaction. The insulating layer must be sufficiently thin to allow the photogenerated charges to be injected into the solution through a tunneling process across the protective layer. For this reason, these films must be very thin ($\leq 3$ nm) and, at the same time, forming a defect-free (conformal) coverage of the semiconductor surface. Usually, these films are obtained by ALD, a technique that usually allows conformal coverage and the formation of pinhole-free coverages. Quite often the insulating layer is coupled with a catalyst film of nanoparticles forming a metal–insulator–semiconductor (MIS) structure or semiconductor insulator semiconductor (SIS) structure. These types of structures have the advantage to protect the photo absorber and, at the same time, prevent the Fermi level pinning [128, 129]. $SiO_2$ was the first insulating and protective layer to be studied since it naturally forms when Si is exposed to air or acidic electrolytes. Esposito et al. prepared, on p-type Si, a 2 nm thick $SiO_2$ layer decorated with Pt/Ti 20–30 nm thick islands to be used as photocathode for the HER. In the obtained MIS structure, Pt was the HER catalyst, while Ti, with its low work function, enhanced the photovoltage. Although 20–30 nm metallic islands are not transparent to light, a careful control of the distance and diameter among the islands allowed to obtain photocurrents of 20 mA cm$^{-2}$ at 0 V vs. RHE and 0.55 V vs. RHE onset potential in 0.5 M $H_2SO_4$ [134]. Similarly, p-InP was protected by a 3 nm thick $TiO_2$ film where a 2 nm Ru film was added as HER catalyst. Stable photocurrents of 37 mA cm$^{-2}$ were achieved with an onset potential of 0.76 V vs. RHE [135]. An alternative way to protect p-InP photocathodes was that of replacing $TiO_2$ with $Al_2O_3$. Layers thicker than 3 nm introduced a resistance lower than that by the $SiO_2$ layer of equal thickness but higher than what was obtained by defect-conductive $TiO_2$. Photovoltages of 490 mV were obtained [136, 137]. In general, the obtained photovoltages were found to depend on the thickness of the insulating layer used for protection. Crystalline

TiO$_2$ reduced this problem in comparison with amorphous TiO$_2$ deposited by ALD, because of dielectric constant increase, obtaining a record photovoltage of 623 mV [138]. Indeed, TiO$_2$ is one of the best candidates to be used as an "insulating" protective layer, and it was proven to be conductive even for thicknesses higher than 10 nm through a hopping mechanism via trap states located ≈1 eV below the CB [34]. This defect band conductivity was proposed as the conduction mechanism in the case of photoanodes by Campet et al. [139]. Sometimes thin passivating and protective layers naturally forms if the photoelectrode works with a bias applied and under illumination, like in the case of copper vanadates Cu$_x$V$_y$O$_z$ [140]. The formation of a thin CuO passivating layer on these ternary oxides photoelectrodes prevents the V leaching in the tetraborate buffer solution used as electrolyte. CuV$_2$O$_6$ nanostripes coated with a thin layer of Co$_3$O$_4$ deposited by CVD were found to be coated by a thin CuO protecting layer after several minutes of electrochemical work with a bias applied and under illumination. In that case, a mechanism involving the formation of CoO, Cu(I) centers, and diffusion of Co(II) ions in the CuV$_2$O$_6$ lattice was verified [87]. Generally, we can say that pin-hole-free, ultrathin metallic, or insulating films are rather difficult to obtain on large area electrodes and especially to last hundreds of hours. Of course, thicker films can last longer, avoiding pinholes and cracks probability, and for this reason, effective protection of photo absorber can be obtained by thick conductive transparent oxides [130]. For films thicker than 5 nm, tunneling is not possible, and therefore the material electronic properties, like interfacial potential barriers, band alignment, charge mobility, band bending, etc., must be carefully considered. Protective layers must be conductive and transparent to radiation so that the choice is reduced to mid-band gap semiconductors (2–5 eV), excluding insulators like Al$_2$O$_3$ or SiO$_2$ ($E_g \approx 8$–9 eV). These protective layers do not need to be as conductive as traditional transparent conductive oxides (FTO, ITO), and only vertical conductivity is needed. Usually, n-type metal oxide semiconductors are used to protect photocathodes, while photoanodes are protected by thick transparent p-type oxides [106, 130, 141, 142]. In both cases, an n–p or p–n heterojunction is formed. The most studied oxides are TiO$_2$, NiO, Co$_3$O$_4$, and CoO$_x$ (Figure 10.12) as well as MnO.

Many works can be found in the literature where tens of nanometers thick TiO$_2$ layer were used as protective coatings. A first example was photocathode protection by a 100 nm thick TiO$_2$ layer obtained by sputtering. In that case, the TiO$_2$ layer was deposited, by reactive sputtering on a 5 nm thick Ti film. A Pt layer was finally added as HER catalyst so that a p-n$^+$-Si/(5 nm)/TiO$_2$ (100 nm)/Pt was obtained. The conductivity through the TiO$_2$ layer was attributed to an electron polaron hopping mechanism through the CB or electron polaron hopping through the Ti$^{3+}$ states close to the CB. The performances of this electrode are expressed as 20 mA cm$^{-2}$ photocurrents at 0.3 V vs. RHE and 70 hours of stability under illumination [143]. The optimal conductivity through 100 nm thick TiO$_2$ ALD layers was found to be related to conductivity through crystalline regions of the film, while amorphous regions were found to dissolve in acidic HER conditions [144, 145]. Since ALD allows to obtain crystalline films at a rather moderate deposition temperature (150–200 °C), this is an ideal technique to avoid the degradation of the photo absorber with thermal treatments at a higher temperature (400–500 °C). Although

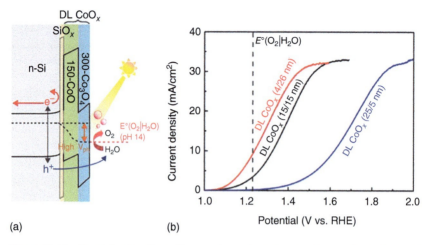

**Figure 10.12** Band diagram for a silicon wafer protected with cobalt oxide. In the article different protective films were studied: CoO, $Co_3O_4$, and the double layer $Co_3O_4$/CoO/$SiO_2$/Si with different CoO and $Co_3O_4$ relative thicknesses. The position of the VB edge of the overlayer compared with the one of silicon determines the overall PEC properties of the material. CoO has a higher VB edge compared with $Co_3O_4$, and this helps the migration of holes from Si. $Co_3O_4$ is instead characterized by very good OER-electrocatalytic properties. Source: Reproduced with permission of Oh et al. [56].

there are some examples of $TiO_2$ thick layers used to protect photoanodes [34, 146], hole conductors and p-type oxides are the ideal candidates to be used in the case of photoanodes protection. NiO-based protective layers have been widely studied because of their abundance, chemical resistance in alkaline media, stability, and catalytic properties. Its band gap is around 3.8–4.0 eV [147, 148], and therefore it does not absorb any visible light. Its p-type conductivity [149] is caused by point defects and $Ni^{2+}$ vacancies as explained by the Mott–Hubbard insulator theory [150, 151]. The $Ni^{2+}$ vacancies introduce extra oxygens in the structure and the compensation of their charge is obtained by the presence of $Ni^{3+}$ sites, that is, electron acceptors. Several examples for the protection of photoanodes by NiO can be found in the literature. For instance, in 2014, Mei et al. used reactive sputtering to deposit 50 nm thick NiO to protect a buried $np^+$-Si homojunction, previously coated by 5 nm of metallic Ni. This structure was not active unless a pretreatment in a Fe-containing electrolyte was performed. Photocurrent as high as 14 mA cm$^{-2}$ at 1.3 V vs. RHE in 1 M KOH was obtained with a 300 hours stability [152]. NiO, grown by ALD at a temperature between 100 and 300 °C, was also used to protect photoanodes. This film was very effective only if periodically depolarized. This reversible degradation mechanism was attributed to the presence of Ni in higher oxidation states, like $NiO_2$, the latter being less conductive and less catalytically active [153]. Films deposited at temperatures higher than 100 °C were more resistive, less defective with a reduction of their p-type behavior. $CoO_x$ is a transition metal oxide, like NiO in many aspects. It is characterized by a p-type conductivity and a high CB edge and is an electron-blocking layer. It is also transparent to the visible part of the spectrum and is characterized by a 2.4 eV band gap [151]. $CoO_x$ in contact with highly alkaline electrolyte is hydroxylated to CoOOH, which is a stable catalyst for $O_2$ evolution [154]. For instance, the deposition of $CoO_x$ by ALD on np+-Si photoanodes allowed the obtainment of 30 mA cm$^{-2}$ at 1.4 V vs. RHE in 1 M NaOH for about 24 hours [155]. Mixed oxides like $NiCoO_x$ [156] or $CoO_x$ with 3–8% vanadium [157] were also used as protective layers. In the first case, 23 mA cm$^{-2}$ at 1.23 V were maintained in 1 M KOH for 72 hours, while the vanadium-doped 70 nm $CoO_x$ layer allowed a photocurrent of 30 mA cm$^{-2}$ from a p+n-Si electrode. MnO layers were used to protect n-Si already in 1987 by Kainthla et al. [158]. The photoanode was stable for 650 hours with a photocurrent of 1 mA cm$^{-2}$ at 1.6 V vs. RHE in a 0.5 M $K_2SO_4$ solution. MnO deposited by ALD with a thickness of 20 nm was used by Strandwitz to protect n-Si obtaining 30 mA cm$^{-2}$ in 1 M KOH solution at 1.5 V vs. RHE.

## 10.6 Evaluation of Photoelectrode Efficiencies

The efficiency of production of solar fuels ($H_2$) starts, in the case of photoelectrodes, from the definition of the mechanism that leads to the creation of minority carriers. Considering an n-type semiconductor, the generation process (illumination) increases the concentration of the free carriers, and if the number of majority carriers is not affected by this process, it is possible to consider only the minority carriers (holes) [159]. If holes are generated far enough from the SCR, they can reach the surface only through diffusion processes. Instead, the holes generated in the SCR interact with an electric field that allows them to migrate rapidly toward the surface. When the holes reach the surface of the semiconductor, they become available for a charge transfer process or can be trapped in defect states. The ideal situation where loss mechanisms are not considered is well described by the Gärtner equation, which calculates the hole flux that reaches the surface of a (in our case an n-type) semiconductor, under illumination. The holes are generated in the field-free region and the space charge zone. In the first case, the holes can migrate only through a diffusion mechanism, while in the second case they migrate toward the surface because of the electric field present in the SCR.

$$\text{EQE} = \frac{j_H}{q\phi} = 1 - \frac{e^{-\alpha W}}{1 + \alpha(\lambda) L_D} \tag{10.21}$$

$$-\ln(1 - \text{EQE}) = \alpha W_{SC} + \ln(1 + \alpha L_D) \tag{10.22}$$

$$W = \sqrt{\frac{2\varepsilon_r \varepsilon_0 (V - V_{FB} - k_b T)}{q N_D}} \tag{10.23}$$

The Gärtner equation (10.21) defines the proportionality between the photogenerated hole flux ($j_H$) and the photon flux ($\phi$). The ratio between these two quantities is defined as the external quantum efficiency (EQE). In the equations, reported in Figure 10.13, $\alpha$ is the absorption coefficient of the photon flux, $L_D$ is the diffusion length of the holes, and $W$ is the width of the SCR in the material. This value

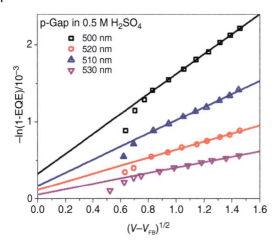

**Figure 10.13** Measurement of the EQE for p-GaP in 0.5 M $H_2SO_4$. The EQE is reported as a function of the band bending $(V-V_{FB})$. Different colors refer to different wavelengths [159, 160]. On the left of the figure, two expressions of the Gärtner equation are reported together with the equation that calculates the width of space charge as a function of the band bending. Sources: Based on Peter [159], Li [160].

depends on the applied potential ($V$) with respect to the flat band potential[1] ($V_{FB}$), the dielectric constant of the material, and the donor concentration ($N_D$). The EQE allows computing the efficiency of the generation process. As explained before, this equation does not consider any loss mechanism, like trapping on defect sites, which reduces the number of holes or electrons that can participate in the redox reaction in the solution.

$$h\nu \rightarrow e^- + h^+ \tag{10.24}$$

$$y + e^- \rightarrow X \tag{10.25}$$

$$R + h^+ \rightarrow O \tag{10.26}$$

$$X + h^+ \rightarrow Y \tag{10.27}$$

$$O + X \rightarrow R + Y \tag{10.28}$$

The overall dynamic of the minority carriers inside the material is schematized in Figure 10.14 (Eqs. (10.24)–(10.28)). It is possible to write a "continuity" set of equations for this system, describing the process in terms of change of carrier concentration with time. Thus, it is possible to link the photon flux with the effective photocurrent that flows in the system [162, 163], and it is also possible to calculate the percentage of the minority carriers that are injected into the solution. This quantity is defined as the charge transfer efficiency $\eta_{CT}$.

For this reason, to evaluate the performances of a photoelectrode correctly, it is of paramount importance not only to consider the effectiveness of conversion between the photon flux and the minority carrier flux (EQE) but also to know how efficient the injection in solution ($\eta_{CT}$) is and the selectivity of the injection toward the desired

---

1 The flat band potential is the potential required to compensate the band bending thus removing the accumulated charge at the interface. This value can be obtained from a Mott–Schottky measurement [161].

## 10.6 Evaluation of Photoelectrode Efficiencies

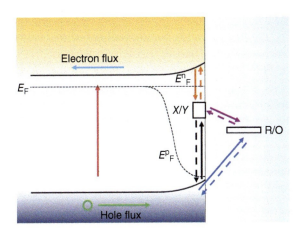

**Figure 10.14** Scheme of the carriers dynamics at the interface between an n-type semiconductor and a solution with a redox couple R/O. The semiconductor has surface states X and Y that represent the trap sites for hole and electrons, respectively. Holes can be injected into the solution form the VB (Eq. (10.26)) or from the surface states (Eq. (10.28)). Equations (10.25) and (10.27) are referred to as recombination processes for electrons and holes, respectively.

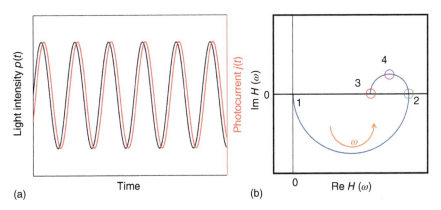

**Figure 10.15** Time-dependent light intensity and photocurrent signals. The phase shift between the light intensity signal (input) and the photocurrent (output) is visible. IMPS plot for a photoanode material, where the imaginary part vs. the real part of $H(\omega)$ are plotted.

reaction ($\eta_F$). This last quantity is defined as the Faradic efficiency. The measure of $\eta_{CT}$ is usually performed by using an intermittent illumination at different frequencies as in the case of intensity-modulated photocurrent spectroscopy (IMPS) [159]. IMPS is a technique that probes the system with a time modulated light intensity ($p(t)$), which corresponds to a time-dependent photocurrent ($j(t)$) (Figure 10.15a).

$$p(t) = p_0 + \tilde{p}_0 \text{sen}(\omega t) \tag{10.29}$$

$$j(t) = j_0 + \tilde{j}_0 \text{sen}(\omega t + \phi) \tag{10.30}$$

From these input and output signals, it is possible to obtain the transfer function in the frequency domain $H(\omega)$ [159, 164, 165]. The system is monitored in a range of frequencies suitable to observe different phenomena (usually between 20 kHz and 0.1 Hz). At high frequency (point 1 in Figure 10.15b), the main response is related to the time constant of the electrochemical cell. At lower frequencies (point 2 in Figure 10.15b), the effect of absorption and migration of minority carriers becomes important, and, in the very low-frequency range (point 3 in Figure 10.15b), the response of the system is related to processes of charge recombination and transfer. The transfer function $H(\omega)$ can be interpreted, in the case of an n-type semiconductor, working as a photoanode, by using the model proposed by Ponoamarev and Peter [165], and a similar analysis can be done on a photocathode [164]. Equation (10.32) is the total expression of the transfer function for a photoanode [165] where it is evident that the values of charge transfer ($k_{CT}$) and recombination ($k_{rec}$) rate constants are the key parameters of the process (from them is possible to define $\eta_{CT}$ with Eq. (10.33)), together with the time constant of the cell and the capacity value of the semiconductor ($C_{SC}$) and that of the Helmholtz layer ($C_H$).

$$H(\omega) = \frac{\mathscr{F}[j(t)]}{\mathscr{F}[p(t)]} = \frac{j(\omega)}{p(\omega)} \tag{10.31}$$

$$H(\omega) = \frac{j_h C_H}{(C_H + C_{SC})} \frac{(k_{CT} + i\omega)}{(k_{CT} + k_{rec} + i\omega)} \left[\frac{1}{1 + i\omega\tau_c}\right] \tag{10.32}$$

$$\eta_{CT} = \frac{LFI}{HFI} = \frac{k_{CT}}{k_{CT} + k_{rec}} \tag{10.33}$$

$$\omega_{ImH \to MAX} = k_{CT} + k_{rec} \tag{10.34}$$

Usually, the data are displayed in a Nyquist plot (Figure 10.15b), which in the case of a photoanode is formed by two semicircles: one in the fourth quadrant and the second in the first quadrant. The negative circle is related to the capacitive response of the cell (recombination [RC] circuit with a time constant $\tau_C$), and the frequency at the minimum of the plot is related to the time constant ($\tau_C$) of the measurement apparatus. The second circle (in the first quadrant) is related to the recombination and charge transfer processes of the minority carriers (in this case holes), and the frequency related to the max (4) of this circle is the sum of the kinetic constants of charge transfer and recombination (Eq. (10.34)). The high-frequency intercept (HFI) is related to the flux of holes generated from the light absorption, as calculated by the Gärtner equation [159, 166]. The value for the HFI is given by

$$HFI = j_h \frac{C_H}{C_H + C_{SC}} \tag{10.35}$$

where $j_h$ is the hole current, without any losses, while $C_H$ and $C_{SC}$ are the capacitances of the Helmholtz layer and the SCR, respectively. Usually (except for nanostructures), $C_H$ is much greater than $C_{SC}$ so that $HFI \approx j_h$.

This maximum theoretical holes current ($j_h$) is reduced due to trapping and the recombination events so that only a lower amount can be transferred to the solution.

The intercept at lower frequency LFI is defined as

$$\text{LFI} = j_h \frac{C_H}{C_H + C_{SC}} \frac{(k_{CT})}{(k_{CT} + k_{rec})} = j_h \frac{C_H}{C_H + C_{SC}} \eta_{CT} \qquad (10.36)$$

From the expressions of low frequency intercept (LFI) and HFI, it is evident that the ratio between these values gives the efficiency of charge transfer, and from the frequency value of the maximum of the imaginary part of $H(\omega)$, it is possible to derive the values of the kinetic constants for the process of charge transfer and recombination.

This technique was used by Zachäus et al. [167] to describe the surface properties of $BiVO_4$, allowing to correlate the limit in the photoelectrocatalytic performance of the material with the recombination mechanism. This aspect was pointed out by depositing on the $BiVO_4$ surface a Co–Pi (amorphous Co–Phosphate) and $RuO_x$ layers. The first one showed a reduction of the recombination kinetic constant, while in the second case, the addition of a well-known electrocatalyst for OER led to a charge transfer efficiency decrease compared to the bare $BiVO_4$. This behavior was assigned to the lower oxidation power of holes in $RuO_x$. The Co–Pi layer passivates the surface states on the $BiVO_4$ surface that cause trapping and recombination phenomena.

In this way, it is possible to quantify the ability of a material to convert the photon flux to a photocurrent ($j$).

The final value of the photocurrent is thus given by the product of the elemental charge of the electron $q_e$, the charge transfer efficiency $\eta_{ct}$, the EQE, and the photon flux $\Phi(\lambda)$, as the following:

$$j = q_e \eta_{CT} \text{EQE} \Phi(\lambda) \qquad (10.37)$$

The evaluation of the Faradic efficiency usually requires a chronoamperometry measurement together with an experimental setup that allows analyzing the formed products (either $H_2$ or $O_2$) like a gas chromatograph or an $O_2$ probe [43] (in the case of photoanodes). In this way, it is possible to verify (by measuring the concentration of the species) the selectivity of the material for the specific reaction. The equation that allows quantifying the faradic efficiency is reported as follows:

$$\eta_F \% = \frac{n° \text{mol produced}}{j * t / nF} * 100 \qquad (10.38)$$

where $j$ is the photocurrent, $t$ is the time (in seconds) for the formation of the desired product, $F$ is the Faraday constant, and $n$ is the number of exchanged electrons in the reaction.

Until now the application of potential is implicit in the definition of EQE, but this aspect has to be carefully considered when defining the efficiency of a process for the generation of solar fuels. For this reason, further definitions of efficiency are based on the ratio between the generated power ($jV$) and the illumination power. In the case of a two-electrode configuration, it is possible to define the overall efficiency for the WS process directly from the $J$–$V$ curve (assuming 100% Faradic efficiency):

$$\eta_{WS}^{2\text{electrode}} = \frac{j(1.23\ V - V_{app})}{P_{in}} \qquad (10.39)$$

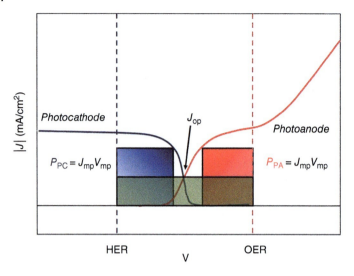

**Figure 10.16** LSV curves for a photocathode and a photoanode tested separately. The $j$ value corresponding to the intercept of the curve defines the performance of a hypothetical WS cell where these materials are used as photoelectrodes. The power produced singularly by the photoanode, and the photocathode are indicated by $P_{PA}$ and $P_{PC}$. The overall device power is obtained at the interception between the two JV curves and is given by $J_{OP} \times 1.23$ V.

In this case, it is important to separate the electrode compartment to avoid cross-reactions that would prevent a correct evaluation of the overall performance. This equation can be used also for a photocatalyst system, where no external bias is needed. In this case, the only variation is to set $V_{app} = 0$.

Usually, photoanodes and photocathodes are studied separately to optimize the material's properties, to create a device that does not require a bias application. A single photoelectrode is studied using a standard potentiostat in a three-electrode cell. It is therefore important to define the efficiency of one single photoelectrode to optimize the final fabrication process (WS cell). This quantification requires the use of the values of current and potential at the maximum power points ($J_{mp}$ and $V_{mp}$ in Figure 10.16).

$$\eta_{WS}^{3\ electrode} = \frac{J_{mp} V_{mp}}{P_{in}} \tag{10.40}$$

For a WS cell with a photocathode and a photoanode electrically connected, it is possible to define also the solar-to-hydrogen efficiency ($\eta_{STH}$). The point where the two JV curves of each photoelectrodes are crossing is $J_{OP}$, so that

$$\eta_{STH} = \frac{J_{op} 1.23\ V}{P_{in}} \tag{10.41}$$

Until now, the evaluation of the properties of the photoelectrodes was defined considering an illumination with a wide spectrum (Figure 10.7) that represents the solar emission filtered by absorption and scattering of the atmosphere [168]. The effect of the atmosphere depends on the path of light in the atmosphere, and this depends on the latitude, season, and time. Usually, it is used as the AM 1.5 standard, which is the condition of mid-latitude (around 37°) and mid-day time with a solar intensity, which is an average value during the year (in summer it is higher, while in winter it is lower) [169].

It is also important to observe how the materials (or devices) work under different types of illumination (different wavelength) and to do that it is necessary to introduce the quantities IPCE and absorbed photon-to-current efficiency (APCE) defined as

$$\text{IPCE (\%)} = \frac{jh\nu}{q_e P(\nu)} 100 \tag{10.42}$$

$$\text{APCE} = \frac{\text{IPCE}}{A(\nu)} \tag{10.43}$$

where $P(\nu)$ is the power of the illumination with a particular frequency ($\nu = c/\lambda$), which corresponds to photons of energy $h\nu$ (where $h$ is the Plank's constant), $q_e$ is the electron charge, $j$ is the photocurrent density, and $A(\nu)$ is the absorbance of the material.

## 10.7 Conclusions and Perspectives

In this chapter, we have presented and discussed several aspects of the PEC-driven WS process for the production of solar fuels (mainly $H_2$). The WS reaction can be easily obtained by using electrolyzers. These devices are simple but require the use of expensive electrocatalysts to keep the overpotentials, on both electrodes, to a minimum value so that an economically efficient process is possible. In a PEC process most of the energy corresponding to the overpotential, for the HER or OER, is given by the sunlight. The electrons and holes necessary to reduce "$H^+$" and oxidize "$O^{2-}$" in the water molecule can be obtained by light absorption in a semiconductor. Electrons are excited in the CB, while holes remain in the VB. The positions of VB and CB edges on the RHE scale depend on the material and can be roughly predicted by using the absolute electronegativity of the semiconductor. When the semiconductor is in contact with the electrolyte and the SLJ is formed, a band bending occurs. Photoanodes are characterized by an upward band bending, while for photocathodes the band bending is in the opposite direction. Many oxides and especially transition metals oxides have the characteristics to be used as photoanodes or photocathodes if excited by UV or visible light. Since the discovery of the SLJ producing few µA cm$^{-2}$ by using a $TiO_2$ photoanode, many other metal oxides such as $Fe_2O_3$, $BiVO_4$, $WO_3$, and $Cu_2O$ have been tested, modified, and improved to maximize the built-in electric field to maximize electron–hole separation and transport. Cocatalysts, doping, surface decoration, or plasmonic nanostructures helped to improve OER and HER performances, but, so far, these materials have lacked significant photocurrent generation efficiency, although they present rather high photovoltages and stability in oxidizing environments. An alternative approach was that of adapting short band gap materials, commonly used in the photovoltaic industry, like Si, GaAs, InP, $CuZnSnS_4$ (CZTS), and CuInGaSe (CIGS) to the PEC process using suitable protective layers. In the beginning, ultrathin metallic films were used to protect the semiconductor surface and work at the same time as catalysts. These very thin metallic films were not passivating enough so that the electrode's life was reduced to a few hours. Thin insulator oxides presented similar problems, and if the thickness exceeds ≈5 nm, the resistances increase dramatically. The best solution consists in protecting short band gap photo absorbers, containing a buried junction, with a semiconductor metal oxide, transparent to a large part of the visible spectrum, and as much conductive as possible. Electrodes with the stability of about 2000 hours have been prepared with photocurrent and photovoltages resembling photovoltaic cells.

Of course, the final goal is that of obtaining complete PEC cells capable of obtaining bias-free solar WS with good efficiency and durability. Only a few works describing stand-alone PEC cells can be found in the literature [105]. Systems based on III–V semiconductors have achieved solar-to-hydrogen (STH) efficiencies up to about 20%. As an example, Cheng obtained an efficiency of 19.3% with a monolithic device based on the GaInAs/GaInP structure protected by a $TiO_2$ layer [170]. Similarly, Licht et al. achieved a very good 18% efficiency with an AlGaAs/Si tandem cell with $RuO_2$/Pt catalyst [171]. Furthermore, recent example of a stand-alone PEC cell was described by Hagfeltd, Grätzel, and coworkers [58]. They proposed a cell consisting of a $Cu_2O$-based photocathode coupled to a perovskite solar cell (PSC) (setup schematized in Figure 10.6d). In that case, the $Cu_2O$ layer was deposited on the solar-blind CuSCN layer that allowed the transmission of unadsorbed solar energy to a rear photo absorber able to provide the necessary extra bias. As the rear photo absorber, PSC can collect the solar radiation up to 850 nm, with two absorbers the visible part of the solar spectrum is well covered. The two basic parts of the cell, the $Cu_2O$ photocathode and PSC, are wired together and in close contact, "back-to-back," to minimize light scattering losses. When the device is illuminated from the $Cu_2O$ side, photogenerated electrons in $Cu_2O$ are injected into the surface $RuO_x$ catalyst to reduce water, while the holes recombine with electrons from PSC through the wire. Finally, holes from PSC are conducted to the CE in the PEC cell for OER on $IrO_x$ deposited on the CE due to its excellent catalytic activity. The obtained STH efficiency was 4.55% and dropped to 4.15% after 10-hour operation, which is one of the highest efficiencies achieved by unbiased solar WS devices using oxides photoelectrodes. PEC WS has been considered for a quite long time a possible way of achieving cheap and scalable hydrogen production directly from sun and water, but lately, the development of photovoltaics and electrolyzers based on Earth-abundant elements seem to allow lower hydrogen production costs. The implementation of PEC WS at an industrial scale requires further efforts in the fabrication of efficient, stable, and scalable photoelectrodes. Stable semiconductor materials that fulfil these requirements have not been found so far and more research is needed with the help of computational design of multielement oxides or nitrides. In conclusion, the best strategy is that of protecting photo absorbers normally used for PV for long-term operations. $TiO_2$, NiO, and $CoO_x$ are the best performing protective layers usually deposited by CVD, ALD, or reactive sputtering to reduce defects and pinhole formation.

# References

1. Rostrup-Nielsen, J.R. and Rostrup-Nielsen, T. (2002). Large-scale hydrogen production. *CATTECH* 6 (4): 150–159.
2. Wagner, N.J., Coertzen, M., Matjie, R.H., and Van Dyk, J.C. (2008). Coal gasification. *Appl. Coal Petrol.* (1): 119–144.
3. Kogan, A. (1998). Direct solar thermal splitting of water and on-site separation of the products - II. Experimental feasibility study. *Int. J. Hydrogen Energy* 23 (2): 89–98.
4. Muhich, C.L., Ehrhart, B.D., Al-Shankiti, I. et al. (2016). A review and perspective of efficient hydrogen generation via solar thermal water splitting. *Wiley Interdiscip. Rev. Energy Environ.* 5 (3): 261–287.
5. Kusumoto, T., Ogawara, R., Ludwig, N. et al. (2020). Water radiolysis with thermal neutrons, fast neutrons and contamination γ rays in the accelerator based thermal neutron field: time dependence of hydroxyl radical yields. *Radiat. Phys. Chem.* 174 (May): 108978.
6. van de Krol, R. (2012). *Photoelectrochemical Hydrogen Production*. Springer US.
7. Van de Krol, R. (2012). Principles of photoelectrochemical cells. In: *Electrochemical and Photoelectrochemical Hydrogen Production* (eds. V. de Krol Roel and M. Grätzel), 14–66. Springer Science+Business Media.
8. Zhou, Z., Pei, Z., Wei, L. et al. (2020). Electrocatalytic hydrogen evolution under neutral pH conditions: current understandings, recent advances, and future prospects. *Energy Environ. Sci.* 13: 3185–3206.
9. Dau, H., Limberg, C., Reier, T. et al. (2010). The mechanism of water oxidation: from electrolysis via homogeneous to biological catalysis. *ChemCatChem* 2 (7): 724–761.
10. Conway, B.E. and Jerkiewicz, G. (2002). Nature of electrosorbed H and its relation to metal dependence of catalysis in cathodic $H_2$ evolution. *Solid State Ionics* 150 (1–2): 93–103.
11. She, Z.W., Kibsgaard, J., Dickens, C.F., Chorkendorff, I., Nørskov, J.K., and Jaramillo, T.F. (2017) Combining theory and experiment in electrocatalysis: insights into materials design. *Science (80-)*, 355 (6321).
12. Fleischer, M. (1954). The abundance and distribution of the chemical elements in the earth's crust. *J. Chem. Educ.* 31 (9): 446–455.
13. Chen, Z., Dinh, H.N., and Miller, E. (2013). *Photoelectrochemical water splitting - standards*. Exp. Methods Protocols.
14. Murthy, A.P., Theerthagiri, J., and Madhavan, J. (2018). Insights on Tafel constant in the analysis of hydrogen evolution reaction. *J. Phys. Chem. C* 122 (42): 23943–23949.
15. Zhang, K., Ma, M., Li, P. et al. (2016). Water splitting progress in tandem devices: moving photolysis beyond electrolysis. *Adv. Energy Mater.* 6 (15), 1600602.
16. Fan, K., Li, F., Wang, L. et al. (2014). Pt-free tandem molecular photoelectrochemical cells for water splitting driven by visible light. *Phys. Chem. Chem. Phys.* 16 (46): 25234–25240.

**17** Song, A., Bogdanoff, P., Esau, A. et al. (2020). Assessment of a W: $BiVO_4$–$CuBi_2O_4$ tandem photoelectrochemical cell for overall solar water splitting. *ACS Appl. Mater. Interfaces* 12 (12): 13959–13970.

**18** Wang, Y., Shi, H., Cui, K. et al. (2020). Reversible electron storage in tandem photoelectrochemical cell for light driven unassisted overall water splitting. *Appl. Catal., B* 275 (November 2019): 119094.

**19** Neumann, B., Bogdanoff, P., and Tributsch, H. (2009). $TiO_2$-protected photoelectrochemical tandem Cu(In,Ga)$Se_2$ thin film membrane for light-induced water splitting and hydrogen evolution. *J. Phys. Chem. C* 113 (49): 20980–20989.

**20** Miller, E.L., Rocheleau, R.E., and Khan, S. (2004). A hybrid multijunction photoelectrode for hydrogen production fabricated with amorphous silicon/germanium and iron oxide thin films. *Int. J. Hydrogen Energy* 29 (9): 907–914.

**21** Fujishima, A. and Honda, K. (1972). Electrochemical photolysis of water at a semiconductor electrode. *Nature* 238 (5358): 38–40.

**22** Würfel, P. and Würfel, U. (2016). Photons. In: *The Physics of Solar Cells*, vol. 3, 11–39. Weinheim: Wiley-VCH Verlag GmbH & Co.

**23** Qian, R., Zong, H., Schneider, J. et al. (2019). Charge carrier trapping, recombination and transfer during $TiO_2$ photocatalysis: an overview. *Catal. Today* 335 (October 2018): 78–90.

**24** Scarongella, M., Gadiyar, C., Strach, M. et al. (2018). Assembly of β-$Cu_2V_2O_7$/$WO_3$ heterostructured nanocomposites and the impact of their composition on structure and photoelectrochemical properties. *J. Mater. Chem. C* 6 (44): 12062–12069.

**25** Zhou, X., Liu, R., Sun, K. et al. (2016). 570 mV photovoltage, stabilized n-Si/$CoO_x$ heterojunction photoanodes fabricated using atomic layer deposition. *Energy Environ. Sci.* 9 (3): 892–897.

**26** Alexander, J.C. (2016). *Surface Modifications and Growth of Titanium Dioxide for Photo-Electrochemical Water Splitting*. Springer.

**27** Young, E.R., Costi, R., Paydavosi, S. et al. (2011). Photo-assisted water oxidation with cobalt-based catalyst formed from thin-film cobalt metal on silicon photoanodes. *Energy Environ. Sci.* 4 (6): 2058–2061.

**28** Barreca, D., Carraro, G., Gasparotto, A., and Maccato, C. (2018). *Metal Oxide Electrodes for Photo-Activated Water Splitting*. Elsevier Ltd.

**29** Zaleska-Medynska, A. (2018). *Metal Oxide-Based Photocatalysis*. Elsevier.

**30** Wang, L., Tong, Y., Feng, J. et al. (2018). G-$C_3N_4$-based films: a rising star for photoelectrochemical water splitting. *Sustain. Mater. Technol.* 17: e00089.

**31** Pei, L., Wang, H., Wang, X. et al. (2018). Nanostructured TaON/$Ta_3N_5$ as a highly efficient type-II heterojunction photoanode for photoelectrochemical water splitting. *Dalton Trans.* 47 (27): 8949–8955.

**32** Strandwitz, N.C., Comstock, D.J., Grimm, R.L. et al. (2013). Photoelectrochemical behavior of n-type Si(100) electrodes coated with thin films of manganese oxide grown by atomic layer deposition. *J. Phys. Chem. C* 117 (10): 4931–4936.

**33** Gao, L., Cui, Y., Vervuurt, R.H.J. et al. (2016). High-efficiency InP-based photocathode for hydrogen production by interface energetics design and photon management. *Adv. Funct. Mater.* 26 (5): 679–686.

**34** Hu, S., Shaner, M.R., Beardslee, J.A. et al. (2014). Amorphous $TiO_2$ coatings stabilize Si, GaAs, and GaP photoanodes for efficient water oxidation. *Science (80-.)* 344 (6187): 1005–1009.

**35** Su, J., Minegishi, T., and Domen, K. (2017). Efficient hydrogen evolution from water using CdTe photocathodes under simulated sunlight. *J. Mater. Chem. A* 5 (25): 13154–13160.

**36** Majumder, S., Quang, N.D., Thi Hien, T. et al. (2021). Nanostructured $\beta$-$Bi_2O_3$/PbS heterojunction as np-junction photoanode for enhanced photoelectrochemical performance. *J. Alloys Compd.* 870: 159545.

**37** Girardi, L., Blanco, M., Agnoli, S. et al. (2020). A DVD-$MoS_2$/$Ag_2$S/Ag nanocomposite thiol-conjugated with porphyrins for an enhanced light-mediated hydrogen evolution reaction. *Nanomaterials* 10 (7): 1–12.

**38** Kosmala, T., Mosconi, D., Giallongo, G. et al. (2018). Highly efficient $MoS_2$/$Ag_2$S/Ag photoelectrocatalyst obtained from a recycled DVD surface. *ACS Sustainable Chem. Eng.* 6 (6): 7818–7825.

**39** Blanco, M., Lunardon, M., Bortoli, M. et al. (2020). Tuning on and off chemical- and photo-activity of exfoliated $MoSe_2$ nanosheets through morphologically selective "soft" covalent functionalization with porphyrins. *J. Mater. Chem. A* 8: 11019–11030.

**40** Zhou, L., Shinde, A., Newhouse, P.F. et al. (2020). Quaternary oxide photoanode discovery improves the spectral response and photovoltage of copper vanadates. *Matter* 3 (5): 1–17.

**41** Newhouse, P.F., Guevarra, D., Zhou, L. et al. (2020). Enhanced bulk transport in copper vanadate photoanodes identified by combinatorial alloying. *Matter* 3 (5): 1601–1613.

**42** Girardi, L., Bardini, L., Michieli, N. et al. (2019). $Co_3O_4$ nanopetals on Si as photoanodes for the oxidation of organics. *Surfaces* 2 (1): 41–53.

**43** Shuang, S., Girardi, L., Rizzi, G. et al. (2018). Visible light driven photoanodes for water oxidation based on novel r-GO/$\beta$-$Cu_2V_2O_7$/$TiO_2$ nanorods composites. *Nanomaterials* 8 (7): 544.

**44** Makimizu, Y., Yoo, J.E., Poornajar, M. et al. (2020). Effects of low oxygen annealing on the photoelectrochemical water splitting properties of $\alpha$-$Fe_2O_3$. *J. Mater. Chem. A* 8 (3): 1315–1325.

**45** Li, W., Li, J., Wang, X., and Chen, Q. (2012). Preparation and water-splitting photocatalytic behavior of S-doped $WO_3$. *Appl. Surf. Sci.* 263: 157–162.

**46** Zachäus, C., Abdi, F.F., Peter, L.M., and Van De Krol, R. (2017). Photocurrent of $BiVO_4$ is limited by surface recombination, not surface catalysis. *Chem. Sci.* 8 (5): 3712–3719.

**47** Mei, B., Mul, G., and Seger, B. (2017). Beyond water splitting: efficiencies of photo-electrochemical devices producing hydrogen and valuable oxidation products. *Adv. Sustain. Syst.* 1 (1–2): 1–6.

**48** Vanka, S., Wang, Y., Ghamari, P. et al. (2018). A high efficiency Si photoanode protected by few-layer MoSe$_2$. *Sol. RRL* 2 (8): 1–6.

**49** Durairasan, M., Karthik, P.S., Balaji, J., and Rajeshkanna, B. (2021). Design and fabrication of WSe2/CNTs hybrid network: A highly efficient and stable electrodes for dye sensitized solar cells (DSSCs). *Diamond Relat. Mater.* 111 (September 2020): 108174.

**50** Si, K., Ma, J., Lu, C. et al. (2020). A two-dimensional MoS2/WSe2 van der Waals heterostructure for enhanced photoelectric performance. *Appl. Surf. Sci.* 507 (December 2019): 145082.

**51** Da, P., Cha, M., Sun, L. et al. (2015). High-performance perovskite photoanode enabled by Ni passivation and catalysis. *Nano Lett.* 15 (5): 3452–3457.

**52** Poli, I., Hintermair, U., Regue, M. et al. (2019). Graphite-protected CsPbBr$_3$ perovskite photoanodes functionalised with water oxidation catalyst for oxygen evolution in water. *Nat. Commun.* 10 (1): 1–10.

**53** Liu, Z., Li, C., Xiao, Y. et al. (2020). Tailored NiFe catalyst on silicon photoanode for efficient photoelectrochemical water oxidation. *J. Phys. Chem. C* 124 (5): 2844–2850.

**54** Cai, Q., Hong, W., Jian, C., and Liu, W. (2020). A high-performance silicon photoanode enabled by oxygen vacancy modulation on NiOOH electrocatalyst for water oxidation. *Nanoscale* 12 (14): 7550–7556.

**55** Yu, Y., Zhang, Z., Yin, X. et al. (2017). Enhanced photoelectrochemical efficiency and stability using a conformal TiO$_2$ film on a black silicon photoanode. *Nat. Energy* 2 (6): 17045.

**56** Oh, S., Jung, S., Lee, Y.H. et al. (2018). Hole-selective CoO$_x$/SiO$_x$/Si heterojunctions for photoelectrochemical water splitting. *ACS Catal.* 8 (10): 9755–9764.

**57** Walter, M.G., Warren, E.L., McKone, J.R. et al. (2010). Solar water splitting cells. *Chem. Rev.* 110 (11): 6446–6473.

**58** Pan, L., Liu, Y., Yao, L. et al. (2020). Cu$_2$O photocathodes with band-tail states assisted hole transport for standalone solar water splitting. *Nat. Commun.* 11 (1): 1–10.

**59** Luo, J., Steier, L., Son, M.K. et al. (2016). Cu$_2$O nanowire photocathodes for efficient and durable solar water splitting. *Nano Lett.* 16 (3): 1848–1857.

**60** Kang, D., Hill, J.C., Park, Y., and Choi, K.S. (2016). Photoelectrochemical properties and photostabilities of high surface area CuBi$_2$O$_4$ and Ag-Doped CuBi$_2$O$_4$ photocathodes. *Chem. Mater.* 28 (12): 4331–4340.

**61** Prévot, M.S., Guijarro, N., and Sivula, K. (2015). Enhancing the performance of a robust Sol-Gel-processed p-type delafossite CuFeO$_2$ photocathode for solar water reduction. *ChemSusChem* 8 (8): 1359–1367.

**62** Wheeler, G.P. and Choi, K.S. (2017). Photoelectrochemical properties and stability of nanoporous p-type LaFeO$_3$ photoelectrodes prepared by electrodeposition. *ACS Energy Lett.* 2 (10): 2378–2382.

**63** Sekizawa, K., Nonaka, T., Arai, T., and Morikawa, T. (2014). Structural improvement of CaFe$_2$O$_4$ by metal doping toward enhanced cathodic photocurrent. *ACS Appl. Mater. Interfaces* 6 (14): 10969–10973.

64 Khaselev, O. and Turner, J.A. (1998). A monolithic photovoltaic-photoelectrochemical device for hydrogen production via water splitting. *Science (80-. )* 280 (5362): 425–427.

65 Fernández, A.M., Dheree, N., Turner, J.A. et al. (2005). Photoelectrochemical characterization of the Cu(In,Ga)S$_2$ thin film prepared by evaporation. *Sol. Energy Mater. Sol. Cells* 85 (2): 251–259.

66 Dominey, R.N., Lewis, N.S., Bruce, J.A. et al. (1982). Improvement of photoelectrochemical hydrogen generation by surface modification of p-type silicon semiconductor photocathodes. *J. Am. Chem. Soc.* 104 (2): 467–482.

67 Heller, A. and Hill, M. (1982). Efficient p-lnP ( Rh-H alloy ) and p-lnP (Re–H alloy) hydrogen evolving photocathodes. *J. Electrochem. Soc.* 129 (9): 2865–2866.

68 Yin, W.J., Tang, H., Wei, S.H. et al. (2010). Band structure engineering of semiconductors for enhanced photoelectrochemical water splitting: the case of TiO$_2$. *Phys. Rev. B: Condens. Matter* 82 (4): 1–6.

69 Huang, X., Gao, X., Xue, Q. et al. (2020). Impact of oxygen vacancies on TiO$_2$ charge carrier transfer for photoelectrochemical water splitting. *Dalton Trans.* 49 (7): 2184–2189.

70 Wang, G., Xiao, X., Li, W. et al. (2015). Significantly enhanced visible light photoelectrochemical activity in TiO$_2$ nanowire arrays by nitrogen implantation. *Nano Lett.* 15 (7): 4692–4698.

71 Wang, G., Wang, H., Ling, Y. et al. (2011). Hydrogen-treated TiO$_2$ nanowire arrays for photoelectrochemical water splitting. *Nano Lett.* 11 (7): 3026–3033.

72 Ros, C., Fàbrega, C., Monllor-Satoca, D. et al. (2018). Hydrogenation and structuration of TiO$_2$ nanorod photoanodes: doping level and the effect of illumination in trap-states filling. *J. Phys. Chem. C* 122 (6): 3295–3304.

73 Wang, F., Di Valentin, C., and Pacchioni, G. (2012). Doping of WO$_3$ for photocatalytic water splitting: hints from density functional theory. *J. Phys. Chem. C* 116 (16): 8901–8909.

74 Wang, F., DiValentin, C., and Pacchioni, G. (2012). Rational band gap engineering of WO$_3$ photocatalyst for visible light water splitting. *ChemCatChem* 4 (4): 476–478.

75 Hamid, S.B.A., Teh, S.J., and Lai, C.W. (2017). Photocatalytic water oxidation on ZnO: a review. *Catalysts* 7 (3): 93.

76 Wang, M., Ren, F., Zhou, J. et al. (2015). N doping to ZnO nanorods for photoelectrochemical water splitting under visible light: engineered impurity distribution and terraced band structure. *Sci. Rep.* 5 (July): 1–13.

77 Chen, Y.S. and Kamat, P.V. (2014). Glutathione-capped gold nanoclusters as photosensitizers. visible light-induced hydrogen generation in neutral water. *J. Am. Chem. Soc.* 136 (16): 6075–6082.

78 Xiao, F.X., Hung, S.F., Miao, J. et al. (2015). Metal-cluster-decorated TiO$_2$ nanotube arrays: a composite heterostructure toward versatile photocatalytic and photoelectrochemical applications. *Small* 11 (5): 554–567.

79 Zhang, Z., Zhang, L., Hedhili, M.N. et al. (2013). Plasmonic gold nanocrystals coupled with photonic crystal seamlessly on TiO$_2$ nanotube photoelectrodes for

efficient visible light photoelectrochemical water splitting. *Nano Lett.* 13 (1): 14–20.

80 Wang, T., Luo, Z., Li, C., and Gong, J. (2014). Controllable fabrication of nanostructured materials for photoelectrochemical water splitting via atomic layer deposition. *Chem. Soc. Rev.* 43 (22): 7469–7484.

81 Fàbrega, C., Andreu, T., Tarancón, A. et al. (2013). Optimization of surface charge transfer processes on rutile $TiO_2$ nanorods photoanodes for water splitting. *Int. J. Hydrogen Energy* 38 (7): 2979–2985.

82 Wu, J.M., Zhang, T.W., Zeng, Y.W. et al. (2005). Large-scale preparation of ordered titania nanorods with enhanced photocatalytic activity. *Langmuir* 21 (15): 6995–7002.

83 Fàbrega, C., Monllor-Satoca, D., Ampudia, S. et al. (2013). Tuning the fermi level and the kinetics of surface states of $TiO_2$ nanorods by means of ammonia treatments. *J. Phys. Chem. C* 117 (40): 20517–20524.

84 Smith, Y.R., Ray, R.S., Carlson, K. et al. (2013). Self-ordered titanium dioxide nanotube arrays: anodic synthesis and their photo/electro-catalytic applications. *Materials (Basel)* 6 (7): 2892–2957.

85 Su, J., Feng, X., Sloppy, J.D. et al. (2011). Vertically aligned $WO_3$ nanowire arrays grown directly on transparent conducting oxide coated glass: synthesis and photoelectrochemical properties. *Nano Lett.* 11 (1): 203–208.

86 Rao, P.M., Cai, L., Liu, C. et al. (2014). Simultaneously efficient light absorption and charge separation in $WO_3/BiVO_4$ core/shell nanowire photoanode for photoelectrochemical water oxidation. *Nano Lett.* 14 (2): 1099–1105.

87 Girardi, L., Rizzi, G.A., Bigiani, L. et al. (2020). Copper vanadate nanobelts as anodes for photoelectrochemical water splitting: influence of $CoO_x$ overlayers on functional performances. *ACS Appl. Mater. Interfaces* 12 (28): 31448–31458.

88 Seabold, J.A. and Choi, K.S. (2012). Efficient and stable photo-oxidation of water by a bismuth vanadate photoanode coupled with an iron oxyhydroxide oxygen evolution catalyst. *J. Am. Chem. Soc.* 134 (4): 2186–2192.

89 Sivula, K., Le Formal, F., and Grätzel, M. (2011). Solar water splitting: progress using hematite ($\alpha$-$Fe_2O_3$) photoelectrodes. *ChemSusChem* 4: 432–449.

90 Yang, X., Liu, R., Lei, Y. et al. (2016). Dual influence of reduction annealing on diffused hematite/FTO junction for enhanced photoelectrochemical water oxidation. *ACS Appl. Mater. Interfaces* 8 (25): 16476–16485.

91 Tang, P., Xie, H., Ros, C. et al. (2017). Enhanced photoelectrochemical water splitting of hematite multilayer nanowire photoanodes by tuning the surface state via bottom-up interfacial engineering. *Energy Environ. Sci.* 10 (10): 2124–2136.

92 Mor, G.K., Varghese, O.K., Wilke, R.H.T. et al. (2008). Erratum: p-type Cu–TiO nanotube arrays and their use in self-biased heterojunction photoelectrochemical diodes for hydrogen generation. *Nano Lett.* 8 (10): 3555.

93 Moniz, S.J.A., Zhu, J., and Tang, J. (2014). 1D Co–Pi modified $BiVO_4$/ZnO junction cascade for efficient photoelectrochemical water cleavage. *Adv. Energy Mater.* 4 (10): 1–8.

94 Paracchino, A. (2011). photoelectrochemical water reduction. *Nat. Mater.* 10 (6): 456–461.

95 Qiu, Y., Liu, W., Chen, W. et al. (2016). Efficient solar-driven water splitting by nanocone $BiVO_4$-perovskite tandem cells. *Sci. Adv.* 2 (6).

96 Barroso, M., Mesa, C.A., Pendlebury, S.R. et al. (2012). Dynamics of photogenerated holes in surface modified $\alpha$-$Fe_2O_3$ photoanodes for solar water splitting. *Proc. Natl. Acad. Sci. U. S. A.* 109 (39): 15640–15645.

97 Hisatomi, T., Le Formal, F., Cornuz, M. et al. (2011). Cathodic shift in onset potential of solar oxygen evolution on hematite by 13-group oxide overlayers. *Energy Environ. Sci.* 4 (7): 2512–2515.

98 Yang, X., Liu, R., Du, C. et al. (2014). Improving hematite-based photoelectrochemical water splitting with ultrathin $TiO_2$ by atomic layer deposition. *ACS Appl. Mater. Interfaces* 6 (15): 12005–12011.

99 Huai, X., Girardi, L., Lu, R. et al. (2019). The mechanism of concentric $HfO_2$/$Co_3O_4$/$TiO_2$ nanotubes investigated by intensity modulated photocurrent spectroscopy (IMPS) and electrochemical impedance spectroscopy (EIS) for photoelectrochemical activity. *Nano Energy* 65 (August): 104020.

100 Benck, J.D., Hellstern, T.R., Kibsgaard, J. et al. (2014). Catalyzing the hydrogen evolution reaction (HER) with molybdenum sulfide nanomaterials. *ACS Catal.* 4 (11): 3957–3971.

101 McCrory, C.C.L., Jung, S., Ferrer, I.M. et al. (2015). Benchmarking hydrogen evolving reaction and oxygen evolving reaction electrocatalysts for solar water splitting devices. *J. Am. Chem. Soc.* 137 (13): 4347–4357.

102 Dutta, A. and Pradhan, N. (2017). Developments of metal phosphides as efficient OER precatalysts. *J. Phys. Chem. Lett.* 8 (1): 144–152.

103 Yang, Y., Niu, S., Han, D. et al. (2017). Progress in developing metal oxide nanomaterials for photoelectrochemical water splitting. *Adv. Energy Mater.* 7 (19), 1700555.

104 Osgood, H., Devaguptapu, S.V., Xu, H. et al. (2016). Transition metal (Fe, Co, Ni, and Mn) oxides for oxygen reduction and evolution bifunctional catalysts in alkaline media. *Nano Today* 11 (5): 601–625.

105 Peerakiatkhajohn, P., Yun, J.-H., Wang, S., and Wang, L. (2016). Review of recent progress in unassisted photoelectrochemical water splitting: from material modification to configuration design. *J. Photonics Energy* 7 (1): 012006.

106 Burke, M.S., Enman, L.J., Batchellor, A.S. et al. (2015). Oxygen evolution reaction electrocatalysis on transition metal oxides and ($O_{xy}$) hydroxides: activity trends and design principles. *Chem. Mater.* 27 (22): 7549–7558.

107 Rahimi, N., Pax, R.A., and Gray, E.M.A. (2016). Review of functional titanium oxides. I: $TiO_2$ and its modifications. *Prog. Solid State Chem.* 44 (3): 86–105.

108 Zhong, D.K., Cornuz, M., Sivula, K. et al. (2011). Photo-assisted electrodeposition of cobalt–phosphate (Co–Pi) catalyst on hematite photoanodes for solar water oxidation. *Energy Environ. Sci.* 4 (5): 1759–1764.

109 Abdi, F.F. and Van De Krol, R. (2012). Nature and light dependence of bulk recombination in Co-Pi-catalyzed $BiVO_4$ photoanodes. *J. Phys. Chem. C* 116 (17): 9398–9404.

110 Pilli, S.K., Furtak, T.E., Brown, L.D. et al. (2011). Cobalt–phosphate (Co–Pi) catalyst modified Mo-doped BiVO$_4$ photoelectrodes for solar water oxidation. *Energy Environ. Sci.* 4 (12): 5028–5034.

111 Ibach, H. and Luth, H. Semiconductors. In: *Solid-State Physics*, 4e, 419–527. Berlin: Springer-Verlag Berlin Heidelberg.

112 Sze, M.S. (1969). *Physics of Semicoonductor Devices*. New York: Wiley-Interscience.

113 Ashcroft, N. and Mermin, N.D. (1976) Solid State Physics, Saunders College Publishing.

114 Li, J., Meng, F., Suri, S. et al. (2012). Photoelectrochemical performance enhanced by a nickel oxide-hematite p–n junction photoanode. *Chem. Commun.* 48 (66): 8213–8215.

115 Liu, Z. and Yan, L. (2016). High-efficiency p–n junction oxide photoelectrodes for photoelectrochemical water splitting. *Phys. Chem. Chem. Phys.* 18 (45): 31230–31237.

116 Zhou, T., Wang, J., Chen, S. et al. (2020). Bird-nest structured ZnO/TiO$_2$ as a direct Z-scheme photoanode with enhanced light harvesting and carriers kinetics for highly efficient and stable photoelectrochemical water splitting. *Appl. Catal., B* 267 (800): 118599.

117 Li, L., Shi, H., Yu, H. et al. (2021). Ultrathin MoSe$_2$ nanosheet anchored CdS–ZnO functional paper chip as a highly efficient tandem Z-scheme heterojunction photoanode for scalable photoelectrochemical water splitting. *Appl. Catal., B* 292 (April).

118 Maeda, K. (2013). *Z - Scheme Water Splitting Using Two Different Semiconductor Photocatalysts.*: 2.

119 Sayama, K., Mukasa, K., Abe, R. et al. (2002). A new photocatalytic water splitting system under visible light irradiation mimicking a Z-scheme mechanism in photosynthesis Kazuhiro. *J. Photochem. Photobiol., A* 148: 71–77.

120 Weinrich, H., Durmus, Y.E., Tempel, H. et al. (2019). Silicon and iron as resource-efficient anode materials for ambient-temperature metal-air batteries: A review. *Materials* 12 (13): 2134.

121 Früh A. (2006). Mn pourbaix diagram. Wikipedia. https://commons.wikimedia.org/wiki/File:Pourbaix_diagram_for_Manganese.svg.

122 Ros, C., Andreu, T., and Morante, J.R. (2020). Photoelectrochemical water splitting: a road from stable metal oxides to protected thin film solar cells. *J. Mater. Chem. A* 8 (21): 10625–10669.

123 Nakato, Y., Yano, H., Nishiura, S. et al. (1987). Hydrogen photoevolution at p-type silicon electrodes coated with discontinuous metal layers. *J. Electroanal. Chem.* 228 (1–2): 97–108.

124 Nakato, Y. and Tsubomura, H. (1985). Structures and functions of thin metal layers on semiconductor electrodes. *J. Photochem.* 29 (1–2): 257–266.

125 Nakato, Y. and Tsubomura, H. (1992). Silicon photoelectrodes modified with ultrafine metal islands. *Electrochim. Acta* 37 (5): 897–907.

126 Nagasubramanian, G. and Bard, A.J. (1981). Photoelectrochemistry of p-type WSe$_2$ in acetonitrile and p-WSe$_2$-nitrobenzene cell. *J. Electrochem. Soc.* 128 (5): 1055–1060.

**127** Nakato, K., Takabayashi, S., Imanishi, A. et al. (2004). Stabilization of n-Si electrodes by surface alkylation and metal nano-dot coating for use in efficient photoelectrochemical solar cells. *Sol. Energy Mater. Sol. Cells* 83 (4): 323–330.

**128** Singh, R., Green, M.A., and Rajkanan, K. (1981). Review of conductor-insulator-semiconductor (CIS) solar cells. *Sol. Cells* 3 (2): 95–148.

**129** Green, M.A. and Blakers, A.W. (1983). Advantages of metal-insulator-semiconductor structures for silicon solar cells. *Sol. Cells* 8 (1): 3–16.

**130** Ran, J., Zhang, J., Yu, J. et al. (2014). Earth-abundant cocatalysts for semiconductor-based photocatalytic water splitting. *Chem. Soc. Rev.* 43 (22): 7787–7812.

**131** Heller, A., Aharon-Shalom, E., Bonner, W.A., and Miller, B. (1982). Hydrogen-evolving semiconductor photocathodes. Nature of the junction and function of the platinum group metal catalyst. *J. Am. Chem. Soc.* 104 (25): 6942–6948.

**132** Harris, L.A., Gerstner, M.E., and Wilson, R.H. (1977). The role of metal overlayers on gallium phosphide photoelectrodes. *J. Electrochem. Soc.* 124 (10): 1511–1516.

**133** Sun, K., Shen, S., Liang, Y. et al. (2014). Enabling silicon for Solar-Fuel production. *Chem. Rev.* 114 (17): 8662–8719.

**134** Esposito, D.V., Levin, I., Moffat, T.P., and Talin, A.A. (2013). $H_2$ evolution at Si-based metal-insulator-semiconductor photoelectrodes enhanced by inversion channel charge collection and H spillover. *Nat. Mater.* 12 (6): 562–568.

**135** Lee, M.H., Takei, K., Zhang, J. et al. (2012). P-type InP nanopillar photocathodes for efficient solar-driven hydrogen production. *Angew. Chem. Int. Ed.* 51 (43): 10760–10764.

**136** Scheuermann, A.G., Kemp, K.W., Tang, K. et al. (2016). Conductance and capacitance of bilayer protective oxides for silicon water splitting anodes. *Energy Environ. Sci.* 9 (2): 504–516.

**137** Choi, M.J., Jung, J.Y., Park, M.J. et al. (2014). Long-term durable silicon photocathode protected by a thin $Al_2O_3/SiO_x$ layer for photoelectrochemical hydrogen evolution. *J. Mater. Chem. A* 2 (9): 2928–2933.

**138** Scheuermann, A.G., Lawrence, J.P., Meng, A.C. et al. (2016). Titanium oxide crystallization and interface defect passivation for high performance insulator-protected schottky junction MIS photoanodes. *ACS Appl. Mater. Interfaces* 8 (23): 14596–14603.

**139** Campet, G., Manaud, J.P., Puprichitkun, C. et al. (1989). Protection of photoanodes against photo-corrosion by surface deposition of oxide films: criteria for choosing the protective coating. *Act. Passiv. Electron. Components* 13 (3): 175–189.

**140** Newhouse, P.F., Boyd, D.A., Shinde, A. et al. (2016). Solar fuel photoanodes prepared by inkjet printing of copper vanadates. *J. Mater. Chem. A* 4 (19): 7483–7494.

**141** Roger, I., Shipman, M.A., and Symes, M.D. (2017). Earth-abundant catalysts for electrochemical and photoelectrochemical water splitting. *Nat. Rev. Chem.* 1: 1–13.

**142** Singh, A. and Spiccia, L. (2013). Water oxidation catalysts based on abundant 1st row transition metals. *Coord. Chem. Rev.* 257 (17–18): 2607.

**143** Seger, B., Pedersen, T., Laursen, A.B. et al. (2013). Using $TiO_2$ as a conductive protective layer for photocathodic $H_2$ evolution. *J. Am. Chem. Soc.* 135 (3): 1057–1064.

**144** Ros, C., Andreu, T., Hernández-Alonso, M.D. et al. (2017). Charge transfer characterization of ALD-grown $TiO_2$ protective layers in silicon photocathodes. *ACS Appl. Mater. Interfaces* 9 (21): 17932–17941.

**145** Ros, C., Carretero, N.M., David, J. et al. (2019). Insight into the degradation mechanisms of atomic layer deposited $TiO_2$ as photoanode protective layer. *ACS Appl. Mater. Interfaces* 11 (33): 29725–29735.

**146** Sivula, K. (2014). Defects give new life to an old material: electronically leaky titania as a photoanode protection layer. *ChemCatChem* 6 (10): 2796–2797.

**147** Peng, T.C., Xiao, X.H., Han, X.Y. et al. (2011). Characterization of DC reactive magnetron sputtered NiO films using spectroscopic ellipsometry. *Appl. Surf. Sci.* 257 (13): 5908–5912.

**148** Sun, K., Saadi, F.H., Lichterman, M.F. et al. (2015). Stable solar-driven oxidation of water by semiconducting photoanodes protected by transparent catalytic nickel oxide films. *Proc. Natl. Acad. Sci. U. S. A.* 112 (12): 3612–3617.

**149** Jang, W.L., Lu, Y.M., Hwang, W.S. et al. (2009). Point defects in sputtered NiO films. *Appl. Phys. Lett.* 94 (6): 1–4.

**150** Schuler, T.M., Ederer, D.L., Itza-Ortiz, S. et al. (2005). Character of the insulating state in NiO: a mixture of charge-transfer and Mott-Hubbard character. *Phys. Rev. B: Condens. Matter* 71 (11): 1–7.

**151** Bae, D., Seger, B., Vesborg, P.C.K. et al. (2017). Strategies for stable water splitting: via protected photoelectrodes. *Chem. Soc. Rev.* 46 (7): 1933–1954.

**152** Mei, B., Permyakova, A.A., Frydendal, R. et al. (2014). Iron-treated NiO as a highly transparent p-type protection layer for efficient Si-based photoanodes. *J. Phys. Chem. Lett.* 5 (20): 3456–3461.

**153** Ros, C., Andreu, T., David, J. et al. (2019). Degradation and regeneration mechanisms of NiO protective layers deposited by ALD on photoanodes. *J. Mater. Chem. A* 7 (38): 21892–21902.

**154** Tung, C.W., Hsu, Y.Y., Shen, Y.P. et al. (2015). Reversible adapting layer produces robust single-crystal electrocatalyst for oxygen evolution. *Nat. Commun.* 6: 1–9.

**155** Yang, J., Walczak, K., Anzenberg, E. et al. (2014). Efficient and sustained photoelectrochemical water oxidation by cobalt oxide/silicon photoanodes with nanotextured interfaces. *J. Am. Chem. Soc.* 136 (17): 6191–6194.

**156** Bae, D., Mei, B., Frydendal, R. et al. (2016). Back-illuminated si-based photoanode with nickel cobalt oxide catalytic protection layer. *ChemElectroChem* 3 (10): 1546–1552.

**157** Xing, Z., Wu, H., Wu, L. et al. (2018). A multifunctional vanadium-doped cobalt oxide layer on silicon photoanodes for efficient and stable photoelectrochemical water oxidation. *J. Mater. Chem. A* 6 (42): 21167–21177.

**158** Kainthla, R.C., Zelenay, B., and Bockris, J.O. (1986). Protection of n-Si photoanode against photocorrosion in photoelectrochemical cell for water electrolysis. *J. Electrochem. Soc.* 133 (2): 248–253.

**159** Peter, L.M. (1990). Dynamic aspects of semiconductor photoelectrochemistry. *Chem. Rev.* 90: 753–769.

**160** Li, J. (1984). Surface recombination at semiconductor electrodes Part II. Photoinduced "near-surface" recombination centres in p-Gap. *J. Electroanal. Chem.* 165 (1–2): 41–59.

**161** Gelderman, K., Lee, L., and Donne, S.W. (2007). Flat-band potential of a semiconductor: using the Mott–Schottky equation. *J. Chem. Educ.* 84 (4): 685.

**162** Bertoluzzi, L. and Bisquert, J. (2017). Investigating the consistency of models for water splitting systems by light and voltage modulated techniques. *J. Phys. Chem. Lett.* 8 (1): 172–180.

**163** Bertoluzzi, L. and Bisquert, J. (2012). Equivalent circuit of electrons and holes in thin semiconductor films for photoelectrochemical water splitting applications. *J. Phys. Chem. Lett.* 3 (17): 2517–2522.

**164** Peter, L. (2019). Photoelectrochemical kinetics: hydrogen evolution on p-type. *J. Electrochem. Soc.* 166 (5).

**165** Ponoamarev, E.A. and Peter, L.M. (1995). A generalized theory of intensity modulated photocurrent spectroscopy (IMPS). *J. Electroanal. Chem.* 396: 219–226.

**166** Thorne, J.E., Jang, J., Liu, E.Y., and Wang, D. (2016). Understanding the origin of photoelectrode performance enhancement by probing surface kinetics. *Chem. Sci.* 7: 3347–3354.

**167** Zachäus, C., Abdi, F.F., Peter, L.M., and van de Krol, R. (2017). Photocurrent of $BiVO_4$ is limited by surface recombination, not surface catalysis. *Chem. Sci.* 8 (5): 3712–3719.

**168** Gueymard, C.A., Myers, D., and Emery, K. (2002). Proposed reference irradiance spectra for solar energy systems testing. *Sol. Energy* 73 (6): 443–467.

**169** Kasten, F. and Young, A.T. (1989). Revised optical air mass tables and approximation formula. *Appl. Opt.* 28 (22): 4735–4738.

**170** Cheng, W.H., Richter, M.H., May, M.M. et al. (2018). Monolithic photoelectrochemical device for direct water splitting with 19% efficiency. *ACS Energy Lett.* 3 (8): 1795–1800.

**171** Licht, S., Wang, B., Mukerji, S. et al. (2001). Over 18% solar energy conversion to generation of hydrogen fuel; theory and experiment for efficient solar water splitting. *Int. J. Hydrogen Energy* 26 (7): 653–659.

# 11

## Tailoring Porous Electrode Structures by Materials Chemistry and 3D Printing for Electrochemical Energy Storage

Sally O'Hanlon[1] and Colm O'Dwyer[1,2,3,4]

[1] University College Cork, School of Chemistry, Cork T12 YN60, Ireland
[2] Tyndall National Institute, Micro-Nano Systems Centre, Lee Maltings, Cork T12 R5CP, Ireland
[3] Trinity College Dublin, AMBER@CRANN, Dublin 2, Ireland
[4] Environmental Research Institute, University College Cork, Lee Road, Cork T23 XE10, Ireland

## 11.1 Strategies for Functional Porosity in Electrochemical Systems

Our planet faces increasingly urgent sustainability challenges that require paradigm shifting approaches to daily life and, consequently, to deliver this, new research priorities in various disciplines. Electrochemical energy storage research and innovation is one such research line of critical importance and impact. Substantial progress in battery technology is essential if we are to succeed in an energy transition toward a more carbon-neutral, sustainable, and cleaner society. We need to move toward more sustainable energy harvesting and storage technologies that become part of a circular economy. Materials sustainability will become an important consideration in the years to come, from mining to refining and to reuse of battery materials conjointly with consumer demands that are very real and ever-increasing. Under such a scenario, the production of Li-ion batteries (LIBs) will grow considerably over the years to come, hence reviving the issue of finite mineral concentrate reserves for some critical raw materials for batteries. This concern has driven researchers to explore new, potentially more sustainable chemistries, including Na-ion, metal–air chemistries Li(Na)–$O_2$, Li–S, multivalent (Mg, Ca), redox flow batteries (RFBs), and aqueous-based technologies. Readers are referred to several books and extensive review article for details of the advanced made using many phases, crystal structures, and stoichiometries of cathode and anode materials that have improved their understanding and application for better LIBs [1, 2]. The accomplishments in materials synthesis and performance-related benefits for LIBs have been extensive in recent years. New forms of electrode material design have shown promise for higher energy density LIBs, including new forms of inorganic material selection [3], and the development of cation disordered and Li-enriched compounds for faster rate and better performing electrode materials [4, 5] is at the forefront in battery materials research currently.

*Tailored Functional Oxide Nanomaterials: From Design to Multi-Purpose Applications*, First Edition.
Edited by Chiara Maccato and Davide Barreca.
© 2022 WILEY-VCH GmbH. Published 2022 by WILEY-VCH GmbH.

High performance LIBs are one way to provide high capacity or energy density, with good, long life cycling stability at high current density, so as to allow fast charging when needed [6, 7]. To achieve these capabilities, new materials with inbuilt features or properties that enhance the ion and electron transport kinetics are always in focus. Controlling material size, arrangement, surface chemistry, and the nature of the reaction process with lithium (alloying, intercalation, conversion, etc.) are all methods used to improve rate response, cycle life, cell voltage, and capacity. These characteristics can be native to the material composition and structure or added during the assembly or electrode manufacturing process. The ultimate aim is to maximize energy density, without negatively affecting charging rate, cycle life, etc., and to keep the capacity stable over hundreds or thousands of cycles [8–14].

Figure 11.1 summarizes a general series of material arrangements that have been used to improve behavior in an LIB. Rational changes to the material arrangement and structure, including the formation of complex composites, have been shown to influence the longevity, efficiency, and stability of the electrode materials in Li-ion cells [16–21]. These modifications were designed to alleviate material breakup during high mole fraction lithium insertion into anodes and cathodes [22] and maintain electrical conductivity between active materials during cycling. Other forms of composite structures encapsulate active material with plastically deformable carbons, again, to alleviate mechanism stress variations on the active material's structural integrity, and more recent efforts with reduced dimensionality (two-dimensional (2D) material) offer ways to improve the rate performance for LIBs [23]. Several generations of micro- and nanostructures have been discovered that they ultimately base their improvement on the physical response to reversible lithiation to enhance cycle

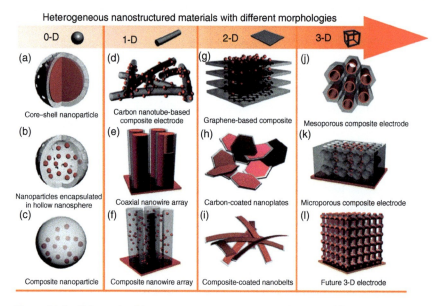

**Figure 11.1** Schematic of heterogeneous nanostructures based on 0D (a)–(c), 1D (d)–(f), 2D (g)–(i), and 3D (j)–(l) structures. Source: Liu et al. [15], Reproduced with permission from Royal Society of Chemistry.

life [24]. A recent example uses the chemistry of the mechanical bond to develop a molecular pulley binder for Si nanostructures in places during high mole fraction reversible lithiation. The free moving of the rings in polyrotaxanes follows the volumetric changes of the silicon particles, efficiently holding Si particles in place without disintegration during continuous volume change cycle after cycle [25].

One approach to tackle the interplay between ion kinetic, ion diffusion in electrolyte and the solid, and electrical conductivity of the electrode material in a way that minimizes contributions from binders or other additives is to fashion a three-dimensional (3D) interconnected network, which is often called an inverse opal (IO) [26] (of which there may be many different types over different length scales), and examples of which are shown in Figure 11.2.

In response to this challenge, bicontinuous porous electrodes were proposed where the electrode active materials were put on preprepared conductive IO structured nickel [18, 27]. These types of electrodes were promising in the early days, showing higher initial capacities and quite remarkable rate response in battery cells. One limitation was that they were unable to provide a long-term cycling stability. Recently, a 3D structured lithiated $MnO_2$ cathode arranged in the normal planar geometry was shown to retain 76% of its capacity when discharged rapidly at 185 C ($nC = 1/n$ hours to full discharge) [18]. Thickening (~150–200 nm) the layer of the IO reduced the high-rate capacity presumably from a change to out of plane electrical conductivity, but the specific capacity remained at 60% at 62 C charging

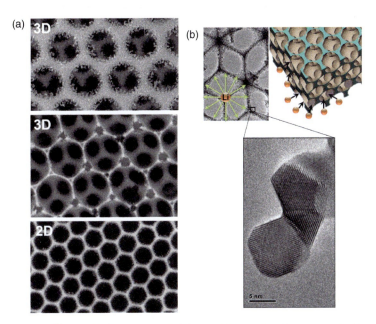

**Figure 11.2** (a) 3D inverse opal $SnO_2$ containing walls of nanocrystals. The 2D version (single IO layer) provides an easier way to assess changes in volume or other modification during cycling without additional complexity of stacked porous layers. (b) Representation of shortened ionic diffusion distance in an IO structure soaked with electrolyte. Source: O'Dwyer [26], Adapted with permission from John Wiley & Sons, Inc.

rates, which showed how engineering the porosity was beneficial to some extent for electrochemical systems such as reversible intercalation processes. $TiO_2$ IOs have also been reported where capacity has retained a >98% efficiency for at least 5000 cycles [28]. As with other reports [27], capacity retention is in part due to a solid solution with Li. As the IO architecture is able to maintain its interconnectivity even after undergoing some swelling from Li uptake, these porous materials provide surprising stability, capacity retention, and cycle life, especially considering that there are no binders or electrical conductivity promoters such as carbons in the electrode formulation. These structures are a good basic to study and evaluate porosity in battery electrode. While porosity has limited volumetric and gravimetric energy density, it can be useful in elevating certain electrochemical responses and prove the effectiveness of binder-free structured macroporous materials.

## 11.2 Benefits and Limitations of Structural Engineering for Electrochemical Performance

Lithium-ion electrodes store their electrochemical charge by several mechanisms, the most common of which is the intercalation reaction, where lithium ions reversibly insert into the cathode or anode material crystal structure [29]. For several materials, such as transition metals (e.g. Sn) and metalloids (e.g. Si, Ge), the mechanism is that of a reversible alloying reaction. The lower voltage reactions typical of negative electrodes (anodes) can be complicated by electrolyte decomposition that forms a so-called secondary electrolyte interphase layers that is generally a stabilizing layer on the anode surface [30, 31]. The positive electrode materials also reversibly intercalate lithium ions. Many of the most prevalent cathode materials today have crystal structures that enable reversible Li ion accommodation to a high mole fraction without deleterious effects on the intrinsic conductivity or structural stability; the material accommodates volume swelling during charge–discharge cycling without pulverization [32]. New materials are now being developed and predicted all the time, each providing incremental improvements over existing materials to maximize energy density, charging rate, or other properties desirable to the application [33].

Battery researchers have for a several years considered using materials with so-called functional porosity that wets a greater volume of electrolyte and, in principle, minimizes the diffusion distance of Li-ions in the electrolyte and in the material that are already small or thin enough to minimize intrinsic solid state diffusion limits [20]. Various forms of porosity from random to hierarchical and organized pores structures have been analyzed and compared in great detail. While their basic premise has been examined and the benefits or otherwise of pores has more or less been extracted, some very instructive findings have emerged that influence more commercial materials and the modes of their deposition, in not only Li-ion but also Li-S technologies, where porosity is very beneficial for other reasons.

Creating battery materials with solid-state diffusion lengths on a nanometer-length scale [34] helps to address efficiency concerns, but for porous materials it is their arrangement that allows the electrolyte wetting and overall shorter diffusion paths. The rate-limiting step in bulk materials is the solid-state diffusion of Li ions into the crystal, hence the advent of nanostructuring to minimize this limitation with the knock-on effect of either unwanted reactivity or in some cases the exposure of certain crystal facets that allow faster lithium insertion [35, 36]. The periodically arranged porous 3D material architecture [37] in IOs gives at least a somewhat definable electron and ion transport lengths compared with a compressed random material composite [18, 27, 38].

Charge rates from several hundred C up to 1000 C or even more are considered possible, enabling fabrication of an Li-ion material that can be fully charged in a matter of minutes or seconds [39]. However, the electrical conductivity is critical at high rates, and porous materials, even interconnected one, often comprise many grain boundaries and tortuous conduction paths that are not a priori shorter than a corresponding bulk material (thickness). Thicker porous electrodes can behave as thick film electrodes in such cases, severely suppressing the total stored charge at high rates even when pore structure is designed to improve many aspects of the electrode–electrolyte interface.

Ordered porous structures can improve rate response in some battery chemistries [14] – the materials, and the relationship between the pore structure and the electrochemical response to reversible lithiation [40] often depends on the interplay between the nature of the pore geometry and the type of material or lithiation reaction mechanism. Alloying materials respond well to fine pore structure, where the active material dimension is on the nanometer scale. Irreversible changes occur to such materials, and they can be packed in reasonably gravimetrically dense (good areal mass loading) electrodes even when internally porous. The primary benefit of these cases is to offset material pulverization and stress build from large volumetric changes, such as in silicon. As we will see in this chapter, some conversion mode reactions or intercalation with metal oxides is less reliant on mitigating volumetric changes, and the porosity should then be designed to minimize another unwanted property, such as solid-state diffusion limits or electrolyte access to active material, among others. 3D porous materials [17, 41] have received recent attention [8], but aside from promising performance as summarized for some materials in IO form in Table 11.1, there are many properties of these ordered materials that can provide very important information on their response to charging and discharging.

## 11.3 Tailoring the Pore Structure of Metal Oxides for Li-ion Battery Cathodes and Anodes

Organized and/or hierarchical porosity with either fixed or variable (but controlled) length scales can be readily made for metal oxide as the precursor solutions are often aqueous solutions and can be formed by drop casting, dip coating, vacuum infiltration, and related methods. Of course, hybrid nanomaterial compositions are also possible with careful tuning of the chemistry of the precursor solution and the method of crystallization [10, 56]. Some of the factors that can be tuned include the

**Table 11.1** Some IO materials and their response metrics as Li-ion battery electrodes.

| Composition | Morphology | Initial capacity (spec. current) | Final capacity (No. of cycles) | References |
|---|---|---|---|---|
| $V_2O_5$ | IO | — | — | [42] |
| $LiNiO_2$ | IO | — | — | |
| $SnO_2$ | IO | — | — | [43] |
| C | IO | 299 (15.2 mA g$^{-1}$) | — | [44] |
| Sn/C | Sn@C IO | 348 (15.2 mA g$^{-1}$) | — | |
| $TiO_2$/C | $TiO_2$@C | 158 (50 µA cm$^{-2}$) | 40 (15) | [45] |
| $Li_4Ti_5O_{12}$ | IO | 152 (6.3 µA cm$^{-2}$) | 145 (70) | [46] |
| $FePO_4$ | IO | 137 (C/10) | 110 (50) | [47] |
| $LiMn_2O_4$/ $Li_{1.5}Al_{0.5}Ti_{1.5}(PO_4)_3$ | IO composite | 64 (–) | — | [48] |
| $LiMn_2O_4$ | IO | — | — | [49] |
| Sn/C | Sn@C | 1500 (100 mA g$^{-1}$) | 96 (100) | [50] |
| Si/C | Si@C | 3180 (–) | — | [51] |
| $LiFePO_4$ | IO | 115 (5C) | 120 (50) | [52] |
| $LiMn_2O_4$ | IO | — | — | [53] |
| $FeF_3$ | IO | 540 (50 mA g$^{-1}$) | 190 (30) | [8] |
| Si @Ni | Bicontinuous IO | 3568 (C/20) | 2660 (100) | [27] |
| $SnO_2$/C | $SnO_2$@C | 1659 (0.5 A g$^{-1}$) | 715 (500) | [54] |
| $TiO_2$ | IO | 442 (25 mA g$^{-1}$) | 117 (200) | [55] |

Source: Osiak et al. [20], Adapted with permission from the Royal Society of Chemistry.

pore size, the periodicity, the variation or blending of pore sizes and length scales. The dimensions of the walls of the interconnected pore structure and the degree of porosity overall in the electrode coating can also be varied. Recent work in our laboratory provides many methods and materials for growing a range of oxides in defined shapes, porosities, arrangement, and assemblies, and as ordered electrode materials. Nanoscale systems for LIBs [57] and Li–O$_2$ batteries [58, 59] also give a range of battery materials that can be fashioned into IO ordered porous morphologies. Cathodes and anodes that contain carbon and oxides, or mixtures of oxides, phosphates, metalloids, or other materials, can have varying degrees of volumetric expansion at a given voltage in a particular battery. Porous structures that contain single or multiple phase materials give, in principle, the advantage of several degrees of freedom for changes in material dimension or volume during intercalation or alloying [60, 61]. Inverted opal oxides such as V$_2$O$_5$ and others have shown good performance and response as LIB electrodes in half-cells, or indeed as full IO cells (with both electrodes in ordered porous interconnected arrangement), with useful specific capacity that remains efficient and stable over hundreds, and, in some cases, thousands of cycles. There are now many systems published using IO materials with various focus on performance,

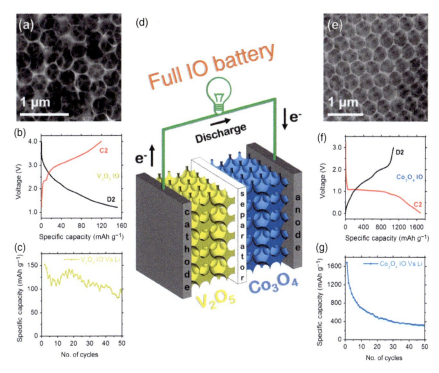

**Figure 11.3** (a)–(g) The behavior of individual $V_2O_5$ IO cathodes and $Co_3O_4$ IO anodes in lithium half-cells over 50 cycles. Source: Xue et al. [62], Adapted with permission from Royal Society of Chemistry.

electrochemical fundamental response, and the examination of process in the materials and the benefit or limitations of these porous architectures.

One example that demonstrates the half-cell response of IO cathode and anodes, as well as an all-IO structures battery cell, is that of $V_2O_5$ paired with $Co_3O_4$, which exploits intercalation in the positive electrode and conversion reactions at the negative electrode [62], using a standard lithium hexafluorophosphate-based electrolyte ($LiFP_6$ in an ethylene carbonate and dimethylcarbonate mixture). Figure 11.3 shows the overall cell and system. Here, the overall response of $V_2O_5$ IO cathode and $Co_3O_4$ IO anode is shown over their respective full voltage ranges, without carbon or added binders, and cycle life response. Once paired to the correct voltage range in a full cell such that both electrodes under reactions with Li mole fractions of ~2, the capacity and efficiency area is stable, keeping 175 mAh g$^{-1}$ after 125 cycles. In Figure 11.4, this cell was shown to be more stable with excess electrolyte, so that the porous network was fully infiltrated. In either 2 electrode or 3-electrode (with additional lithium metal reference electrodes), the response was nearly identical in an all-IO LIB using porous materials with two different reaction modes.

On the anode side, germanium-based materials are also interesting since, like silicon, it is a metalloid with a high theoretical capacity (1384 mAh g$^{-1}$), and it can be readily processed in nanoscale form much like silicon [63–69]. Pure Ge in

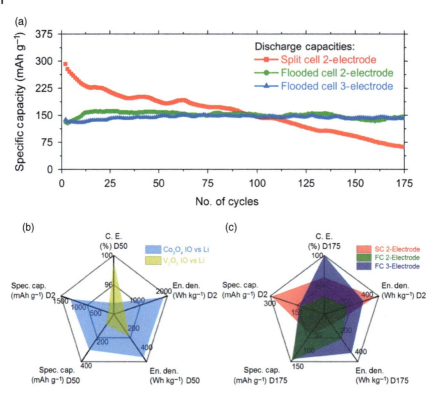

**Figure 11.4** (a–c) A comparison of the discharge capacities for $V_2O_5$ IO/$Co_3O_4$ IO cells in a 2-electrode split cell and 2 and 3-electrode excess electrolyte flooded cells. Source: McNulty et al. [62], Adapted with permission from the Royal Society of Chemistry.

precursor form can be expensive for quantity scaling, and synthetic methods are often specific, requiring high-temperature solvent or systems. Low-temperature routes to electrodeposited Ge microwires for battery anodes have been recently demonstrated using liquid Ga seed sources in water electrolytes and show some promises [70]. Using $GeO_2$ as the initial electrode formulation may provide opportunities for Ge-based anodes. In IO structure, the material has the inbuilt porosity that is often shown to be beneficial for high capacity anode materials that through alloying reactions undergo large volume changes during initial cycling. The capacity values from the oxide are obviously lower than those from pure Ge [71–73], but when made in a highly ordered, porous, 3D IO structure, the measured capacity approaches those found using more complex Ge nanostructured anodes in similar battery cell assemblies [21].

A Ge nanocrystal-containing $GeO_2$ IO LIB anode was created from a germanium(IV) ethoxide ($Ge(OC_2H_5)_4$) precursor that was able to provide state-of-the-art capacity, efficiency, stability of voltage, and good charging rate response over 200 cycles as a binder and conductive additive-free electrode, and aspects of the IO structure and electrode are shown in Figure 11.5. As with most IO structures, the morphology at electrode scale is that of large islands, separated by cracks. While

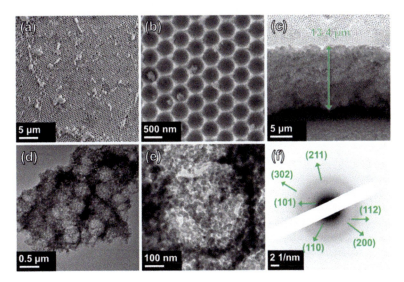

**Figure 11.5** Scanning electron microscopy (SEM) images showing (a) and (b) a top down view and (c) the cross-sectional thickness of a typical $GeO_2$ IO. Transmission electron microscopy (TEM) images of a typical $GeO_2$ IO showing (d) the porous structure and (e) the nanoparticles, which comprise the walls of the IO. (f) Electron diffraction pattern of a typical $GeO_2$ IO. Source: McNulty et al. [21], Adapted with permission from Elsevier.

cracks can reduce the areal mass loading, the mechanical and electrical adhesion to the current collector matters and is generally maintained. Within each island is an ordered arrangement of $GeO_2$. The walls of this particular structure comprise small nanocrystals. In LIB cells, these $GeO_2$ anode IO materials were capable of delivering high capacities at relatively large specific current (1000 mA g$^{-1}$). Even under asymmetric galvanostatic charge–discharge (Figure 11.6a–c) (discharging at 1000 mA g$^{-1}$, 3× faster than charge), the $GeO_2$ IO maintained charge capacities of 524 and 508 mAh g$^{-1}$ after the 50th and 150th cycles, respectively. Analysis of differential capacity plots obtained from charge–discharge galvanostatic curves from these IOs showed that the dominant phase for charge storage switched from c-$Li_{15}Ge_4$ to a-$Li_{15}Ge_4$. The oxide was reduced during the initial lithiation cycles to provide amorphous Ge for alloying reactions.

Rutile $TiO_2$ made in IOs has also been investigated as an LIB anode material [28]. This naturally occurring polymorph of $TiO_2$ is less commonly observed compared with anatase $TiO_2$. Very often the phase is a mixed anatase and rutile structure, especially in powders involving low-dimensional structures of the material [74] and oxidized bulk titanium metals [75]. Since the early report of ordered porous $TiO_2$ materials by Kavan et al. in 2004 [76], there have been many other investigations involving $TiO_2$ and other materials for LIB electrodes [18, 19, 34, 62, 77–80].

Until recently, the long-term cycling behavior of IO structured $TiO_2$ was not known. McNulty et al. tested these forms of anodes in Li-ion half-cells and reported long-term cycle life behavior over a 5000 cycle tests at 450 mA g$^{-1}$, as shown in Figure 11.7e. Between 1000 and 5000 cycles, an electrode consisting of only $TiO_2$ in

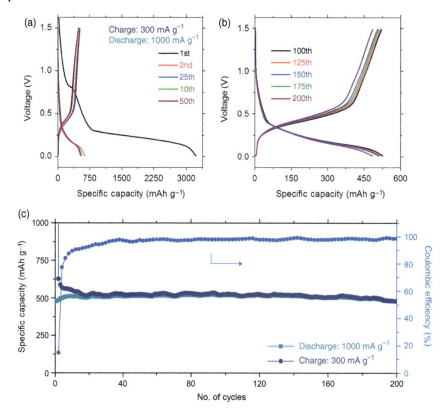

**Figure 11.6** (a), (b) Charge and discharge voltage profiles for charging at 300 mA g$^{-1}$ and discharging at 1000 mA g$^{-1}$ between 1.50 and 0.01 V. (c) Specific capacity and Coulombic efficiency from asymmetric cycling. Source: McNulty et al. [21], Adapted with permission from Elsevier.

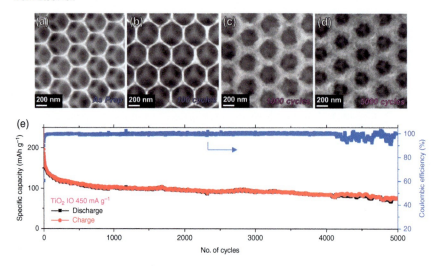

**Figure 11.7** (a)–(d) SEM images of the TiO$_2$ electrode as-prepared, and after 100, 1000, and 5000 cycles, are shown, where swelling occurs without disconnection. (e) Response of a rutile TiO$_2$ Li-ion battery anode in a half-cell over 5000 cycles at a specific current of 450 mA g$^{-1}$. Source: McNulty et al. [28], Adapted with permission from John Wiley & Sons, Inc.

IO form varied from ~103 to 76 mAh g$^{-1}$ at a reasonably high effective C-rate. Part of the stability was due to the interconnected structure being maintained. While the intrinsic electronic conductivity changes due to the Li mole fraction in successive cycling, the IO structure prevented disconnection and local increases in resistance over time, shown in Figure 11.7a–d. Modifications to this can be envisioned, where composites of materials are used, or highly conductive trusses improve electrical and mechanical connectivity within the porous electrode.

## 11.4 Developments in 3D Printing of Porous Electrodes for Electrochemical Energy Storage

Energy storage systems are more important today than ever before. With an ever-increasing reliance on mobile devices, as well as the emergence of wearable and flexible technologies, our energy storage systems constantly face new challenges, to keep up with modern demands [23, 81]. Unsurprisingly, many of these efforts focus on developing better and more sustainable active materials for electrodes and electrolytes [81–83]. Recently there has been a shift in focus on the manufacturing processes, material refinement, and the manner in which electrochemical energy storage devices can be made. Many of these approaches are forward looking, seeking to evaluate the likelihood of alternative form factors for batteries and supercapacitors, and teasing out the issues that accompany new ways of integrating materials, manufacturing cells, and seeing how they perform.

3D printing, or additive manufacturing, is a revolutionary manufacturing method where a virtual model designed and optimized on computer is converted into a 3D physical object by a printing technique that usually forms the object shaped directly, layer-by-layer [84]. This disruptive technology has found immense success across many different fields, including aerospace, biotechnology, bioengineering, construction, food, fashion, and energy storage [84–87]. 3D printing has developed quite rapidly in the area of energy storage device and materials research. New materials formulations and ways of site-selectively depositing composites of materials with different properties in intricate designs and shapes, which can be manipulated and controlled in all dimensions, is providing new opportunities to investigate alternative form factor energy storage devices that was not possible before. 3D printing has provided a simplified process for many complex manufacturing procedures such as the fabrication of interdigitated electrode designs [88, 89].

The nature of an electrode and the active material and its geometry dictates many of the electrochemical response and, in turn, the form factor of the battery, for example. These factors also influence ion transport distances, which then dictate energy density, power density, and rate capability [90–92]. The precision and control offered by 3D printing allows complete design freedom for almost any geometric shape, which then allows assessment and comparison of the electrochemical response of materials and the influence of design and structure on the requirement for high performance anodes and cathodes [93, 94]. Figure 11.8 summarizes a vision for 3D printable batteries and supercapacitors, the materials, electrodes, and

**Figure 11.8** Basis and motivation for additive manufacturing and 3D printing materials, casings, electrodes, electrolytes, and designs for electrochemical energy storage devices such as batteries and supercapacitors. Source: Egorov et al. [95].

printing innovation required to realize printable electrochemical energy storage device (EESD) technology.

When implementing a 3D printed version of a "normal" electrode (materials on a metal substrate), control over some obvious parameters such as electrode thickness is important especially for batteries. Out of plane conductivity reduction for thicker nonmetallic materials does increase areal mass loading and energy density but limits power density and the overall capacity and cycle life [96, 97]. Due to the layer-by-layer material deposition method, 3D printing offers the user strict control and precision over electrode thickness, potentially optimizing important performance parameters.

## 11.5 Porous Current Collectors by 3D Printing

The current collector has several functions; the most important of which is the electrical connection between the active material and the outer circuit. One option for current collectors, especially for natively low conductivity materials, is to infuse the current collection capability through the active material matrix. This of course sacrifices overall active mass but reduces the thickness-induced resistance of thick films by having the metal placed throughout the material. Metals foams (steel, copper, and nickel foams as common examples) are used in some cases as active material supports in several electrochemical technologies such as water splitting, supercapacitor, and, sometimes, battery electrodes [27].

Figure 11.9 shows various approaches involving 3D printing and modification of self-assembly and synthetic chemistry to create functional porous metals or metal

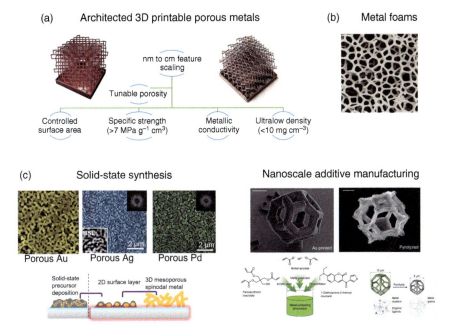

**Figure 11.9** (a) 3D printed porous metals and associated characteristics by metallization of photopolymerized resins using stereolithographic printing. (b) SEM of a metal foam structure of Ni. Source: Egorov and O'Dwyer [98], Licensed under CC BY 4.0. (c) Solid-state precursor synthetic method to create porous Au, Ag, and Pd as a porous surface film directly by pyrolysis. Source: Valenzuela et al. [60], Licensed under CC BY NC SA 3.0. Nanoscale additive manufacturing using chemical precursors and photoinitiators is capable of producing printed architected metals with nanoscale features. Source: Vyatskikh et al. [99], Licensed under CC BY 4.0.

lattices that are potential useful for electrochemical energy storage. Nanoscale metal lattices can now be produced from precursor chemistry and pyrolysis (Figure 11.9). Porous metallic materials are well known to have enhanced electrochemical activity that is intrinsic either to the high surface area of the porous metal [42, 100] or from the active material subsequently coated onto the porous current collector [28, 36, 40, 42, 77, 78, 101, 102]. For batteries and supercapacitors, this intrinsic activity is not always wanted, but the integrated conductivity of the current collector in addition to the pore structure that is design to improve the response of the active material can be beneficial. 3D printing of porous metals, and their oxides by subsequent oxidation, are unique since they can be rationally designed prior to manufacture and made into exactly the complex structure that is predicted to be useful for application. One benefit of trying such an approach is for materials that have spurious electrochemical response or capacities where it can be difficult to decouple effects from high surface area intrinsic reactivity from the geometric enhancement only at specific voltages that are under various specific currents. By coding and printing porosity with different degrees of order, pores sizes, geometries, etc., these effects can be tested.

3D printing will likely become commonplace for laboratory-scale examination of electrodes where porosity is important. In particular, for metallic lattices and more

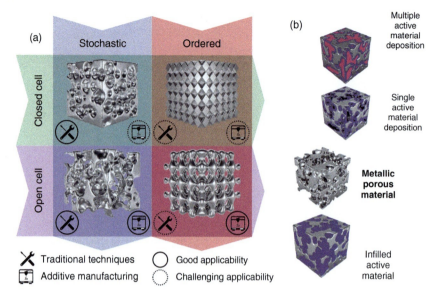

**Figure 11.10** (a) Various porous metal geometries that can be produced by 3D printing. (b) Filling or coating these types of printed porous metals with active materials can provide a range of unique active material formulations in almost any structure. Source: Egorov and O'Dwyer [98], Licensed under CC BY 4.0.

complex current collectors, 3D printing can print metals directly or make 3D supports that contain active materials. Some methods use decomposable materials that induced a hierarchical porosity (porosity in the trusses with a larger length scale porous material), and the porosity can be ordered, disordered, or be closed cell, open cell, or even optimized for space filling in two or three dimensions (Figure 11.10). In addition to porosity, the reduction in ionic tortuosity when filled with electrolyte, pore sizes optimized for volumetric swelling, and mechanical strength or flexibility can be coded from the boom up and material incorporated directly into the printing process [103]. With many notional benefits already stated, 3D printing offers one more obvious but very important capability – reproducibility of very complex pore systems and their scaling in dimension, which can be difficult even with the most controlled self-assembly or top-down etching or decomposition processes, which is again different depending in whether the material is metal, oxide, polymer, carbonaceous, or ceramic.

## 11.6 Battery and Supercapacitor Materials from 3D Printing

Accurate modeling and printing of porous structures may allow for new ways to interrogate electrochemical processes such as surface reactivity, intercalation, pseudocapacitance, capacitance, secondary electrolyte interphase (SEI) formation, and other processes in supercapacitors, batteries, and related technologies and also

## 11.6 Battery and Supercapacitor Materials from 3D Printing

allow for their comparison without issues associated with exact complex electrode replication. Many metallic trusses and lattices are also examined in the mechanical engineering disciplines due to their extreme toughness and compression tolerance, which is also useful for electrochemical energy storage technologies where conventional coin cell-type battery assembly can compress and eradicate porosity designed to improve response or performance.

There have been some interesting preludes to the current state-of-the-art 3D printed materials and cells in supercapacitors and batteries. Many of these investigations focused on sensors, capacitors, and some energy storage chemistries that were initially more amenable to 3D printing apparatus such as fused deposition modeling (the typical acrylonitrile butadiene styrene (ABS) plastic hobbyist technique). Palmer [104] used direct printing to produce 3D printed shapes with electrical components embedded into them. Lopez et al. [105] furthered this idea by combining stereolithographic apparatus (SLA) and a dispensing system, to produce simple temperature sensor prototypes. Sun et al. [89] 3D interdigitated microbattery architectures, using $Li_4Ti_5O_{12}$ (LTO) and $LiFePO_4$ (LFP), as the anode and cathode materials. Then, Ho et al. [106] showed a zinc microbattery with an ionic liquid gel electrolyte using direct write dispenser printing. In this work, a developed dispenser printed a microbattery comprising of zinc and manganese dioxide electrodes, which sandwich an ionic liquid gel electrolyte. Zhao et al. [88] produced a 3D printed electrode and used selective laser melting (SLM) to fabricate interdigitated supercapacitors as a 3D printed electrode. Pseudocapacitors were produced using a 3D interdigitated $Ti_6Al_4V$ electrode, which is fine metal powder. Then, Zhu et al. [107] reported fabrication of 3D periodic graphene composite aerogel (3D-GCA) microlattices for supercapacitors using a direct-ink-writing technique. In this study, the 3D-GCA was produced and fabricated as an electrode.

Traditional electrochemical energy storage devices including batteries typically involve a series of steps that evolve from a current collector with coated active materials to assembly, electrolyte additional, and casing as a prismatic cell, cylindrical cell, coin or pouch cell, etc. Many of these processes are commercially optimized for high throughput production; 3D printed cells are being investigated because of their ability to be fashioned into almost any form factor, which provides options for cell customization. The initial development were based on fused deposition modeling (FDM) methods, involving PLA and ABS plastics. This approach is still the most common due to being the most accessible for laboratory infrastructure, but newer methods such as SLA and others offer better choice in materials and often finer printing fidelity, among other attributes.

One issue that is endemic in FDM printing for technologies in areas of surface active as batteries and related electrochemical systems is that of impurities. What has been found recently is that lower grade poly(lactic acid) and ABS plastics contain significant impurities that can change and sometimes dominate the electrochemical response. This can modify effective (useful) surface area in these thermoplastics when they are used in certain solvents [108] or by thermal decomposition [109] during the very printing process. While several papers have addressed why this happens and what the root cause was in some cases, the issue remains in mass-produced

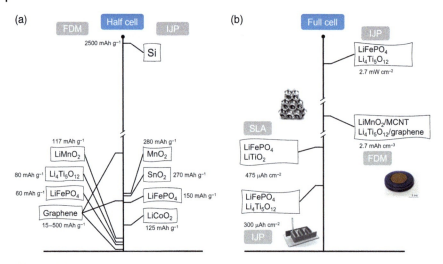

**Figure 11.11** Summary of reported areal power density, volumetric capacity, and areal capacity performance metrics for Li-ion (a) half-cells and (b) full cells. These data were taken from papers where cells/materials were printed using fused deposition modeling, inkjet printing, or stereolithographic photopolymerization printing. Source: Egorov et al. [95], John Wiley & Sons, Inc.

raw materials. One approach to alleviate this results in home-made ABS and PLA together with in situ extrusion to create high purity base materials for printing. However, the intrinsic electronic and ionic conductivity in any printed electrode support material or active material composite is critical when these materials are used to create current collector, for example.

Nevertheless, advances have been made in 3D printed systems such as LIBs, which is noteworthy given the chemical sensitivity of various thermoplastics and related materials to organic electrolytes that are common place in these devices. The first attempts reported in the literature used deposition modeling (FDM), inkjet printing (IJP), and SLA. These techniques were used to make electrodes and electrolytes in half-cell and full-cell LIB configurations. In Figure 11.11, many of recent reports using half-cell and full-cell Li batteries are summarized in terms of common metrics of areal power density, volumetric capacity, and areal capacity. So far, the advances used graphene [110–112], $Li_4Ti_5O_{12}$ [113], $SnO_2$ [114], $MnO_2$ [115], and Si [116] as negative electrodes, with reports of $LiFePO_4$ [113, 117, 118] and $LiCoO_2$ positive electrodes [119, 120]. Lithium iron phosphate and lithium titanate are the materials most commonly used in the full cell configurations so far [89, 121, 122], although new versions are being investigated with water-in-salt and other aqueous or gel-type electrolytes.

## 11.7 Conclusions and Outlook

Adding porosity that is useful to materials and electrodes does offer some performance advantages for batteries, supercapacitors, and other electrochemical systems and technologies. For electrochemical energy storage, the concepts of random and order porosity, where this can be to a large extent controlled, have proven its effectiveness in Li-ion, Li-air, and Li-S batteries, for example. A careful choice should be made when considering modifying or adding porosity to materials when performance improvement is the primary motivation. Can the Li reaction mechanism be kept consistent without exacerbating side reactions? Will energy density sacrifice be useful for additional power density and are the additional degrees of freedom for volumetric expansion in alloying materials useful for voltage and cycling stability? There will be some scope for functional porosity for beyond LIB material involving larger cations in intercalation reactions or other forms of electrochemical energy storage.

Additive manufacturing has revolutionized the building of materials, and 3D-printing has become a useful tool for complex electrode assembly for batteries and supercapacitors. What is the motivation for continuing to investigate porous materials in batteries and 3D printing or other additive manufacturing methods for different battery or supercapacitor form factors? Depending on the application, cells with porosity that improves rate capability at the expense of volumetric energy density are useful, specifically for miniature power sources and wearables where continued charging and power delivery are more important than high capacity battery power-only approaches. In such cases cycle life stability and rapid charging for some materials are improved by engineering the materials arrangement on the electrodes. A case in point is the Internet of Everything, where trillions of sensors deployed to monitor as much as possible in real time. Many of these deployments are wearable, medical technologies that do not want wired connections to power, remotely deployed modules, and many others. Batteries and supercapacitors are useful power solutions for these technologies with microbatteries and new forms of printable power sources being developed. There are many reviews and reports available on stretchable and wearable textile-based capacitors and some 3D printed batteries that the reader is referred to, but new form factor batteries for off-grid telecommunications for IoT are important. Small battery cells, remotely deployed, printed to be "invisible" within the environment, and designed by the requirement can reduced the need for large cells. As an example, the total data traffic growth is exponential. In order to meet this capacity growth, the telecommunications industry is introducing small cell technology – small cells are miniature base stations that transmit over much shorter distances and hence consume less spectrum and less power. As a consequence, moving to small cells implies that many more products will be deployed in denser networks, and in this situation, access to power for the devices will become a challenge. This requires more cells, closer to devices, and the on-grid electrical infrastructure would be impractical at this level – off-grid power is the way forward for opening up the free higher frequency bandwidths in telecoms. These technological challenges may benefit from energy storage and battery technologies for powering the IoT infrastructure.

New form factor cells with better internal active materials and electrodes, together with printable form factors and cell casings, offer possibilities also. For most wearable and IoT technologies, the battery is the bulkiest, heaviest, and least green component. Batteries have been limited to "classic" form factors (rectangular cells, cylindrical containers, button cells, etc.) so far for good reason, including safety, reproducibility, and production all governed by certification requirements. Fixed form factors, in principle, limit the form factor of products such as wearable devices, which must compromise their shape to accommodate the batteries. A 3D printed battery could be built to fit into the device, and not the other way round. Also, because the batteries are separate, they must possess an additional layer of packaging, often metal, which increases the overall weight and size of a small device. The concept of miniaturized, conformal batteries, with energy density sufficient for their need, which can be adapted to any form factor or shape, will completely open the design space for future technology, devices, peripherals, and wearables for enhanced quality of life and connected living.

## References

1 Song, M.-K., Park, S., Alamgir, F.M. et al. (2011). Nanostructured electrodes for lithium-ion and lithium-air batteries: the latest developments, challenges, and perspectives. *Mater. Sci. Eng., R* 72: 203.
2 Choi, J.W. and Aurbach, D. (2016). Promise and reality of post-lithium-ion batteries with high energy densities. *Nat. Rev. Mater.* 1: 16013.
3 Croguennec, L. and Palacin, M.R. (2015). Recent achievements on inorganic electrode materials for lithium-ion batteries. *J. Am. Chem. Soc.* 137: 3140–3156.
4 Seo, D.H., Lee, J., Urban, A. et al. (2016). The structural and chemical origin of the oxygen redox activity in layered and cation-disordered Li-excess cathode materials. *Nat. Chem.* 8 (7): 692–697.
5 Lee, J., Papp, J.K., Clement, R.J. et al. (2017). Mitigating oxygen loss to improve the cycling performance of high capacity cation-disordered cathode materials. *Nat. Commun.* 8 (1): 981.
6 Lee, S.W., Lee, S.W., Lee, S.W. et al. (2010). High-power lithium batteries from functionalized carbon-nanotube electrodes. *Nat. Nanotechnol.* 5: 531–537.
7 Tarascon, J.M. and Armand, M. (2001). Issues and challenges facing rechargeable lithium batteries. *Nature* 414: 359.
8 Ma, D.-L., Cao, Z.-Y., Wang, H.-G. et al. (2012). Three-dimensionally ordered macroporous $FeF_3$ and its in situ homogenous polymerization coating for high energy and power density lithium ion batteries. *Energy Environ. Sci.* 5 (9): 8538–8542.
9 Sakamoto, J.S. and Dunn, B. (2002). Hierarchical battery electrodes based on inverted opal structures. *J. Mater. Chem.* 12: 2859–2861.
10 Armstrong, M.J., Burke, D.M., Gabriel, T. et al. (2013). Carbon nanocage supported synthesis of $V_2O_5$ nanorods and $V_2O_5/TiO_2$ nanocomposites for Li-ion batteries. *J. Mater. Chem. A* 1: 12568–12578.
11 Li, H.Q. and Zhou, H.S. (2012). Enhancing the performances of Li-ion batteries by carboncoating: present and future. *Chem. Commun.* 48: 1201–1217.

**12** Kang, D.-Y., Kim, S.-O., Chae, Y.J. et al. (2013). Particulate inverse opal carbon electrodes for lithium-ion batteries. *Langmuir* 29 (4): 1192–1198.

**13** McSweeney, W., Lotty, O., Mogili, N.V.V. et al. (2013). Doping controlled roughness and defined mesoporosity in chemically etched silicon nanowires with tunable conductivity. *J. Appl. Phys.* 114: 034309.

**14** Stein, A. (2011). Energy storage: batteries take charge. *Nat. Nanotechnol.* 6 (5): 262–263.

**15** Liu, R., Duay, J., and Lee, S.B. (2011). Heterogeneous nanostructured electrode materials for electrochemical energy storage. *Chem. Commun.* 47 (5): 1384–1404.

**16** McSweeney, W., Geaney, H., and O'Dwyer, C. (2015). Metal assisted chemical etching of silicon and the behaviour of nanoscale silicon materials as Li-ion battery anodes. *Nano Res.* 8: 1395–1442.

**17** Shaijumon, M.M., Perre, E., Daffos, B. et al. (2010). Nanoarchitectured 3D cathodes for Li-ion microbatteries. *Adv. Mater.* 22: 4978–4981.

**18** Zhang, H., Yu, X., and Braun, P.V. (2011). Three-dimensional bicontinuous ultrafast-charge and discharge bulk battery electrodes. *Nat. Nanotechnol.* 6: 277–281.

**19** Vu, A., Qian, Y.Q., and Stein, A. (2012). Porous electrode materials for lithium-ion batteries – how to prepare them and what makes them special. *Adv. Energy Mater.* 2: 1056–1085.

**20** Osiak, M., Geaney, H., Armstrong, E., and O'Dwyer, C. (2014). Structuring materials for lithium-ion batteries: advancements in nanomaterial structure, composition, and defined assembly on cell performance. *J. Mater. Chem. A* 2: 9433–9460.

**21** McNulty, D., Geaney, H., Buckley, D., and O'Dwyer, C. (2018). High capacity binder-free nanocrystalline $GeO_2$ inverse opal anodes for Li-ion batteries with long cycle life and stable cell voltage. *Nano Energy* 43: 11–21.

**22** Liu, N., Lu, Z., Zhao, J. et al. (2014). A pomegranate-inspired nanoscale design for large-volume-change lithium battery anodes. *Nat. Nanotechnol.* 9 (3): 187–192.

**23** Lukatskaya, M.R., Dunn, B., and Gogotsi, Y. (2016). Multidimensional materials and device architectures for future hybrid energy storage. *Nat. Commun.* 7: 12647.

**24** McDowell, M.T., Lee, S.W., Nix, W.D., and Cui, Y. (2013). 25th anniversary article: understanding the lithiation of silicon and other alloying anodes for lithium-ion batteries. *Adv. Mater.* 25 (36): 4966–4985.

**25** Choi, S., Kwon, T.-W., Coskun, A., and Choi, J.W. (2017). Highly elastic binders integrating polyrotaxanes for silicon microparticle anodes in lithium ion batteries. *Science* 357 (6348): 279–283.

**26** O'Dwyer, C. (2016). Colour-coded batteries – inverse opal materials circuitry for enhanced electrochemical energy storage and optically encoded diagnostics. *Adv. Mater.* 28: 5681–5688.

**27** Zhang, H. and Braun, P.V. (2012). Three-dimensional metal Scaffold supported bicontinuous silicon battery anodes. *Nano Lett.* 12 (6): 2778–2783.

**28** McNulty, D., Carroll, E., and O'Dwyer, C. (2017). Rutile $TiO_2$ inverse opal anodes for Li-ion batteries with long cycle life, high-rate capability and high structural stability. *Adv. Energy Mater.* 7 (12): 1602291.

**29** Lee, J., Urban, A., Li, X. et al. (2014). Unlocking the potential of cation-disordered oxides for rechargeable lithium batteries. *Science* 343: 519–522.

**30** Kang, B. and Ceder, G. (2009). Battery materials for ultrafast charging and discharging. *Nature* 458: 190–193.

**31** Kang, K.S., Meng, Y.S., Breger, J. et al. (2006). Electrodes with high power and high capacity for rechargeable lithium batteries. *Science* 311: 977–980.

**32** Roberts, M., Johns, P., Owen, J. et al. (2011). 3D lithium ion batteries-from fundamentals to fabrication. *J. Mater. Chem.* 21 (27): 9876–9890.

**33** Armstrong, E., Khunsin, W., Osiak, M. et al. (2014). Light scattering investigation of 2D and 3D opal template formation on hydrophilized surfaces. *ECS Trans.* 58 (47): 9–18.

**34** Pikul, J.H., Zhang, H.G., Cho, J. et al. (2013). High-power lithium ion microbatteries from interdigitated three-dimensional bicontinuous nanoporous electrodes. *Nat. Commun.* 4: 1732–1736.

**35** Goodenough, J.B. and Kim, Y. (2010). Challenges for rechargeable Li batteries. *Chem. Mater.* 22: 587–603.

**36** Lytle, J.C. (2013). Inverse opal nanoarchitectures as lithium-ion battery materials. In: *Nanotechnology for Lithium-Ion Batteries*, Nanostructure Science and Technology (ed. Y. Abu-Lebdeh and I. Davidson), 13–41. Springer Science+Business Media, LLC.

**37** Gough, D.V., Juhl, A.T., and Braun, P.V. (2009). Programming structure into 3D nanomaterials. *Mater. Today* 12: 28.

**38** Li, X., Dhanabalan, A., Gu, L., and Wang, C. (2012). Three-dimensional porous core-shell Sn@carbon composite anodes for high-performance lithium-ion battery applications. *Adv. Energy Mater.* 2 (2): 238–244.

**39** Ergang, N.S., Lytle, J.C., Yan, H.W., and Stein, A. (2005). Effect of a macropore structure on cycling rates of $LiCoO_2$. *J. Electrochem. Soc.* 152: A1989–A1995.

**40** Ergang, N.S., Lytle, J.C., Lee, K.T. et al. (2006). Photonic crystal structures as a basis for a three-dimensionally interpenetrating electrochemical-cell system. *Adv. Mater.* 18: 1750–1753.

**41** Rolison, D.R. (2009). Multifunctional 3D nanoarchitectures for energy storage and conversion. *Chem. Soc. Rev.* 38: 226–252.

**42** Long, J.W., Dunn, B., Rolison, D.R., and White, H.S. (2004). Three-dimensional battery architectures. *Chem. Rev.* 104 (10): 4463–4492.

**43** Lytle, J.C., Yan, H.W., Ergang, N.S. et al. (2004). Structural and electrochemical properties of three-dimensionally ordered macroporous tin(IV) oxide films. *J. Mater. Chem.* 14 (10): 1616–1622.

**44** Lee, K.T., Lytle, J.C., Ergang, N.S. et al. (2005). Synthesis and rate performance of monolithic macroporous carbon electrodes for lithium-ion secondary batteries. *Adv. Funct. Mater.* 15 (4): 547–556.

**45** Bing, Z., Yuan, Y., Wang, Y., and Fu, Z.W. (2006). Electrochemical characterization of a three dimensionally ordered macroporous anatase $TiO_2$ electrode. *Electrochem. Solid-State Lett.* 9 (3): A101–A104.

**46** Sorensen, E.M., Barry, S.J., Jung, H.K. et al. (2006). Three-dimensionally ordered macroporous $Li_4Ti_5O_{12}$: effect of wall structure on electrochemical properties. *Chem. Mater.* 18 (2): 482–489.

**47** W-j, C., H-j, L., Wang, C.-x., and Y-y, X. (2008). Highly ordered three-dimensional macroporous $FePO_{(4)}$ as cathode materials for lithium-ion batteries. *Electrochem. Commun.* 10 (10): 1587–1589.

**48** Nakano, H., Dokko, K., Hara, M. et al. (2008). Three-dimensionally ordered composite electrode between $LiMn_2O_4$ and $Li_{1.5}Al_{0.5}Ti_{1.5}(PO_4)_{(3)}$. *Ionics* 14 (2): 173–177.

**49** Tonti, D., Torralvo, M.J., Enciso, E. et al. (2008). Three-dimensionally ordered macroporous lithium manganese oxide for rechargeable lithium batteries. *Chem. Mater.* 20 (14): 4783–4790.

**50** Wang, Z., Fierke, M.A., and Stein, A. (2008). Porous carbon/tin(IV) oxide monoliths as anodes for lithium-ion batteries. *J. Electrochem. Soc.* 155 (9): A658–A663.

**51** Wang, Z., Li, F., Ergang, N.S., and Stein, A. (2008). Synthesis of monolithic 3D ordered macroporous carbon/nano-silicon composites by diiodosilane decomposition. *Carbon* 46 (13): 1702–1710.

**52** Doherty, C.M., Caruso, R.A., Smarsly, B.M., and Drummond, C.J. (2009). Colloidal crystal templating to produce hierarchically porous $LiFePO_4$ electrode materials for high power lithium ion batteries. *Chem. Mater.* 21 (13): 2895–2903.

**53** Qu, Q., Fu, L., Zhan, X. et al. (2011). Porous $LiMn_2O_4$ as cathode material with high power and excellent cycling for aqueous rechargeable lithium batteries. *Energy Environ. Sci.* 4 (10): 3985–3990.

**54** Huang, X., Chen, J., Lu, Z. et al. (2013). Carbon inverse opal entrapped with electrode active nanoparticles as high-performance anode for lithium-ion batteries. *Sci. Rep.* 3: 1736–1745.

**55** Jiang, H., Yang, X., Chen, C. et al. (2013). Facile and controllable fabrication of three-dimensionally quasi-ordered macroporous TiO2 for high performance lithium-ion battery applications. *New J. Chem.* 37 (5): 1578–1583.

**56** O'Dwyer, C., Lavayen, V., Tanner, D.A. et al. (2009). Reduced surfactant uptake in three-dimensional assemblies of $VO_x$ nanotubes improves reversible $Li^+$ intercalation and charge capacity. *Adv. Funct. Mater.* 19: 1736.

**57** McNulty, D., Buckley, D.N., and O'Dwyer, C. (2014). Polycrystalline vanadium oxide nanorods: growth, structure and improved electrochemical response as a Li-ion battery cathode material. *J. Electrochem. Soc.* 161: A1321–A1329.

**58** Geaney, H. and O'Dwyer, C. (2015). Electrochemical investigation of the role of $MnO_2$ nanorod catalysts in water containing and anhydrous electrolytes for $Li-O_2$ battery applications. *Phys. Chem. Chem. Phys.* 17: 6748–6759.

**59** Bhatt, M.D., Geaney, H., Nolan, M., and O'Dwyer, C. (2014). Key scientific challenges in current rechargeable non-aqueous $Li-O_2$ batteries: experiment and theory. *Phys. Chem. Chem. Phys.* 16: 12093–12130.

60 Valenzuela, C.D., Carriedo, G.A., Valenzuela, M.L. et al. (2013). Solid state pathways to complex shape evolution and tunable porosity during metallic crystal growth. *Sci. Rep.* 3: 2642.

61 Diaz, C., Valenzuela, M.L., Bravo, D. et al. (2008). Synthesis and characterization of cyclotriphosphazene containing silicon as single solid state precursors for the formation of silicon/phosphorus nanostructured materials. *Inorg. Chem.* 47: 11561–11569.

62 McNulty, D., Geaney, H., Armstrong, E., and O'Dwyer, C. (2016). High performance inverse opal Li-ion battery with paired intercalation and conversion mode electrodes. *J. Mater. Chem. A* 4: 4448–4456.

63 Hu, Z., Zhang, S., Zhang, C., and Cui, G. (2016). High performance germanium-based anode materials. *Coord. Chem. Rev.* 326: 34–85.

64 Chan, C.K., Zhang, X.F., and Cui, Y. (2008). High capacity Li-ion battery anodes using Ge nanowires. *Nano Lett.* 8 (1): 307–309.

65 Xue, D.-J., Xin, S., Yan, Y. et al. (2012). Improving the electrode performance of Ge through Ge@C core–shell nanoparticles and graphene networks. *J. Am. Chem. Soc.* 134 (5): 2512–2515.

66 Xiao, X., Li, X., Zheng, S. et al. (2017). Nanostructured germanium anode materials for advanced rechargeable batteries. *Adv. Mater. Inter.* 4 (6): 1600798.

67 Li, D., Wang, H., Liu, H.K., and Guo, Z. (2016). A new strategy for achieving a high performance anode for lithium ion batteries—encapsulating germanium nanoparticles in carbon nanoboxes. *Adv. Energy Mater.* 6 (5): 1501666.

68 Li, D., Wang, H., Zhou, T. et al. (2017). Unique structural design and strategies for germanium-based anode materials toward enhanced lithium storage. *Adv. Energy Mater.* 7, 1700488. https://doi.org/10.1002/aenm.201700488

69 Park, M.-H., Cho, Y., Kim, K. et al. (2011). Germanium nanotubes prepared by using the kirkendall effect as anodes for high-rate lithium batteries. *Angew. Chem.* 123 (41): 9821–9824.

70 Ma, L., Fahrenkrug, E., Gerber, E. et al. (2017). High-performance polycrystalline ge microwire film anodes for Li ion batteries. *ACS Energy Lett.* 2 (1): 238–243.

71 Son, Y., Park, M., Son, Y. et al. (2014). Quantum confinement and its related effects on the critical size of $GeO_2$ nanoparticles anodes for lithium batteries. *Nano Lett.* 14 (2): 1005–1010.

72 Lin, Y.-M., Klavetter, K.C., Heller, A., and Mullins, C.B. (2013). Storage of lithium in hydrothermally synthesized $GeO_2$ nanoparticles. *J. Phys. Chem. Lett.* 4 (6): 999–1004.

73 Jahel, A., Darwiche, A., Matei Ghimbeu, C. et al. (2014). High cycleability nano-$GeO_2$/mesoporous carbon composite as enhanced energy storage anode material in Li-ion batteries. *J. Power Sources* 269: 755–759.

74 Su, R., Bechstein, R., Sø, L. et al. (2011). How the anatase-to-rutile ratio influences the photoreactivity of $TiO_2$. *J. Phys. Chem. C* 115 (49): 24287–24292.

75 Nam, I., Park, J., Park, S. et al. (2017). Observation of crystalline changes of titanium dioxide during lithium insertion by visible spectrum analysis. *Phys. Chem. Chem. Phys.* 29: 13140.

76 Kavan, L., Zukalová, M., Kalbáč, M., and Graetzel, M. (2004). Lithium insertion into anatase inverse opal. *J. Electrochem. Soc.* 151 (8): A1301–A1307.

77 Collins, G., Armstrong, E., McNulty, D. et al. (2016). 2D and 3D photonic crystal materials for photocatalysis and electrochemical energy storage and conversion. *Sci. Technol. Adv. Mater.* 17 (1): 563–582.

78 Armstrong, E., McNulty, D., Geaney, H., and O'Dwyer, C. (2015). Electrodeposited structurally stable $V_2O_5$ inverse opal networks as high performance thin film lithium batteries. *ACS Appl. Mater. Interfaces* 7 (48): 27006–27015.

79 McNulty, D., Geaney, H., and O'Dwyer, C. (2017). Carbon-coated honeycomb Ni–Mn–Co–O inverse opal: a high capacity ternary transition metal oxide anode for Li-ion batteries. *Sci. Rep.* 7: 42263.

80 McNulty, D., Geaney, H., Carroll, E. et al. (2017). The effect of particle size, morphology and C-rates on 3D structured $Co_3O_4$ inverse opal conversion mode anode materials. *Mater. Res. Express* 4 (2): 025011.

81 Zhang, F., Wei, M., Viswanathan, V.V. et al. (2017). 3D printing technologies for electrochemical energy storage. *Nano Energy* 40 (May): 418–431.

82 Mendoza-Sánchez, B. and Gogotsi, Y., Oak Ridge National Lab ORTN(2016). Synthesis of two-dimensional materials for capacitive energy storage. *Adv. Mater.* 28 (29): 6104–6135.

83 Peng, X., Liu, H., Yin, Q. et al. (2016). A zwitterionic gel electrolyte for efficient solid-state supercapacitors. *Nat. Commun.* 7 (1): 11782.

84 Ligon, S.C., Liska, R., Stampfl, J. et al. (2017). Polymers for 3D printing and customized additive manufacturing. *Chem. Rev.* 117 (15): 10212–10290.

85 Gross, B.C., Erkal, J.L., Lockwood, S.Y. et al. (2014). Evaluation of 3D printing and its potential impact on biotechnology and the chemical sciences. *Anal. Chem.* 86 (7): 3240–3253.

86 Zhu, W., Ma, X., Gou, M. et al. (2016). 3D printing of functional biomaterials for tissue engineering. *Curr. Opin. Biotechnol.* 40: 103–112.

87 Sun, J., Zhou, W., Huang, D. et al. (2015). An overview of 3D printing technologies for food fabrication. *Food Bioprocess Technol.* 8 (8): 1605–1615.

88 Zhao, C., Wang, C., Gorkin, R. et al. (2014). Three dimensional (3D) printed electrodes for interdigitated supercapacitors. *Electrochem. Commun.* 41: 20–23.

89 Sun, K., Wei, T.S., Ahn, B.Y. et al. (2013). 3D printing of interdigitated Li-ion microbattery architectures. *Adv. Mater.* 25 (33): 4539–4543.

90 Qi, D., Liu, Y., Liu, Z. et al. (2017). Design of architectures and materials in in-plane micro-supercapacitors: current status and future challenges. *Adv. Mater.* 29 (5): 1602802.

91 Pech, D., Brunet, M., Dinh, T.M. et al. (2013). Influence of the configuration in planar interdigitated electrochemical micro-capacitors. *J. Power Sources* 230: 230–235.

92 Liu, W., Lu, C., Wang, X. et al. (2015). High-performance microsupercapacitors based on two-dimensional graphene/manganese dioxide/silver nanowire ternary hybrid film. *ACS Nano* 9 (2): 1528–1542.

93 Wei, T.S., Ahn, B.Y., Grotto, J., and Lewis, J.A. (2018). 3D printing of customized Li-ion batteries with thick electrodes. *Adv. Mater.* 30 (16): 1–7.

**94** Bergmann, C., Lindner, M., Zhang, W. et al. (2010). 3D printing of bone substitute implants using calcium phosphate and bioactive glasses. *J. Eur. Ceram. Soc.* 30 (12): 2563–2567.

**95** Egorov, V., Gulzar, U., Zhang, Y. et al. (2020). Evolution of 3D printing methods and materials for electrochemical energy storage. *Adv. Mater.* 32: 2000556.

**96** Zheng, H., Li, J., Song, X. et al. (2012). A comprehensive understanding of electrode thickness effects on the electrochemical performances of Li-ion battery cathodes. *Electrochim. Acta* 71: 258–265.

**97** Zhao, R., Liu, J., and Gu, J. (2015). The effects of electrode thickness on the electrochemical and thermal characteristics of lithium ion battery. *Appl. Energy* 139: 220–229.

**98** Egorov, V. and O'Dwyer, C. (2020). Architected porous metals for electrochemical energy storage. *Curr. Opin. Electrochem.* 21: 201–208.

**99** Vyatskikh, A., Delalande, S., Kudo, A. et al. (2018). Additive manufacturing of 3D nano-architected metals. *Nat. Commun.* 9 (1): 593.

**100** Taberna, P.L., Mitra, S., Poizot, P. et al. (2006). High rate capabilities $Fe_3O_4$-based Cu nano-architectured electrodes for lithium-ion battery applications. *Nat. Mater.* 5 (7): 567–573.

**101** McNulty, D., Lonergan, A., O'Hanlon, S., and O'Dwyer, C. (2018). 3D open-worked inverse opal $TiO_2$ and $GeO_2$ materials for long life, high capacity Li-ion battery anodes. *Solid State Ionics* 314: 195–203.

**102** Armstrong, E. and O'Dwyer, C. (2015). Artificial opal photonic crystals and inverse opal structures - fundamentals and applications from optics to energy storage. *J. Mater. Chem. C* 3: 6109–6143.

**103** Gulzar, U., Glynn, C., and O'Dwyer, C. (2020). Additive manufacturing for energy storage: methods, designs and materials selection for customizable 3D printed batteries and supercapacitors. *Curr. Opin. Electrochem.* 20: 46–53.

**104** Palmer, J.A., Yang, P., Davis, D.W. et al. (2004). Rapid prototyping of high density circuitry. In: *Rapid Prototyping & Manufacturing 2004 Conference Proceedings, , Rapid Prototyping Association of the Society of Manufacturing Engineers*, 24. Dearborn: Society of Manufacturing Engineers.

**105** Joe Lopes, A., MacDonald, E., and Wicker, R.B. (2012). Integrating stereolithography and direct print technologies for 3D structural electronics fabrication. *Rapid Prototyping J.* 18 (2): 129–143.

**106** Ho, C.C., Evans, J.W., and Wright, P.K. (2010). Direct write dispenser printing of a zinc microbattery with an ionic liquid gel electrolyte. *J. Micromech. Microeng.* 20 (10): 104009.

**107** Zhu, C., Liu, T., Qian, F. et al. (2016). Supercapacitors based on three-dimensional hierarchical graphene aerogels with periodic macropores. *Nano Lett.* 16 (6): 3448–3456.

**108** Browne, M.P., Novotný, F., Sofer, Z., and Pumera, M. (2018). 3D printed graphene electrodes' electrochemical activation. *ACS Appl. Mater. Interfaces* 10 (46): 40294–40301.

**109** Novotný, F., Urbanová, V., Plutnar, J., and Pumera, M. (2019). Preserving fine structure details and dramatically enhancing electron transfer rates in graphene

3D-printed electrodes via thermal annealing: toward nitroaromatic explosives sensing. *ACS Appl. Mater. Interfaces* 11 (38): 35371–35375.

110 Foster, C.W., Down, M.P., Zhang, Y. et al. (2017). 3D printed graphene based energy storage Devices. *Sci. Rep.* 7: 42233.

111 Foster, C.W., Zou, G.-Q., Jiang, Y. et al. (2019). Next-generation additive manufacturing: tailorable graphene/polylactic(acid) filaments allow the fabrication of 3D printable porous anodes for utilisation within lithium-ion batteries. *Batteries Supercaps* 2 (5): 448–453.

112 Vernardou, D., Vasilopoulos, K.C., and Kenanakis, G. (2017). 3D printed graphene-based electrodes with high electrochemical performance. *Appl. Phys. A* 123 (10): 623.

113 Ragones, H., Menkin, S., Kamir, Y. et al. (2018). Towards smart free form-factor 3D printable batteries. *Sustainable Energy Fuels* 2 (7): 1542–1549.

114 Zhao, Y., Zhou, Q., Liu, L. et al. (2006). A novel and facile route of ink-jet printing to thin film $SnO_2$ anode for rechargeable lithium ion batteries. *Electrochim. Acta* 51 (13): 2639–2645.

115 Xu, F., Wang, T., Li, W., and Jiang, Z. (2003). Preparing ultra-thin nano-$MnO_2$ electrodes using computer jet-printing method. *Chem. Phys. Lett.* 375 (1): 247–251.

116 Lawes, S., Sun, Q., Lushington, A. et al. (2017). Inkjet-printed silicon as high performance anodes for Li-ion batteries. *Nano Energy* 36: 313–321.

117 Gu, Y., Wu, A., Sohn, H. et al. (2015). Fabrication of rechargeable lithium ion batteries using water-based inkjet printed cathodes. *J. Manuf. Processes* 20: 198–205.

118 Delannoy, P.E., Riou, B., Brousse, T. et al. (2015). Ink-jet printed porous composite $LiFePO_4$ electrode from aqueous suspension for microbatteries. *J. Power Sources* 287: 261–268.

119 Huang, J., Yang, J., Li, W. et al. (2008). Electrochemical properties of $LiCoO_2$ thin film electrode prepared by ink-jet printing technique. *Thin Solid Films* 516 (10): 3314–3319.

120 Lee, J.-H., Wee, S.-B., Kwon, M.-S. et al. (2011). Strategic dispersion of carbon black and its application to ink-jet-printed lithium cobalt oxide electrodes for lithium ion batteries. *J. Power Sources* 196 (15): 6449–6455.

121 Delannoy, P.E., Riou, B., Lestriez, B. et al. (2015). Toward fast and cost-effective ink-jet printing of solid electrolyte for lithium microbatteries. *J. Power Sources* 274: 1085–1090.

122 Cohen, E., Menkin, S., Lifshits, M. et al. (2018). Novel rechargeable 3D-microbatteries on 3D-printed-polymer substrates: feasibility study. *Electrochim. Acta* 265: 690–701.

# 12

## Ferroic Transition Metal Oxide Nano-heterostructures: From Fundamentals to Applications

*G. Varvaro[1], A. Omelyanchik[1,2,3], and D. Peddis[1,2]*

[1] National Research Council, Institute of Structure of Matter, nM²-Lab, Research Area Roma 1, Via Salaria km 29.300, 00015 Monterotondo Scalo (Roma), Italy
[2] University of Genova, Department of Chemistry and Industrial Chemistry, nM²-Lab, Via Dodecaneso 31, 16146 Genova, Italy
[3] Immanuel Kant Baltic Federal University, REC "Smart Materials and Biomedical Applications", Nevskogo 14, 236041, Kaliningrad, Russia

## 12.1 Introduction

Transition metal oxides (TMOs) represent a very interesting class of materials consisting of transition metals bonded to oxygen atoms. Their rich variety of crystal chemistry turns into a unique plethora of physical properties related mainly to spin and electronic structures, making TMOs applicable in many technological fields of industry and interesting objects for fundamental research [1, 2].

Such flexible materials have provided a fertile playground for the discovery and application of different and novel forms of magnetism. In magnetic materials, the atomic magnetic moments can mutually act together (i.e. cooperative magnetism) and lead to a different behavior from what it would be observed if all the magnetic moments were reciprocally isolated (i.e. noncooperative magnetism). This effect, coupled with the different types of magnetic interaction, leads to a rich variety of magnetic properties. Two classes of interactions can be distinguished, namely, *direct exchange interaction* (i.e. interactions occur between moments close enough to have significant overlap of their wave functions) and *indirect exchange interaction* (i.e. interactions *arising* when the atomic magnetic moments are coupled over relatively large distances). The latter can also be mediated by a nonmagnetic (NM) ion (e.g. oxygen atoms in TMOs), which is placed in between the magnetic ions, and, in this case, it is called *super-exchange interactions*. From this picture, it is clear that the magnetic properties of TMOs are strongly related, and they are controlled by super-exchange interactions, giving rise, below a certain temperature, to several types of magnetic orders such us ferri(Fi)-, antiferro(AF)-, and ferromagnetism(F) [3–5] (Figure 12.1a). In F materials, individual magnetic moments ($\mu$) are aligned in

---

*Tailored Functional Oxide Nanomaterials: From Design to Multi-Purpose Applications*, First Edition.
Edited by Chiara Maccato and Davide Barreca.
© 2022 WILEY-VCH GmbH. Published 2022 by WILEY-VCH GmbH.

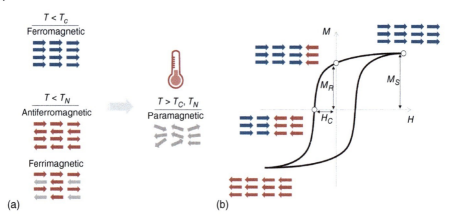

**Figure 12.1** Schematic representation of the (a) main magnetic order and (b) field-dependent magnetization loop of a ferro(i)magnetic material.

one direction in a certain volume of material called "magnetic domain." In contrast, in AF and Fi materials, neighboring individual magnetic moments are aligned antiparallelly; thus they cancel each other's in AF materials and partially in Fi materials [6]. Macroscopically, the magnetic properties of Fi materials are like those of an F: below a critical temperature (i.e. Curie temperature, $T_C$), they both have pronounced saturation magnetization ($M_s$), non-zero coercivity ($H_C$), and remanence magnetization ($M_R$), showing a typical M-H hysteresis loop (Figure 12.1b). Above $T_C$ (called Néel temperature for antiferromagnetic, $T_N$), the energy of thermal fluctuations became higher than the exchange energy, bringing the system in a paramagnetic state.

Like the ferromagnetic properties, ferroelectricity (FE) is characterized by the presence of a polarization under an applied electric field. Ferroelectricity depends on the whole structure and symmetry of the compounds, and the mechanism of this phenomenon is related to the order, disorder, and displacement of ions within the crystalline structure. Again, in analogy with ferromagnetism, the ferroelectric property results in the hysteresis loop related to the formation of electric domain (Figure 12.2) with saturation polarization ($P_s$, maximum polarization), remanent polarization ($P_r$, residual polarization in the material), and coercive field ($E_c$, electric field that must be applied to get zero polarization) [7]. Above a certain critical temperature, depending on the nature of the material, a phase change from ferroelectric to paraelectric is observed.

Changing the chemical composition and crystal structure strongly affects the ferroic behaviour and elctronic properties of TMOs (i.e. insulator, metallic, and semiconductors). In addition, in the last decades the interplay of charge, spin, and orbital degrees has become a hot topic in spintronics [8–10]. In this view, a key example is given by the change of resistivity in the magnetic field related to phase transition from paramagnetic to ferromagnetic state (colossal magnetoresistance) observed in $LaMnO_3$ [11].

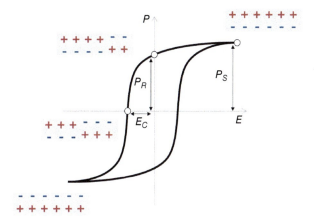

**Figure 12.2** Hysteresis (P–E) curve in ferroelectric materials.

Materials possessing two or more ferroic properties (i.e. multiferroic materials [12]) have become the subject of an intense research aimed at investigating the chemical nature and at studying the solid-state physics [13]. The new research frontiers on these materials are the design by chemical engineering and nano-architecturing of complex TMO heterostructures in the form of nanoparticles, nanowires/tubes, nanocomposites, and thin films. Indeed, it is well known that physical properties, especially spin-dependent ones, are strongly affected by the size of material constituents. In particular, when the dimension of magnetic materials decrease below a certain size, it overcomes from multi-domain to single-domain state [14, 15], due to the tendency of the system to minimize its magnetostatic energy (Figure 12.2), with a significant change of the overall magnetic properties. The energy of a magnetic particle is generally dependent on the magnetization direction, and for particles with uniaxial anisotropy and negligible interparticle interactions, it can be considered proportional to magnetic anisotropy ($K$) and the nanoparticle volume ($V$) (i.e. $\Delta E_B = KV$). Above a certain temperature (the blocking temperature, $T_B$) and on a certain time scale, the particle moment can produce a thermally activated transition, behaving like a paramagnet but with a different time and magnetization scale; for this reason, this phenomenon is called "superparamagnetism." Below $T_B$, the particle moment is blocked and unable to rotate over the barrier in the time of a measurement [16]. The magnetic behavior of a nanoparticle assembly can also be strongly affected by interparticle interactions that lead to an increase in the energy barrier, evolving from ferromagnetic (FM)-like including a spin glass-like behavior. The magnetism of nanoparticle assembly has been often called supermagnetism. In this general framework, it appears clear that tuning the magnetic anisotropy and magnetic interaction represents a key step to design nanostructured materials with suitable magnetic properties for specific applications. Magnetic anisotropy at the nanoscale is related by the interplay of several sources: spin–orbit coupling, shape, surface/interface, and others. The spin–orbit coupling gives rise to magnetocrystalline anisotropy, which is mainly dependent on the chemical composition of the material. In magnetic nanocomposites (e.g. magnetic particles in the magnetic matrix or core–shell structures) or multilayers, surface spins at the interface of one material interact with those of its counterpart via exchange interactions if they are in direct contact. This interaction leads to several interesting phenomena (i.e. exchange bias) representing an additional tool to tune the magnetic properties of TMO nano-heterostructures. For example, if the magnetic anisotropy of two counterparts is significantly different or in ferro(i-)magnetic/antiferromagnetic composites, such phenomena as exchange bias can be observed. Thus, complex TMO materials represent an exciting playground to design new materials with tuneable magnetic properties by fabrication of complex multimagnetic nano-heterostructures with various shape, size, and chemical composition of materials. In this chapter, a brief overview on the magnetic and ferroic properties of nanostructured complex TMOs will be given. Then, recent examples of different TMO nano-heterostructures will be discussed, showing the possibility to design new multifunctional materials. Finally, underexplored and emerging research directions in the field will be briefly discussed.

## 12.2 Ferroic Properties of Complex Transition Metal Oxides

Ferroic metal oxides incorporating 3d transition metals exhibit manifold physical properties, such as spontaneous magnetization, electric polarization, and strain, which make them of great interest for both fundamental studies and applications in different fields, including information storage/processing, sensors, biomedicine, and energy. The functional properties of ferroic TMOs are strictly related to the characteristic electronic states arising from the hybridization of 3d transition metals with oxygen, and the specific crystal symmetry of these structures, where cations are located in the lattice sites of a closely packed substructure of oxygen atoms. The main properties of the most investigated TMOs, i.e. $AB_2O_4$ spinels and $ABO_3$ perovskites consisting of A and B cations distributed in a closely packed oxygen lattice with a face-centered cubic (fcc) structure, will be illustrated while also discussing in brief the effect of size on their features. In addition, the key properties of other TMOs, including $AB_{12}O_{19}$ hexaferrites, $A_2(BB')O_6$ double perovskites, and AO rock-salt monoxides, will be briefly reviewed.

### 12.2.1 Spinel Ferrites

Spinel oxides ($M^{2+}M^{3+}_2O_4$) containing 3d metals represent one of the most important classes of TMOs due to their rich crystal chemistry, allowing for an excellent fine-tuning of the magnetic properties. The spinel ferrite structure can be described as a cubic close-packed arrangement of $O^{2-}$ anions, with $M^{2+}$ and $M^{3+}$ ions occupying tetrahedral and octahedral coordinated sites, termed as A- and B-sites, respectively. When the (A)-sites are occupied by $M^{2+}$ cations and the [B]-sites by $M^{3+}$ cations, the structure is referred to as *normal* spinel, $(M^{2+})[M^{3+}]$. However, if A-sites are completely occupied by $M^{3+}$ and B-sites are occupied by

$M^{2+}$ and $M^{3+}$, the structure is referred to *inverse* spinel, $(M^{3+})[M^{3+}\ M^{2+}]$. In general, the cationic distribution in octahedral and tetrahedral sites is quantified by the "inversion degree" ($\gamma$), which is defined as the fraction of divalent ions in octahedral sites [17]. Super-exchange interactions between atomic magnetic moments in A (A–O–A) and B (B–O–B) interstices lead to a ferromagnetic order between the ions located in the two sites, giving rise to two magnetic sublattices. On the other hand, interactions between magnetic ions in the A- and B-sites (A–O–B) induce antiferromagnetic order, and they are tenfold stronger than the intra-lattice interactions. The dominant inter-lattice interactions induce a non-compensated antiferromagnetic order (ferrimagnetism), and the saturation magnetization ($M_s$) is given by the difference between the A and B sublattice magnetizations [17].

Among spinel oxides, magnetite ($Fe_3O_4$) and maghemite ($\gamma\text{-}Fe_2O_3$) ferrites are the most investigated and still represent an interesting class of materials both for fundamental research and technological applications [18]. While magnetite contains both $Fe^{2+}$ and $Fe^{3+}$ ions, the maghemite only contains $Fe^{3+}$ with some vacancies in the spinel structure [19, 20]. The absence of catalytically active $Fe^{2+}$ ions in $\gamma\text{-}Fe_2O_3$ avoids the formation of reactive oxygen species in biological media, making maghemite of great interest for biomedical applications [21]. Magnetite is a common soft ferrimagnetic material with a high Curie temperature ($\sim 860$ K), characterized by a metal–insulator transition occurring at the Verwey temperature ($T_V = 120–125$ K for bulk magnetite) [22]. Above $T_V$, the magnetite is conductive with a half-metallic character due to the fluctuation of valences of $Fe^{2+}$ and $Fe^{3+}$ over octahedral sides in the inverted spinel structure. Below $T_V$, magnetite distorts into a monoclinic superstructure with $Cc$ space group symmetry. This first-order structural transition is caused by charge and orbital ordering through electrostatic repulsion and Jahn–Teller compression [23]. In this charge-ordered state, the resistivity of magnetite sharply increases by two orders of magnitude. At the nanoscale (i.e. thin films and nanoparticles), size effects, as well as strains and structure defects, lead $T_V$ to shift to lower temperatures in the 25–125 K range [24, 25]. In some cases, due to the oxidation of magnetite to maghemite (i.e. the disappearance of $Fe^{2+}$), the observation of Verwey transition is hindered. The half-metallicity and the charge-ordering effect make magnetite potentially interesting for many applications including spintronic and high-frequency devices [26].

Efficient control of the magnetic structure and properties can be obtained by chemical engineering of spinels with suitable ions. For example, substituting the $Fe^{2+}$ ions with other magnetic (e.g. Co, Ni, Mn) and/or non-magnetic (e.g. Zn) bivalent cations allows the electronic and magnetic properties (e.g. insulating/conductive state, magnetic anisotropy, Curie temperature, and saturation magnetization) to be finely tuned. In Table 12.1, the main physical properties of the most common magnetic spinel ferrites in the bulk form are listed. All of them show ferrimagnetic properties, except zinc and magnesium ferrites that are antiferromagnetic or weakly ferrimagnetic depending on the inversion degree. Magnetite and manganese ferrites show the highest values of saturation magnetization at 0 K, owing to a combination of the highest magnetic moments of divalent metal ions and the inversion degree. $CoFe_2O_4$ has the highest value of magnetic anisotropy and

**Table 12.1** Common spinel ferrites and their main properties: density, inversion degree ($\gamma$), magnetic Curie (Néel) temperature ($T_{C(N)}$), saturation magnetization ($M_S$), and magnetic anisotropy constant ($K$).

| Compound | Density (kg m$^{-3}$) | $\gamma$ | $T_{C(N)}$ (K) | $M_S$ (Am² kg$^{-1}$) 0 K | $M_S$ (Am² kg$^{-1}$) 300 K | $K$ (×10⁴ J m$^{-3}$) |
|---|---|---|---|---|---|---|
| $Fe_3O_4$ | 5240 | 1 | 858 | 97 | 91 | 1.2 |
| $\gamma\text{-}Fe_2O_3$ | 4900 | 1 | 948 | 81 | 73 | 0.46 |
| $MnFe_2O_4$ | 5000 | 0.2 | 573 | 112 | 80 | 0.3 |
| $CoFe_2O_4$ | 5290 | 0.8–1 | 793 | 90 | 80 | 20 |
| $NiFe_2O_4$ | 5380 | 1 | 858 | 56 | 50 | 0.62 |
| $ZnFe_2O_4$ | 5200 | 0 | 9 | AF | P | — |
| $MgFe_2O_4$ | 5280 | 0.9 | 713 | 27 | 23 | 1.1[a] |

AF, antiferromagnetic; P, paramagnetic.
a) K value for $MgFe_2O_4$ is for a 110 nm thin film [27].
Source: Data are adapted from Dionne [2], Cullity and Graham [3], and Sanchez-Lievanos et al. [28].

can be considered as a magnetically semi-hard material. Moreover, it has one of the largest room temperature magnetostriction among spinel ferrites ($\lambda_{100} \sim -650 \times 10^6$, $\lambda_{111} \sim 120 \times 10^6$, $\lambda_{s,\text{polycrystalline}} \sim -110 \times 10^6$) [29], which allows for a fine tuning of the physical properties by lattice strain. Further control of magnetic properties can be achieved in mixed spinel ferrites containing two different bivalent cations. For example, when magnetite is doped with non-magnetic zinc, $Zn^{2+}$ preferentially occupies tetrahedral sites pushing part of the $Fe^{3+}$ cations into octahedral positions, leading to a change of the saturation magnetization. A similar effect was observed in cobalt and nickel ferrites doped by zinc [30–33]. The increase in the concentration of non-magnetic zinc leads to a weakening of the super-exchange interaction between A- and B-sites, and the magnetization decreases due to a spin canting phenomenon. Even with the significant influence of the spin canting, the saturation magnetization at low temperatures is significantly higher than that of undoped ferrite. A side effect of this approach, which may be advantageous for specific applications, such as the self-regulated magnetic hyperthermia, is the reduction of the Curie temperature and then of the saturation magnetization at high temperatures [31]. It should be finally mentioned that the overall properties are strongly affected by the system size. For example, in nanostructured materials, the inversion degree is often observed to be significantly different from the thermodynamical equilibrium, and it is strongly related to the synthesis method, temperature treatment, particles size (Figure 12.3), and shape [34–36]. As a consequence, a different saturation magnetization can be observed [37].

In nanocrystalline materials, the inversion degree strongly depends on the synthesis method and temperature treatment. Elias Ferreiro-Vila with colleagues investigated thin film of cobalt ferrite prepared via a chemical solution and pulsed laser deposition methods. By changing the deposition temperature and epitaxial

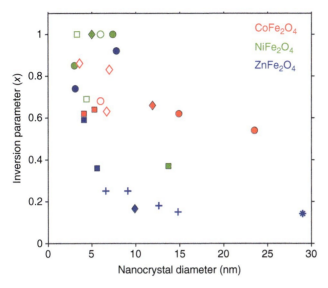

**Figure 12.3** Dependence of inversion degree of representative spinel ferrites (CoFe$_2$O$_4$, NiFe$_2$O$_4$, and ZnFe$_2$O$_4$) as a function of nanocrystal size. Source: Reprinted with permission from Sanchez-Lievanos et al. [28].

stress, Co$^{2+}$ cations can migrate from octahedral to tetrahedral sites in strained layers. This leads to a drastic drop of the magnetic anisotropy as confirmed by the appearance of two-phase-like M-H hysteresis loops [38].

## 12.2.2 Perovskites

Perovskites are among the most investigated TMOs, owing to their potential interest for different applications, such as catalysis, solar cells, semiconductors, and magnetoelectric devices [39]. The general chemical formula used to describe the perovskite is ABO$_3$, where A is an alkaline (e.g. La), rare-earth (e.g. Sm), transition (e.g. Y), or post-transition (e.g. Bi) metal and B is a transition metal (e.g. Fe, Co, Mn, Ti). In a typical structure, the A-site and O ions form an *fcc* lattice, while 1/4 of the B-site is occupied by TM ions forming BO$_6$ octahedra. Very interesting are also double perovskites (discussed in the next paragraph) with a general formula A$_2$BB'O$_3$, where B is magnetic TM and B' is nonmagnetic TM or lanthanide ion with different nominal charges. The B–O–B super-exchange bridging by oxygen favors the antiferromagnetic order if ions in B-sites are magnetic. Only in a few compounds, ferro(i)magnetic order is observed at low temperatures, such as YTiO$_3$ [40] and BiBO$_3$ (e.g. B = Fe, Mn) perovskites [41]. Nevertheless, many common antiferromagnetic perovskites, such as LaMnO$_3$, CaMnO$_3$, and LaFeO$_3$, can exhibit ferromagnetic properties caused by oxygen vacancies, strains, or by doping with other elements. A paramount example is represented by LaMnO$_3$ perovskites doped by Sr, Ba, or Pb, which undergoes a phase transformation from rhombohedral to orthorhombic structure accompanied by the appearance of an ferromagnetic order. For example, in La$_x$Sr$_{1-x}$MnO$_3$, the saturation magnetization can exceed 4.2 $\mu_B$/f.u., and the Curie temperature can vary in the 145–370 K range depending on the $x$ values (0.1–0.3) [42]. Due to the high saturation magnetization and spin polarization, this material exhibits colossal magnetoresistance, and is of great interest for spintronic applications [11, 43]. A peculiar property of perovskites, and in particular of manganites, which is related to their electronic configuration, is the so-called *phase separation phenomenon* [44]. The strong coupling between electronic and elastic degrees of freedom leads to the formation of spontaneous regions with local nano- or microscale electronic configuration. This can result, for example, in the formation of conducting islands in an insulating matrix or the appearance of ferromagnetic clusters in an antiferromagnetic matrix [45–47]. Some perovskites with relatively high resistivity (e.g. $PbTiO_3$ [48] and $BaTiO_3$ [49]) exhibit a ferroelectric order, i.e. a spontaneous electric polarization that can be manipulated by means of an external electric field. Ferroelectricity of these materials is usually explained by displacement of the $Ti^{4+}$ cations in the O octahedron due to the second order Jahn–Teller effect. Both $PbTiO_3$ and $BaTiO_3$ also exhibit piezoelectricity (i.e. applying mechanical stresses leads to the appearance of electrical polarization, and, in the opposite, by switching electrical field it is possible to induce mechanical stresses). This correlation is widely exploited in strain-coupled artificial multiferroic composites consisting of ferroelectric/piezoelectric and ferro(i)magnetic/magnetostrictive materials in intimate contact, where the ferroic property (i.e. electric polarization, magnetization) can be manipulated with the conjugate field (magnetic, electric) of the other. Multiferroic behavior is also observed in single phase materials with a perovskite structure, e.g. $BiFeO_3$, $TbMnO_3$, $PbXO_3$, $EuTiO_3$, $YMnO_3$, and $RMO_3$–R = rare-earth, M = Fe, Cr, Mn – [50–53]. Among them, bismuth ferrite ($BiFeO_3$) is the only room-temperature single-phase multiferroic material with a ferroelectric Curie temperature $T_C = 1123$ K and a Néel antiferromagnetic temperature $T_N = 643$ K. $BiFeO_3$ exhibits a high value of ferroelectric polarization ($\sim$90–100 $\mu C\,cm^{-2}$) and uncompensated magnetic moment reaching $\sim$1 $\mu_B$/f.u. [54, 55]. The coexistence of the two ferroic orders has been exploited to control the magnetic state through an electric field [56] or to regulate the electric polarization with an external magnetic field [57], thus paving the way for the development of novel functional devices [58–61]. $BiMnO_3$ is the only multiferroic material showing a ferromagnetic order with a large saturation magnetization of $\sim$4 $\mu_B$/f.u. and a Curie temperature in the 99–105 K range. On the other hand, it shows a relative low ferroelectric polarization with a theoretical value of 0.52 $\mu C\,cm^{-2}$, which has been, however, never observed experimentally.

### 12.2.3 Other Magnetic Oxides

Double perovskites with a general formula $A_2BB'O_3$, where B is magnetic TM and B′ is nonmagnetic TM or lanthanide ion with different nominal charges, have recently attracted a great deal of attention, owing to the possibility to finely tune the physical properties by chemical engineering. In B-site-ordered perovskites, $BO_6$ and $B'O_6$ octahedra form 3D structural order, which via indirect B–O–B′–O–B exchange interaction provide anti- or ferromagnetic (or even its coexistence) properties above room temperature [62, 63]. After discovering magnetoresistance in $Sr_2FeMoO_6$, this material has become the most studied among double perovskites [64]. Magnetic properties of this material are primary governed by anti-site disorder, i.e. by the percentage of misplaced B ions at B'-sites ($w$), according to the expression:

$$M_S = (4 - 8w)\mu_B/\text{f.u.} \quad (12.1)$$

From this equation, the saturation magnetization can reach a value as high as $4\,\mu_B$/f.u. The synthesis of new types of double perovskites with controlled structural order is underemphasized because of interest in fundamental science and a high potential for several applications.

Monoxides such as NiO, CoO, FeO, and MnO represent another interesting class of TMOs showing antiferromagnetic order with $T_N$ of 523, 291, 198, and 122 K, respectively [65]. They assume a cubic NaCl type (rock salt) unit cell at temperatures above Néel temperatures, while below $T_N$ the magnetic order induces a transition from cubic to a rhombohedral symmetry in NiO, FeO, and MnO and a tetragonal symmetry in CoO.

M-type hexagonal ferrites with chemical formula $MFe_{12}O_{19}$ (where $M$ = Ba or Sr) show outstanding properties compared with spinel ferrites in terms of magnetic anisotropy [66] and dielectric properties. Their large uniaxial magnetic anisotropy has made hexaferrites of great interest for the development of permanent magnets [67, 68].

## 12.3 Magnetic Oxide Heterostructures

The interfacial effect in composite heterostructures combining two or more different oxide materials has emerged as a powerful tool to extend the already reach phenomenology of TMOs. Novel functional materials with tailored chemical and physical properties matching the requirements for specific applications can be designed by changing the nature of each component, their relative size, or by tuning the interface interactions. *Hard/soft ferro(i)magnetic exchange-coupled* composites, *antiferromagnetic/ferro(i)magnetic exchange bias* systems, and *synthetic antiferromagnets* (SAFs) combining two or more ferro(i)magnetic phases separated by a non-magnetic spacer have been the subject of an intense activity involving the fabrication and investigation of heterostructures with different architectures including core–shell nanoparticles, thin film multilayers, and granular nanocomposites.

### 12.3.1 Hard/Soft Exchange-Coupled Systems

The nanoscale interaction occurring at the interface between hard ($h$) and soft ($s$) ferro(i)magnetic (F/Fi) materials has been widely exploited to design novel functional materials, which benefit from the attributes of the constituent phases, i.e., the high magnetic anisotropy $K$ (and coercivity) of the $h$-phase and the high saturation magnetization ($M_s$) of the $s$-phase [69–71]. By changing the constituent materials, the size of the two phases, and the shape of the interface, the overall properties can

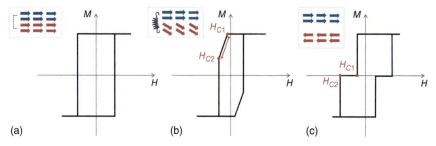

**Figure 12.4** Schematic representation of the field-dependent magnetization loop of exchange-coupled hard/soft composites for different magnetic regimes. (a) *Rigid coupling regime*: the two phases reverse simultaneously at a critical field $H_c$. (b) *Exchange spring regime*: the soft phase reverses first at $H_{c1}$ in a reversible way (indicated by the red arrow) and supports the switching of the hard phase occurring at $H_{c2}$. (c) *Decoupled regime*: the two phases reverse separately at $H_{c1}$ and $H_{c2}$.

be finely modulated. The critical parameters determining the magnetic behavior of hard/soft exchange-coupled composites are the soft layer thickness ($t_s$) and the exchange length $l_w = \sqrt{A/K_h}$, where $A$ is the exchange stiffness constant and $K_h$ the magnetic anisotropy of the hard phase (Figure 12.4). For $t_s < l_w$, the spins of the soft and the hard phases are strongly coupled, and the system behaves as a single magnetic phase with the properties averaged between the two materials (*rigid coupled regime*). When $t_s$ is of the order of a few $l_w$, the two phases are still coupled, but the soft phase reverses first in a reversible way and supports the switching of the hard phase, which occurs at a field lower than that expected for the single hard material (*exchange spring regime*). With increasing $t_s$ well above $l_w$, the two phases become magnetically decoupled and reverse independently and irreversibly at different critical fields (*decoupled regime*). Depending on the strength of the interphase exchange interaction, hard/soft exchange-coupled composites can be exploited for different applications. Rigid-coupled magnets can be designed to increase the remanent magnetization while limiting the reduction of coercivity for the development of high-performance permanent magnets with an enhanced maximum energy product [72–77]. Exchange-spring systems are on the contrary of great interest for magnetic recording applications as they allow reducing the coercivity of the *h*-phase without significantly affecting the thermal stability [71, 78–83].

Despite most of the studies have been focused on metallic systems, a growing attention has been devoted to all-oxide hard/soft composites for both fundamental studies and technological applications [68, 72, 73, 84–97].

Hard/soft exchange-coupled core–shell magnetic nanoparticles (MNPs) have attracted a significant attention, owing to the great advantage of such structures for several applications (e.g., magnetic hyperthermia and permanent magnets). An example of these systems is reported in Figure 12.5, showing the microstructural features and the magnetic behavior of $CoFe_2O_4/MnFe_2O_4$ core–shell MNPs with a narrow size distribution. The smooth shape of the hysteresis loop of the core–shell MNPs indicates that the two phases are rigid coupled giving rise to a material whose properties are averaged between the two phases [98]. The peculiar properties

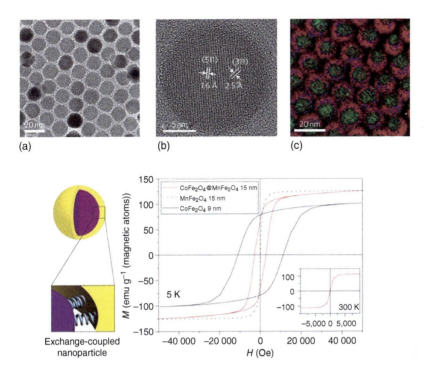

**Figure 12.5** (a) Transmission electron microscopy (TEM) and (b) high-resolution TEM images, (c) electron energy loss spectroscopy elemental map (Co – green, Fe – red, and Mn – blue) of 15 nm $CoFe_2O_4@MnFe_2O_4$. (d) M–H curve of single-phase 15 nm $MnFe_2O_4$, 9 nm $CoFe_2O_4$ MNPs, and exchange-coupled core–shell 15 nm $CoFe_2O_4@MnFe_2O_4$ MNPs. Source: Lee et al. [98], Reproduced with permission from Springer Nature.

of these systems result in a great increment of the specific loss power (up to one order of magnitude) with respect to single-phase iron oxides MNPs, with significant implication for applications based on the magnetic hyperthermia [98].

Varying the composition, the shell thickness, and the sequence of layers yields to further degrees of freedom to tune the magnetic properties and designing multifunctional magnetic materials. Figure 12.6 shows simple linear dependence of coercivity versus the volume fraction of hard phase for hard@soft nanoparticles. However, this linear trend is oversimplified, and for this reason an important deviation is observed in inverted soft/hard systems [99] (Figure 12.6). In this view, more efforts should be done to comparatively investigate normal (hard/soft) and inverted (soft/hard) core–shell systems [101].

A very important, but relatively poorly studied phenomenon in core–shell MNPs systems is the interplay between inter- and intra-particle interactions [102] and its influence on the magnetic properties and performance of the materials. For example, it was demonstrated that modulating interparticle interactions in core–shell systems allows improving either the stability of particles in a liquid solution or the heat performance for hyperthermia application as well as the R2 relaxivity for contrast enhancement in magnetic resonance imaging (MRI) [102].

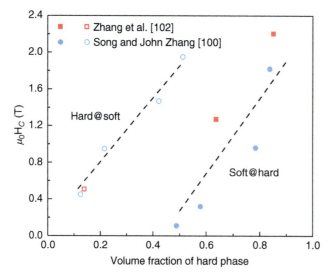

**Figure 12.6** Dependence of coercivity on the volume fraction of the hard phase in $CoFe_2O_4$@$MnFe_2O_4$ and $MnFe_2O_4$@$CoFe_2O_4$ nanoparticles. Source: Data are adapted from Song and Zhang [99] and Zhang et al. [100].

Hard/soft thin film heterostructures consisting of complex TMOs are of great interest as both model systems and components for novel electronic devices. Despite their high potential, a limited number of systems have been studied so far, including all-ferrite (e.g. s-$CoFe_2O_4$/h-$BaFe_{12}O_{19}$ [90], s-$Fe_3O_4$/h-$CoFe_2O_4$ [88, 92], s-$NiFe_2O_4$/h-$CoFe_2O_4$ [84]) and all-perovskite (e.g. s-$La_{0.7}Sr_{0.3}MnO_3$/h-$La_{0.7}Sr_{0.3}CoO_3$) [89] composite films as well as mixed heterostructures combining the two different materials (e.g. s-$La_{0.7}Sr_{0.3}MnO_3$/h-$CoFe_2O_4$) [85, 93]. An illustrative example of these studies is represented by the work of Lavorato et al. [88], which clearly proved that the magnetic behavior of $Fe_3O_4$/$CoFe_2O_4$ spinel ferrite bilayers grown on $MgAl_2O_4$ (001) substrates can be finely modulated by changing the thickness of the softer $Fe_3O_4$ phase (Figure 12.7). A rigid-coupling regime can be indeed achieved for $Fe_3O_4$ thicknesses lower than the exchange length of the hard layer (~8 nm), while two separated reversals are observed for larger $Fe_3O_4$ thicknesses, indicating a partial decoupling of the two layers.

### 12.3.2 Ferro(i)magnetic/Antiferromagnetic Systems

The magnetic coupling between ferro(i)magnetic (F/Fi) and antiferromagnetic (AF) materials is a fundamental interfacial phenomenon of considerable scientific interest and technological importance. Since its discovery in oxidized Co fine particles [103], F(Fi)/AF heterostructures have become an integral part of modern magnetism with implications for both fundamental research and technological applications [104–108]. One of the major effects arising from the exchange interaction at the F(Fi)/AF interface is the development of a unidirectional magnetic anisotropy when the composite is field-cooled through the Néel temperature ($T_N$)

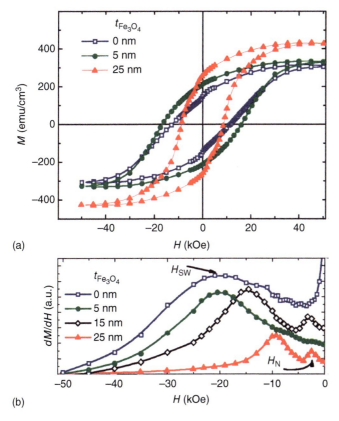

**Figure 12.7** Low-temperature (10 K) (a) field-dependent magnetization loops and (b) corresponding first derivative curves of $Fe_3O_4$ ($t$ nm)/$CoFe_2O_4$ (25 nm) bilayers as a function of the soft layer thickness. $H_{sw}$ and $H_N$ were associated with the irreversible switching field of the $h$-phase ($CoFe_2O_4$) and the nucleation field of the $s$-phase ($Fe_3O_4$), respectively. Source: Reprinted with permission from Lavorato et al. [88].

of the AF phase from a temperature $T_N < T < T_c$, where $T_c$ is the Curie temperature of the F(Fi) phase. This phenomenon, called *exchange bias* effect, manifests with a shift of the F(Fi) hysteresis loop along the field axis by an amount called exchange bias field $H_{EB} = |H_1 + H_2|/2$, where $H_1$ and $H_2$ are the left and right coercive fields (Figure 12.8).

A simplified picture of the spin configurations in the F(Fi) and the AF phases, before and after a field cooling process, is reported in Figure 12.8 (top panel). When a magnetic field is applied at a temperature $T_N < T < T_C$ (point 1), all the spins in the F(Fi) phase align parallel to $H$, i.e. the F(Fi), are in the saturated state. Meanwhile, the AF remains in the paramagnetic state as $T > T_N$. When the F(Fi)/AF system is cooled through $T_N$, the magnetic order in the AF is set up (point 2). During the cooling, the spins at the F(Fi)/AF interface interact with each other by exchange coupling; the first layer of spins in the AF will tend to align parallel to the spins in the F(Fi) phase (assuming a ferromagnetic interaction at the interface), while the adjacent layers in the AF will orient antiparallel to each other. The coupling between the

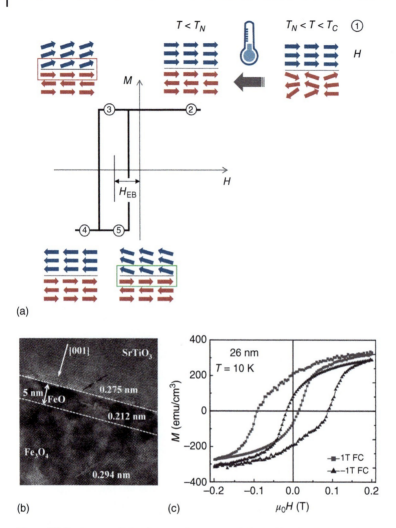

**Figure 12.8** Top panel. (a) Schematic representation of the field-dependent magnetization loop of exchange bias F(Fi)/AF heterostructures. Bottom panel. (b) High-resolution TEM image of Fi/AF interface and (c) field cooled (FC) M–H hysteresis loops of $Fe_3O_4$ (26 nm)/FeO (5 nm) thin films epitaxially grown on an $SrTiO_3$ substrate. Source: Zhu et al. [109], Reproduced with permission from AIP Publishing LLC.

AF and F(Fi) spins at the interface exerts an additional torque on the F spins, which opposes/adds (point 3/point 5) to the reversal field, thus resulting in the shift of the loop. This shift is commonly accompanied by an increase of the coercivity of the F(Fi) phase [104]. The high tunability of the magnetic properties of complex TMOs, which can manifest both F(Fi) and AF behaviors to changes in compositions, made them common building blocks for the realization of exchange bias heterostructures. A paramount example of the exchange bias phenomenon is reported in Figure 12.8 (bottom panel), showing the positive (negative) shift of hysteresis loop in $Fe_3O_4$/FeO bilayers after field cooling under a negative (positive) external field.

AF/F(Fi) core–shell MNPs and nanogranular composites based on TMOs are of great interest due to their increased anisotropy and thermal stability, which make them potential candidates for rare-earth-free permanent magnets [110] and advanced magnetic memory devices [111]. Fabrication of nanocomposites containing TM monoxides as the AF phase is particularly interesting owing to their relatively high magnetic anisotropy, resulting in a large increase in coercivity and exchange bias field [111]. Skumryev et al. observed a large exchange bias field and a drastic increase of the blocking temperature from 10 to 290 K in 4 nm ferromagnetic Co particles when they are embedded in an AF CoO matrix [111]. It is worth mentioning that the exchange bias effect does not always manifest with a shift of the hysteresis loop after field cooling; this generally happens in the case of very thin AF layers or if the two phases have a similar magnetic anisotropy constant [112]. For example, the exchange bias effect was not observed in CoO@CoFe$_2$O$_4$ MNPs in the size range of 5–11 nm and with a core diameter ranging between ∼2.5 and 6.0 nm. Nevertheless, an increase of the coercivity from ∼2.2 T to ∼3.1 T at 5 K was observed when the particle size is reduced. The exchange bias effect can be also observed at the interface between a F(Fi) magnetic layer and a spin glass system [113]. In very small MNPs with a high fraction of frustrated surface spins, at temperatures lower than the glassy temperature ($T_g$), these spins became frozen and act as a spin glass. Such spin glass-like frozen surface spins may act as a magnetically hard phase increasing anisotropy and leading to the intrinsic exchange bias effect in chemically single-phase systems [114, 115].

### 12.3.3 All-Oxide Synthetic Antiferromagnets

Antiferromagnets have long been considered an interesting but useless class of materials with little practical applications due to their intrinsic zero magnetic moment and the need for high magnetic fields to manipulate the antiferromagnetic order. However, after the discovery of the exchange bias phenomenon in F(Fi)/AF heterostructures (see Section 12.3.2) [103, 104] and the exploitation of this effect in spintronic thin film stacks (e.g. spin valves and magnetic tunneling junctions), AF materials have acquired increasing relevance over the past years as passive elements in spintronic devices for data storage/processing technologies and sensors [106, 107, 116]. To expand the range of applications of AF materials, great efforts have been performed in the last few years to manipulate the AF spins with the aim to develop ferromagnetic-free functional devices based on AF materials as active spin-dependent elements [117, 118]. Compared with conventional ferromagnetic spintronics, AF spintronic devices have several advantages including a high robustness against perturbation from external magnetic fields, the absence of parasitic stray fields, and ultrafast operation speeds (THz regime). However, the characteristic insensitivity of AF materials to magnetic perturbations also limits the control of the AF order by using external magnetic fields easily accessible in a laboratory, thus requiring complex, and not always efficient strategies that still hinder the use of crystal AFs for ferromagnetic-free devices. To overcome this limitation, synthetic antiferromagnets (SAFs) consisting of two or more F(Fi) layers indirectly coupled

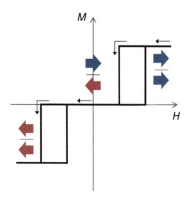

**Figure 12.9** Schematic illustration of the field-dependent magnetization loop of an SAF system with in-plane uniaxial magnetic anisotropy. Thick arrows indicate the direction of the magnetization in the two F(Fi) layers, depending on the external magnetic field, while thin arrows indicate the direction of the field sweep.

through a non-magnetic (NM) metallic or insulating spacer (Figure 12.9) [119–121] have gained increasing attention, owing to the easy manipulation and control of the magnetic configuration (parallel/antiparallel alignment) by using relatively small and easily accessible magnetic fields. Moreover, the overall magnetic properties of the SAFs can be finely tuned by changing the single layers features (e.g. type of material, thickness) [119], thus enabling for extra degrees of freedom for the optimization of the material performance. All these features have made the SAFs of particular interest for the development of both conventional and novel spintronic devices [122–130] being also of interest for biomedical applications [131–133]. For SAFs with metallic spacers, the interlayer exchange coupling (IEC) between the F layers is essentially a spin-dependent *Ruderman–Kittel–Kasuya–Yosida* (RKKY) coupling with an oscillatory behavior inducing an ferromagnetic or an antiferromagnetic coupling depending on the NM layer thickness [120]. For insulating spacers, the IEC depends on the spin-polarized quantum tunneling between the F(Fi) layers and the strength decays exponentially with the spacer layer thickness [121].

So far, most of the theoretical and experimental works on SAFs have been focused on multilayer systems based on transition metals or alloys, while SAFs consisting of correlated oxide multilayers have rarely been investigated [134–140], owing to the stringent constraints required to achieve a robust IEC, i.e. the epitaxial growth with atomic layer control and the retainment of a ferromagnetic behavior in ultrathin layers. In this scenario, only few attempts have been reported in the literature exploring the use of half-metal oxides as the ferromagnetic layers (e.g. $Fe_3O_4$, $La_{0.67}Ca_{0.33}MnO_3$, and $La_{0.67}Sr_{0.33}MnO_3$) and both conductive and insulating oxides as spacers (e.g. MgO, $CaRu_{1-x}Ti_xO_3$, and $LaNiO_3$) [134–140]. The most significant results have been achieved so far in $La_{0.67}Ca(Sr)_{0.33}MnO_3/CaRu_{1-x}Ti_xO_3$ superlattices [134–137], which display a strong IEC with layer-resolved magnetic switching that can be finely tuned by changing the thickness of the layers, the doping of the insulating spacer, and the crystallographic orientation (Figure 12.10). These results are the effects of a number of conditions occurring in these systems, i.e. (i) the interaction at the interface between $La_{0.67}Ca(Sr)_{0.33}MnO_3$ and $CaRu_{1-x}Ti_xO_3$, which allows retaining a robust ferromagnetism in the manganite layers even at very low thicknesses; (ii) the epitaxial growth with atomically flat interfaces (Figure 12.11), which are fundamental prerequisites for obtaining AF-IEC; and

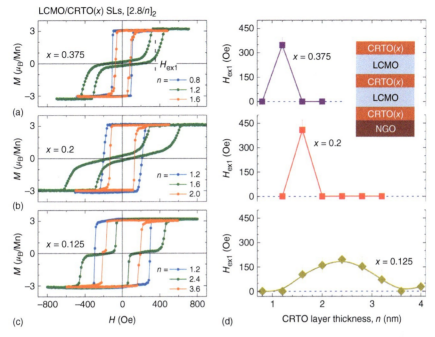

**Figure 12.10** (a)–(c) Low-temperature (100 K) field-dependent magnetization loops of [La$_{0.67}$Ca$_{0.33}$MnO$_3$ (2.8 nm)/CaRu$_{1-x}$Ti$_x$O$_3$ (n nm)]$_2$ (LCMO/CRTO) superlattices deposited on NdGaO$_3$ (001) single crystal substrates as a function of the CaRu$_{1-x}$Ti$_x$O$_3$ spacer thickness (n) and the Ti doping level (x). The external field was applied along the in-plane easy axis. (d) Corresponding trend of the interlayer exchange field, defined as the field shift of the minor loop, as a function of the CaRu$_{1-x}$Ti$_x$O$_3$ spacer thickness (n). Source: Reprinted with permission from Lan et al. [134]. CC BY 4.0.

(iii) the strain induced by the substrate on the manganite, which induces a significant uniaxial anisotropy in the plane. Despite the concept of all-oxide SAFs was clearly demonstrated, most of the results have been achieved at low temperatures, and further work will be necessary to increase the operating temperature and add novel functionalities to devices with correlated-oxide interfaces.

## 12.4 Artificial Multiferroic Oxide Heterostructures

Magnetoelectric multiferroic (ME-MF) materials hosting ferroelectric (FE) and ferro(i)/antiferromagnetic (FM) orders show a very rich and fascinating physics arising from the coupling between FE and FM ordering parameters, which allows one ferroic property (electric polarization, magnetization) to be manipulated with the conjugate field (magnetic, electric) of the other [141–143]. Since the first observation of magnetoelectricity in Cr$_2$O$_3$ oxides in the early 1960s [144], the magnetoelectric effect has been extensively explored in single-phase multiferroic compounds. However, due to the inherent incompatibility between ferroelectricity, requiring transition metals with empty $d$ orbitals and magnetism, which needs

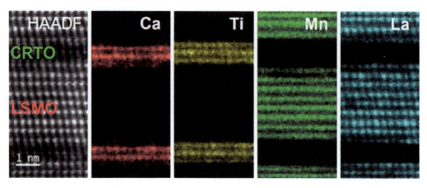

**Figure 12.11** Cross-section scanning transmission electron microscopy (STEM) images with elemental mapping showing the element-resolved interfaces between LSMO and CRTO layers. Source: Xu et al. [135], Reproduced with permission from American Physical Society.

partially filled $d$ orbitals [145], only a few examples of single-phase magnetoelectric oxides have been discovered (e.g. $BiFeO_3$, $TbMnO_3$, $PbXO_3$, $EuTiO_3$, $YMnO_3$, $RMO_3$–R = rare-earth, M = Fe, Cr, Mn–) [50–53]. Moreover, most of them show a week magnetoelectric coupling resulting in a poor manipulation of the magnetic (electric) polarization by using an electric (magnetic) field and low operating temperature limiting their integration into practical devices. Compared to single-phase ME-MFs, artificial hybrid heterostructures combining dissimilar ferroic phases at the interface (e.g. FE/FM, MF/FM, MF/FE, MF/MF) are more likely to be used in magnetoelectric devices as they show: (i) robust magnetic and ferroelectric properties at room temperature, which can be independently optimized, (ii) highly tunable and strong magnetoelectric responses exceeding those of single-phase multiferroic materials by several orders of magnitude, and (iii) a high flexibility for device designs [146–150]. Owing to the advantages of hybrid heterostructures over single phases, nano-heterostructures with different geometries and microstructures have been largely investigated over the last few years. They include core–shell spherical particles, nanorods/wires dispersed in a continuous matrix, and layered thin film stacks, where $XFe_2O_4$ (X = Fe, Co, Ni, Zn, etc.), $La_{1-x}Sr(Ca)_xMnO_3$, $BiFeO_3$, or $RMO_3$ (R = rare-earth, M = Fe, Cr, Mn, …) have been often used as the magnetic phase and $BaTiO_3$ or $Pb(ZrTi)O_3$ perovskites as the ferroelectric material (Figure 12.12). Like the single-phase ME-MFs, the magnetoelectric coupling in hybrid nano-heterostructures relies on the interplay among the four fundamental degrees of freedom (spin, orbit, charge, and lattice) but across the interface between the constituent phases rather than within one crystal lattice as in the single phase. Three main coupling mechanisms have been investigated so far: (i) strain transfer, (ii) charge modulation, and (iii) exchange interaction. The strain-mediated mechanism is currently the only one available for the control of electric polarization by an external magnetic field (*direct magnetoelectric coupling*); while all three mechanisms have been successfully exploited for the electrical control of magnetism (*converse magnetoelectric effect*).

(a)  (b)  (c)

**Figure 12.12** Schematic of magnetoelectric hybrid heterostructures: (a) core–shell spherical particles, (b) nanorods/wires in a continuous matrix, and (c) layered thin film stacks.

### 12.4.1 Strain Transfer Mechanism

In most of the hybrid heterostructures, the magnetoelectric coupling is obtained through the elastic interactions between a ferro(i)magnetic/magnetostrictive phase in intimate contact with an ferroelectric/piezoelectric material. In the direct process, an external magnetic field induces a strain in the F(Fi)/magnetostrictive phase, which is then transferred to the FE/piezoelectric component, thus inducing an electric polarization. The converse process exploits the same strain-mediated mechanism to induce a magnetization in the F(Fi) phase by applying an electric field. Nowadays, the most studied system is the $CoFe_2O_4$/$BaTiO_3$ hybrid heterostructure consisting of a Fi/magnetostrictive $CoFe_2O_4$ spinel ferrite (CFO) and a FE/piezoelectric $BaTiO_3$ (BTO) perovskite. The large magnetostriction of CFO ($\lambda_{100}$: $-250$ to $-590 \times 10^{-6}$) [151] combined with the high piezoelectric coefficient of BTO ($d_{33}$ up to 191 pC N$^{-1}$) [152] and the good crystallographic coherence between the two oxides leads to a colossal magnetoelectric effect (Figure 12.13) [153], which can be exploited for several applications, ranging from energy-efficient data storage and processing to personalized nanomedicine.

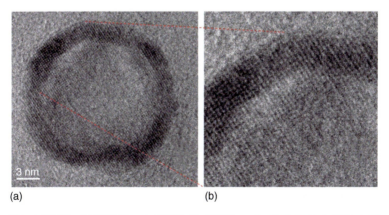

(a)  (b)

**Figure 12.13** (a) TEM image of $CoFe_2O_4$/$BaTiO_3$ core–shell nanoparticles and (b) corresponding magnification showing the lattice matching between the crystal lattices of the two phases. Source: Wang et al. [153], Reproduced with permission from American Chemical Society.

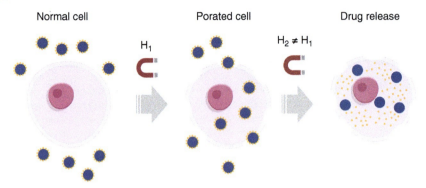

**Figure 12.14** Schematic representation of the nano-electroporation effect using magnetoelectric nanoparticles (blue circles) decorated with drug molecules (yellow circles) and the subsequent transport and release of drugs into targeted cells. The external field $H_1$ depends on the electric properties of the cell membrane, while $H_2$ is related to the binding force between the drug molecules and the nanoparticles.

A strong correlation between magnetic and electrical orders was observed in $CoFe_2O_4$@$BaTiO_3$ core–shell nanoparticles obtained by combining thermal decomposition methods and sol–gel process [153], showing a colossal magnetoelectric coefficient up 5 V cm$^{-1}$/Oe. Such a huge effect has been recently exploited to develop smart ME drug nanocontainers for the treatment of several human diseases, such as the ovarian and glioblastoma brain cancers and HIV [154–158]. Under the applications of DC or AC magnetic fields, the strain-mediated magnetoelectric coupling allows generating localized, intense, and tunable electric fields (*direct ME effect*), which can modify the porosity of the membrane of cells in contact with the particles (Figure 12.14). Owing to the different sensitivity to the electric field of healthy and malignant cells, pores with different sizes can be generated into the membrane, thus allowing ME particles with proper sizes to selectively enter the malignant cells. Magnetic-induced electric fields of higher intensities can be finally used for a precise and on-demand release of drug molecules previously attached to the particles, thus increasing the efficiency of the treatment while reducing the effect on healthy cells. In addition to drug delivery, core–shell ME nanoparticles can play a major role in many other applications, including noninvasive stimulation of brains or other tissues, treatment of neurological disorders, and imaging of neuronal activities.

A strong magnetoelectric coupling has been also observed in nanocomposite systems consisting of magnetic particles embedded in an ferroelectric/piezoelectric matrix [97, 159] and in three-component composites based on piezopolymers [160–162]. Figure 12.15a,b shows a typical TEM image of $CoFe_2O_4$/$LaFeO_3$ nanocomposites prepared via a modified sol–gel autocombustion method [163]. They consist of an $LaFeO_3$ matrix with embedded nanocrystals of $CoFe_2O_4$. Sarkar et al. developed an effective chemical method to synthesize and tune the magnetic properties of $AFeO_3$–$MFe_2O_4$ (where A is $Bi^{3+}$ or $La^{3+}$, M is $Co^{2+}$ or $Ni^{2+}$) nanocomposites [97]. The nanocomposites were prepared via a two-step sol–gel self-combustion process: presynthesized $MFe_2O_4$ (M: $Co^{2+}$, $Ni^{2+}$) spinel ferrite nanoparticles used as seeds to grow the ferroelectric/piezoelectric $AFeO_3$

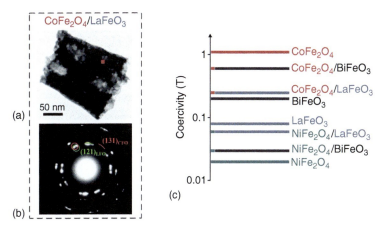

**Figure 12.15** (a) TEM bright-field image and (b) corresponding SAED pattern of CoFe$_2$O$_4$/LaFeO$_3$ nanocomposites. Source: Sayed et al. [163], Reproduced with permission Royal Society of Chemistry / CC BY-NC 3.0. (c) Schematic representation of the coercivity measured at $T = 5$ K for pure spinel and perovskite nanoparticles and nanocomposites. Source: Sarkar et al. [97], Reproduced with permission Royal Society of Chemistry / CC BY-NC 3.0.

(A:Bi$^{2+}$, La$^{2+}$) phase. Since CoFe$_2$O$_4$ is magnetically hard while NiFe$_2$O$_4$ is magnetically soft, the magnetic anisotropy of composites can be fine-tuned by changing the concentration and ratio of spinel phases. Figure 12.15 shows the change of coercivity of pure materials and composites with 10 wt% of magnetic phase.

An efficient and scalable one-step mixed precursor approach was also used by the same group to synthesize LaFeO$_3$–CoFe$_2$O$_4$ [95] and PbZr$_{0.52}$Ti$_{0.48}$O$_3$(PZT)–CoFe$_2$O$_4$ [164] nanocomposites. This synthetic approach is characterized by an interesting symbiotic effect since spinel phase combusting at a lower temperature promotes the reaction of perovskite phase precursor. Namely, in a typical sol–gel self-combustion process reaction of spinel ferrite, phase formation occurs at about 200 °C, while the temperature of perovskite phase formation is about 500 °C. For the synthesis of LaFeO$_3$–CoFe$_2$O$_4$ nanocomposites, the minimal concentration of 50% of spinel ferrite precursor was enough to obtain a fully crystalized LaFeO$_3$ phase [95] at 300 °C.

The indirect effect has been extensively studied in thin film systems to achieve an easy and energy-efficient control of magnetism using an electric field, which may be exploited for several applications, such as memory elements and sensors. The peculiar properties of these systems have triggered an increasing research effort in the field, and the combinations of different oxide materials have been explored, including magnetic spinel ferrites (e.g. MeFe$_2$O$_4$; Me = Fe, Co, Ni) and perovskites (e.g. La$_{1-x}$Sr(Ca)$_x$MnO$_3$) as the magentic layer and piezoelectric perovskites (e.g. BaTiO$_3$ or Pb(ZrTi)O$_3$) as the FE phase [165–169]. Despite the electric-field control of the magnetic state has been clearly demonstrated in many systems, as reported, for example, in Figure 12.16 for a Fe$_3$O$_4$/Pb(Zn$_{0.33}$Nb$_{0.66}$)O$_3$-PbTiO$_3$ hybrid thin film stack [169], the strength of the magnetoelectric coupling is generally reduced with respect to what is expected because of the substrate clamping effect. To overcome this

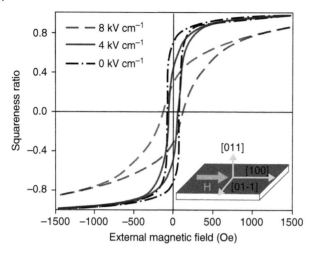

**Figure 12.16** Field-dependent magnetization loops of $Fe_3O_4$/Pb($Zn_{0.33}Nb_{0.66}$)$O_3$-$PbTiO_3$ hybrid thin film stack under various electric fields applied across the thickness of ferroelectric substrate. Source: Reprinted with permission from Liu et al. [169].

issue, several strategies have been proposed, including the thickness/stacking order variation, the strain engineering, and the use of patterned structures [166, 170–172].

### 12.4.2 Charge Modulation Mechanism

The magnetoelectric coupling at the FM/FE interface may originate from pure electronic mechanisms. These effects have been generally observed in layered heterostructures containing a thin ferromagnetic film coupled to a ferroelectric layer or a dielectric spacer. Upon the application of an external electric field, the magnetic properties of the ferromagnetic phase can be tuned through different mechanisms occurring at the interface, such as the accumulation/depletion of charges that modifies the charge density of the magnetic layer via charge screening, the electronic hybridization between 3D transition metal atoms, and ionic displacements at the interface [146, 173–175]. Among the different materials that have been investigated so far, doped manganites showed the largest effects, owing to their strong lattice–spin–charge coupling. The hole doping concentration can be modulated by the accumulation/depletion of charge carriers near the interface. Therefore, a change of the polarization of a ferroelectric or a dielectric phase in close contact may alter the electronic ground state and the magnetic properties of a manganite layer when it is close to its phase transition. Theoretical studies have demonstrated that changing the polarization direction of the ferroelectric layer in $BaTiO_3$/$La_{0.5}X_{0.5}MnO_3$ (X = Sr, Ca, or Ba) layered heterostructures can induce a change of the magnetization state of the manganite from ferromagnetic to antiferromagnetic [176]. The effects of an external electric field were also demonstrated experimentally. For example, in $La_{0.8}Sr_{0.2}MnO_3$/$PbZr_{0.2}Ti_{0.8}O_3$ bilayers, changes of the polarization state of the $PbZr_{0.2}Ti_{0.8}O_3$ layer induce the modification of either the temperature of magnetic phase transitions or the switching between two magnetization states likely because of an electrostatic modulation of the valence state of Mn ions [177].

### 12.4.3 Exchange Interaction Mechanism

Multiferroic materials showing an antiferromagnetic order (e.g. $BiFeO_3$ – $T_N \sim 643$ K, $LuMnO_3$ – $T_N \sim 90$ K, $YMnO_3$ – $T_N \sim 80$ K) can be exploited to manipulate the magnetization of an exchange-coupled magnetic layer by controlling the exchange bias coupling through the application of an electric field [178–180]. Among all-oxide heterostructures, the $BiFeO_3/La_{0.7}Sr_{0.3}MnO_3$ stack has been widely investigated either experimentally or theoretically [146, 178, 181, 182]. Upon the application of an external electric field, the switching of the ferroelectric polarization of the $BiFeO_3$ phase causes ionic displacements in the ferroelectric phase that alters the F(Fi)/FE interface, thus leading to a change of the exchange bias coupling, which results in a modification of the magnetic properties of the ferromagnetic phase as reported in Figure 12.17 as a representative example.

## 12.5 All-Oxide Spintronic Heterostructures

Spin electronics or spintronics is a rapidly developing area that exploits the spin of the electrons to carry, manipulate, and store information, in contrast to conventional electronics based on the electron charge [106, 107, 116, 183–186]. Since the pioneering works by Tedrow and Meservey on the spin-dependent nature of tunneling current [187] and, a few years later, by M. Julliere on the magnetoresistance in

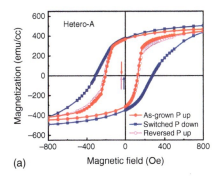
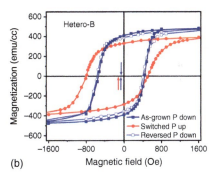

**Figure 12.17** Low-temperature (10 K) field-dependent magnetization loops of $BiFeO_3/La_{0.7}Sr_{0.3}MnO_3$ thin film heterostructures measured after a field cooling of 1 T for different ferroelectric polarization states: as-grown state, oppositely switched state, and reversible switched state. Heterostructures A (a) and B (b) differ for the atomic sequence across the interface: (Hetero A) $FeO_2/BiO/MnO_2/La_{0.7}Sr_{0.3}O$ and (Hetero B) $BiO/FeO_2/MnO_2/La_{0.7}Sr_{0.3}O/MnO_2$. Changing the polarization state of the piezoelectric phase allows the magnetization and the $H_{EB}$ field of the magnetic phase to be reversible changed, thus indicating that the magnetic properties of the magnetic component can be finely modulated by using an external electric field. Source: Reprinted with permission from Yi et al. [178].

Fe/Ge/Co heterostructures [188], a significant acceleration in the field of spintronics came in the 1980s after the discovery of the *giant magnetoresistance* (GMR) effect in Fe/Cr multilayers consisting of alternating ferromagnetic (Fe) and non-magnetic (Cr) layers [189, 190]. The relative alignment of the magnetization in the ferromagnetic layers affects the conduction of spin-polarized electrons, thus determining a significant enhancement of the electrical resistance when the ferromagnetic layers are magnetically aligned antiparallel. A larger change of the electric resistance can be obtained by using a thin nonmagnetic insulating spacer, owing to the spin-dependent tunneling effect, from which the name *tunneling magnetoresistance* (TMR). Both the GMR and TMR effects have been widely exploited to develop spintronic elements for sensing and data storage/processing technologies. They include, among others, the spin valve and magnetic tunneling junction structures consisting of two ferromagnetic layers separated by a thin metallic or insulating spacer, respectively. The magnetization of one of the two ferromagnetic layers (named *reference electrode*) is set along a fixed direction through the coupling with an antiferromagnetic pinning layer (exchange bias effect) or by using a synthetic antiferromagnetic stack, while the magnetization of the second ferromagnetic layer (named *free electrode*) can be freely switched/rotated under the action of external stimulus, thus leading to a change of the resistance depending on the mutual alignment of the magnetization in the two layers. In addition to the original procedure to manipulate the magnetization of the free electrode relying on an external magnetic field, alternative and more efficient strategies have been developed over the years exploiting external electric fields or the current-induced spin-transfer torque (STT) and spin–orbit torque (SOT) phenomena.

Despite most of the efforts have been devoted to the development of spintronic systems based on metallic ferromagnetic layers, a growing attention has been paid to spintronic devices based on complex TMOs, owing to the rich variety of physical properties they exhibit, the strong environmental stability, and the high compatibility with other oxides (with different properties), which allow for the fabrication of multifunctional devices [185, 191–193]. As the resistance change in spintronic devices is proportional to the spin polarization of the conducting charge carriers, a great attention has been dedicated to half-metal ferro(i)magnetic oxides, showing an ideal 100% spin polarization, which can be used as the electrodes of spintronic devices to improve their efficiency. Magnetic oxides that have been predicted to present this unusual half-metal character include $Fe_3O_4$ spinel ferrite, $La_{1-x}A_xMnO_3$ (A = Sr, Ca, Ba, etc.) manganites, and some double perovskites (e.g. $Sr_2FeMoO_6$), among others [185]. $Fe_3O_4$ is a ferrimagnetic oxide with a Curie temperature $T_c \sim 850\,K$, which was predicted to be half-metal at room temperature. Despite the high expectations, when $Fe_3O_4$ is prepared in the form of thin films, the presence of crystal imperfections, such as antiphase boundaries, atomic-site disorder, and dislocations, induced by the high lattice mismatch and/or the different crystal symmetry of the commonly used substrates (e.g. MgO, $SrTiO_3$, $MgAl_2O_4$) [192], leads to magnetoresistance effects lower than the predicted ones [191]. MgO has indeed a good lattice match but a different crystal symmetry, $MgAl_2O$ is isostructural but has a large mismatch, while $SrTiO_3$ has both a higher lattice

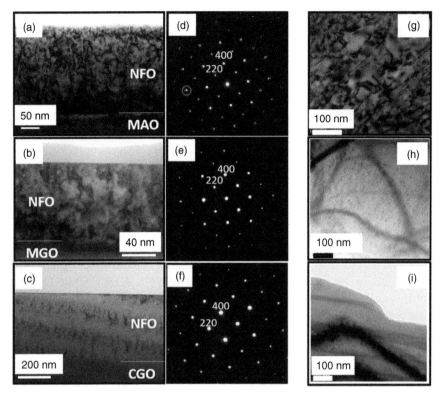

**Figure 12.18** TEM bright field images and corresponding diffraction pattern of $NiFe_2O_4$ (NFO) films deposited on (a) and (d) $MgAl_2O_4$ (MAO), (b) and (e) MgO (MGO), and (c) and (f) $CoGa_2O_4$ (CGO) substrates. Films deposited on MAO show strain relaxation as indicated by the splitting of intensities at high g vector diffractions (red circle in (d)), while no relaxation is present in films grown on MGO and CGO. Plan view images (g), (h), and (i) indicate the presence of antiphase grain boundaries only when the film is deposited on MAO. The results indicate that using an isostructural CGO substrate matching the lattice parameter of the NFO allows improving the quality of the films. Preliminary evidence that similar improvements can also be obtained in other members of the ferrite family, such as $Fe_3O_4$ and $CoFe_2O_4$, were reported by the authors. Source: Singh et al. [195], Reproduced with permission John Wiley & Sons, Inc.

mismatch and a different crystal symmetry. Recent studies have demonstrated that spinel ferrites thin films with a high-quality crystalline structure can be obtained by using spinel substrates with a closer lattice parameter (e.g. $CoGa_2O_4$, $MgGa_2O_4$, and $Co_2TiO_4$) [194–196] (Figure 12.18). However, the limited availability of such single-crystal spinel substrates, which are not commercially offered, still hinders the use of $Fe_3O_4$ for the development of all-oxide spintronic devices.

Among the family of perovskites, $La_{1-x}A_xMnO_3$ (A = Sr, Ca, Ba, etc.) manganites represent a very interesting class of oxides, which exhibit both strong ferromagnetism and metallic conductivity [26]. Despite high magnetoresistance values have reported in magnetic tunneling junctions using manganite thin films and $SrTiO_3$ oxide tunnel barrier, the effect is generally observed at low temperatures [197], thus triggering considerable research efforts to individuate perovskite oxides with room temperature half-metallicity. In this regard, the family of double perovskites represents a promising choice as some of them are half-metal at room temperature (e.g. $Sr_2FeMoO_6$, $T_c \sim 420$ K) [64]. Despite the promising properties of this class of materials, a week magnetoresistance effect has been observed so far at room temperature, likely because of the high sensitivity of the half-metal behavior to small structural and stoichiometric variations. Besides conventional magnetic tunneling junctions using dielectric oxides as the tunnel barrier, additional functionalities can be obtained by using insulating spacers with intrinsic functionality such as ferro(i)magnetic, ferroelectric, and multiferroic oxides. Insulating ferromagnetic oxides (e.g. $CoFe_2O_4$, $BiMnO_3$, $La_{0.1}Bi_{0.9}MnO_3$) can be used, for example, as spin filters to artificially reproduce the half-metal behavior [198–200]. In such a case, the magnetic oxide is used as tunnel barrier by sandwiching it between a non-magnetic metal, which acts as an electron source, and a ferromagnetic layer, acting as receiving electrode, to produce a strong spin polarization. A further degree of freedom can be added by using active materials, such as ferroelectric oxides, to form the tunnel barrier. Such structures, called *multiferroic tunnel junctions* [201] and consisting of two ferromagnetic films separated by a ferroelectric spacer, allow a ferroelectric control of the spin polarization as well as multiple resistance states by using both external electric and magnetic fields, thus providing a powerful strategy to develop multifunctional spintronic devices (Figure 12.19).

## 12.6 Conclusion and Perspectives

TMOs represent an extremely rich class of materials in terms of chemical composition and crystal structure, which result in a unique variety of ferroic and multiferroic properties. The new research frontiers on this field are focused on the rational design of complex TMO nano-architectures (i.e., nanocomposites, thin

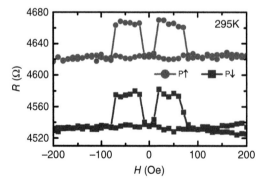

**Figure 12.19** Room-temperature change of the electrical resistance as a function of the external magnetic field for two different polarization state of the ferroelectric barrier in all-perovskite $La_{0.7}Sr_{0.3}MnO_3/Ba_{0.95}Sr_{0.05}TiO_3/La_{0.7}Sr_{0.3}MnO_3$ multiferroic tunnel junctions. After switching the polarization direction of the ferroelectric barrier by applying positive and negative voltage pulses, the resistance curves shift, and four resistance states are obtained. Source: Reprinted with permission from Yin et al. [201].

film, multilayers). The recent progress in synthesis methods opens the possibility to control morpho-structural and chemical properties at the nanoscale, allowing a fine-tuning of the physical properties of TMOs as well as to engineer new nano-heterostructures. In this view, a careful tuning of inversion degree in nanostructured spinel ferrites or the preservation of crystalline phase of some perovskite structures still represents a challenge, catalyzing the interest of many researchers. Starting from this exciting playground, many interesting results have been obtained by synthesis on new nano-heterostructures showing several exotic physical properties, such as exchange bias, magneto transport, multiferroic, and others. Even if this huge variety of properties make complex TMO nano-heterostructures very interesting for potential applications in many fields (e.g., technolgies for sensing and data storage/processing, energy and biomedicine), the physical models used to describe these materials are often oversimplified, and further efforts need to be done to fully understand the physics behind these phenomena.

## References

1 Coey, J.M.D., Venkatesan, M., and Xu, H. (2013). *Functional Metal Oxides*, 1–49. Weinheim, Germany: Wiley-VCH Verlag GmbH & Co. KGaA.
2 Dionne, G.F. (2009). *Magnetic Oxides*, vol. 8. Boston, MA: Springer US.
3 Cullity, B.D. and Graham, C.D. (2009). *Introduction to Magnetic Materials*. Wiley.
4 Jiles, D. (1991). *Introduction to Magnetism and Magnetic Materials*, vol. 1. Boston, MA: Springer US.
5 Spaldin, N.A. (2010). *Magnetic materials: Fundamentals and applications*. Cambridge University Press.
6 Néel, L. (1948). *Ann. Phys. (Paris)* 12: 137–198.
7 Lone, I.H. et al. (2019). *Nanoscale Research Letters* 14: 142.
8 Biswas, A., Talha, M., Kashir, A., and Jeong, Y.H. (2019). *Curr. Appl Phys.* 19: 207–214.
9 Opel, M. (2012). *J. Phys. D: Appl. Phys.* 45: 033001.
10 Bader, S.D. and Parkin, S.S.P. (2010). *Annu. Rev. Condens. Matter Phys.* 1: 71–88.
11 Ramirez, A.P. (1997). *J. Phys. Condens. Matter* 9: 8171–8199.
12 Fiebig, M., Lottermoser, T., Meier, D., and Trassin, M. (2016). *Nat. Rev. Mater.* 1: 16046.
13 Spaldin, N.A. and Ramesh, R. (2019). *Nat. Mater.* 18: 203–212. https://doi.org/10.1063/1.5050217.
14 Frenkel, J., Doefman, J., and Dorfman, J. (1930). *Nature* 126: 274–275.
15 Chikazumi, S. (2009). *Physics of Ferromagnetism*. Oxford: Oxford University Press.
16 Peddis, D., Orrù, F., Ardu, A. et al. (2012). *Chem. Mater.* 24: 1062–1071.
17 West, A.R. (1984). *Solid State Chemistry and Its Applications*. New York: Wiley.
18 García, J. and Subías, G. (2014). *J. Phys. Condens. Matter* 16: R145–R178. https://doi.org/10.1088/0953-8984/16/7/R01.

19 Frison, R., Cernuto, G., Cervellino, A. et al. (2013). *Chem. Mater.* 25: 4820–4827.
20 Schwaminger, S.P., Syhr, C., and Berensmeier, S. (2020). *Crystals* 10: 214.
21 Maldonado, A.C.M., Winkler, E.L., Raineri, M. et al. (2019). *J. Phys. Chem. C* 123: 20617–20627.
22 Walz, F. (2002). *J. Phys. Condens. Matter* 14: R285–R340.
23 Pinto, H.P. and Elliott, S.D. (2006). *J. Phys. Condens. Matter* 18: 10427–10436.
24 Hamed, M.H., Hinz, R.A., Lömker, P. et al. (2019). *ACS Appl. Mater. Interfaces* 11: 7576–7583.
25 Liu, X.H., Rata, A.D., Chang, C.F. et al. (2014). *Phys. Rev. B* 90: 125142.
26 Coey, J.M.D., Viret, M., and Von Molnár, S. (1999). *Adv. Phys.* 48: 167–293.
27 Ade, R., Chen, Y.S., Huang, C.H., and Lin, J.G. (2020). *J. Appl. Phys.* 127: 113904. https://doi.org/10.1063/5.0003542.
28 Sanchez-Lievanos, K.R., Stair, J.L., and Knowles, K.E. (2021). *Inorg. Chem.* 60: 4291–4305.
29 Coey, J.M.D. (2010). *Magnetism and Magnetic Materials*. New York: Cambridge University Press.
30 Mameli, V., Musinu, A., Ardu, A. et al. (2016). *Nanoscale* 8: 10124–10137.
31 Albino, M., Fantechi, E., Innocenti, C. et al. (2019). *J. Phys. Chem. C* 123: 6148–6157.
32 Omelyanchik, A., Singh, G., Volochaev, M. et al. (2019). *J. Magn. Magn. Mater.* 476: 387–391.
33 Omelyanchik, A., Levada, K., Pshenichnikov, S. et al. (2020). *Materials (Basel)* 13: 1–13.
34 Peddis, D. (2014). *Magnetic Nanoparticle Assemblies*, vol. 7 (ed. K.N. Trohidou), 978–981. Singapore: Pan Stanford Publishing.
35 Peddis, D., Cannas, C., Musinu, A. et al. (2006). *J. Chem. Phys.* 125: 164714. https://doi.org/10.1063/1.2354475.
36 Peddis, D., Mansilla, M.V., Mørup, S. et al. (2008). *J. Phys. Chem. B* 112: 8507–8513.
37 Peddis, D., Cannas, C., Piccaluga, G. et al. (2010). *Nanotechnology* 21: 125705.
38 Ferreiro-Vila, E., Iglesias, L., Lucas Del Pozo, I. et al. (2019). *APL Mater.* 7: 031109. https://doi.org/10.1063/1.5087559.
39 Petrović, M., Chellappan, V., and Ramakrishna, S. (2015). *Sol. Energy* 122: 678–699.
40 Varignon, J., Bibes, M., and Zunger, A. (2019). *Nat. Commun.* 10: 1–11.
41 Belik, A.A., Iikubo, S., Yokosawa, T. et al. (2007). *J. Am. Chem. Soc.* 129: 971–977.
42 Urushibara, A., Moritomo, Y., Arima, T. et al. (1995). *Phys. Rev. B* 51: 14103–14109.
43 Jin, S., Tiefel, T.H., McCormack, M. et al. (1994). *Science (80-.)* 264: 413–415.
44 Markovich, V., Wisniewski, A., and Szymczak, H. (2014). *Handbook of Magnetic Materials*, 1ste, vol. 22, 1–201. Elsevier B.V.
45 Niebieskikwiat, D. and Salamon, M.B. (2005). *Phys. Rev. B* 72: 174422–174426.
46 Giri, S., Patra, M., and Majumdar, S. (2011). *J. Phys. Condens. Matter* 23: 073201. https://doi.org/10.1088/0953-8984/23/7/073201.

47 Smari, M., Hamdi, R., Slimani, S. et al. (2020). *J. Phys. Chem. C* 124: 23324–23332.
48 Wang, Y., Zhao, H., Zhang, L. et al. (2017). *Phys. Chem. Chem. Phys.* 19: 17493–17515.
49 Acosta, M., Novak, N., Rojas, V. et al. (2017). *Appl. Phys. Rev.* 4: 041305. https://doi.org/10.1063/1.4990046.
50 Khomskii, D. (2009). *Physics (College. Park. Md).* 2: 20. https://doi.org/10.1103/Physics.2.20.
51 Ramesh, R. and Spaldin, N.A. (2009). *Nanosci. Technol. A Collect. Rev. Nat.. J.* 3: 20–28.
52 Eerenstein, W., Mathur, N.D., and Scott, J.F. (2006). *Nature* 442: 759–765.
53 Prellier, W., Singh, M.P., and Murugavel, P. (2005). *J. Phys. Condens. Matter* 17: R803.
54 Catalan, G. and Scott, J.F. (2009). *Adv. Mater.* 21: 2463–2485.
55 Ederer, C. and Spaldin, N.A. (2005). *Phys. Rev. B: Condens. Matter Mater.* 71: 060401. https://doi.org/10.1103/PhysRevB.71.060401.
56 Lebeugle, D., Mougin, A., Viret, M. et al. (2009). *Phys. Rev. Lett.* 103: 257601. https://doi.org/10.1103/PhysRevLett.103.257601.
57 Tokunaga, M., Akaki, M., Ito, T. et al. (2015). *Nat. Commun.* 6: 1–5.
58 Catalan, G., Seidel, J., Ramesh, R., and Scott, J.F. (2012). *Rev. Mod. Phys.* 84: 119–156.
59 Rojac, T., Bencan, A., Drazic, G. et al. (2017). *Nat. Mater.* 16: 322–327. https://doi.org/10.1038/nmat4799.
60 Sharma, P., Zhang, Q., Sando, D. et al. (2017). *Sci. Adv.* 3: e1700512. https://doi.org/10.1126/sciadv.1700512.
61 Parkin, S.S.P., Hayashi, M., and Thomas, L. (2008). *Science (80-.)* 320: 190–194.
62 Madhogaria, R.P., Clements, E.M., Kalappattil, V. et al. (2020). *J. Magn. Magn. Mater.* 507: 166821.
63 Thompson, C.M., Chi, L., Hayes, J.R. et al. (2015). *Dalton Trans.* 44: 10806–10816.
64 Serrate, D., De Teresa, J.M., and Ibarra, M.R. (2007). *J. Phys. Condens. Matter* 19: 023201. https://doi.org/10.1088/0953-8984/19/2/023201.
65 Roth, W.L. (1958). *Physiol. Rev.* 110: 1333–1341.
66 Pullar, R.C. (2012). *Prog. Mater Sci.* 57: 1191–1334.
67 Petrecca, M., Muzzi, B., Oliveri, S.M. et al. (2021). *J. Phys. D: Appl. Phys.* 54: 134003.
68 Maltoni, P., Sarkar, T., Barucca, G. et al. (2021). *J. Magn. Magn. Mater.* 535: 168095.
69 López-Ortega, A., Estrader, M., Salazar-Alvarez, G. et al. (2014). *Phys. Rep.* 553: 1–32.
70 Laureti, S., Gerardino, A., D'Acapito, F. et al. (2021). *Nanotechnology* 32: 205701. https://doi.org/10.1088/1361-6528/abe260.
71 Suess, D., Lee, J., Fidler, J., and Schrefl, T. (2009). *J. Magn. Magn. Mater.* 321: 545–554.

72 Maltoni, P., Sarkar, T., Varvaro, G. et al. (2021). *J. Phys. D: Appl. Phys.* 54: 124004. https://doi.org/10.1088/1361-6463/abd20d.
73 De Julián Fernández, C., Sangregorio, C., de la Figuera, J. et al. (2021). *J. Phys. D: Appl. Phys.* 54: 153001.
74 Jiang, J.S. and Bader, S.D. (2014). *J. Phys. Condens. Matter* 26: 064214.
75 Skomski, R., Manchanda, P., Kumar, P.K. et al. (2013). *IEEE Trans. Magn.* 49: 3215–3220.
76 Balamurugan, B., Sellmyer, D.J., Hadjipanayis, G.C., and Skomski, R. (2012). *Scr. Mater.* 67: 542–547.
77 Gutfleisch, O., Willard, M.A., Brck, E. et al. (2011). *Adv. Mater.* 23: 821–842.
78 Varvaro, G., Laureti, S., and Fiorani, D. (2014). *J. Magn. Magn. Mater.* 368: 415–420.
79 Barucca, G., Speliotis, T., Giannopoulos, G. et al. (2017). *Mater. Des.* 123: 147–153.
80 Di Bona, A., Luches, P., Albertini, F. et al. (2013). *Acta Mater.* 61: 4840–4847.
81 Varvaro, G., Albertini, F., Agostinelli, E. et al. (2012). *New J. Phys.* 14: 073008.
82 Alexandrakis, V., Speliotis, T., Manios, E. et al. (2011). *J. Appl. Phys.* 109: 07B729.
83 Asti, G., Ghidini, M., Pellicelli, R. et al. (2006). *Phys. Rev. B* 73: 094406 1–094406 16.
84 Kang, Y.M., Lee, S.H., Kim, T.C. et al. (2017). *Appl. Phys. A* 123, 123–648.
85 Tang, X., Wei, R., Hu, L. et al. (2017). *J. Appl. Phys.* 121: 245305.
86 Jenuš, P., Topole, M., McGuiness, P. et al. (2016). *J. Am. Ceram. Soc.* 99: 1927–1934.
87 Muscas, G., Kumara, P.A., Barucca, G. et al. (2016). *Nanoscale* 8: 2081–2089.
88 Lavorato, G., Winkler, E., Rivas-Murias, B., and Rivadulla, F. (2016). *Phys. Rev. B* 94: 054405.
89 Li, B., Chopdekar, R.V., Arenholz, E. et al. (2014). *Appl. Phys Express* 105: 202401.
90 Yoon, S.D., Oliver, S.A., and Vittoria, C. (2002). *J. Appl. Phys.* 91: 7379–7381.
91 Maltoni, P., Sarkar, T., Barucca, G. et al. (2021). *J. Phys. Chem. C* 125: 5927–5936.
92 Hu, L., Sun, X., Zhou, F. et al. (2021). *Ceram. Int.* 47: 2672–2677.
93 Tang, X., Zhu, S., Wei, R. et al. (2020). *Compos. Part B Eng.* 186: 107801.
94 Saeedi Afshar, S.R., Masoudpanah, S.M., and Hasheminiasari, M. (2020). *J. Electron. Mater.* 49: 1742–1748.
95 Sayed, F., Kotnana, G., Muscas, G. et al. (2020). *Nanoscale Adv.* 2: 851–859.
96 Sayed, F., Muscas, G., Jovanovic, S. et al. (2019). *Nanoscale* 11: 14256–14265.
97 Sarkar, T., Muscas, G., Barucca, G. et al. (2018). *Nanoscale* 10: 22990–23000.
98 Lee, J.-H., Jang, J., Choi, J. et al. (2011). *Nat. Nanotechnol.* 6: 418–422.
99 Song, Q. and Zhang, Z.J. (2012). *J. Am. Chem. Soc.* 134: 10182–10190.
100 Zhang, Q., Castellanos-Rubio, I., Munshi, R. et al. (2015). *Chem. Mater.* 27: 7380–7387.
101 Omelyanchik, A., Villa, S., Vasilakaki, M. et al. (2021). *Nanoscale Adv.* 3(24): 6912–6924. https://doi.org/10.1039/d1na00312g.

102 Da Yang, M., Ho, C.H., Ruta, S. et al. (2018). *Adv. Mater.* 30: 1802444. https://doi.org/10.1002/adma.201802444.
103 Meiklejohn, W.H. and Bean, C.P. (1957). *Physiol. Rev.* 105: 904–913.
104 Nogués, J. and Schuller, I.K. (1999). *J. Magn. Magn. Mater.* 192: 203–232.
105 Nogués, J., Sort, J., Langlais, V. et al. (2005). *Phys. Rep.* 422: 65–117.
106 Dieny, B., Prejbeanu, I.L., Garello, K. et al. (2020). *Nat. Electron.* 3: 446–459.
107 Joshi, V.K. (2016). *Eng. Sci. Technol. Int. J.* 19: 1503–1513.
108 Tokunaga, Y., Taguchi, Y., and Arima, T. (2014). *Phys. Rev. Lett.* 112: 037203 1–037203 5.
109 Zhu, Q.X., Zheng, M., Yang, M.M. et al. (2014). *Appl. Phys. Lett.* 105: 241604. https://doi.org/10.1063/1.4904471.
110 Lottini, E., López-Ortega, A., Bertoni, G. et al. (2016). *Chem. Mater.* 28: 4214–4222.
111 Skumryev, V., Stoyanov, S., Zhang, Y. et al. (2003). *Nature* 423: 850–853.
112 Lavorato, G.C., Lima, E., Tobia, D. et al. (2014). *Nanotechnology* 25: 355704.
113 Ali, M., Adie, P., Marrows, C.H. et al. (2007). *Nat. Mater.* 6: 70–75.
114 Phan, M.-H., Alonso, J., Khurshid, H. et al. (2016). *Nanomaterials* 6: 1–31.
115 Peddis, D., Cannas, C., Piccaluga, G. et al. (2010). *Nanotechnology* 21, 125705 1–10.
116 Dieny, B., Goldfarb, R.B., and Lee, K.-J. (ed.) (2017). *Introduction to Magnetic Random-Access Memory*. Wiley-IEEE Press.
117 Baltz, V., Manchon, A., Tsoi, M. et al. (2018). *Rev. Mod. Phys.* 90: 15005.
118 Jungwirth, T., Marti, X., Wadley, P., and Wunderlich, J. (2016). *Nat. Nanotechnol.* 11: 231–241.
119 Parkin, S.S.P. (1991). *Phys. Rev. Lett.* 67: 3598.
120 Bruno, P. and Chappert, C. (1992). *Phys. Rev. B* 46: 261–270.
121 Faure-Vincent, J., Tiusan, C., Bellouard, C. et al. (2002). *Phys. Rev. Lett.* 89: 107206.
122 Duine, R.A., Lee, K., Parkin, S.S.P., and Stiles, M.D. (2018). *Nat. Phys.* 14: 217–219.
123 Makushko, P., Oliveros Mata, E.S., Cañón Bermúdez, G.S. et al. (2021). *Adv. Funct. Mater.* 31: 2101089.
124 Hassan, M., Laureti, S., Rinaldi, C. et al. (2021). *Nanoscale Adv.* 3: 3076–3084.
125 Back, C., Cros, V., Ebert, H. et al. (2020). *J. Phys. D: Appl. Phys.* 53: 363001.
126 Legrand, W., Maccariello, D., Ajejas, F. et al. (2020). *Nat. Mater.* 19: 34–42. https://doi.org/10.1038/s41563-019-0468-3.
127 Yu, Z., Shen, M., Zeng, Z. et al. (2020). *Nanoscale Adv.* 2: 1309–1317.
128 Dohi, T., Duttagupta, S., Fukami, S., and Ohno, H. (2019). *Nat. Commun.* 10: 5153.
129 Yang, Q., Wang, L., Zhou, Z. et al. (2018). *Nat. Commun.* 9: 991.
130 Wang, X., Yang, Q., Wang, L. et al. (2018). *Adv. Mater.* 30, 1803612 1–38.
131 Welbourne, E.N., Vemulkar, T., and Cowburn, R.P. (2021). *Nano Res.* 14: 3873–3878. https://doi.org/10.1007/s12274-021-3307-1.
132 Varvaro, G., Laureti, S., Peddis, D. et al. (2019). *Nanoscale* 11: 21891–21899.
133 Mansell, R., Vemulkar, T., Petit, D.C.M.C. et al. (2017). *Sci. Rep.* 7: 4257. (7pp).

134 Lan, D., Chen, B., Qu, L. et al. (2019). *APL Mater.* 7: 031119.
135 Xu, H., Chen, F., Chen, B. et al. (2018). *Phys. Rev. Appl.* 10: 024035.
136 Xu, H., Wan, S., Chen, B. et al. (2017). *Appl. Phys. Lett.* 110: 082402.
137 Chen, B., Xu, H., Ma, C. et al. (2017). *Science (80-.)* 194: 191–194.
138 Gibert, M., Viret, M., Zubko, P. et al. (2016). *Nat. Commun.* 7: 11227.
139 Nikolaev, K.R., Dobin, A.Y., Krivorotov, I.N. et al. (2000). *Phys. Rev. Lett.* 85: 3728–3731.
140 Van der Heijden, P., Bloemen, P., Metselaar, J. et al. (1997). *Phys. Rev. B* 55: 11569–11575.
141 Pradhan, D.K., Kumari, S., and Rack, P.D. (2020). *Nanomaterials* 10: 2072.
142 Speliotis, T., Giannopoulos, G., Niarchos, D. et al. (2016). *J. Appl. Phys.* 119: 233904.
143 Hu, J.M., Chen, L.Q., and Nan, C.W. (2016). *Adv. Mater.* 28: 15–39.
144 Folen, V.J., Rado, G.T., and Stalder, E.W. (1961). *Phys. Rev. Lett.* 6: 607–608.
145 Hill, N.A. (2000). *J. Phys. Chem. B* 104: 6694–6709.
146 Vaz, C.A.F. and Staub, U. (2013). *J. Mater. Chem. C* 1: 6731–6742.
147 Ma, J., Hu, J., Li, Z., and Nan, C.-W. (2011). *Adv. Mater.* 23: 1062–1087.
148 Vaz, C.A.F., Hoffman, J., Ahn, C.H., and Ramesh, R. (2010). *Adv. Mater.* 22: 2900–2918.
149 Srinivasan, G. (2010). *Annu. Rev. Mater. Res.* 40: 153–178.
150 Nan, C.W., Bichurin, M.I., Dong, S. et al. (2008). *J. Appl. Phys.* 103: 031101.
151 Gao, M., Viswan, R., Tang, X. et al. (2018). *Sci. Rep.* 8: 323.
152 Gao, J., Xue, D., Liu, W. et al. (2017). *Actuators* 6: 24. https://doi.org/10.3390/act6030024.
153 Wang, P., Zhang, E., Toledo, D. et al. (2020). *Nano Lett.* 20: 5765–5772.
154 Stewart, T.S., Nagesetti, A., Guduru, R. et al. (2018). *Nanomedicine* 13: 423–438.
155 Rodzinski, A., Guduru, R., Liang, P. et al. (2016). *Sci. Rep.* 6: 1–14.
156 Kaushik, A., Jayant, R.D., Sagar, V., and Nair, M. (2014). *Expert Opin. Drug Deliv.* 11: 1635–1646.
157 Guduru, R., Liang, P., Runowicz, C. et al. (2013). *Sci. Rep.* 3: 1–8.
158 Nair, M., Guduru, R., Liang, P. et al. (2013). *Nat. Commun.* 4: 1707.
159 Walther, T., Straube, U., Köferstein, R., and Ebbinghaus, S.G. (2016). *J. Mater. Chem. C* 4: 4792–4799.
160 Martins, P. and Lanceros-Méndez, S. (2013). *Adv. Funct. Mater.* 23: 3371–3385.
161 Feng, Y., Zhang, Y., Sheng, J. et al. (2021). *J. Phys. Chem. C* 125: 8840–8852.
162 Omelyanchik, A., Antipova, V., Gritsenko, C. et al. (2021). *Nanomaterials* 11: 1154.
163 Sayed, F., Kotnana, G., Barucca, G. et al. (2020). *J. Magn. Magn. Mater.* 503: 166622.
164 Kotnana, G., Sayed, F., Joshi, D.C. et al. (2020). *J. Magn. Magn. Mater.* 166792.
165 Wang, L., Petracic, O., Kentzinger, E. et al. (2017). *Nanoscale* 9 (35): 12957–12962.
166 Wang, J., Li, Z., Wang, J. et al. (2015). *J. Appl. Phys.* 117: 044101.
167 Geprägs, S., Mannix, D., Opel, M. et al. (2013). *Phys. Rev. B* 88: 054412.
168 Alberca, A., Munuera, C., Tornos, J. et al. (2012). *Phys. Rev. B* 86: 144416.
169 Liu, M., Obi, O., Lou, J. et al. (2009). *Adv. Funct. Mater.* 19: 1826–1831.

170 Kim, D., Rossell, M.D., Campanini, M. et al. (2021). *Appl. Phys. Lett.* 119: 012901.
171 Lorenz, M., Lazenka, V., Schwinkendorf, P. et al. (2014). *J. Phys. D: Appl. Phys.* 47: 135303.
172 Fina, I., Dix, N., Rebled, J.M. et al. (2013). *Nanoscale* 5: 8037–8044.
173 Hu, J.M., Duan, C.G., Nan, C.W., and Chen, L.Q. (2017). *NPJ Comput. Mater.* 3: 18.
174 Niranjan, M.K., Velev, J.P., Duan, C.G. et al. (2008). *Phys. Rev. B* 78: 104405.
175 Radaelli, G., Petti, D., Plekhanov, E. et al. (2014). *Nat. Commun.* 5: 3404.
176 Yin, Y.W., Burton, J.D., Kim, Y.M. et al. (2013). *Nat. Mater.* 12: 397–402.
177 Vaz, C.A.F., Hoffman, J., Segal, Y. et al. (2010). *Phys. Rev. Lett.* 104: 127202.
178 Yi, D., Yu, P., Chen, Y.C. et al. (2019). *Adv. Mater.* 31: 1806335.
179 White, J.S., Bator, M., Hu, Y. et al. (2013). *Phys. Rev. Lett.* 111: 037201.
180 Zandalazini, C., Esquinazi, P., Bridoux, G. et al. (2011). *J. Magn. Magn. Mater.* 323: 2892–2898.
181 Vafaee, M., Finizio, S., Deniz, H. et al. (2016). *Appl. Phys. Lett.* 108: 072401.
182 Wu, S.M., Cybart, S.A., Yu, P. et al. (2010). *Nat. Mater.* 9: 756–761.
183 Fert, A. (2008). *Angew. Chem.* 47: 5956–5967.
184 Hirohata, A., Yamada, K., Nakatani, Y. et al. (2020). *J. Magn. Magn. Mater.* 509: 166711.
185 Hirohata, A., Sukegawa, H., Yanagihara, H. et al. (2015). *IEEE Trans. Magn.* 51: 0800511.
186 Bertacco, R. and Cantoni, M. (2016). *Ultra-High Density Magnetic Recording-Storage Materials and Media Designs* (ed. G. Varvaro and F. Casoli). Pan Stanford Publishing.
187 Tedrow, P.M. and Meservey, R. (1971). *Phys. Rev. Lett.* 26: 192–195.
188 Julliere, M. (1975). *Phys. Lett. A* 54: 225–226.
189 Baibich, M.N., Broto, J.M., Fert, A. et al. (1988). *Phys. Rev. Lett.* 61: 2472–2475.
190 Binasch, G., Grünberg, P., Saurenbach, F., and Zinn, W. (1989). *Phys. Rev. B* 39: 4828.
191 Bibes, M. and Barthélémy, A. (2007). *IEEE Trans. Electron Devices* 54: 1003–1023.
192 Emori, S. and Li, P. (2021). *J. Appl. Phys.* 129: 020901. https://doi.org/10.1063/5.0033259.
193 Yin, Y. and Li, Q. (2017). *J. Mater.* 3: 245–254.
194 Budhani, R.C., Emori, S., Galazka, Z. et al. (2018). *Appl. Phys. Lett.* 113: 082404.
195 Singh, A.V., Khodadadi, B., Mohammadi, J.B. et al. (2017). *Adv. Mater.* 29: 1701222.
196 Liu, X., Chang, C.-F., Rata, A.D. et al. (2016). *NPJ Quantum Mater.* 1: 16027.
197 Bowen, M., Bibes, M., Barthélémy, A. et al. (2003). *Appl. Phys. Lett.* 82: 233–235.
198 Matzen, S., Moussy, J.B., Mattana, R. et al. (2012). *Appl. Phys. Lett.* 101: 042409.
199 Gajek, M., Bibes, M., Barthélémy, A. et al. (2005). *Phys. Rev. B* 72: 020406.
200 Gajek, M., Bibes, M., Barthlmy, A. et al. (2005). *J. Appl. Phys.* 97: 103909.
201 Yin, Y.W., Raju, M., Hu, W.J. et al. (2011). *J. Appl. Phys.* 109: 2010–2012.

# 13

## Metal-Oxide Nanomaterials for Gas-Sensing Applications

*Pritamkumar V. Shinde[1], Nanasaheb M. Shinde[1], Shoyebmohamad F. Shaikh[2], and Rajaram S. Mane[1]*

[1] Swami Ramanand Teerth Marathwada University, School of Physical Science, Center for Nanomaterial & Energy Devices, Nanded, Maharashtra 431606, India
[2] King Saud University, College of Science, Chemistry Department, PO Box 2455, Riyadh 11451, Saudi Arabia

## 13.1 Introduction

Urbanization and industrial growth have resulted in severe health issues for the living and nonliving things because of increased pollution. There is a need for developing low-cost, reliable, and eco-friendly detection devices, which can monitor pollution level before actual damage to environment. Continuous respiration inside the stale cabin can induce far-reaching negative effects. Due to continuous industrial production, the greenhouse gases are emitted mainly from the burning of fossil fuels, which are responsible for a global warming effect [1]. The American Conference of Government Industrial Hygienists reported a source of production, toxicity, and threshold limit value (TLV) of various environmentally harmful gases including toxic and greenhouse gases present in the atmosphere [2–13].

According to Jacob Fraden [14], a sensor is a device that receives a signal or stimulus and responds with an electrical signal. The reason for the output of a sensor to be limited to electrical signals is related to the present development of signal processing, which is almost exclusively performed using electronic devices. Hence, a sensor should be a device that detects physical, chemical, or biological signal and converts it into an electric signal that should be compatible with electronic circuits. Sensor seems to have come from the word "sense" given that usually sensor devices try to mimic human senses characteristics (Figure 13.1).

In the biological sense, the output is also an electrical signal that is transmitted to the nervous system. Usually, sensors are part of larger complex systems, made by several transducers, signal conditioners, signal processors, memory devices, and actuators. Gas sensors are the heart of the system as they transduce the physical signal in the presence of toxic gases to the monitoring system that can recognize and send an alert signal to humans, both qualitatively and quantitatively or can give the feedback to maintain a congenial atmosphere. In recent years, numerous gases and volatile organic compounds (VOCs) (refer to Table 13.1) are being envisaged

*Tailored Functional Oxide Nanomaterials: From Design to Multi-Purpose Applications*, First Edition.
Edited by Chiara Maccato and Davide Barreca.
© 2022 WILEY-VCH GmbH. Published 2022 by WILEY-VCH GmbH.

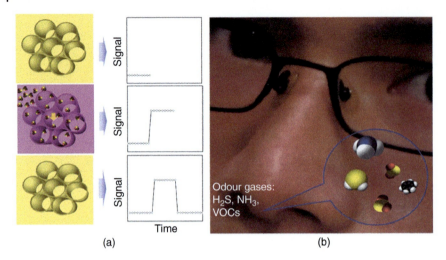

**Figure 13.1** (a) Carton of a general gas sensor, in which the adsorption of gaseous molecules to the sensing materials leads to the change in readout signals (physical or chemical properties). (b) A nose as an odor gas sensor. Source: Hoa et al. [1]. https://www.hindawi.com/journals/jnm/2015/972025/ (last accessed 12/07/2021) CC BY 3.0.

in industrial production and research applications [15]. The efficient and reliable control and monitoring of these harmful species is a main concern considering a huge risk of damage to assets and lives. There are certain gases that are toxic to human and environment as they cause big explosions. Accordingly, there is a need of gas sensors with the intention of continuous and effective detection of these gases to evade predominantly the said dangers. Researchers are trying to develop devices for early detection and/or alarm of certain flammable, explosive, and toxic gases. For this purpose, gas sensors were invented for monitoring of pollution gases [16, 17]. Moreover, recent developments in the field of nanotechnology and materials science have paved the way for synthesis of numerous novel materials with desired morphologies and special physical–chemical properties. In particular, several nanomaterials including metal-oxide nanomaterials (MONs) have received considerable attention for their use as promising sensing materials [18, 19]. The MONs exhibits several unique features such as controllable size, functional biocompatibility, biosafety, chemical stability, and catalytic properties with enhanced electron transfer kinetics and strong adsorption capability [20–22]. In fact, MONs are preferred as optical emitters, electronic conductors, catalysts, and carriers for amplified detection signal and biosensing interfaces [23]. Moreover, the amalgamation of conducting and semiconducting nanoparticles (NPs)/quantum dots (QDs) such as gold, silver, platinum, carbon nanotubes, graphene, etc. has been reported to enhance the optical, electrical, and magnetic properties of the MONs with improved selectivity, chemical stability, and increased sensing performance [21, 22]. Hence, smart selection, design, and application of MONs will lead to a new generation of sensing devices with enhanced signal amplification and coding strategies that may address future economic and social needs.

Table 13.1 A brief summary of the toxic gases.

| Sr. no. | Gas | Toxicity to human | TLV (ppm) | Ref. |
| --- | --- | --- | --- | --- |
| 1 | $Cl_2$ | It causes distress in respiration as it communicates with the mucous membrane of the lungs and otherwise can be a serious problem after a few breaths at 1000 ppm | 1000 | [2] |
| 2 | $NH_3$ | Eye, nose, and throat irritation | 35 | [3] |
| 3 | $C_2H_5OH$ | Nausea, fatigue, and visual disorders | 1000 | [4] |
| 4 | $CH_3OH$ | Methanol poisoning causes blindness, organ failure, or even death when recognized too late | 200 | [5] |
| 5 | $H_2S$ | Eye irritation and lung irritation, can paralyze the nerves and cause tissue hypoxia, and may cause death over 500 ppm | 10 | [6] |
| 6 | $C_7H_8$ | Toluene can negatively affect the nervous system, causing disturbances in brain function and causes weakening of sight, hearing, and speech. Furthermore, it can damage the kidneys and liver | 100 | [7] |
| 7 | $C_6H_6$ | Narcosis, eye irritation, and skin irritation | 2.5 | [7] |
| 8 | $C_3H_6O$ | Nose and throat irritation | 500 | [8] |
| 9 | Liquefied petroleum gas (LPG) | The blasts caused by leaks | — | [9] |
| 10 | $H_2$ | $H_2$ is nontoxic but can lead to an extreme decrease of oxygen concentration in human body, a condition called asphyxia | — | [10] |
| 11 | CO | Headache, discomfort, possibility of collapse, and can even cause a death; its limit is over 1600 ppm | 25 | [11] |
| 12 | $CO_2$ | High concentration exposure causes problems such as dizziness or headache, which on long-term exposure results in breathing complications and oblivion | — | [12] |
| 13 | $NO_2$ | Eye, nose, and throat irritation may cause death over 100 ppm | 5 | [13] |

In this chapter, recent research developments made over last five years in synthesis of MONs and their use in fabricating sensing devices by reviewing published works are presented. Basically, MONs are grown in many different nanoscale forms such as nanoparticles, nanowires, nanosheets, nanorods, nanotubes, nanofilms, etc. for developing various novel sensor devices to detect various hazardous gases at low and high temperatures [8–13]. Different gas-sensing mechanisms are delineated with possible limitations. Future perspectives and real-time usages for economic and social benefits of MON sensors are also explored.

## 13.2 Types of Gas Sensors

A sensor consists of a transducer and an active layer for converting the chemical information into electronic signal like resistance change, current change, or voltage change. Fundamentally, conventional detection methods construct systems that ignite an aural alarm to inform people when there is seepage of dangerous or toxic gases. However, it is not very consistent because it is required to get precise real-time measurements of the concentration of a gas. Gas sensors are typically classified into various types based on the type of the sensing elements. The classification of the various types (Figure 13.2 and Table 13.1) of the gas sensors is based on the sensing element. There are no evidences for assertion that all materials are equally effective for gas sensors applications. Therefore, at such a big variety of materials, which can be used, the selection of optimal sensing material becomes key problem in both designing and manufacturing gas sensors with an expected performance [24, 25]. We began to understand that for implementation of all requirements, a material for solid-state gas sensors should have a specific combination of its physical and chemical properties. According to [26] in order to be used in practice, a gas sensor should fulfill many requirements, which depend on the purposes, locations, and conditions of sensor operation. Among the requirements, primarily important are the sensing performance related to sensitivity, selectivity, response rate, and reliability related to drift, stability, and interfering gases. In this regard, the selection and processing of the sensing materials are of utmost importance in research and development of gas sensors. In recent years, the sensing technology has evolved and continued to develop for betterment of human well-being, safety, quality of air, and for environment protection [11].

A variety of MONs (ZnO, $MnO_2$, $SnO_2$, $TiO_2$, $WO_3$, $CeO_2$, CuO, $Co_3O_4$, $In_2O_3$, $Fe_2O_3$, and $Bi_2O_3$ composite oxides) are being used in gas sensors because of their distinctive oxide characteristic properties [8, 9, 13, 27–35]. Low fabrication

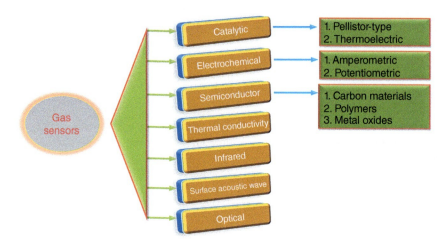

**Figure 13.2** Types of gas sensors.

cost, miniaturized size, and ease of integration have made the semiconductor metal-oxide-based sensors as attractive cost-effective sensing materials [36]; metal-oxide materials in nano-thin films [37], nanoparticles [38], nanotubes [39], nanobelts [40], and nanowires [41] play a vital role by virtue of their large surface area in sensor application. Thanks to the variety of nanostructures, non-toxicity, high refractive index, dielectric permittivity, and ionic conductivity, they also are preferred in different opto-electrochemical devices [42, 43]. At the nanoscale level, unique optical, electrical, and magnetic properties, and even performance enhancement, are typically observed [13, 44].

## 13.3 Metal-Oxide Nanomaterial-Based Gas Sensors

Nanomaterials with dimensions between 1 and 100 nm have demonstrated steadily growing interest due to their fascinating chemical and physical properties as compared to their bulk counterparts [45]. Nanosized materials of higher surface area and surface-to-volume ratio are useful for catalytic reaction [46]. Actually, as the particle size decreases, the surface-to-volume ratio increases. The reduction in particle size of material gives rise to quantum confinements [47]. The MONs sensors are widely investigated, due to small size, cost, eco-friendly nature, earth abundant, chemical stability, and appreciating their chemico-physical characteristics [48]. In fact, MONs sensors are widely accepted in technology like dye-sensitized solar cells [49]. Metal oxides exhibit metalic, semiconducting, or insulating characteristic as they are composed of positive metalic ions and negative oxygen ions as ionic compounds [50]. Moreover s-shell in metal oxides is filled by electrons, and their d shell may not be completely filled. The physical, chemical, and electronic properties can be engineered by changing their structure, size, composition, stoichiometry, and doping [11]. They are known for detecting combustible, reducing, or oxidizing gases by conductive measurements. Since few decades, the oxides like ZnO, $MnO_2$, $SnO_2$, $TiO_2$, $WO_3$, $CeO_2$, CuO, $Co_3O_4$, $In_2O_3$, $Fe_2O_3$, and $Bi_2O_3$ with their composites are being used in sensors [8, 9, 13, 27–35]. There are two main types of semiconducting metal-oxide-based sensors, n-type (whose majority carrier is an electron) and p-type (whose majority carrier is a hole). The working principle of MON-based gas sensors is based on an equilibrium shift of the surface reactions associated with the target analyte.

Commonly, reducing gases like $NH_3$, CO, $H_2$, HCHO, etc. lead to an increase in the conductivity for n-type semiconductor and a decrease for p-type semiconductors, while the effect of oxidizing gases ($NO_2$, $O_3$, $Cl_2$, etc.) is vice versa. The p-type MON-based gas sensors have availed relatively little attention compared with n-type MONs [51]. The response of the p-type MON-based gas sensors is only equal to the square root of that of an n-type MONs to the same gas when the morphological configurations of both sensor materials are identical [52, 53]. The p-type MONs have their own advantages, such as low humidity dependence [54] and high catalytic

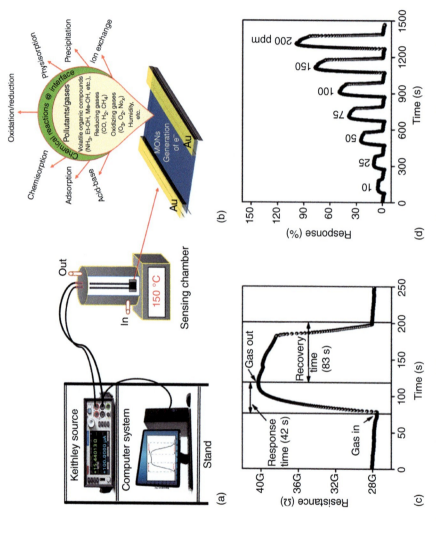

**Figure 13.3** (a) Experimental setup of a gas sensor [13], (b) sketch showing the working principles of chemical sensors, (c) transient vs. time (at 100 ppm), and (d) response vs. $H_2$ concentration of the $Bi_2O_3$ film sensor. Source: Shinde et al. [10]. Reproduced with permission of Elsevier.

properties [55]. Considering that a single MON cannot satisfy the requirements for monitoring in harsh environments, a combinatorial approach needs to be employed for the selection of sensing materials. As is known, the fabrication of gas sensors is composed of the selection and the preparation of sensing materials and the material sensor integration. Therefore, the sensing properties of metal oxides to a special gas are dependent on the whole manufacturing process, including material type, configuration, temperature, and the integration mode for the sensing materials and devices (see Figure 13.3a,b).

For commercial applications, a gas sensor should fulfill many requirements, which depend on the purpose, location, and sensor operation conditions. Primarily requirements of the gas-sensing performance include sensitivity, selectivity, detection limit, and response time/recovery time (refer to Figure 13.3c,d and Table 13.2), which are eventually connected with the sensing materials used so that selection and processing of the sensing materials are essentially important in research and development of gas sensors [10, 56, 57]. The authors believe that sensing materials synthesis is a base for new sensors enabling significant innovations in gas sensor technology.

Since the sensing properties of the MON-based gas sensors improve through the morphological or microstructure manipulation, the sensing mechanism and the influencing factor need to be comprehensively analyzed to offer a guideline for the preparation of sensing materials for higher sensing performance [52, 56–61]. Factors including the particle size, pore size, grain boundary, heterointerface, etc. are essentially important in the sensing mechanism. After the optimization of sensing materials in material selection and material preparation, the material sensor

Table 13.2 Definitions of various gas-sensing parameters.

| Sr. no. | Gas-sensing parameter | Definition |
|---|---|---|
| 1. | Sensitivity | The sensitivity ($S$) of a sensor is calculated from the mathematical relation, $S(\%) = \frac{R_a - R_g}{R_a} \times 100$, where $R_a$ is the stabilized resistance of the sensor material in presence of air and $R_g$ is the stabilized resistance in presence of the target gas. Conveniently, $S$ is expressed in terms of percentage of sensitivity |
| 2. | Selectivity | Selectivity refers to the characteristics that determine whether a sensor can respond selectively to a gas or a group of gases |
| 3. | Detection limits | Detection limit is the lowest concentration of the gas that can be detected by the sensor under given conditions |
| 4. | Response time | The time for the resistance reaches the 90% of the final change in resistance after the sensor is exposed to the target gas |
| 5. | Recovery time | The time for the resistance to recover the 90% of the maximum change in resistance once the target gas is turned off |

integrating mode should be taken into account. As is known, the microstructure of sensing materials is damaged in the process of posttreatment [59, 60]. The film thickness [62], the binding between the sensing materials and the devices [63], the electrode distance [64], and the morphological evolution are major concerns in the sensor fabrication process [65].

## 13.4 Preparation of Metal-Oxide Gas Sensors

The choice of a proper metal-oxide semiconductor is significant for gas sensing; the methods and conditions for preparing gas-sensing nanomaterials have profound influence on the gas-sensing properties as morphologies, crystalline sizes, contacting geometry between crystals, packing densities, packing thicknesses, defect densities, and so on are different [66]. Additionally, the sensing properties are augmented by promoting the sensing reaction of a specific gas through loading the sensing material with foreign metal-oxide semiconductors [67, 68] or noble metal catalysts [69–72]. Therefore, to obtain gas-sensing materials with excellent sensing properties, the sensing mechanism and the effect of structural parameters of metal oxides on the sensing properties are outlined in Sections 13.4.1–13.4.6.

### 13.4.1 Operation Mechanism

The sensing mechanism of metal-oxide gas sensors has a key importance for designing and fabricating novel gas-sensing materials with excellent performances. The fundamental sensing mechanism of metal-oxide-based gas sensors is based on a shift of equilibrium of the surface chemisorbed oxygen reaction due to the presence of target gas. Based on the conduction type, the metal-oxide semiconductors are divided into n- and p-types, which exhibited different sensing behaviors to the same detecting gas. Upon exposure to oxidizing gases, the gas species act as acceptors, leading to a resistance increase for n-type semiconductors and a decrease for p-type semiconductors [73]. As for the reducing gases, the gas species act as donors, leading to a resistance decrease for n-type semiconductors and an increase for p-type semiconductors. At present, the sensing mechanism based on metal-oxide semiconductors is mainly described by two different models: the ionosorption model and the oxygen vacancy model. The first model considers the space charge effects or the changes in the electric surface potential that result from the gas adsorption, ionization, and redox reactions. The second model centers on the reaction between oxygen vacancies and gas molecules, and the variation of the amount of the subsurface/surface oxygen vacancies and their reduction–reoxidation mechanism leads to changes in the oxygen stoichiometry [74].

In order to facilitate the understanding of the sensing mechanism, the sensing properties are assumed to be determined by three factors: receptor function, transducer function, and utility factor [56] as shown in Figure 13.4. The receptor function is concerned with how each constituent responds to the surrounding atmosphere containing oxygen and the target gases. The amount of oxygen adsorbed in this

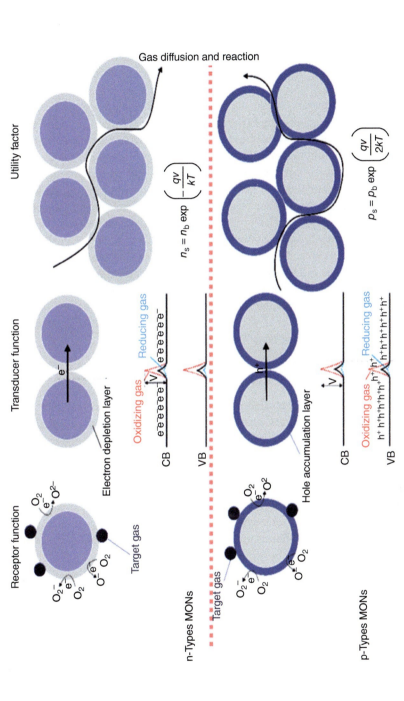

**Figure 13.4** Diagram description of the gas-sensing mechanism and the conduction model based on n-type MONs and p-type MONs. Source: Zhang et al. [75]. Reproduced with permission of Royal Society of Chemistry.

process direclty determines sensing properties, which depends on the specific surface area of the sensing materials used. The specific surface area of the sensing materials is altered significantly with the change in the particle size, the pore size, and the surface morphology. As for the transducer function, it is concerned with how the response of each particle is transformed into that of the whole device, which is based on the free carrier transport mechanism between the adjacent crystals. Commonly, a double Schottky barrier model is used to reveal this process. For n-type semiconductors, oxygen is adsorbed on the grains in air to attract electrons from the grains forming oxygen species ($O^-$, $O^{2-}$, $O_2^-$, etc.), inducing an electron-depleted layer to increase the surface potential. Thus, a potential barrier is formed between adjacent grains, which hinder the migration of electrons from one grain to another. After exposure to a reducing gas, the oxygen species are consumed, leading to a lower potential barrier height, which is beneficial for electron transport between grains. A drastic influence on electron migration between grains is played not only by the potential barrier produced by gas adsorption (determining the free carrier transport), but also by the trapping and defect states [76–78]. As for the utility factor, it is concerned with the attenuation of the response due to the effects of diffusion and the reaction of reactive target gases through the pores of the assembly of crystals. Therefore, to obtain higher sensing performance of the metal-oxide-based sensors, the crystal structure, surface morphology, and the defect structure are in Sections 13.4.2.1–13.4.2.2, which would offer a guideline for researchers while preparing sensing materials.

### 13.4.2 Morphology-Related Structural Parameters

The microstructural parameters of metal oxides are very important factors to control the response parameters of resistive-type gas sensors. As discussed earlier, the sensing properties are determined by the pore size, specific surface area, particle size, film thickness, etc. The stacking effect or the opposite effect occurs in the gas-sensing process, and adjusting one factor improves performance metric by degrading another [79]. Therefore, considering the complexity of the gas-sensing mechanism and numerous factors, various morphological and structural parameters, the grain size, and pore size, which are correlated with the other parameters, such as the specific surface area, morphological evolution, etc. of metal oxides with gas-sensing properties should be separately investigated.

#### 13.4.2.1 Grain Size

The grain size, an important parameter, controls the sensing properties of metal-oxide semiconductors. The effect of the grain size on the gas-sensing properties has widely been studied with respect to theory and experiment. Xu et al. [80] synthesized ZnO with different particle sizes using the chemical precipitation method and found that the gas sensitivity of ZnO is affected by the grain size. Zhang and Liu [38] reported that the reduction of the particle size to nanometers, or to dimension comparable with the thickness of charge depletion layers, leads to a dramatic improvement in the sensitivity and the response rate of $SnO_2$-based gas sensors. Xu et al. [81] studied the effect of the grain size on sensitivity of porous $SnO_2$-based gas sensors and concluded that the sensitivity increases steeply as the grain size decreases to be comparable with or less than twice of the depth of the space charge layer. Theoretical research based on the numerical simulations and semiquantitative model has also been employed. A. Rothschild and Y. Komem [82] calculated the effective carrier concentration as a function of the surface state density (Eq. (13.1)) for nanosized $SnO_2$ crystallites with different grain sizes from 5 to 80 nm. Accordingly, the carrier concentration significantly decreases, and the surface state density reaches a critical value proportional to the grain size ($d$), suggesting the sensitivity is proportional to $1/d$. For the semiquantitative model, the analysis is mainly based on a comparison of the grain size ($d$) with the Debye length ($L_D$). In semiconductors, the extent of the space charge layer ($L_S$) depends on the Debye length [82]:

$$L_S = L_D \sqrt{\frac{eV_S^2}{KT}} \quad (13.1)$$

where $K$ is the Boltzmann constant, $T$ is the absolute temperature, and $V_S$ is the surface potential barrier. For large crystallites with grain size diameter $d \gg 2L_S$, the conductance of sensing materials is determined by Schottky barriers at the grain boundary, which is independent of the grain size. It is assumed that the gas sensitivity is practically independent of the grain size. However, the specific surface area of the sensing material increases with decrease in grain size, leading to enhanced gas adsorption and higher sensitivity. If $d < 2L_S$, the depletion region extends throughout the whole grain, implying that each grain is fully involved in the space charge layer. As a result, the conductance of sensing materials decreases sharply, which essentially is controlled by the intra-crystalline conductivity (grain controlled). In this case, the sensitivity depends critically on the grain size, and the highest gas sensitivity is obtained. The sensing materials with a smaller grain size demonstrate a larger sensing response [57]. However, the former discussion is based on the changes in depletion layers and potential barriers at grain boundaries, which is usually used for the analysis of the n-type metal-oxide semiconductors; as for the p-type metal oxides, the hole-accumulation layer (HAL) acts as a charge carrier transport channel, and there is no need for the holes to overcome the potential barriers across the grain boundaries. Therefore, the extent of grain size impact of the sensing properties in the p-type semiconductors is lower than that in the n-type semiconductors. For example, Choi et al. [83] synthesized p-type NiO nanofibers using the electrospinning method, and the effect of the grain size on the sensing ability is investigated where the size of the grains demonstrates an influence on the response to only a small extent. Therefore, to obtain an appropriate grain size for a better sensing performance of the metal oxide, the type of metal oxide and the different sensing mechanisms should also be comprehensively considered.

#### 13.4.2.2 Pore Size
Apart from the particle size effect, the pore size is also an important factor in sensing performance. In some cases, the effect of porosity associated with the pore size on

the sensitivity is even more powerful than that of the grain size [84]. As mentioned previously, the sensing properties of metal-oxide semiconductors are determined by the receptor function, the transducer function, and the utility factor. And the utility factor is concerned with the effects of the diffusion and the reaction of reactive target gases through the pores of the assembly of crystals. During the response and recovery process, the gas diffuse process determines the sensing properties, which depends on pore size. Gas diffusion through a porous material consists of surface diffusion, Knudsen diffusion, and molecule diffusion [85] whose constants increase with the pore size. Gas diffusion, determined by temperature ($T$), the pore radius ($r$), the molecular weight ($M$), and $R$, is the gas constant of the diffusing gas and is dominated by Knudsen diffusion in mesoporous sensing materials. The diffusion coefficient ($D_K$) is expressed as [85]

$$D_K = \frac{4r}{3}\sqrt{\frac{2RT}{\pi M}} \tag{13.2}$$

A big pore size leads to a larger diffusion coefficient. Concerning the effect of the pore size on the sensing properties, extensive research studies have been carried out. The variation in the pore size tunes the sensing properties, such as sensitivity, selectivity, response/recovery time, etc. The sensing materials with a porous structure exhibit enhanced sensitivity and therefore envisaged for detecting gases. Tian et al. prepared hierarchical $SnO_2$ mesoporous structures with different pore sizes by templating cottons where the sensor response of the as-prepared sensors to formaldehyde gradually increased with increasing pore size [86]. According to Eq. (13.2), a larger pore size means a higher diffusion constant, suggesting higher interfacial contact area that allows most detected gas molecules to diffuse easily inside the deeper region with an enhanced sensor response. Sun et al. prepared porous $SnO_2$ nanosheets with enhanced sensor response to ethanol as compared with $SnO_2$ nanoparticles [87]. Shinde et al. [12] reported that a room-temperature, direct, and superfast chemical bath deposition method has successfully been applied for synthesizing various morphologies of $Bi_2O_3$ nanosensors on a glass substrate. A direct and rapid synthesis of $Bi_2O_3$ nanostructures of different surface area used poly-ethylene glycol (PEG), ethylene glycol (EG), and ammonium fluoride (AF) surfactants by a soft chemical method at room temperature on glass substrate and named as $Bi_2O_3$ (BO), PEG@$Bi_2O_3$ (PBO), EG@$Bi_2O_3$ (EBO), and AF@$Bi_2O_3$ (ABO), respectively. These nanosensors were exploited as gas sensors that demonstrated excellent sensing performances for $CO_2$ gas at room temperature. The results (see Figure 13.5) emphasize the potential applications of the room-temperature-synthesized $Bi_2O_3$ nanosensors of the woolen globe, nanosheet, flower-rose, and spongy square-plate-type surface appearances on a glass substrate and that they would find commercial benefits of detecting $CO_2$ gas at room temperature. Figure 13.5b displays the dynamic response of the $Bi_2O_3$ nanosensors for $CO_2$ gas at 10, 20, 40, 60, 80, and 100 ppm concentrations where the graphs are plotted with an identical scale for quickly comparing the sensing responses. It is evident that the response amplitude is highly dependent on gas concentration. The gas-sensing transients of $Bi_2O_3$ nanosensors exhibit a

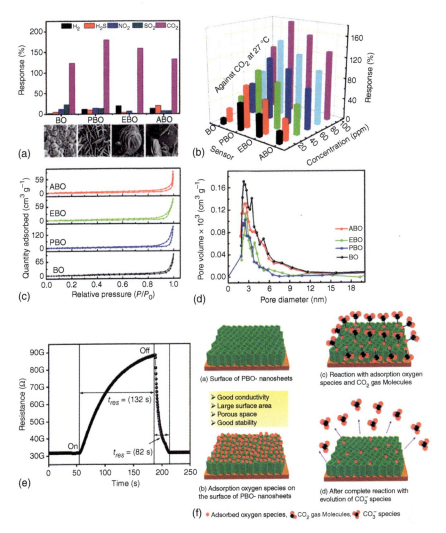

**Figure 13.5** (a) Gas selectivity (at 100 ppm) and (b) response vs. $CO_2$ concentration of the BO, PBO, EBO, and ABO nanosensors at 27 °C. (c) Nitrogen adsorption–desorption isotherms and (d) Barrett–Joyner–Halenda (BJH) pore size distribution plots of the BO, PBO, EBO, and ABO nanosensors. (e) Transient vs. time (at 100 ppm) of the PBO nanosensor, and (f) gas-sensing mechanism of the PBO, nanosensor in air and $CO_2$ gas. Source: Shinde et al. [12]. Reproduced with permission of Royal Society of Chemistry.

smooth response/recovery behavior. Apparently, the PBO nanosheet-type sensor reveals a better $CO_2$ gas-sensing response than that of the others. When PBO nanosheet-type sensor was operated under 10 ppm $CO_2$ gas, the response value is about 52%. With increasing $CO_2$ gas concentration from 10 to 100 ppm, the recorded responses were 52%, 80%, 101%, 118%, 146%, and 179%, respectively. The gas response of PBO nanosheet-type sensor to 10 ppm of $CO_2$ gas (52%) was higher than that of BO woolen-globe (12%), EBO rose-flower (42%), and ABO square-plate (23%)-type gas sensors. Nevertheless, there are performance differences among

BO woolen-globe, PBO nanosheet, EBO rose-flower, and ABO square-plate-type $Bi_2O_3$ nanosensors for $CO_2$ gas. In all cases, with the concentration of $CO_2$ gas, the response of $CO_2$ gas increases, suggesting that the surface morphology of $Bi_2O_3$ nanosensor has an impact on $CO_2$ gas-sensing performance at room temperature. To confirm the relationship between the morphology of $Bi_2O_3$ nanosensor and gas-sensing performance, nitrogen adsorption and desorption measurements of these four nanosensors were carried out to estimate the properties. The nitrogen adsorption and desorption cyclic curves as in Figure 13.5c for BO, PBO, ABO, and EBO nanosensors exhibited 31, 58, 44, and 37 $m^2\ g^{-1}$, surface areas, respectively, indicating a downtrend of the active surface area, so it can be concluded that the moderate nanosheet diameter and a uniform distribution contribute to a large surface area and hence lead to high sensitivity. Pore size distribution curves (Figure 13.5d) of these nanosensors suggest that the moderate nanostructures diameter (~50 nm) and density (~285 $mm^{-2}$) link to uniform and proper pore size, which are important factors for mass transport and effective surface area. The average pore size distribution maxima of BO, PBO, ABO, and EBO nanosensors were, respectively, 5.16, 2.53, 2.87, and 3.89 nm. Compared with the $Bi_2O_3$ nanosensors, thin and sparse woolen-globe-type sensor produces a larger pore volume but a less effective surface area, while thick and dense nanosheet-type sensor contributes a smaller pore space and high surface area. Only the moderate diameter and uniform distribution of these nanosheets would lead to produce an appropriate pore space with a large surface area and, thus, can be more beneficial for their use as sensing materials with a high sensing performance. In a nutshell, the surface area and pore size of nanosensor structures also contribute in gas-sensing performance. The PBO nanosensor exhibits a higher sensing performance as compared with those of the BO, ABO, and EBO nanosensors owing to the moderate diameter and uniform distribution of the nanosheets, resulting in the production of an appropriate pore space with a large surface area [12]. When BO, PBO, EBO, and ABO nanosensors are exposed to $CO_2$ gas, $Bi_2O_3$ catalyzes the oxidation of $CO_2$ where the electrons get trapped by the dissociated oxygen species that changes the electrical resistance to create a better sensing response. The transient gas response time periods for $CO_2$ gas-sensing are shown in Figure 13.5e (show only higher sensing performance). On the exposure of $CO_2$ gas to the PBO nanosensor, the resistance is increased from 31.68 to 88.46 G$\Omega$. The PBO nanosensors demonstrate a maximum response of 179% @100 ppm $CO_2$ gas at room temperature. The response time/recovery time depends on the rate of diffusion of gas molecules onto the sensor surface (adsorption and desorption) and/or associated reaction rate between the target gas molecules with the sensor elements. The response time/recovery time depends on the rate of diffusion of gas molecules onto the sensor surface (adsorption and desorption) and/or associated reaction rate between the target gas molecules with the sensor elements to moderate the fast response time/recovery time value of 132/82 seconds, respectively. With this reaction, many extracted electrons can be released to the PBO nanosensor surface, leading to a decrease of Schottky surface barrier on thinning a space charge layer. Therefore, the electrical conductivity layers increase at room temperature itself. Moreover, pores are regularly distributed over the PBO $Bi_2O_3$ nanosensor surface with several channels facilitating $CO_2$ gas diffusion, making the contact of $CO_2$ gas with inner $Bi_2O_3$ grains more easily in the reduction reaction process and degassing equally during the recovery process as porous surface provides a sufficient gas absorption surface sites; therefore, more gas molecules would absorb at a relatively low temperature, which causes an enhancement of the $CO_2$ gas sensitivity (Figure 13.5f). Consequently, the mesoporous PBO nanostructures are essentially important in the fast response/recovery process. However, the pore size does not always maintain a linear relationship with sensitivity. For instance, mesoporous NiO nanosheets with different pore sizes prepared by a simple hydrothermal method for room-temperature $NO_2$ sensing demonstrate different sensitivities at room temperature. The monotonic changes in the pore size cannot figure out the non-monotonic change rule of the sensitivity [88].

In addition to the sensitivity, materials with porous structure exhibit enhanced response/recovery rate and selectivity. The Knudsen diffusion constant depends significantly on the molecular weight; the heavier the gas molecules is, the slower will be their motion in the porous sensing films (Eq. (13.2)). According to this principle, the selectivity to different gas molecules is achieved by modulating their pore size.

Shimizu et al. [89] investigated effects of gas diffusivity and reactivity on gas-sensing properties of porous $SnO_2$-based sensors. The interior of a phase pure $SnO_2$ sensor endows enhanced sensitivity to $H_2$ compared to the surface, leading to excellent selectivity to $H_2$. The gas sensors with higher porosity adduce better gas permeability for faster response. Varghese et al. [90] prepared highly ordered nanoporous alumina films as humidity sensors where sensors with a large pore size confirm shortest response/recovery time. The pore-widening treatment was also emphasized to obtain a short response time, which is attributed to the short diffusion time aroused by the porous structure, suggesting an importance of pore size in response time [91]. In addition to pore size, the film thickness and film quality need to be taken into consideration as pore size effect is valid only for thick films [92].

### 13.4.3 Crystallographic Defective and Heterointerface Structures

Several researchers have explored the effect of the morphology and crystallographic structure on the sensing properties of metal oxides [57, 79]. However, the comprehensive coverage based on the defective and the heterointerface structures is scare. In addition to the microstructural parameters, effect due to the defect structure and the heterointerface like the grain boundary [93], defect concentration [88, 94], p–n/n–n heterojunction [95–97], noble metal sensitization [98–100], etc. is important. Therefore, the effect of the defect and the heterointerface structures on the sensing properties should also be comprehensively analyzed for better understanding.

#### 13.4.3.1 Defect Structure
The defect structure in semiconductors, grain boundary structures, are usually produced by some interaction between neighboring grains or some other additives,

which can change the electrical properties of semiconductors with different sensing properties. The sensing properties of metal oxides are usually described by the grain boundary model [56, 101–103], implying the important role of grain boundaries. The grain boundaries contain a great number of surface states that can trap or scatter free carriers, leading to a reduction in the mobility and an increase in the effective resistivity of the semiconductor [104, 105]. Zhang et al. [93] grain boundary scattering theory was employed to detail the different room-temperature $NO_2$ sensitivity of single crystalline and polycrystalline NiO nanostructures where based on the photoluminescence (PL) and the electrochemical impedance spectroscopy (EIS) spectra, the polycrystalline NiO possesses more grain boundaries compared with the single-crystalline NiO. For the p-type NiO, the conductivity is expressed in Eq. (13.3) as follows:

$$\sigma = pq\mu_p \tag{13.3}$$

where $p$ is the hole density and $\mu_p$ is the carrier mobility. The sensitivity is not only dependent on the hole density but also relies on the carrier mobility. As for the single-crystalline and polycrystalline NiO, the extent of surface band bending associated with the amount of nickel vacancies is nearly equal to each other, suggesting higher hole density in single-crystalline NiO (SC NiO) and polycrystalline NiO (PC NiO) after exposure to $NO_2$ was also equal. In this case, the different sensitivity of SC NiO and PC NiO is explained on the basis of the differences in carrier mobility, which depends on the carrier scattering, and the relationship between the carrier mobility and the scattering potential [106] is expressed in Eq. (13.4):

$$\mu_p = \mu_o \exp\left(-\frac{\varphi}{kT}\right) \tag{13.4}$$

where $\varphi$ is the scattering potential. Therefore, a larger number of extra grain boundaries in polycrystalline NiO lead to a larger scattering potential, significantly reducing its carrier mobility. The grain boundary scattering mechanism based on SC NiO and PC NiO may enhance the sensing mechanism, accounting for a better sensitivity. Point defects, responsible for control and optimization of the sensing properties, represent a deviation from the long-range periodicity of the crystal lattice [107]. There are many types of point defects, such as cation vacancies, anion vacancies, interstitial impurities, substitutional impurities, etc. Concerning the effect of point defects on the sensing properties of metal oxides, many researchers put the emphasis on the oxygen vacancies, especially for the n-type metal oxides [94, 108–110]. For the oxygen ionosorption model, the larger number of oxygen vacancies is beneficial for the formation of oxygen species on the surface of sensing materials so that more oxygen species participate in the gas-sensing reaction for obtaining an enhanced sensitivity, whereas for the oxygen vacancy model, the increased oxygen vacancies offer enough active sites so as to increase the gas adsorption for enhanced sensitivity. The effect of oxygen vacancies on the sensing properties of metal oxides has been widely reported [111], which is attributed to the larger number of oxygen vacancies based on the former two models. For example, Epifani et al. [112] prepared $SnO_2$ nanocrystals for $NO_2$ sensing and carefully analyzed the effect of oxygen vacancies on the sensing properties. The oxygen vacancy sites that interact with $NO_2$ in presence of bridging oxygen vacancies strongly promote the charge transfer process from the surface to $NO_2$. Ahn et al. [113] fabricated network-structure ZnO nanowires and investigated the correlation between the oxygen vacancy-related defects and gas sensitivity where the gas sensitivity is proportional to the oxygen vacancy density. However, despite the many researchers who have paid attention in finding the relationship between the oxygen vacancies and the gas-sensing properties, the effect of metal cation defects on the sensing properties is rate, especially due to the fact that the metal cation defects are the main defect types in p-type metal oxides.

Zhang et al. [88] prepared mesoporous NiO with different amounts of nickel vacancies as room-temperature $NO_2$ sensor where NiO with a higher concentration of nickel vacancies exhibits a higher sensitivity. The nickel vacancies act as the dominant active sites participating in the gas–solid reaction. Due to the equivalent defect ionization rate at room temperature, NiO with a higher concentration of nickel vacancies facilitates the ionization of neutral nickel vacancies followed by participation of charged nickel vacancies in the gas-sensing reaction. Due to a reaction between the charged nickel vacancies and the $NO_2$ molecules, charged nickel vacancies concentration decreases. More neutral nickel vacancies are facilitated to ionize into charged nickel vacancies, leading to an increase in the hole concentration followed sensitivity (Figure 13.6).

Except for the effect of metal cation defect, the sensing properties of metal oxides can also be influenced by introducing dopants. When doping with elements, such as transition or non-transition metal ions, the catalytic properties, electric conductivity, or the defect concentration of metal oxides are changed with an improved sensitivity and selectivity. Kaur et al. prepared indium-doped $SnO_2$ by the sol–gel spin coating process where on indium doping the sensor response and selectivity toward $NO_2$ gas are improved [114]. Therefore, the manipulation of the defect structure, such as grain boundary, oxygen vacancies, cation vacancies, and dopants, improves the sensing properties of metal oxides.

### 13.4.3.2 Heterointerface Structure

Introducing another phase to form a heterointerface structure, such as loading noble metals, compositing with another metal oxide, is also an efficient method to improve the sensing properties of metal oxides. As for the noble metal additives, their role is to increase the reaction rate of gas molecules. The effect of the addition of many different noble metals on the gas-sensing properties of metal oxides has been widely studied [115–117]. Platinum (Pt)-modified NiO demonstrates an enhancement of the response toward $H_2$ as compared with the unmodified NiO samples due to the catalytic properties of Pt toward $H_2$ [69]. Effect of gold (Au) sensitization on the liquefied petroleum gas-sensing properties for ZnO nanorods has been reported by Nakate et al. [118]. The gas response of ZnO is greatly improved on Au doping, which is attributed to (i) the spillover mechanism, (ii) the synergistic effects induced by the coupling, and (iii) the heterointerface between two different classes of nanomaterials, suggesting the importance of heterostructured sensors over single one.

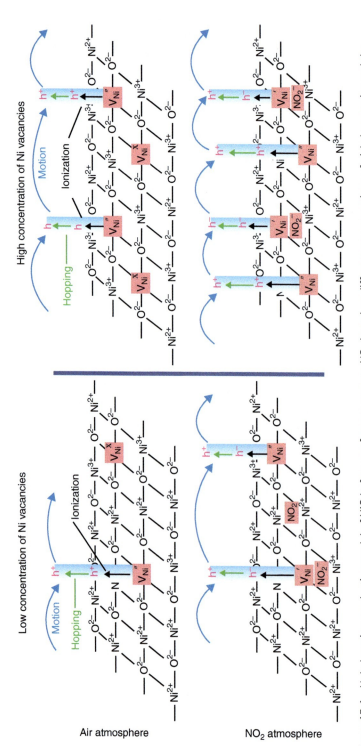

**Figure 13.6** Hole hopping conduction model of NiO before or after exposure to $NO_2$ based on different concentrations of nickel vacancies and the corresponding defect ionization and gas-sensing reaction equation. Source: Zhang et al. [88]. Reproduced with permission of American Chemical Society.

Therefore, the construction of a p–n heterojunction or an n–n/p–p homo-type heterojunction has become an efficient strategy to improve the sensing properties of metal oxides. In the p–n heterojunction, electrons move from the n-type MONs to the p-type MONs, and holes are transferred from the p-type MONs to the n-type MONs until the Fermi level of the nanocomposite reaches to equilibrium. Therefore, the electron depletion layer (EDL) at the interface of the heterojunction expands, and the resistance increases. In addition, when the material is in the environment of a reducing gas, the target gas reacts with the adsorbed oxygen on the surface of the material and transports the electrons back to the conduction band of the n-type MONs. Some electrons enter into the conduction band of p-type MONs. This process causes electron–hole recombination, resulting in a decrease in hole concentration and an increase in electrons in the p-type MONs. Therefore, the carrier concentration on both sides of the p–n junction is reduced. The barrier at the interface is reduced due to the limitation of carrier diffusion. The EDL shrinks more than without doping, and at the macroscopic level, the performance improvement manifests as an enhancement in the material conductivity. In fact, both n–n junctions and p–p junctions are due to the difference in work functions between the two MONs. In order to maintain balance in the system, the movement of electrons or holes occurs, so the amount of change in the electrical signal is amplified [119–121]. The presence of heterojunctions helps in providing more adsorption and reaction sites, resulting in better catalytic activity than that obtained with monomers [122].

For instance, novel NiO@ZnO heterostructured nanotubes fabricated by the co-electrospinning method exhibits highly improved sensing performance compared with pristine ZnO and NiO [123]. The heterostructured NiO–SnO$_2$ nanocomposites endow an enhanced sensing property toward toluene due to combination of p-type NiO and n-type ZnO [124].

The two competing mechanisms, one change in the potential energy barrier encountered by an electron crossing a heterojunction interface and another narrowing of the charge conduction channel, are accepted [61]. To understand further details of the sensing mechanism of heterostructure-based sensors, the contact potential, which is an electrostatic potential generated locally across the contact to compensate the Fermi level difference having existed between the isolated grains, is measured [59, 60]. The generation of contact potential attenuates drift mobility of electrons running against it and thus influencing the relevant contact resistance. The resistance of heterostructured sensing materials is influenced dually by the ambient gases, first through the electron density term and second through the mobility term, yielding an enhancement of the transducer function up to the square of that of the sensing materials composed of the only one kind of semiconductor. Moreover, this guideline has also been extracted for designing a contact potential-promoted gas sensor with higher sensing performance, such as the larger difference in the work function between two grains, easy electron motion through hetero-contacts, and the larger specific surface area ensuring the sensitive receptor function. Self-assembled NiO–SnO$_2$ heterojunction nanocomposite sensors were constructed as room-temperature gas sensors [125]. The carrier transport at the heterointerface via the interface bonds and the charge transfer between the adsorbed $NO_2$ molecules and the heterostructure is involved.

### 13.4.4 Chemical Composition

Semiconducting metal oxides are being investigated extensively at elevated temperatures for the detection of simple gases [126]. There are many parameters of materials for gas sensor applications, for example, adsorption ability, catalytic activity, sensitivity, thermodynamic stability, etc. Different metal-oxide materials demonstrate these properties, but very few of them are suitable to all requirements. For this situation, research on composite materials, such as $SnO_2$–ZnO [127, 128], $Fe_2O_3$–ZnO [129], ZnO–CuO [130], etc. is at the front line. In addition to binary oxides, numerous ternaries, quaternary, and composite metal oxides are envisaged in sensor application [131, 132]. The composite ZnO–$SnO_2$ sensors exhibit higher sensitivity than sensors constructed solely from $SnO_2$ or ZnO when tested under identical experimental conditions [129]. Due to a synergistic effect between the two components, sensors based on the two components mixed together are more sensitive than the individual components. Details about the synergistic effect is still unknown, but de Lacy Costello et al. [128] have suggested a possible mechanism. Taking $SnO_2$–ZnO binary oxide butanol sensors as an example, they hypothesize that butanol is more effectively dehydrogenated to butanol by $SnO_2$, but that $SnO_2$ is relatively ineffective in the catalytic breakdown of butanol, i.e. ZnO catalyzes the breakdown of butanol. A combination of the two materials would effectively dehydrogenate butanol and then subsequently catalyze the breakdown of butanol. It is to be noted that not all composite gas sensors adduce better performance than their individuals. Only when the catalytic action of the components complements each other, the performance of gas sensors is increased. However, composite sensors comprising mixtures of ZnO and $In_2O_3$ demonstrate a reduction in sensitivity over equivalent single oxide sensors. In addition to the synergistic effect, heterojunction interface between two or more components also contributes to the enhancement of the composite gas sensor performance [133–138]. The principle of formation of heterojunction barriers in air ambient and their disruption on exposure to target gas is employed. So the resistance and proportion of p–n heterojunctions in the composite gas sensor becomes a control factor to the gas sensor performance. Furthermore, it has been shown that changing the proportions of each material in the composite yields a wide range of sensor materials with very different sensing characteristics.

### 13.4.5 Addition of Noble Metal Particles

In many gas sensors, the conductivity response is determined by the efficiency of catalytic reactions with detected gas participation, taking place at the surface of gas-sensing material. Therefore, control of catalytic activity of gas sensor material is one of the most commonly used means to enhance the performance. However, in practice, the widely used gas-sensing metal-oxide materials such as $TiO_2$, ZnO, $SnO_2$, $Cu_2O$, $Ga_2O_3$, and $Fe_2O_3$ are the least active with catalytic point of view [26]. The pure $SnO_2$ sensor without any catalyst exhibits a very poor sensitivity (~3) [139]. Noble metals are high-effective oxidation catalysts, and this ability can be used to enhance the reactions on gas sensor surface. A wide diversity of methods, including impregnation, sol–gel, sputtering, and thermal evaporation, are being used for introducing noble metal additives into oxide semiconductors. Different doping states can be obtained by different methods. Mixture of noble metal particles and metal oxides are obtained by sol–gel method, while metal oxides modified by noble metal particles on the surface are obtained by sputtering or thermal evaporation. In several reports, an enhancement of sensitivity modified by noble metals, such as Pt, Au, Pd, Ag, etc. is reported [140–145].

Kim et al. [7] reported a ZnO nanowires (NWs) synthesized by using a VLS method and then functionalized with Pt and Pd NPs by means of sputtering and UV irradiation reduction methods, respectively. Gas-sensing studies indicated that the annealing temperature should be optimized to achieve the best sensing performance of the Pt-functionalized gas sensor. Further, for the Pd functionalized gas sensor, the UV irradiation time used for the reduction of Pd precursor had a significant effect on the final gas response. In the self-heating mode, optimized Pt- and Pd-functionalized gas sensors selectively detected toluene and benzene gases, respectively [7]. As shown in Figure 13.7, the Pt/ZnO or Pd/ZnO sensors are exposed to benzene or toluene gases; electrons will return to the ZnO surface owing to the reaction between the reducing gas and oxygen, thereby increasing the surface electron density. The elevation of the ZnO Fermi level will increase the height of the Schottky barrier. Accordingly, an electron cannot easily escape to Pt or Pd, and a further increase in the resistance of ZnO cannot contribute to any enhancement of the sensor response. Moreover, the Pt and Pd NPs can play a catalytic role in decreasing the energy barrier for the adsorption of the toluene and benzene gases to the sensor surface, thereby facilitating the adsorption of these gases, and then by means of the so-called spillover effect [146] transfer the gas molecules onto the ZnO NWs. The oxidation of benzene and toluene with oxygen is enhanced owing to the catalytic activities of Pd and Pt. For instance, in the case of sensing of toluene with Pt/ZnO, toluene is first chemisorbed onto the Pt/ZnO surface sites and then reacts with the adsorbed oxygen species to form benzaldehyde, which is further converted to benzoate species, and the increased decomposition of benzoate species is the main contributor to the catalytic activity of Pt in toluene oxidation. Subsequently, the benzoate species are decomposed to carboxylates and carboxylic acid, and these are further converted to anhydrides and carboxylates. The anhydrides are subsequently converted to adsorbed $H_2O$ and CO species, and the latter can react with adsorbed oxygen to produce $CO_2$ [147]. One of the main parameters that leads to the good response of a gas sensor toward a specific gas is the strength of chemisorption/adsorption between the catalytic component (noble metal) and the target gas. When there is a strong bonding between the catalyst surface and the target gas molecules, the catalytic activity decreases, since high energy is required to break the bonds. When the bonding is weak, the catalytic activity is once again decreased, since the gas molecules do not stick to the surface long enough to promote the subsequent reaction. When the bonding strength is moderate, the molecules have enough flexibility to move around and form transition states [148]. Moreover, the reasons for

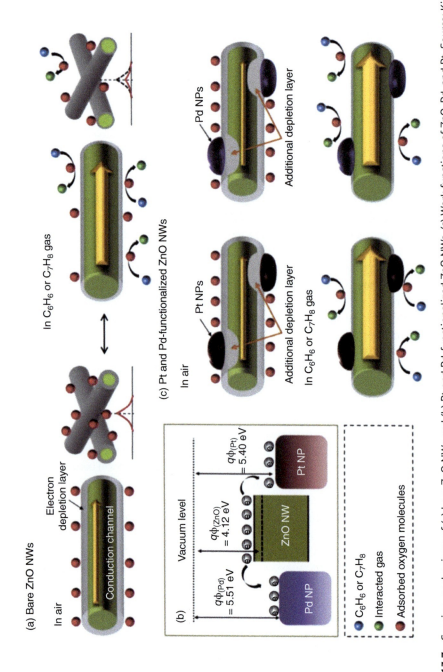

**Figure 13.7** Sensing mechanisms of (a) bare ZnO NWs and (b) Pt and Pd-functionalized ZnO NWs. (c) Work functions of ZnO, Pd, and Pt. Source: Kim et al. [7]. Reproduced with permission of Elsevier.

the good selectivity of these gas sensors toward toluene and benzene is attributed to the unique catalytic effects of Pt and Pd to toluene and benzene, respectively. By the careful selection of noble metals for functionalizing the active sensor material, selective gas sensors that can operate with low power consumption can be fabricated for the detection of VOCs [7].

### 13.4.6 Humidity and Temperature

Environmental humidity is an important factor influencing the performance of metal-oxide gas sensors, as many humidity gas sensors based on metal oxides are commercialized. However, mechanism of sensing water vapor and other pollution gases such as CO, $NO_2$, and $H_2S$ is different. For metal-oxide humidity gas sensors, ionic-type humidity sensors are the most common patterns. The conduction mechanism depends on $H^+$ or $H_3O^+$, from dissociation of adsorption water, which hops between adjacent hydroxyl groups. Details about the adsorption of water on metal-oxide surfaces and mechanism of sensing water vapor are given in [149] and [150] references. Water adsorption on the metal-oxide surface does not donate electrons to sensing layers. Moreover, as explained in [151] and [152] references, it will lower the sensitivity of metal-oxide sensors. This is attributed to the reaction between the surface oxygen and the water molecules responsible to a decrease in baseline resistance of the gas sensor [152], and the adsorption of water molecules leads to less chemisorption of oxygen species on the $SnO_2$ surface due to the decrease of the surface area that is responsible for the sensor response. On the other hand, water molecules also act as a barrier against $C_2H_2$ adsorption. The superficial migration of the $C_2H_2$ on the $SnO_2$ surface becomes difficult; thus the sensitivity decreases and the response and recovery times increases. Figure 13.8 shows the schematic of the effect of humidity on the $C_2H_2$ sensing properties.

Water adsorption significantly lowers the sensitivity of metal-oxide gas sensors (Figure 13.9a). Prolonged exposure to humid environments leading to the gradual formation of stable chemisorbed $OH^-$ on the surface [149] causes a progressive deterioration of the sensitivity of gas sensors. However, surface hydroxyls start to desorb at about 400 °C [154], and the hydroxyl ions are removed by heating to temperatures higher than 400 °C. As seen in [155], after several humidity pulses the sensor resistance does not recover to its initial level; however, subsequent heating up the sensor to a temperature of 450 °C for several minutes leads to a full recovery of the signal. In the conclusions of [145], water strongly inhibits methane combustion even at low partial pressures. The temperature is also an important factor for the metal-oxide gas sensors. Typical curves of gas response vs. temperature are shown in Figure 13.9b. Gas sensors with different compositions reveal a similar shape. The responses increase and reach their maximum levels at a certain temperature, and then decreases rapidly with increasing the temperature, as typically observed [153, 156–159].

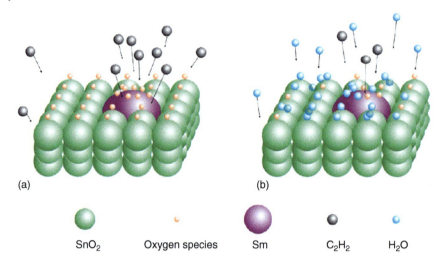

**Figure 13.8** Gas-sensing mechanism of $Sm_2O_3$-doped $SnO_2$ in the atmosphere of (a) $C_2H_2$ and (b) $C_2H_2$ and humidity. Source: Qi et al. [151]. Reproduced with permission of Elsevier.

## 13.5 Gas-Sensing Mechanisms

The common MON-based gas sensors detect a gas because of the change in electrical signal caused by the gas. Gas-sensing mechanisms explain why the gas can cause changes in the electrical properties of a sensor in presence of gas molecules. The widely used gas-sensing mechanism is based on the changes in electrical properties from a relatively microscopic perspective including Fermi level control theory, grain boundary barrier control theory, and EDL/HAL theory, which are basically theoretical, and, in any application, changes in electrical properties are necessarily accompanied by changes in physical properties such as energy bands and work functions. The adsorption/desorption model, the bulk resistance control mechanism and the gas diffusion control mechanism, belong to this kind of theory. These theories enable better analysis of the process of gas-sensing reactions by modern material analysis techniques based on the actual physical phenomena. These latter models can also be used to explain some of the physics principles mentioned in the previous theory, such as why bands move and bend or why the grain boundary barrier changes.

### 13.5.1 Adsorption/Desorption Model

As early as the 1930s, researchers discovered that the conductance and work function of a MONs change when the material is in contact with gases such as $O_2$ and CO. In fact, the adsorption/desorption model is the basis of most of the current main stream gas-sensing mechanisms. This model is based on the chemical or physical adsorption/desorption of the gas on the surface of a material, resulting in a change in resistance due to the change in charge carrier concentration. The most common oxygen adsorption model, chemical adsorption/desorption without considering oxygen adsorption conditions, and physical adsorption/desorption are briefed in succinct.

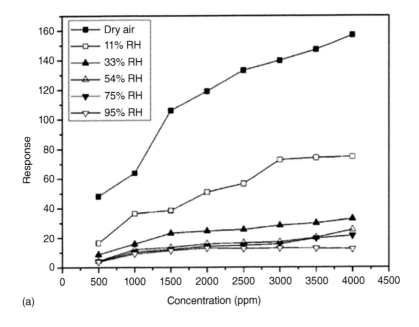

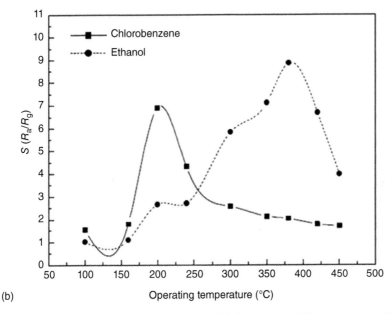

**Figure 13.9** (a) Response of the $Sm_2O_3$-doped $SnO_2$ sensor to different concentrations of $C_2H_2$ at different relative humidity (RH). Source: Qi et al. [151]. Reproduced with permission of Elsevier. (b) Gas response vs. operating temperature of porous ZnO nanoplate sensor to 100 ppm chlorobenzene and ethanol. Source: Jind and Zhan [153]. Reproduced with permission of John Wiley and Sons.

### 13.5.1.1 Oxygen Adsorption Model

The oxygen adsorption model is for almost all MON-based gas-sensing materials, and EDL and HAL theories are extensions of the oxygen adsorption model. When MONs are exposed to air, oxygen molecules are adsorbed on the surface of the material. As electron donor, an EDL will generate in an n-type oxide semiconductor, while an HAL in a p-type oxide semiconductor. As a result, high-resistance EDL exists at the shells of the particles in Figure 13.10a, and low-resistance HAL exists near the surface of the particles in Figure 13.10b. Accordingly, n-type and p-type semiconductors yield different performances due to their diverse charge-flowing paths [160].

Li et al. [161] used nanosheet-assembled hierarchical $SnO_2$ nanostructures to fabricate a simple integrated device that was effective at monitoring the presence of

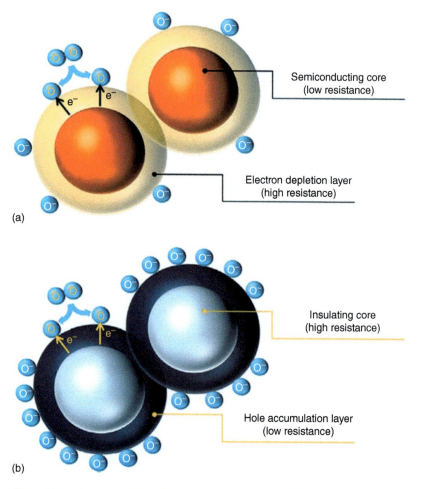

**Figure 13.10** Formation of electronic core–shell structures in (a) n-type and (b) p-type oxide semiconductors. Source: Cui et al. [160] https://ietresearch.onlinelibrary.wiley.com/doi/10.1049/hve.2019.0130 (last accessed 12/07/2021) CC BY 3.0.

beer. They noted that the surface of $SnO_2$, which should theoretically exist in the inert form, acted as a highly doped semiconductor with oxygen vacancies at high temperatures and in reducing environments.

$$O_0 = V_{\ddot{O}} + 2e' + \frac{1}{2}O_2(g) \tag{13.5}$$

As shown in Figure 13.11a, every vacancy is associated with a filled donor state that is below the conduction band edge. When oxygen molecules are adsorbed on the surface of $SnO_2$, they will occupy the previously formed oxygen vacancies and

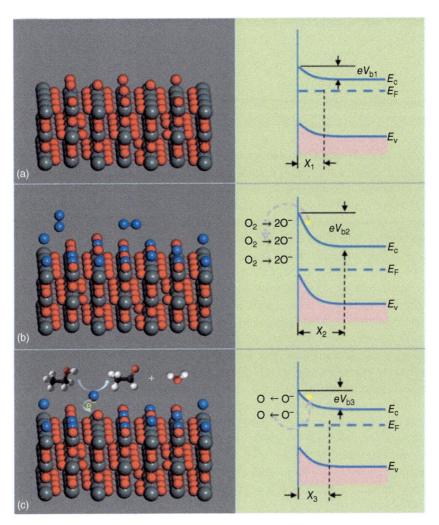

**Figure 13.11** (a–c) Schematic illustration and corresponding energy band diagram of $SnO_2$ during the gas-sensing reaction process. Source: Li et al. [161]. Reproduced with permission of Elsevier.

take electrons from the conduction band to form negative oxygen species as follows:

$$O_2(g) = O_2(ad) \tag{13.6}$$

$$O_2(ad) + e^- = O_2^-(ad) \tag{13.7}$$

$$O_2^-(ad) + e^- = 2O^-(ad) \tag{13.8}$$

$$O^-(ad) + e^- = O^{2-} \tag{13.9}$$

Since the electrons in the conduction band are consumed, an EDL is formed. When the thickness of nanoscale $SnO_2$ is less than twice the Debye length ($L_D$), the EDL will extend to the entire material, and thus the electrical signal changes will reach a maximum.

$$L_D = \sqrt{\frac{\varepsilon k_B T}{q^2 N_D}} \tag{13.10}$$

Furthermore, as shown in Figure 13.11b, oxygen adsorption causes the Fermi level to move down and the conduction band to bend upward. When the target gas enters and undergoes a redox reaction with the negative oxygen species, the electrons are rereleased back into the conduction band, the EDL shrinks, the oxygen vacancies reform, the Fermi level and the conduction band return to normal, and eventually the resistivity is restored (Figure 13.11c).

Moreover, Xiao et al. [162] synthesized ordered mesoporous $SnO_2$ exhibits excellent $H_2S$ sensing properties and superior stability. For interpretation a grain boundary potential barrier control model, based on an oxygen adsorption model, is used. In air, oxygen molecules begin to adsorb on the grain surface at a small scale. Due to the porous structure of the material, the oxygen molecules gradually diffuse and eventually cover all the grains. In addition, the oxygen molecules take electrons from the grains during adsorption and are converted into oxygen anions. The loss of electrons naturally causes the generation of an EDL and the formation of a potential barrier between adjacent grains, which in turn leads to an increase in the resistance toward electrons flowing at grain boundaries. When this sensor is exposed to $H_2S$, the oxyanions react with $H_2S$ and deliver a large number of electrons to the grains. Therefore, the EDL shrinks, the barrier between adjacent crystal grains is lowered, the electron flow becomes smooth again, and the electric resistance is also decreased. The previously mentioned process is clearly shown in Figure 13.12. However, they also noted that the reason why $SnO_2$-based gas sensors undergo a change in resistance during the $H_2S$ sensing process is due to not only oxygen adsorption but also chemical adsorption between $SnO_2$ and $H_2S$. This aspect will be discussed in Section 13.5.1.2. In addition to n-type MONs, the oxygen adsorption model is also applicable to p-type MONs.

Shinde et al. [13] prepared 3D α-$MnO_2$ mesoporous cubes composed of interconnected nanocrystallites via a hydrothermal method and found that the resulting α-$MnO_2$ sensor demonstrated good selectivity to $NO_2$ at 150 °C operating temperature. The basic sensing mechanism of α-$MnO_2$-based sensors is the same as that in most MON-based devices due to the resistance change caused by the

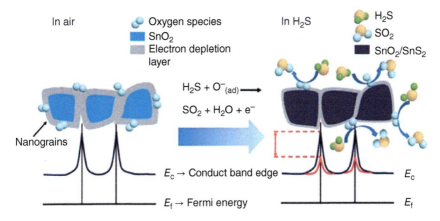

**Figure 13.12** Schematic diagram of the $H_2S$-sensing mechanism of sensors based on mesoporous $SnO_2$. Source: Xiao et al. [162]. Reproduced with permission of American Chemical Society.

adsorption/desorption process on the sensor surface and that the sudden decrease in resistance of α-$MnO_2$ upon the interaction of $NO_2$ was observed, which is mainly attributable to the adsorption/desorption procedure of $NO_2$ molecules on the surface of α-$MnO_2$ sensing material. The observed sensing performance of α-$MnO_2$ sensor was due to its distinguishing morphology 3D mesoporous interconnected cubes. This could provide high free surface as well as interconnected crystallites support for direct transformation. When a characteristic n-type $MnO_2$ sensor was exposed to air, oxygen molecules from air could adsorb onto the sensor surfaces by capturing electrons from α-$MnO_2$ conduction band, and consequently forming adsorbed oxygen species such as $O^-$, $O_2^-$, and $O^{2-}$ [163] on the α-$MnO_2$ surface (refer to Eqs. (13.6)–(13.9)). This could lead to the formation of EDL (Figure 13.13). When reducing $NO_2$ gas molecules that are in contact with α-$MnO_2$ surface, the formerly adsorbed surface oxygen species interact with the $NO_2$ molecules by releasing the captured electrons back into the conduction band of α-$MnO_2$. As a result, resistance of α-$MnO_2$ sensor decreases due to the electron donating nature of $NO_2$, which decreases a thickness of EDL. The schematic of gas-sensing mechanism between α-$MnO_2$ sensor and $NO_2$ is shown in Figure 13.13, and possible reaction mechanism is as follows:

$$NO_2 \text{ (gas)} + e^- \rightarrow NO_2^- \text{ (ads)} \quad (13.11)$$

$$NO_2^- \text{ (ads)} + O^- \text{ (ads)} + 2e^- \rightarrow NO \text{ (gas)} + 2O^{2-} \text{ (ads)} \quad (13.12)$$

### 13.5.1.2 Chemical Adsorption/Desorption

According to the chemical adsorption/desorption model, the gas is in direct contact with the crystal grains and undergoes a chemical reaction, inducing a change in the electrical signal. In general, the oxygen adsorption model is a kind of chemical adsorption/desorption model, but oxygen adsorption has already been widely described. To better elucidate these models, this chapter will explain chemical

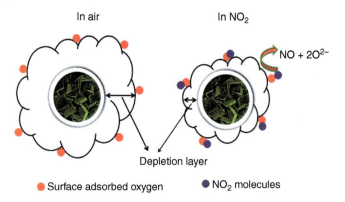

**Figure 13.13** Schematic of gas-sensing mechanism between α-MnO$_2$ sensor and NO$_2$. Source: Shinde et al. [13]. Reproduced with permission of Elsevier.

adsorption/desorption separately. Most current articles discuss the gas-sensing mechanism of MONs-based sensors but simply use the direct oxygen adsorption model. In reality, however, in many cases, such analysis is incomplete. Many materials inevitably undergo chemical adsorption/desorption when exposed to a specific gas, and this phenomenon also affects the gas sensitivity of the material [164]. For example, the SnO$_2$-based gas sensor mentioned in Section 13.5.1.1 undergoes chemical adsorption during the detection of H$_2$S. To better monitor the sensing mechanism between SnO$_2$ and H$_2$S, the inductive coating of a sensor after two minutes of reaction with 50 ppm H$_2$S was analyzed by X-ray photoelectron spectroscopy (XPS), where during the sensing measurement process, in addition to reacting with the adsorbed oxygen anions, H$_2$S directly chemically adsorbs onto SnO$_2$ to produce SnS$_2$.

$$2H_2S + SnO_2 \rightarrow SnS_2 + 2H_2O \tag{13.13}$$

Due to its narrower band gap, the presence of SnS$_2$ can effectively reduce the bulk resistance of the grains, thereby increasing sensitivity. However, because it takes a certain time to convert SnS$_2$ back to SnO$_2$, this process affects the sensor's recovery properties. Zhu et al. [165] used SnO$_2$ as a gas-sensing material to measure the gas response of H$_2$ under ambient atmosphere conditions and in a vacuum environment and compared the two different gas-sensing mechanisms by a first-principle calculations. The experimental results showed that the gas-sensing properties in the two environments are very similar, even though the gas response value under vacuum was slightly higher (Figure 13.14a,b). Through the first-principle approach, a theoretical model of H$_2$ and the sample surface was established to calculate the atomic and electronic structure changes during the sensing process, and then the gas-sensing mechanism of SnO$_2$ in the different environments was analyzed. According to the density of states (DOS) of SnO$_2$ before and after H$_2$ adsorption in the two environments (Figure 13.14c,d), only the relationship between H$_2$ and adsorbed oxygen has been considered. The adsorption of H$_2$ only causes a small band shift, which has little effect on the conductivity. However, the valence band undergoes a shift to some extent when H$_2$ is directly chemisorbed onto

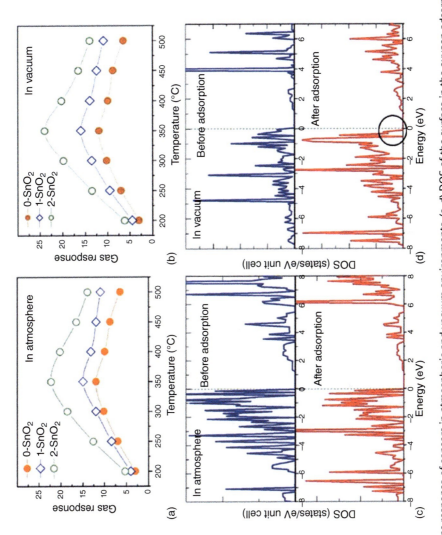

**Figure 13.14** (a, b) gas response of sensors in atmospheric and vacuum environments, (c, d) DOS of the surface in the oxygen adsorption and chemical adsorption models. Source: Zhu et al. [165]. Reproduced with permission of Elsevier.

$SnO_2$. In addition, based on the Mulliken population analysis, $H_2$ can transfer more electrons to $SnO_2$ when chemical adsorption occurs. Therefore, the chemical adsorption/desorption model is a gas-sensing mechanism that is generally overlooked [165].

### 13.5.1.3 Physical Adsorption/Desorption

Physical adsorption is the adsorption of gas molecules onto MONs crystals by Coulomb forces, hydrogen bonding, and other intermolecular forces without chemical changes. Such adsorption is a very common physical phenomenon in gas-sensing processes but is rarely been used to explain the gas-sensing mechanism. The change in conductivity of MON materials caused by pure physical adsorption is negligible. The most common type of MONs-based sensor with physical adsorption/desorption as the main gas-sensing mechanism is a humidity sensor. However, even for humidity sensors, although physical adsorption/desorption is a dominant mechanism, in most cases, the sensing mechanism is still initiated by chemical (oxygen) adsorption [33, 166, 167].

For example, quasi-cubic $Fe_2O_3$ nanostructures display a latent ability to detect low humidity as humidity sensors [168]. The sensing mechanism can be divided into two steps (Figure 13.15); (i) chemical adsorption: $H_2O$ molecules contact and interact with the surface of the sample and are dissociated to form $OH^-$, and, finally, a hydroxyl layer is obtained that occurs at very low humidity, and does not change further with increasing humidity [169] and (ii) physical adsorption: at a high humidity

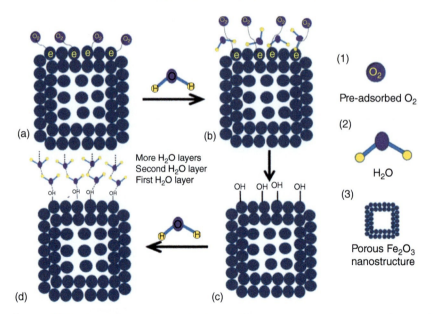

**Figure 13.15** Humidity sensing mechanism at different stages: (a) pre-adsorbed oxygen's deprive free electrons from the surface of $Fe_2O_3$ nanostructures, (b) water molecules replace the pre-adsorbed oxygen's, (c) hydroxyl groups form on the surface of nanostructures, and (d) water molecular layers adsorb on the surface of nanostructures. Source: Yu et al. [168]. Reproduced with permission of Elsevier.

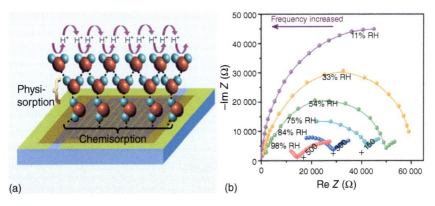

**Figure 13.16** (a) Water adsorption process on the CeO$_2$ NPs surface, (b) the complex impedance modulus spectra (Nyquist plot) based on CeO$_2$ NPs-based sensor, where RH varies from 11% to 98%. Source: Poonia et al. [31]. Reproduced with permission of Royal Society of Chemistry.

level in the environment, H$_2$O molecules are physically adsorbed onto the hydroxyl layer by hydrogen bonding to form a first H$_2$O layer. After that, new H$_2$O molecules are physically adsorbed onto the previous H$_2$O layer, and a new H$_2$O layer is formed. This cycle will increase the number of H$_2$O layers. Proton hopping occurs between adjacent H$_2$O molecules in each continuous H$_2$O layer (Figure 13.15b–d), and the eventual result is an increase in both conductivity and impedance.

To distinguish the effects of the two adsorption mechanisms in humidity sensing, Poonia et al. [31] used a CeO$_2$-based humidity sensor to perform complex impedance measurements under different humidity conditions to prove the dominant role of physical adsorption in the sensing process (Figure 13.16). In most of the cases, physical adsorption has little effect on gas-sensing performance, and is not considered for discussion. However, even for the oxygen adsorption model, at lower temperatures, oxygen molecules are present on the surface of the sample in a physically adsorbed form. Although such physical adsorption has no direct effect on the electrons, it has been shown to interfere with the directional motion of surface carriers, resulting in a small increase in resistance [170, 171].

### 13.5.2 Bulk Resistance Control Mechanism

The core point of the bulk resistance control mechanism is that the resistance change in some MON-based gas sensors is caused by a phase transformation of the gas-sensing material. However, the mechanism is relatively narrow in scope, and is only applicable for the analysis of the gas-sensing processes of materials such as metal oxides, binary, ternary, and chemical composite oxides of MONs [172]. Wang et al. [173] prepared Fe$_2$O$_3$ with different phase compositions of α-Fe$_2$O$_3$ and γ-Fe$_2$O$_3$ by heating Fe-MOF (FeFe(CN)$_6$) templates. Gas-sensing tests were carried out based on the two materials, and their mixed phases and the different gas-sensing mechanisms of the two iron oxides were analyzed. Since α-Fe$_2$O$_3$ has a relatively stable phase structure, its sensing mechanism is explained by a typical oxygen adsorption model [174–176]. Zhang and coworkers [177] proposed the effect of phase transition on the stability of $WO_3$ hydrogen sensing films. By analyzing the results of infrared (IR) and Raman spectroscopy with first-principle calculations, the changes in the $WO_3$ film in the $H_2$ sensing process are screened. Initially, the material exists in the form of $W_3O_{12}$ clusters and is assembled to form $(W_3O_{12})_3$ three-member rings by sharing the W—O—W bonds at the corners. Finally, these three-member ring clusters collapse and produce a monoclinic phase of $O_3$. During the previously mentioned transitions, the highest occupied molecular orbital (HOMO)–lowest unoccupied molecular orbital (LUMO) gap and hydrogen desorption energy also change with changing the bulk resistance for the $H_2$ sensing performance.

### 13.5.3 Gas Diffusion Control Mechanism

There are two components, materials and gases, in the process of gas-sensing. While the first two mechanisms focus more on the physical and chemical properties of the material, this mechanism is more focused on the gas diffusion process, and the most important factor affecting this process is the morphology of the material. As early as the 1990s, scientists proposed the theory that gas diffusion controls the sensitivity of semiconductor gas sensors [85, 178, 179]. In fact, this theory can be used as a supplement to the first two mechanisms to fully explain the changes in physical parameters during the gas-sensing process. Wang et al. [180] used experiments with hierarchical porous $SnO_2$ microspheres to construct a pore canal model and a hollow sphere model, which comprehensively explained the relationship between the surface chemical reaction (SCR) and gas diffusion and their effects on gas-sensing properties. According to Knudsen diffusion theory and reaction kinetics, the gas diffusion rate and the rate of the SCR are represented by Eqs. (13.14) and (13.15), respectively, where $r$ is the pore radius, $R$ is the gas constant [8.314 J (mol K)$^{-1}$], $T$ is the temperature, $M$ is the molecular weight, $A$ is the pre-exponential factor, and $E_a$ is the activation energy.

$$D_k = \frac{4r}{3}\sqrt{\frac{2RT}{\pi M}} \qquad (13.14)$$

$$k = A\exp\left(-\frac{Ea}{RT}\right) \qquad (13.15)$$

According to these two formulas, $k$ is primarily affected by the temperature, so the gas-sensing process can be divided into three stages; (i) $T < 180\,°C$: the SCR is the rate determining step (RDS), and the external surface area is the main factor that affects the gas-sensing properties, (ii) $180\,°C < T < 260\,°C$: the effect of the two phenomena on gas sensitivity is similar, and (c) $260\,°C < T$: gas diffusion is the RDS, and, at this stage, the pore size has a greater impact on the gas-sensing properties. Pore canal models with different pore sizes (Figure 13.17a) can be used to analyze the gas diffusion process within $SnO_2$ hollow spheres. Gas diffusion is controlled by the gas molecular mean free path (GMFP, $\lambda$), which $\lambda$ decreases upon increasing the operation temperature. Therefore, at a lower temperature, the gas is mainly adsorbed

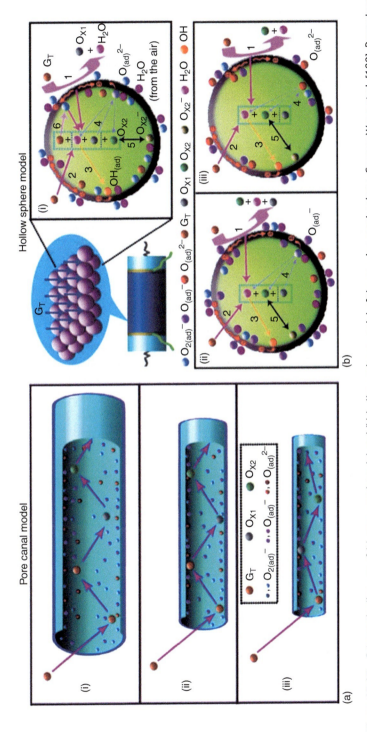

**Figure 13.17** Schematic diagram of (a) pore canal model and (b) hollow sphere model of the sensing mechanisms. Source: Wang et al. [180]. Reproduced with permission of American Chemical Society.

onto the outer surface. As the temperature increases, the probability of gas molecules entering the inner-pore canals is enhanced. This relationship explains why the gas sensitive reaction is divided into the three stages, as mentioned earlier. A hollow sphere model as shown in Figure 13.17b is proposed for a better understanding.

At lower temperatures ($T < 150\,°C$), the adsorbed oxygen on the surface of $SnO_2$ is mainly poorly reactive $O_2^-$, so the SCR is inactive, limiting the gas-sensitive reaction. Some of the target gas is partially oxidized on the outer surface, and the remaining gas diffuses into the interior of the pores to undergo chemical reaction, including partial oxidation, complete oxidation, and ionization of a small amount of oxidation products. In the transition ($150\,°C < T < 200\,°C$) and high-temperature ($T > 200\,°C$) ranges, oxygen anions are mainly present in the form of $O^-$ and $O^{2-}$, and the SCR is more active, making the target gas completely oxidized and ionized inside the hollow sphere. In this process, gas diffusion gradually becomes the RDS. Although the gas diffusion control mechanism still needs to be optimized in terms of quantitative analysis of multiple factors, it has widely been used to evaluate the effect of the morphology of MONs gas-sensing materials on their gas-sensing properties [28, 181, 182].

## 13.6 Conclusions and Future Perspectives

Due to the high stability, the low production cost, and the simplicity of their use, MONs have attracted great attention during the past few decades as promising gas-sensing materials. In this chapter the recent development of MONs-based gas sensors, including the screening, the preparation of sensing materials, the sensing mechanism, and material–sensor integration, has thoroughly been briefed. The gas-sensing process is strongly related to the surface reactions. Different metal-oxide-based materials demonstrate different reaction activation energies to the target gases. Composite metal oxides usually endow a better gas response than the single component if the catalytic actions of the components complement each other. Noble metal additives with high-effective oxidation catalytic activity are preferred to enhance the sensitivity of pure metal oxides due to the "spillover effect." Although a great success has been achieved with MONs-based gas sensors over the past few decades, the design of sensors, the sensing properties, and the sensing mechanism need to be further explored to satisfy the actual requirements in various fields, such as the following:

1. More experiments with modern techniques of materials analysis such as in situ analysis to quantify the effects of different gas-sensing mechanisms on gas-sensing measurement will help the researchers to select an appropriate gas-sensing mechanism more accurately and comprehensively.
2. Through a combination of various gas-sensing mechanisms, the selectivity and repeatability of MONs gas-sensing materials need to be judged to some extent.

3. The mechanism is the essence of the entire gas-sensitive process, so by evaluating the mechanisms involved, one can better understand the existing strategies for enhancing gas sensitivity.
4. Similarly, one can also find new strategies or improve existing strategies by analyzing the mechanisms to balance the sacrifices in other parameters caused by improving a kind of gas-sensing indicator.
5. There are still some missing phenomena in the gas-sensing process that need to be addressed with the current gas-sensing mechanisms, for example, why the response–recovery curves of target gases are not always level. It is necessary to propose a new gas-sensing mechanism or to conduct more in-depth research on existing gas-sensing mechanisms for better understanding.
6. In the field of medical diagnosis based on human breath, gas detection in harsh environments, etc., the excellent selectivity, the low power consumption, the fast response/recovery rate, the low humidity dependence, and the low limit of detection concentration should be fulfilled, and sensors satisfying these requirements need be to be developed.

In addition to the strategies detailed previously, the gas sensitivity of a sensor material needs to be improved by various methods, such as light activation, conductive polymer doping, and morphology control. Furthermore, through viewing the essence of the phenomenon, one can easily understand the basic ideas of this design strategy from several aspects of the gas-sensing mechanism.

## References

1 Hoa, N.D., Duy, N.V., El-Safty, S.A., and Hieu, N.V. (2015). Meso-/nanoporous semiconducting metal oxides for gas sensor applications. *J. Nanomater.* 2015: 1–14.
2 Navale, S.T., Jadhav, V.V., Tehare, K.K. et al. (2017). Solid-state synthesis strategy of ZnO nanoparticles for the rapid detection of hazardous $Cl_2$. *Sens. Actuators, B* 238: 1102–1110.
3 Kulandaisamy, A.J., Reddy, J.R., Srinivasan, P. et al. (2016). Room temperature ammonia sensing properties of ZnO thin films grown by spray pyrolysis: effect of Mg doping. *J. Alloys Compd.* 688: 422–429.
4 Yang, X., Li, H., Li, T. et al. (2019). Highly efficient ethanol gas sensor based on hierarchical $SnO_2$/$Zn_2SnO_4$ porous spheres. *Sens. Actuators, B* 282: 339–346.
5 Van den Broek, J., Abegg, S., Pratsinis, S.E., and Güntner, A.T. (2019). Highly selective detection of methanol over ethanol by a handheld gas sensor. *Nat. Commun.* 10 (1): 1–8.
6 Na, H.B., Zhang, X.F., Deng, Z.P. et al. (2019). Large-scale synthesis of hierarchically porous ZnO hollow tubule for fast response to ppb-level $H_2S$ gas. *ACS Appl. Mater. Interfaces* 11 (12): 11627–11635.

**7** Kim, J.H., Lee, J.H., Park, Y. et al. (2019). Toluene-and benzene-selective gas sensors based on Pt-and Pd-functionalized ZnO nanowires in self-heating mode. *Sens. Actuators, B* 294: 78–88.

**8** Shinde, P.V., Ghule, B.G., Shinde, N.M. et al. (2018). Room-temperature successive ion transfer chemical synthesis and the efficient acetone gas sensor and electrochemical energy storage applications of $Bi_2O_3$ nanostructures. *New J. Chem.* 42 (15): 12530–12538.

**9** Ghule, B.G., Shaikh, S.F., Shinde, N.M. et al. (2018). Promoted room-temperature LPG gas sensor activities of graphene oxide@ $Fe_2O_3$ composite sensor over individuals. *Mater. Res. Express* 5 (12): 125001.

**10** Shinde, P.V., Ghule, B.G., Shaikh, S.F. et al. (2019). Microwave-assisted hierarchical bismuth oxide worm-like nanostructured films as room-temperature hydrogen gas sensors. *J. Alloys Compd.* 802: 244–251.

**11** Mahajan, S. and Jagtap, S. (2020). Metal-oxide semiconductors for carbon monoxide (CO) gas sensing: a review. *Appl. Mater. Today* 18: 100483.

**12** Shinde, P.V., Shinde, N.M., Shaikh, S.F. et al. (2020). Room-temperature synthesis and $CO_2$-gas sensitivity of bismuth oxide nanosensors. *RSC Adv.* 10 (29): 17217–17227.

**13** Shinde, P.V., Xia, Q.X., Ghule, B.G. et al. (2018). Hydrothermally grown α-$MnO_2$ interlocked mesoporous micro-cubes of several nanocrystals as selective and sensitive nitrogen dioxide chemoresistive gas sensors. *Appl. Surf. Sci.* 442: 178–184.

**14** Fraden, J. (2010). Acoustic sensors. In: *Handbook of Modern Sensors* (ed. J. Fraden), 431–443. New York: Springer.

**15** Miramirkhani, F. and Navarchian, A.H. (2017). Morphology, structure, and gas sensing performance of conductive polymers and polymer/carbon black composites used for volatile compounds detection. *IEEE Sens. J.* 17 (10): 2992–3000.

**16** Hoa, N.D., An, S.Y., Dung, N.Q. et al. (2010). Synthesis of p-type semiconducting cupric oxide thin films and their application to hydrogen detection. *Sens. Actuators, B* 146 (1): 239–244.

**17** Van Tong, P., Hoa, N.D., Van Duy, N., and Van Hieu, N. (2015). Micro-wheels composed of self-assembled tungsten oxide nanorods for highly sensitive detection of low-level toxic chlorine gas. *RSC Adv.* 5 (32): 25204–25207.

**18** Pandey, P., Datta, M., and Malhotra, B.D. (2008). Prospects of nanomaterials in biosensors. *Anal. Lett.* 41 (2): 159–209.

**19** Doong, R.A. and Shih, H.M. (2010). Array-based titanium dioxide biosensors for ratiometric determination of glucose, glutamate and urea. *Biosens. Bioelectron.* 25 (6): 1439–1446.

**20** Wang, J., Thomas, D.F., and Chen, A. (2008). Nonenzymatic electrochemical glucose sensor based on nanoporous PtPb networks. *Anal. Chem.* 80 (4): 997–1004.

**21** Feng, D., Wang, F., and Chen, Z. (2009). Electrochemical glucose sensor based on one-step construction of gold nanoparticle–chitosan composite film. *Sens. Actuators, B* 138 (2): 539–544.

22 Xiao, F., Zhao, F., Mei, D. et al. (2009). Nonenzymatic glucose sensor based on ultrasonic-electrodeposition of bimetallic PtM (M = Ru, Pd and Au) nanoparticles on carbon nanotubes–ionic liquid composite film. *Biosens. Bioelectron.* 24 (12): 3481–3486.

23 Hahn, Y.B., Ahmad, R., and Tripathy, N. (2012). Chemical and biological sensors based on metal oxide nanostructures. *Chem. Commun.* 48 (84): 10369–10385.

24 Sensors, G. (1992). Gas sensors, principles, operation and developments. In: *Principles, Operation, and Developments* (ed. G. Sberveglieri), 1–409. Netherlands: Springer.

25 Korotcenkov, G. (2005). Gas response control through structural and chemical modification of metal oxide films: state of the art and approaches. *Sens. Actuators, B* 107 (1): 209–232.

26 Korotcenkov, G. (2007). Metal oxides for solid-state gas sensors: what determines our choice? *Mater. Sci. Eng., B* 139 (1): 1–23.

27 Zhu, L. and Zeng, W. (2017). Room-temperature gas sensing of ZnO-based gas sensor: a review. *Sens. Actuators, A* 267: 242–261.

28 Mohammad-Yousefi, S., Rahbarpour, S., and Ghafoorifard, H. (2019). Describing the effect of Ag/Au modification on operating temperature and gas sensing properties of thick film $SnO_2$ gas sensors by gas diffusion theory. *Mater. Chem. Phys.* 227: 148–156.

29 Shaikh, S.F., Ghule, B.G., Nakate, U.T. et al. (2018). Low-temperature ionic layer adsorption and reaction grown anatase $TiO_2$ nanocrystalline films for efficient perovskite solar cell and gas sensor applications. *Sci. Rep.* 8 (1): 1–11.

30 Shaikh, S.F., Ghule, B.G., Shinde, P.V. et al. (2020). Continuous hydrothermal flow-inspired synthesis and ultra-fast ammonia and humidity room-temperature sensor activities of $WO_3$ nanobricks. *Mater. Res. Express* 7 (1): 015076.

31 Poonia, E., Mishra, P.K., Kiran, V. et al. (2019). Aero-gel based $CeO_2$ nanoparticles: synthesis, structural properties and detailed humidity sensing response. *J. Mater. Chem. C* 7 (18): 5477–5487.

32 Patil, L.A. and Patil, D.R. (2006). Heterocontact type CuO-modified $SnO_2$ sensor for the detection of a ppm level $H_2S$ gas at room temperature. *Sens. Actuators, B* 120 (1): 316–323.

33 Andre, R.S., Pereira, J.C., Mercante, L.A. et al. (2018). ZnO-$Co_3O_4$ heterostructure electrospun nanofibers modified with poly (sodium 4-styrenesulfonate): evaluation of humidity sensing properties. *J. Alloys Compd.* 767: 1022–1029.

34 Sun, L., Fang, W., Yang, Y. et al. (2017). Highly active and porous single-crystal $In_2O_3$ nanosheet for $NO_x$ gas sensor with excellent response at room temperature. *RSC Adv.* 7 (53): 33419–33425.

35 Sangale, S.S., Jadhav, V.V., Shaikh, S.F. et al. (2020). Facile one-step hydrothermal synthesis and room-temperature $NO_2$ sensing application of α-$Fe_2O_3$ sensor. *Mater. Chem. Phys.* 246: 122799.

36 Arya, S.K., Krishnan, S., Silva, H. et al. (2012). Advances in materials for room temperature hydrogen sensors. *Analyst* 137 (12): 2743–2756.

**37** Ishihara, T., Higuchi, M., Takagi, T. et al. (1998). Preparation of CuO thin films on porous $BaTiO_3$ by self-assembled multibilayer film formation and application as a $CO_2$ sensor. *J. Mater. Chem.* 8 (9): 2037–2042.

**38** Zhang, G. and Liu, M. (2000). Effect of particle size and dopant on properties of $SnO_2$-based gas sensors. *Sens. Actuators, B* 69 (1–2): 144–152.

**39** Sayago, I., Santos, H., Horrillo, M.C. et al. (2008). Carbon nanotube networks as gas sensors for $NO_2$ detection. *Talanta* 77 (2): 758–764.

**40** Comini, E., Faglia, G., Sberveglieri, G. et al. (2002). Stable and highly sensitive gas sensors based on semiconducting oxide nanobelts. *Appl. Phys. Lett.* 81 (10): 1869–1871.

**41** Francioso, L., Taurino, A.M., Forleo, A., and Siciliano, P. (2008). $TiO_2$ nanowires array fabrication and gas sensing properties. *Sens. Actuators, B* 130 (1): 70–76.

**42** Korotcenkov, G., Brinzari, V., and Cho, B.K. (2018). $In_2O_3$-and $SnO_2$-based ozone sensors: design and characterization. *Crit. Rev. Solid State Mater. Sci.* 43 (2): 83–132.

**43** Gujar, T.P., Shinde, V.R., Lokhande, C.D. et al. (2005). Bismuth oxide thin films prepared by chemical bath deposition (CBD) method: annealing effect. *Appl. Surf. Sci.* 250 (1–4): 161–167.

**44** Bhande, S.S., Mane, R.S., Ghule, A.V., and Han, S.H. (2011). A bismuth oxide nanoplate-based carbon dioxide gas sensor. *Scr. Mater.* 65 (12): 1081–1084.

**45** Khan, I., Saeed, K., and Khan, I. (2019). Nanoparticles: properties, applications and toxicities. *Arabian J. Chem.* 12 (7): 908–931.

**46** Zhou, W., Shao, Z., and Jin, W. (2006). Synthesis of nanocrystalline conducting composite oxides based on a non-ion selective combined complexing process for functional applications. *J. Alloys Compd.* 426 (1–2): 368–374.

**47** Roduner, E. (2006). Size matters: why nanomaterials are different. *Chem. Soc. Rev.* 35 (7): 583–592.

**48** Eranna, G. (2011). *Metal Oxide Nanostructures as Gas Sensing Devices*. CRC Press.

**49** Yu, X., Marks, T.J., and Facchetti, A. (2016). Metal oxides for optoelectronic applications. *Nat. Mater.* 15 (4): 383–396.

**50** Devan, R.S., Patil, R.A., Lin, J.H., and Ma, Y.R. (2012). One-dimensional metal-oxide nanostructures: recent developments in synthesis, characterization, and applications. *Adv. Funct. Mater.* 22 (16): 3326–3370.

**51** Kim, H.J. and Lee, J.H. (2014). Highly sensitive and selective gas sensors using p-type oxide semiconductors: overview. *Sens. Actuators, B* 192: 607–627.

**52** Barsan, N., Simion, C., Heine, T. et al. (2010). Modeling of sensing and transduction for p-type semiconducting metal oxide-based gas sensors. *J. Electroceram.* 25 (1): 11–19.

**53** Hübner, M., Simion, C.E., Tomescu-Stănoiu, A. et al. (2011). Influence of humidity on CO sensing with p-type CuO thick film gas sensors. *Sens. Actuators, B* 153 (2): 347–353.

**54** Kim, H.R., Haensch, A., Kim, I.D. et al. (2011). The role of NiO doping in reducing the impact of humidity on the performance of $SnO_2$-based gas sensors:

synthesis strategies, and phenomenological and spectroscopic studies. *Adv. Funct. Mater.* 21 (23): 4456–4463.

55 Kaye, S.S. and Long, J.R. (2005). Hydrogen storage in the dehydrated prussian blue analogues $M_3[Co(CN)_6]_2$ (M = Mn, Fe, Co, Ni, Cu, Zn). *J. Am. Chem. Soc.* 127 (18): 6506–6507.

56 Yamazoe, N. and Shimanoe, K. (2009). Receptor function and response of semiconductor gas sensor. *J. Sensors* 2009: 1–21.

57 Yamazoe, N. and Shimanoe, K. (2008). Roles of shape and size of component crystals in semiconductor gas sensors: I. Response to oxygen. *J. Electrochem. Soc.* 155 (4): J85.

58 Barsan, N., Koziej, D., and Weimar, U. (2007). Metal oxide-based gas sensor research: how to? *Sens. Actuators, B* 121 (1): 18–35.

59 Yamazoe, N. and Shimanoe, K. (2011). Basic approach to the transducer function of oxide semiconductor gas sensors. *Sens. Actuators, B* 160 (1): 1352–1362.

60 Yamazoe, N. and Shimanoe, K. (2013). Proposal of contact potential promoted oxide semiconductor gas sensor. *Sens. Actuators, B* 187: 162–167.

61 Miller, D.R., Akbar, S.A., and Morris, P.A. (2014). Nanoscale metal oxide-based heterojunctions for gas sensing: a review. *Sens. Actuators, B* 204: 250–272.

62 Sakai, G., Baik, N.S., Miura, N., and Yamazoe, N. (2001). Gas sensing properties of tin oxide thin films fabricated from hydrothermally treated nanoparticles: dependence of CO and $H_2$ response on film thickness. *Sens. Actuators, B* 77 (1–2): 116–121.

63 Landau, O., Rothschild, A., and Zussman, E. (2009). Processing-microstructure-properties correlation of ultrasensitive gas sensors produced by electrospinning. *Chem. Mater.* 21 (1): 9–11.

64 Capone, S., Siciliano, P., Quaranta, F. et al. (2001). Moisture influence and geometry effect of Au and Pt electrodes on CO sensing response of $SnO_2$ microsensors based on sol–gel thin film. *Sens. Actuators, B* 77 (1–2): 503–511.

65 Tian, S., Zeng, D., Peng, X. et al. (2013). Processing–microstructure–property correlations of gas sensors based on ZnO nanotetrapods. *Sens. Actuators, B* 181: 509–517.

66 Qin, Y., Shen, W., Li, X., and Hu, M. (2011). Effect of annealing on microstructure and $NO_2$-sensing properties of tungsten oxide nanowires synthesized by solvothermal method. *Sens. Actuators, B* 155 (2): 646–652.

67 Kaur, J., Vankar, V.D., and Bhatnagar, M.C. (2010). Role of surface properties of $MoO_3$-doped $SnO_2$ thin films on $NO_2$ gas sensing. *Thin Solid Films* 518 (14): 3982–3987.

68 Xu, L., Song, H., Dong, B. et al. (2010). Preparation and bifunctional gas sensing properties of porous $In_2O_3$-$CeO_2$ binary oxide nanotubes. *Inorg. Chem.* 49 (22): 10590–10597.

69 Hotovy, I., Huran, J., Siciliano, P. et al. (2004). Enhancement of $H_2$ sensing properties of NiO-based thin films with a Pt surface modification. *Sens. Actuators, B* 103 (1–2): 300–311.

**70** Zheng, W., Lu, X., Wang, W. et al. (2009). Assembly of Pt nanoparticles on electrospun $In_2O_3$ nanofibers for $H_2S$ detection. *J. Colloid Interface Sci.* 338 (2): 366–370.

**71** Cho, N.G., Whitfield, G.C., Yang, D.J. et al. (2010). Facile synthesis of Pt-functionalized $SnO_2$ hollow hemispheres and their gas sensing properties. *J. Electrochem. Soc.* 157 (12): J435.

**72** Cho, N.G., Woo, H.S., Lee, J.H., and Kim, I.D. (2011). Thin-walled NiO tubes functionalized with catalytic Pt for highly selective $C_2H_5OH$ sensors using electrospun fibers as a sacrificial template. *Chem. Commun.* 47 (40): 11300–11302.

**73** Liu, Z., Yamazaki, T., Shen, Y. et al. (2007). Room temperature gas sensing of p-type $TeO_2$ nanowires. *Appl. Phys. Lett.* 90 (17): 173119.

**74** Gurlo, A. and Riedel, R. (2007). In situ and operando spectroscopy for assessing mechanisms of gas sensing. *Angew. Chem. Int. Ed.* 46 (21): 3826–3848.

**75** Zhang, J., Qin, Z., Zeng, D., and Xie, C. (2017). Metal-oxide-semiconductor based gas sensors: screening, preparation, and integration. *Phys. Chem. Chem. Phys.* 19 (9): 6313–6329.

**76** Blatter, G. and Greuter, F. (1986). Carrier transport through grain boundaries in semiconductors. *Phys. Rev. B* 33 (6): 3952.

**77** Tsurekawa, S., Kido, K., and Watanabe, T. (2005). Measurements of potential barrier height of grain boundaries in polycrystalline silicon by Kelvin probe force microscopy. *Philos. Mag. Lett.* 85 (1): 41–49.

**78** Lin, Y.H., Li, M., Nan, C.W. et al. (2006). Grain and grain boundary effects in high-permittivity dielectric NiO-based ceramics. *Appl. Phys. Lett.* 89 (3): 032907.

**79** Korotcenkov, G. (2008). The role of morphology and crystallographic structure of metal oxides in response of conductometric-type gas sensors. *Mater. Sci. Eng., R* 61 (1–6): 1–39.

**80** Xu, J., Pan, Q., and Tian, Z. (2000). Grain size control and gas sensing properties of ZnO gas sensor. *Sens. Actuators, B* 66 (1–3): 277–279.

**81** Xu, C., Tamaki, J., Miura, N., and Yamazoe, N. (1991). Grain size effects on gas sensitivity of porous $SnO_2$-based elements. *Sens. Actuators, B* 3 (2): 147–155.

**82** Rothschild, A. and Komem, Y. (2004). The effect of grain size on the sensitivity of nanocrystalline metal-oxide gas sensors. *J. Appl. Phys.* 95 (11): 6374–6380.

**83** Choi, J.M., Byun, J.H., and Kim, S.S. (2016). Influence of grain size on gas-sensing properties of chemiresistive p-type NiO nanofibers. *Sens. Actuators, B* 227: 149–156.

**84** Korotcenkov, G., Brinzari, V., Cerneavschi, A. et al. (2004). $In_2O_3$ films deposited by spray pyrolysis: gas response to reducing (CO, $H_2$) gases. *Sens. Actuators, B* 98 (2–3): 122–129.

**85** Sakai, G., Matsunaga, N., Shimanoe, K., and Yamazoe, N. (2001). Theory of gas-diffusion controlled sensitivity for thin film semiconductor gas sensor. *Sens. Actuators, B* 80 (2): 125–131.

**86** Tian, S., Ding, X., Zeng, D. et al. (2013). Pore-size-dependent sensing property of hierarchical $SnO_2$ mesoporous microfibers as formaldehyde sensors. *Sens. Actuators, B* 186: 640–647.

**87** Sun, P., Zhao, W., Cao, Y. et al. (2011). Porous $SnO_2$ hierarchical nanosheets: hydrothermal preparation, growth mechanism, and gas sensing properties. *CrystEngComm* 13 (11): 3718–3724.

**88** Zhang, J., Zeng, D., Zhu, Q. et al. (2016). Effect of nickel vacancies on the room-temperature $NO_2$ sensing properties of mesoporous NiO nanosheets. *J. Phys. Chem. C* 120 (7): 3936–3945.

**89** Shimizu, Y., Maekawa, T., Nakamura, Y., and Egashira, M. (1998). Effects of gas diffusivity and reactivity on sensing properties of thick film $SnO_2$-based sensors. *Sens. Actuators, B* 46 (3): 163–168.

**90** Varghese, O.K., Gong, D., Paulose, M. et al. (2002). Highly ordered nanoporous alumina films: effect of pore size and uniformity on sensing performance. *J. Mater. Res.* 17 (5): 1162–1171.

**91** Wu, S., Zhou, H., Hao, M. et al. (2016). Fast response hydrogen sensors based on anodic aluminum oxide with pore-widening treatment. *Appl. Surf. Sci.* 380: 47–51.

**92** Tiemann, M. (2007). Porous metal oxides as gas sensors. *Chem. Eur. J.* 13 (30): 8376–8388.

**93** Zhang, J., Zeng, D., Zhu, Q. et al. (2015). Effect of grain-boundaries in NiO nanosheet layers room-temperature sensing mechanism under $NO_2$. *J. Phys. Chem. C* 119 (31): 17930–17939.

**94** Pati, S., Majumder, S.B., and Banerji, P. (2012). Role of oxygen vacancy in optical and gas sensing characteristics of ZnO thin films. *J. Alloys Compd.* 541: 376–379.

**95** Wang, Z., Li, Z., Sun, J. et al. (2010). Improved hydrogen monitoring properties based on p-NiO/n-$SnO_2$ heterojunction composite nanofibers. *J. Phys. Chem. C* 114 (13): 6100–6105.

**96** Xu, K., Li, N., Zeng, D. et al. (2015). Interface bonds determined gas-sensing of $SnO_2$-$SnS_2$ hybrids to ammonia at room temperature. *ACS Appl. Mater. Interfaces* 7 (21): 11359–11368.

**97** Rai, P., Jeon, S.H., Lee, C.H. et al. (2014). Functionalization of ZnO nanorods by CuO nanospikes for gas sensor applications. *RSC Adv.* 4 (45): 23604–23609.

**98** Manjula, P., Arunkumar, S., and Manorama, S.V. (2011). Au/$SnO_2$ an excellent material for room temperature carbon monoxide sensing. *Sens. Actuators, B* 152 (2): 168–175.

**99** Choi, S.W. and Kim, S.S. (2012). Room temperature CO sensing of selectively grown networked ZnO nanowires by Pd nanodot functionalization. *Sens. Actuators, B* 168: 8–13.

**100** Park, S., Kim, H., Jin, C. et al. (2012). Enhanced CO gas sensing properties of Pt-functionalized $WO_3$ nanorods. *Thermochim. Acta* 542: 69–73.

**101** Barsan, N., Hubner, M., and Weimar, U. (2011). Conduction mechanisms in $SnO_2$ based polycrystalline thick film gas sensors exposed to CO and $H_2$ in different oxygen backgrounds. *Sens. Actuators, B* 157 (2): 510–517.

**102** Castro-Hurtado, I., Malagu, C., Morandi, S. et al. (2013). Properties of NiO sputtered thin films and modelling of their sensing mechanism under formaldehyde atmospheres. *Acta Mater.* 61: 1146–1153.

**103** Yang, L., Zhang, S., Li, H. et al. (2013). Conduction model of coupled domination by bias and neck for porous films as gas sensor. *Sens. Actuators, B* 176: 217–224.

**104** Levinson, J., Shepherd, F.R., Scanlon, P.J. et al. (1982). Conductivity behavior in polycrystalline semiconductor thin film transistors. *J. Appl. Phys.* 53 (2): 1193–1202.

**105** Wu, C.M. and Yang, E.S. (1982). Physical basis of scattering potential at grain boundary of polycrystalline semiconductors. *Appl. Phys. Lett.* 40 (1): 49–51.

**106** Nalage, S.R., Chougule, M.A., Sen, S. et al. (2012). Sol-gel synthesis of nickel oxide thin films and their characterization. *Thin Solid Films* 520 (15): 4835–4840.

**107** Tuller, H.L. and Bishop, S.R. (2011). Point defects in oxides: tailoring materials through defect engineering. *Annu. Rev. Mater. Res.* 41: 369–398.

**108** Nandy, S., Saha, B., Mitra, M.K., and Chattopadhyay, K.K. (2007). Effect of oxygen partial pressure on the electrical and optical properties of highly (200) oriented p-type $Ni_{1-x}O$ films by DC sputtering. *J. Mater. Sci.* 42 (14): 5766–5772.

**109** Chang, C.M., Hon, M.H., and Leu, I.C. (2010). Preparation of ZnO nanorod arrays with tailored defect-related characterisitcs and their effect on the ethanol gas sensing performance. *Sens. Actuators, B* 151 (1): 15–20.

**110** Chen, M., Wang, Z., Han, D. et al. (2011). Porous ZnO polygonal nanoflakes: synthesis, use in high-sensitivity $NO_2$ gas sensor, and proposed mechanism of gas sensing. *J. Phys. Chem. C* 115 (26): 12763–12773.

**111** Wu, J., Huang, Q., Zeng, D. et al. (2014). Al-doping induced formation of oxygen-vacancy for enhancing gas-sensing properties of $SnO_2$ NTs by electrospinning. *Sens. Actuators, B* 198: 62–69.

**112** Epifani, M., Prades, J.D., Comini, E. et al. (2008). The role of surface oxygen vacancies in the $NO_2$ sensing properties of $SnO_2$ nanocrystals. *J. Phys. Chem. C* 112 (49): 19540–19546.

**113** Ahn, M.W., Park, K.S., Heo, J.H. et al. (2008). Gas sensing properties of defect-controlled ZnO-nanowire gas sensor. *Appl. Phys. Lett.* 93 (26): 263103.

**114** Kaur, J., Kumar, R., and Bhatnagar, M.C. (2007). Effect of indium-doped $SnO_2$ nanoparticles on $NO_2$ gas sensing properties. *Sens. Actuators, B* 126 (2): 478–484.

**115** Rai, P., Kim, Y.S., Song, H.M. et al. (2012). The role of gold catalyst on the sensing behavior of ZnO nanorods for CO and $NO_2$ gases. *Sens. Actuators, B* 165 (1): 133–142.

**116** Du, N., Zhang, H., Ma, X., and Yang, D. (2008). Homogeneous coating of Au and $SnO_2$ nanocrystals on carbon nanotubes via layer-by-layer assembly: a new ternary hybrid for a room-temperature CO gas sensor. *Chem. Commun.* 46: 6182–6184.

**117** Kim, B., Lu, Y., Hannon, A. et al. (2013). Low temperature $Pd/SnO_2$ sensor for carbon monoxide detection. *Sens. Actuators, B* 177: 770–775.

**118** Nakate, U.T., Bulakhe, R.N., Lokhande, C.D., and Kale, S.N. (2016). Au sensitized ZnO nanorods for enhanced liquefied petroleum gas sensing properties. *Appl. Surf. Sci.* 371: 224–230.

**119** Li, Z., Li, H., Wu, Z. et al. (2019). Advances in designs and mechanisms of semiconducting metal oxide nanostructures for high-precision gas sensors operated at room temperature. *Mater. Horiz.* 6 (3): 470–506.

**120** Zhou, X., Cheng, X., Zhu, Y. et al. (2018). Ordered porous metal oxide semiconductors for gas sensing. *Chin. Chem. Lett.* 29 (3): 405–416.

**121** Liu, J., Dai, M., Wang, T. et al. (2016). Enhanced gas sensing properties of $SnO_2$ hollow spheres decorated with $CeO_2$ nanoparticles heterostructure composite materials. *ACS Appl. Mater. Interfaces* 8 (10): 6669–6677.

**122** Cao, Y., He, Y., Zou, X., and Li, G.D. (2017). Tungsten oxide clusters decorated ultrathin $In_2O_3$ nanosheets for selective detecting formaldehyde. *Sens. Actuators, B* 252: 232–238.

**123** Xu, L., Zheng, R., Liu, S. et al. (2012). NiO@ZnO heterostructured nanotubes: coelectrospinning fabrication, characterization, and highly enhanced gas sensing properties. *Inorg. Chem.* 51 (14): 7733–7740.

**124** Liu, L., Zhang, Y., Wang, G. et al. (2011). High toluene sensing properties of NiO-$SnO_2$ composite nanofiber sensors operating at 330° C. *Sens. Actuators, B* 160 (1): 448–454.

**125** Zhang, J., Zeng, D., Zhu, Q. et al. (2016). Enhanced room temperature $NO_2$ response of NiO-$SnO_2$ nanocomposites induced by interface bonds at the p–n heterojunction. *Phys. Chem. Chem. Phys.* 18 (7): 5386–5396.

**126** Gopel, W. and Schierbaum, K.D. (1995). $SnO_2$ sensors: current status and future prospects. *Sens. Actuators, B* 26 (1–3): 1–12.

**127** Yu, J.H. and Choi, G.M. (1998). Electrical and CO gas sensing properties of ZnO-$SnO_2$ composites. *Sens. Actuators, B* 52 (3): 251–256.

**128** de Lacy Costello, B.P.J., Ewen, R.J., Jones, P.R.H. et al. (1999). A study of the catalytic and vapour-sensing properties of zinc oxide and tin dioxide in relation to 1-butanol and dimethyldisulphide. *Sens. Actuators, B* 61 (1–3): 199–207.

**129** Zhu, C.L., Chen, Y.J., Wang, R.X. et al. (2009). Synthesis and enhanced ethanol sensing properties of α-$Fe_2O_3$/ZnO heteronanostructures. *Sens. Actuators, B* 140 (1): 185–189.

**130** Yoon, D.H., Yu, J.H., and Choi, G.M. (1998). CO gas sensing properties of ZnO-CuO composite. *Sens. Actuators, B* 46 (1): 15–23.

**131** Teterycz, H., Klimkiewicz, R., and Licznerski, B.W. (2001). A new metal oxide catalyst in alcohol condensation. *Appl. Catal., A* 214 (2): 243–249.

**132** Meixner, H. and Lampe, U. (1996). Metal oxide sensors. *Sens. Actuators, B* 33 (1–3): 198–202.

**133** de Lacy Costello, B.P.J., Ewen, R.J., Ratcliffe, N.M., and Sivanand, P.S. (2003). Thick film organic vapour sensors based on binary mixtures of metal oxides. *Sens. Actuators, B* 92 (1–2): 159–166.

**134** Ivanovskaya, M., Kotsikau, D., Faglia, G., and Nelli, P. (2003). Influence of chemical composition and structural factors of $Fe_2O_3$/$In_2O_3$ sensors on their selectivity and sensitivity to ethanol. *Sens. Actuators, B* 96 (3): 498–503.

**135** Hu, Y., Zhou, X., Han, Q. et al. (2003). Sensing properties of CuO-ZnO heterojunction gas sensors. *Mater. Sci. Eng., B* 99 (1–3): 41–43.

**136** Ling, Z. and Leach, C. (2004). The effect of relative humidity on the $NO_2$ sensitivity of a $SnO_2/WO_3$ heterojunction gas sensor. *Sens. Actuators, B* 102 (1): 102–106.

**137** Aygun, S. and Cann, D. (2005). Hydrogen sensitivity of doped CuO/ZnO heterocontact sensors. *Sens. Actuators, B* 106 (2): 837–842.

**138** Herran, J., Mandayo, G.G., and Castano, E. (2007). Physical behaviour of $BaTiO_3$-CuO thin-film under carbon dioxide atmospheres. *Sens. Actuators, B* 127 (2): 370–375.

**139** Haridas, D., Gupta, V., and Sreenivas, K. (2008). Enhanced catalytic activity of nanoscale platinum islands loaded onto $SnO_2$ thin film for sensitive LPG gas sensors. *Bull. Mater. Sci.* 31 (3): 397–400.

**140** Hyodo, T., Baba, Y., Wada, K. et al. (2000). Hydrogen sensing properties of $SnO_2$ varistors loaded with $SiO_2$ by surface chemical modification with diethoxydimethylsilane. *Sens. Actuators, B* 64 (1–3): 175–181.

**141** Lu, Y., Li, J., Han, J. et al. (2004). Room temperature methane detection using palladium loaded single-walled carbon nanotube sensors. *Chem. Phys. Lett.* 391 (4–6): 344–348.

**142** Wang, D., Ma, Z., Dai, S. et al. (2008). Low-temperature synthesis of tunable mesoporous crystalline transition metal oxides and applications as Au catalyst supports. *J. Phys. Chem. C* 112 (35): 13499–13509.

**143** Haridas, D., Sreenivas, K., and Gupta, V. (2008). Improved response characteristics of $SnO_2$ thin film loaded with nanoscale catalysts for LPG detection. *Sens. Actuators, B* 133 (1): 270–275.

**144** Shimizu, Y., Matsunaga, N., Hyodo, T., and Egashira, M. (2001). Improvement of $SO_2$ sensing properties of $WO_3$ by noble metal loading. *Sens. Actuators, B* 77 (1–2): 35–40.

**145** Ruiz, A.M., Cornet, A., Shimanoe, K. et al. (2005). Effects of various metal additives on the gas sensing performances of $TiO_2$ nanocrystals obtained from hydrothermal treatments. *Sens. Actuators, B* 108 (1–2): 34–40.

**146** Kolmakov, A., Klenov, D.O., Lilach, Y. et al. (2005). Enhanced gas sensing by individual $SnO_2$ nanowires and nanobelts functionalized with Pd catalyst particles. *Nano Lett.* 5 (4): 667–673.

**147** Rui, Z., Tang, M., Ji, W. et al. (2017). Insight into the enhanced performance of $TiO_2$ nanotube supported Pt catalyst for toluene oxidation. *Catal. Today* 297: 159–166.

**148** Mirzaei, A., Kim, J.H., Kim, H.W., and Kim, S.S. (2018). Resistive-based gas sensors for detection of benzene, toluene and xylene (BTX) gases: a review. *J. Mater. Chem. C* 6 (16): 4342–4370.

**149** Traversa, E. (1995). Ceramic sensors for humidity detection: the state-of-the-art and future developments. *Sens. Actuators, B* 23 (2–3): 135–156.

**150** McCafferty, E. and Zettlemoyer, A.C. (1971). Adsorption of water vapour on α-$Fe_2O_3$. *Discuss. Faraday Soc.* 52: 239–254.

**151** Qi, Q., Zhang, T., Zheng, X. et al. (2008). Electrical response of $Sm_2O_3$-doped $SnO_2$ to $C_2H_2$ and effect of humidity interference. *Sens. Actuators, B* 134 (1): 36–42.

**152** Gong, J., Chen, Q., Lian, M.R. et al. (2006). Micromachined nanocrystalline silver doped $SnO_2$-$H_2S$ sensor. *Sens. Actuators, B* 114 (1): 32–39.

**153** Jing, Z. and Zhan, J. (2008). Fabrication and gas-sensing properties of porous ZnO nanoplates. *Adv. Mater.* 20 (23): 4547–4551.

**154** Egashira, M., Kawasumi, S., Kagawa, S., and Seiyama, T. (1978). Temperature programmed desorption study of water adsorbed on metal oxides. I. Anatase and rutile. *Bull. Chem. Soc. Jpn.* 51 (11): 3144–3149.

**155** Tischner, A., Maier, T., Stepper, C., and Kock, A. (2008). Ultrathin $SnO_2$ gas sensors fabricated by spray pyrolysis for the detection of humidity and carbon monoxide. *Sens. Actuators, B* 134 (2): 796–802.

**156** Malyshev, V.V. and Pislyakov, A.V. (2008). Investigation of gas-sensitivity of sensor structures to hydrogen in a wide range of temperature, concentration and humidity of gas medium. *Sens. Actuators, B* 134 (2): 913–921.

**157** Kim, I.D., Rothschild, A., Lee, B.H. et al. (2006). Ultrasensitive chemiresistors based on electrospun $TiO_2$ nanofibers. *Nano Lett.* 6 (9): 2009–2013.

**158** Cao, M., Wang, Y., Chen, T. et al. (2008). A highly sensitive and fast-responding ethanol sensor based on $CdIn_2O_4$ nanocrystals synthesized by a nonaqueous sol–gel route. *Chem. Mater.* 20 (18): 5781–5786.

**159** Van Duy, N., Van Hieu, N., Huy, P.T. et al. (2008). Mixed $SnO_2$/$TiO_2$ included with carbon nanotubes for gas-sensing application. *Physica E* 41 (2): 258–263.

**160** Cui, H., Zhang, X., Zhang, J., and Zhang, Y. (2019). Nanomaterials-based gas sensors of $SF_6$ decomposed species for evaluating the operation status of high-voltage insulation devices. *High Voltage* 4 (4): 242–258.

**161** Li, T., Zeng, W., Long, H., and Wang, Z. (2016). Nanosheet-assembled hierarchical $SnO_2$ nanostructures for efficient gas-sensing applications. *Sens. Actuators, B* 231: 120–128.

**162** Xiao, X., Liu, L., Ma, J. et al. (2018). Ordered mesoporous tin oxide semiconductors with large pores and crystallized walls for high-performance gas sensing. *ACS Appl. Mater. Interfaces* 10 (2): 1871–1880.

**163** Navale, S.T., Liu, C., Gaikar, P.S. et al. (2017). Solution-processed rapid synthesis strategy of $Co_3O_4$ for the sensitive and selective detection of $H_2S$. *Sens. Actuators, B* 245: 524–532.

**164** Ji, H., Zeng, W., and Li, Y. (2019). Gas sensing mechanisms of metal oxide semiconductors: a focus review. *Nanoscale* 11 (47): 22664–22684.

**165** Zhu, L., Zeng, W., and Li, Y. (2019). A non-oxygen adsorption mechanism for hydrogen detection of nanostructured $SnO_2$ based sensors. *Mater. Res. Bull.* 109: 108–116.

**166** Singh, S., Gupta, G., Yadav, S. et al. (2019). Highly-sensitive potassium-tantalum-niobium oxide humidity sensor. *Sens. Actuators, A* 295: 133–140.

**167** Wang, L., Zhao, F., Han, Q. et al. (2015). Spontaneous formation of $Cu_2O$-$gC_3N_4$ core-shell nanowires for photocurrent and humidity responses. *Nanoscale* 7 (21): 9694–9702.

**168** Yu, H., Gao, S., Cheng, X. et al. (2019). Morphology controllable $Fe_2O_3$ nanostructures derived from Fe-based metal-organic frameworks for enhanced humidity sensing performances. *Sens. Actuators, B* 297: 126744.

**169** Dey, K.K., Bhatnagar, D., Srivastava, A.K. et al. (2015). $VO_2$ nanorods for efficient performance in thermal fluids and sensors. *Nanoscale* 7 (14): 6159–6172.

**170** Miao, Z. (2016). Study on the preparation of nano $TiO_2$ gas sensors by flame synthesis with a rotating surface. PhD thesis. Beijing, China: Tsinghua University (in Chinese).

**171** Hong, S., Shin, J., Hong, Y. et al. (2018). Observation of physisorption in a high-performance FET-type oxygen gas sensor operating at room temperature. *Nanoscale* 10 (37): 18019–18027.

**172** Deng, Y. (2019). Sensing mechanism and evaluation criteria of semiconducting metal oxides gas sensors. In: *Semiconducting Metal Oxides for Gas Sensing* (ed. Y. Deng), 23–51. Singapore: Springer.

**173** Wang, M., Hou, T., Shen, Z. et al. (2019). MOF-derived $Fe_2O_3$: phase control and effects of phase composition on gas sensing performance. *Sens. Actuators, B* 292: 171–179.

**174** Yang, M., Liu, X., Zhao, X. et al. (2014). Acetone sensing characteristics of γ-$Fe_2O_3$ nanofibers. *Chem. Res. Chin. Univ.* 35: 1615–1619.

**175** Ming, J., Wu, Y., Wang, L. et al. (2011). $CO_2$-assisted template synthesis of porous hollow bi-phase γ-/α-$Fe_2O_3$ nanoparticles with high sensor property. *J. Mater. Chem.* 21 (44): 17776–17782.

**176** He, J., Rao, X., Yang, C. et al. (2014). Glucose-assisted synthesis of mesoporous maghemite nanoparticles with enhanced gas sensing properties. *Sens. Actuators, B* 201: 213–221.

**177** Gao, G., Wu, J., Wu, G. et al. (2012). Phase transition effect on durability of $WO_3$ hydrogen sensing films: an insight by experiment and first-principle method. *Sens. Actuators, B* 171: 1288–1291.

**178** Williams, D.E., Henshaw, G.S., Pratt, K.F.E., and Peat, R. (1995). Reaction-diffusion effects and systematic design of gas-sensitive resistors based on semiconducting oxides. *J. Chem. Soc., Faraday Trans.* 91 (23): 4299–4307.

**179** Lu, H., Ma, W., Gao, J., and Li, J. (2000). Diffusion-reaction theory for conductance response in metal oxide gas sensing thin films. *Sens. Actuators, B* 66 (1–3): 228–231.

**180** Wang, X., Wang, Y., Tian, F. et al. (2015). From the surface reaction control to gas-diffusion control: the synthesis of hierarchical porous $SnO_2$ microspheres and their gas-sensing mechanism. *J. Phys. Chem. C* 119 (28): 15963–15976.

**181** Selvaraj, K., Kumar, S., and Lakshmanan, R. (2014). Analytical expression for concentration and sensitivity of a thin film semiconductor gas sensor. *Ain Shams Eng. J.* 5 (3): 885–893.

**182** Yang, S., Wang, Z., Zou, Y. et al. (2017). Remarkably accelerated room-temperature hydrogen sensing of $MoO_3$ nanoribbon/graphene composites by suppressing the nanojunction effects. *Sens. Actuators, B* 248: 160–168.

# Index

## a

$A_2(BB')O_6$ double perovskites  408
$AB_{12}O_{19}$ hexaferrites  408
$ABO_3$ perovskites  408
$AB_2O_4$ spinels  408
absorbed photon to current efficiency (APCE)  365
adsorption/desorption model  462
advanced oxidation processes (AOPs)  281, 317
AF/F(Fi) core/shell MNPs  419
Ag-SrTiO$_3$ nanocomposites  238
Ag/AgCl-(BiO)$_2$CO$_3$ system  242
air and water cleaning photocatalyst  317
Al$_2$O$_3$/p-GaSb interface  23
all-oxide spintronic heterostructures  427
all-oxide synthetic antiferromagnets  419
α-Fe$_2$O$_3$ nanofibres  249
American Conference of Government Industrial Hygienists  439
amorphous α-nickel-cobalt hydroxide nanodendrites  47
anatase  287
anatase/rutile mixture system  326
anatase TiO$_2$ mesocrystals  236
anthropogenic greenhouse gas emissions  277
anti-fogging surfaces  281
AO rock-salt monoxides  408
artificial multiferroic oxide heterostructures  421
artificial photosynthesis (AP) processes  277
atomic layer deposition (ALD)  2, 356
Au-Mn$_3$O$_4$ nanosystems  115
azo-dye (Plasmocorinth B, PB)  327

## b

ball milling  148
bandgap ($E_g$)  317
barium titanate (BaTiO$_3$)
    dissolution-reprecipitation pathway  50–51
    hydrothermal crystallization mechanisms  50–51
    in organic solvents  51
    ligand-free synthesis  53
BaTiO$_3$ core/shell nanoparticles  424
battery and supercapacitor materials  392
Bi$_2$MoO$_6$ photocatalyst  241
Bi$_2$O$_2$CO$_3$/ZnFe$_2$O$_4$ composite  250
bicontinuous porous electrodes  381
binder-free structured macroporous materials  382
BiOBr/Bi$_{12}$O$_{17}$Br$_2$ systems  242
BiOIO$_3$/BiOI system  242
bismuth ferrate (BiFeO$_3$)  412
    Pechini process  54
    phase selectivity  54
    sonochemical synthesis  54
bismuth oxyhalides  241
BiVO$_4$  363
blue-moon ensemble  114
Bouduard-like reactions  286
Bragg diffraction  121
brookite  288

## c

CaAl$_2$O$_4$(Eu,Nd)/(Ta,N)-co-doped TiO$_2$/Fe$_2$O$_3$ composite  249
CaO/TiO$_2$/Al$_2$O$_3$ ternary oxides, photoactivity of  237

capacitance-voltage (CV) measurements 23
carbon capture and storage (CCS) 278
carbon nanotubes (CNTs) 154
  array synthesis strategy 90–91
  varieties 154–155
carbon quantum dots/zeolite imidazole frameworks (CQDs/ZIF-8) 247
$CaRu_{1-x}Ti_xO_3$ 420
cation doping 290
$CdIn_{1-z}Ga_zSe_2$ (CIGS) 350
$CeO_2$ 292
charge modulation mechanism 426
charge-ordering effect 409
chemical adsorption/desorption model 467
chemical vapor deposition (CVD) 2, 3, 90, 112, 115, 116
chemical warfare agents (CWA) 112
$CNQDs/GO-InVO_4$ aerogel composite 250
CNT/dielectrics 95
$Co_{0.6}Fe_{2.4}O_4$ nanoparticles 180, 181
$Co_2FeO_4$/N-doped CNTs 156, 157
$Co_3O_4$ microtube arrays ($Co_3O_4$-MTA) 154
$Co_3O_4$-coated Ni powder preparation 147
$Co_xFe_{3-x}O_4$ (CFO) nanoislands 163
co-catalysts, for $CO_2$ reduction 299
cobalt hydroxides 47
cobalt oxyhydroxides 47
coercive field 406
$CoFe_2O_4$ 409
$CoFe_2O_4$ (CFO) nanorod-containing nanocomposite 152
$CoFe_2O_4$/graphene oxides hybrids 162
colloidal synthesis 76
  for tungsten bronzes 72
colossal magnetoresistance 406
combustion synthesis 147
commercial $TiO_2$-white cement 254
$CoMn_2O_4$ (CMO) spinel nanotubes 157
computational-guided structural resolution 122
conduction band (CB) 317, 319
contact-resonance piezoresponse force microscopy (CR-PFM) 14
conventional detection methods 442
converse magnetoelectric effect 422
$CoFe_2O_4$ MNPs 419
cooperative magnetism 405
copper oxides 292, 298
core-ring structured $NiCo_2O_4$ nanoplatelets 163
crystalline phase composition 327
C-$TiO_2$ mesoporous anatase nanocrystals 234
$Cu_{1.5}Mn_{1.5}O_4$ spinel ceramic pigment 147
Cu(hfa)2TMEDA 118
Cu-Mn-citric xerogel 148
Curie Temperature 406

## d

decoupled regime 414
$DeNO_x$ actions 229
$DeNO_x$ activity, of $TiO_2$ photocatalyst 236
density functional theory (DFT) 113, 114
di(propyleneglycol) monomethyl ether (DPGME) 115, 116
diamond anvil cell (DAC) 124
dielectric oxide materials 18
Diels-Alder cycloaddition reaction 282
diffusion constant ($D_K$) 450
direct exchange interaction 405
direct liquid injection chemical vapour deposition (DLI-CVD) 10, 11
direct magneto-electric coupling 422

## e

effective oxide thickness (EOT) 19
electrocatalysts, for water splitting 45
electron energy loss spectra (EELS) 331
electron mediators 297
electron paramagnetic resonance (EPR) 323
electronic excitations 131
electrospinning technique 288
energy conversion process 346
energy storage systems 389
environmental humidity 461
EPR spectra 331
European Union (EU) 278
Evonik P25 287
exchange-bias effect 417
exchange bias field 417
exchange interaction mechanism 427
exchange-spring regime 414
exchange-spring systems 414
exfoliation, of layered materials 49
exhaustive surface cleaning protocols 286
external quantum efficiency (EQE) 359

## f

Faradic efficiency  363
$Fe_2O_3$ photoanodes  352
$Fe_2O_3/SiO_2$ composites  249
$Fe_2O_3/TiO_2$ nanocomposites  249
$Fe_3O_4/CoFe_2O_4$ spinel ferrite bilayers  416
$Fe_3O_4/Pb(Zn_{0.33}Nb_{0.66})O_3$-$PbTiO_3$ hybrid thin film stack  425
Fe(hfa)2TMEDA  119
Fe/NiOOH cocatalyst  352
FeCo/graphene nanosheet hybrids  163
ferro(i)magnetic/antiferromagnetic systems  416
ferroelectric materials  4
ferroelectricity  406
ferroic properties, of complex transition metal oxides
  magnetic oxides  412–413
  perovskites  411–412
  spinel ferrites  408–411
ferroic thin films
  $Bi_2O_3$ buffer layers  9
  $BiFeO_3$ films on $IrO_2$/Si substrates  8
  competitive wafer throughputs  18
  CR-PFM  14
  CVD  6, 7, 11, 13, 18
  data storage applications  13
  data-storage technologies  14
  DLI-CVD  10, 11, 16
  ferroelectric and multiferroic properties  8
  ferroelectrics and multiferroics  7
  layer-by-layer growth method  17
  layered Aurivillius structures  15
  metalorganic precursors  9
  microstructural defects  7
  MOCVD-grown PZT  12
  OPB defects  17
  $SrBi_2Ta_2O_9$ samples  9
  2D ultrathin B6TFMO films  15
F(Fi)/AF heterostructures  419
first principles molecular dynamics (FPMD)  114
flexible single-crystalline LFO thin films  159
fluorine doped tin oxide (FTO)  346
Frank-van der Merve-type growth  14
free electrode  428
free-standing flexible $TiO_2$/CNT films  102, 103
functional porosity in electrochemical systems  379
functionalized titania  319

## g

$g$-$C_3N_4/TiO_2$ composites  231
$g$-$C_3N_4/TiO_2$ heterojunction photo-luminescence studies  232
Gärtner equation  359
gas sensing parameters  445
gas sensors
  adsorption/desorption model  462
    chemical adsorption/desorption model  467–470
    oxygen adsorption model  464–467
    physical adsorption/desorption model  470–471
  bulk resistance control mechanism  471–472
  commercial practices  445
  gas diffusion control mechanism  472–474
  mechanisms  462
  metal oxide
    chemical composition  458
    crystallographic defective and hetero-interface structures  453–458
    humidity and temperature  461–462
    morphology-related structural parameters  448
    noble metal particles  458–461
    operation mechanism  446
    preparation  446
  metal oxide nanomaterial-based  443
  types  442
Ge nanocrystal-containing $GeO_2$ inverse opal Li-ion battery anode  386
generalized gradient (GGA) approximations  114
germanium based materials  385
germanium (IV) ethoxide ($Ge(OC_2H_5)_4$) precursor  386
giant magnetoresistance (GMR) effect  428
gold nanoparticles  300
good water oxidation catalyst  295
grafting  319

graphene oxide/TiO$_2$ photocatalysts  232
graphene/TBZ fibre preparation  233
graphene/TiO$_2$ photocatalysts  232
graphitic carbon nitride  298

## h

half-metallicity  409
hard/soft exchange-coupled core shell magnetic nanoparticles (MNPs)  414
hard/soft exchange-coupled systems  413
hard/soft thin film heterostructures  416
hard systems  415
Helmholtz layer  362
heterogeneous photocatalysis
 applications  281–283
 fundamentals  279–281
heterojunction  295
hexagonal Co$_3$O$_4$ nanoplatelets  163
high performance computing (HPC)  111
high performance Li ion batteries (LIBs)  380
high-frequency intercept (HFI)  362
high-pressure thermal plasma synthesis  142
high-resolution transmission electron microscopy (HRTEM)  10
high-temperature processes  339
high-temperature solid solution method  148
holes-current ($j_h$)  362
HOPG  26
H$_{0.07}$WO$_3$ hexagonal bronze nanoplatelets  72
hybrid functionals  114
hybrid supercapacitor electrodes  91, 94
hybrid TiO$_2$/CNT systems  99
hydrogen (H$_2$) production  99, 100
hydrogen economy  339
hydrogen efficiency  364
hydrogen evolution reaction (HER)  341
hydrothermal method  143, 290
hydroxyl  281
hyperfine interaction  322

## i

incident photon to current efficiency (IPCE)  365
indirect exchange interaction  405
indium tin oxide (ITO)  346
InGaAs/Al$_2$O$_3$ MOS structures  23
innovative photocatalysts  258
inorganic molten salts  53
in-silico methodology  111
in-situ diffuse reflectance infrared technique (DRIFTS)  251, 252
insulating protective layer  357
intensity modulated photocurrent spectroscopy (IMPS)  361
inverse opal (IO)  381
inverse spinel  409
inversion degree  409
IrO$_2$/CNT nanocomposites  91, 95

## l

La$_{0.67}$Ca(Sr)$_{0.33}$MnO$_3$  420
La$_{0.67}$Ca(Sr)$_{0.33}$MnO$_3$/CaRu$_{1-x}$Ti$_x$O$_3$ superlattices  420
La$_{0.67}$Sr$_{0.33}$MnO$_3$ nanocubes  56
La$_2$NiO$_4$ and LaSrNiO$_4$ layered perovskite Ruddlesden-Popper nanoparticles  58
La$_x$Sr$_{1-x}$MnO$_3$  411
LaFeO$_3$–CoFe$_2$O$_4$  425
laser pyrolysis  141
laser pyrolysis synthesis, of iron/iron oxide nanoparticles  141
LaTiO$_3$-SrTiO$_3$ composites  238
layered nickelates  45
layered NiO$_6$-based hybrid material synthesis  46
layered oxide-based materials  45
layered structured ferroelectrics  5
Li$_4$Ti$_5$O$_{12}$ (LTO)  393
Li-ion battery electrodes  384
LiFePO$_4$ (LFP)  393
LiMn$_2$O$_4$ nanorods  152, 153
linear sweep voltammetry (LSV)  343
LiNi$_{0.5}$Mn$_{1.5}$O$_4$ spinel cathode  148, 149
liquid-phase synthesis  50, 53
localized surface plasmon resonance (LSPR)  299

## m

M(hfa)2TMEDA   121
M-type hexagonal ferrites   413
maghemite ($\gamma$-Fe$_2$O$_3$)   409
Magnéli phases   63
  tungsten   73
magnetic anisotropy   407
magnetic domain   406
magnetic nanocomposites   407
magnetic oxide heterostructures
  all-oxide synthetic antiferromagnets
    419–421
  ferro(i)magnetic/antiferromagnetic
    systems   416–419
  hard/soft exchange-coupled systems
    413–416
magnetic oxide nanomaterials   112, 132
magnetic oxides   412
magnetically soft epitaxial spinel
  NiZnAl-ferrite film properties   159,
  160
magnetite (Fe$_3$O$_4$)   407
magneto-electric multiferroic (ME-MF)
  materials   421
magneto-optical activity, of WO$_{2.72}$
  nanoparticles   71
magnetocrystalline anisotropy   407
magnetostrictive/ferrimagnetic CoFe$_2$O$_4$
  spinel ferrite (CFO)   423
manganese oxide nanomaterials   114–116
materials sustainability   379
maximum polarization   406
mesoporous MnCo$_2$O$_4$ catalysts   167–168
mesoporous NiCo$_2$O$_4$ nanowire arrays   154
mesoporous silica   320
mesoporous SiO$_2$ structure   323
metal alkylamides   21
metal dopants   319
metal doping   290
metal organic chemical vapour deposition
  (MOCVD) deposition parameters
  8, 9
metal-organic frameworks (MOF)
  crystallization   45
metal oxide dielectrics, band gap vs.
  dielectric constant for   98
metal oxide gas sensors
  crystallographic defective and
    hetero-interface structures   453
  chemical composition   458
  defect structure   453–455
  heterointerface structure   455–458
  noble metal particles   458–461
  humidity and temperature   461
  morphology-related structural parameters
    448
    grain size   448–449
    pore size   449–453
  operation mechanism   446–448
  preparation   446
metal oxide heterostructures   353
metal oxide nanomaterials (MONs)   440,
  441
metal oxides   43
metal-oxide synergies   299
metals foams   390
metastable oxides isolation, templating
  strategy for   74
mixed nickel-cobalt oxyhydroxides   47
mixed valence Ti(III)/Ti(IV) oxides   63
mixed W(V)/W(VI) oxides   68
Mn-Ni-Co ternary oxide nanowires   155
Mn-Ni-Co ternary spinel oxide (MNCO)
  nanowires   154
MnCo$_2$O$_4$ (MCO) spinel nanotubes   157
MnO and Fe$_3$O$_4$ monodispersed
  nanocrystals   175, 179
Mo-O bonds   27
molten salts   55
  inorganic   53
  synthesis, versatility of   56
monomeric tert-butoxide [(OBu$^t$)] group
  20
multicationic oxides
  layered oxide-based materials   45–50
  perovskites   50–53
multicomponent oxides   243
multiferroic tunnel junctions   430

## n

nano-hetero structures   422
nano-objects   43, 44, 47
nanocrystals, La$_{0.67}$Sr$_{0.33}$MnO$_3$   56
nanogranular composites   419
nanomaterials classification   137
nanoparticles, Ti$_4$O$_7$ and Ti$_6$O$_{11}$   65

nanoscaled tungsten bronzes  69
nanostructuration, for oxides  43
nanostructured oxide systems  111
N-doped $TiO_2/g\text{-}C_3N_4$ composites  232
Néel temperature for antiferromagnetic  406
$Ni_xCo_{3-x}O_4$ hierarchical nanostructures  175
Ni-Co oxide hierarchical nanosheets (NCO-HNSs)  159–161
nickel-cobalt layered double hydroxides  47
$NiCo_2O_4$ core-sheath nanowires  171, 172
$NiCo_2O_4$ micro-urchins  171, 174
NiO-based protective layers  358
$NO_x$ gases photochemical oxidation mechanism  251
non-cooperative magnetism  405
non-metal doping  290
non-zero coercivity  406
normal spinel  408
$N\text{-}TiO_2/ZrO_2$ photocatalysts  234
n-type metal oxide semiconductors  357
n-type semiconductor  359

## o

octahedral and cubic 3D magnetite ($Fe_3O_4$) nanocrystals  165, 166
octahedral-to-pyramidal isomerization mechanism  120
ordered porous structures  383
out-of-phase boundary (OPB) defects  17
oxygen evolution reaction (OER)  45, 154, 341
oxide ferroelectrics  5
oxide nanomaterial
  classical force-field techniques  113
  CVD  116
  DFT  113–114
  geometrical approaches  113
  molecular dynamics (MD) approaches  113
  molecular interactions, manganese  114–116
  Monte Carlo (MC)  113
  oxide porous materials  121–131
  synthesis/characterization of  111
oxide porous materials
  high-pressure conditions  124–127

hybrid microporous functional materials  127–131
structural properties  121–123
oxide-based gate dielectrics  89
oxide-based heterojunctions  295
oxides bearing titanium(III) species, synthesis of  59
oxidizing radicals  281
oxygen adsorption model  464
oxygen evolution reaction (OER) pathway  45, 342
oxyhydroxides of Ni(III)  45

## p

palladium  303
Paris Agreement  277
partial oxidation reactions  282
$PbZr_{0.52}Ti_{0.48}O_3$(PZT)–$CoFe_2O_4$  425
Pechini method  146, 147
perovskite solar cell (PSC)  366
perovskites  411
3,4,9,10-perylenetetracarboxylic acid (PTCA)  26
phase separation phenomenon  412
photo-electrochemical reactions  339
photobiorefinery  283
photocatalysis  229, 256, 279, 282
  titania-based coatings  317
photocatalyst  279
photocatalytic $CO_2$ reduction  283
photocatalytic $CO_2$ reduction
  current issues  283–286
  $TiO_2$-based  286–291
photocatalytic pathway  283
photocatalytic processes  278
photocatalytic reactions  280
photocatalytic removal, of nitrogen oxide gases  229
photocatalytically active films  327
photocathodes  349
photochemical $DeNO_x$ action  229
photochemical oxidation pathways, for $NO_x$ molecules  252
photochemical process  230
photocurrents  351, 352
photoelectrochemical (PEC)  278
  carbon-free  339
  conduction band (CB)  344

electrolyzer   344
metal oxide heterostructures   353
metal oxides candidate materials
    photoanodes   349
    photocathodes   349–350
    properties   348–349
photoelectrode efficiencies evaluation   359–365
protective anti-corrosion layer   354–359
renewable energy sources   339
solar fuels   339
tailoring surface catalytic sites and catalyst use   350–353
valence band (VB)   344
water-splitting   339
photoelectrodes   352
photogenerated charge carriers   326
photonic efficiency (PE)   281
photonic yield (PY)   281
photoreduction process   284
photosynthetic organisms   277
photovoltaic cell   344
photovoltaics   288
physical adsorption/desorption   470
piezoelectric/ferroelectric $BaTiO_3$ (BTO) perovskite   423
pinning effects   7
Plank's constant   365
plasma methods   142
platinum   300
polaron hopping   69
polyrotaxanes   381
porous current collector   390
porous $ZnAl_2O_4$ monoliths   143
Pourbaix diagrams   355
pressure-induced hydration   124
pressure-transmitting fluid   124
protective anti-corrosion layer   354
Pt nanoparticles   302

## q

quantum efficiency (QE)   281
quantum yield (QY)   280

## r

radiolysis   339
recombination   318
reference electrode (RE)   343, 428
remanence magnetization   406

remanent polarization   406
Resazurin ink   330
residual polarization in the material   406
reversible hydrogen electrode   343
rigid-coupled magnets   414
rigid-coupled regime   414
Ruderman-Kittel-Kasuya-Yosida (RKKY) coupling   420
rutile   287
rutile $TiO_2$   387

## s

sacrificial agents   285
saturation magnetization   406
saturation polarization   406
SBA-15   323
Schottky junction   356
Schrödinger equation   113
second order Jahn–Teller effect   412
secondary electrolyte interphase layers   382
self-assembled monodisperse magnetite   180, 182
self-cleaning reactions   25
semiconductor core-shell structures   167
semiconductors   278
sensor   439
siliceous zeolite ferrierite (Si-FER)   124–127
single-phase iron oxides MNPs   415
Sn-modified $TiO_2$ photocatalysts   325
$SnO_2$-$Zn_2SnO_4$/graphene composite   244
$SnO_2$/$Zn_2SnO_4$ system   243
soda-lime glass   321
solar energy   282
solar fuels   278, 339, 365
solar-to-hydrogen (STH) efficiency   278
sol-gel method   143, 288
sol-gel process   44, 55
solid phase synthesis methods, of spinel nanomaterials
    ball milling   148
    combustion synthesis   147–148
    high-temperature solid solution method   148–150
    solid-state thermal decomposition   146–147
solid-state gelation   55
solid-state thermal decomposition   146

solution phase synthesis methods, of spinel nanomaterials
   hydrothermal methods   143
   sol-gel methods   143
   solvothermal methods   145–146
   thermal decomposition   143–145
solvothermal methods   145
sonochemical method   291
spherical hierarchical morphology   167
spin-glass-like frozen surface spins   419
spin–orbit coupling   407
spin-orbit torque (SOT)   428
spin-transfer torque (STT)   428
spinel ferrite $LiFe_5O_8$ (LFO) nanofilms   159
spinel ferrites   408
spinel nanomaterials   138
   one and three-dimensional structures   171
   one and two-dimensional structures   170
   one-dimensional (1D) structure   151
     nanorods   151–154
     nanotubes   154–159
     nanowires   154
   self-assembled structures   175
   solid phase synthesis methods
     ball milling   148
     combustion synthesis   147–148
     high-temperature solid solution method   148–150
     solid-state thermal decomposition   146–147
   solution phase synthesis methods
     hydrothermal methods   143
     sol-gel methods   143
     solvothermal methods   145–146
     thermal decomposition   143–145
   structure-effect applications   150–151
   three-dimensional structures   165
   two and three dimensional structure   173
   two-dimensional (1D) structure
     nanofilms   159
     nanoplatelets   163–165
     nanosheets   159–163
   vapor phase synthesis methods   138
     laser pyrolysis process   141–142
     plasma methods   142–143
     spray pyrolysis   140–141

spray pyrolysis   140
$Sr_2FeMoO_6$   413
standard lithium hexafluorphosphate-based electrolyte   385
strain transfer mechanism   423
strain-coupled artificial multiferroic composites   412
structural engineering for electrochemical performance   382
structure-property-application relationships, for functional spinel oxide nanomaterials   183
sulphur doping   291
super-exchange interactions   405
supermagnetism   407
superparamagnetism   407
surface heterojunction   290
surface scattering and reflecting effect (SSR)   240
surface treatments   24
surfactant wrapping sol–gel method   101
synthetic antiferromagnets (SAFs)   419

## t

tailoring surface catalytic sites and catalyst   350
TDMAHf   25
terephthalic acid   330
tetrakis(diethylamido)hafnium (TDEAHf)   26
3D interdigitated $Ti_6Al_4V$ electrode   393
3D interdigitated microbattery architectures   393
3D periodic graphene composite aerogel microlattices   393
3D-printed graphene composite aerogel (3D-GCA)   393
3D printing
   battery and supercapacitor materials   392–394
     porous current collector   390–392
     of porous electrodes   389–390
3D spinel nanostructures   169
thermal decomposition   143, 167, 339
thermal-CVD   117
thin films of photocatalytic $DeNO_x$ materials   259

thioglycolic acid (TGA)-capped CdTe colloidal quantum dots  235
three-dimensionally structured lithiated MnO$_2$ cathode  381
Ti$_2$O$_3$ nanostructures  61
Ti$_4$O$_7$ nanocomposite  64
TiO$_2$ based materials  230
   dopant elements and quantum dots  234–235
   NO$_x$ abatement in real environments  254–256
   photocatalytic construction materials in urban areas  253–254
TiO$_2$ nanostructure
   defects  235–236
   electronic band structure  237
   oxygen vacancies  235–236
TiO$_2$-containing cementitious materials  229, 230
TiO$_2$-nanocarbon interface engineering  99
TiO$_2$-on-CNT systems
   vacuum techniques  103–107
   wet chemistry  100–103
TiO$_2$/Al$_2$O$_3$ binary oxides  237
TiO$_2$/ZnO mixture  247
TiO$_2$-NPs  323
titania-based coatings
   air and water cleaning photocatalyst  317–319
   characterization of  321–323
   fabrication of  320–321
   functionalization  319–320
   SiO$_2$ supported TiO$_2$  329–334
   Sn-modified TiO$_2$ photocatalysts  325–329
   TiO$_2$-SiO$_2$  323–324
titanium  286
titanium based oxides  237
titanium based perovskite semiconductors  237
titanium dioxide (TiO$_2$)  317
titanium Magnéli nanostructures  65
titanium oxide (TiO$_2$)  96
   crystal structures  98
   nanosizing  98
toxic gasses  441
transition metal  319
transition metal complex  117
transition metal dichalcogenides (TMDs)  27
transition metal oxide DeNO$_x$-photocatalyst alternatives  238
   bismuth oxides  238–242
   tin and zinc based oxides  242–247
   transition metal oxides  247–250
transition metal oxides (TMOs)  349–350, 358, 365, 405
tunneling magnetoresistance (TMR)  428
turnover frequencies (TOF)  324

## u

ultrathin Ni(0)-embedded α-Ni(OH)$_2$ heterostructured nanosheets  45
urchin-like mesoporous TiO$_2$ hollow spheres (UMTHS)  236

## v

vacuum scale  354
valence band (VB)  317, 319
vanadium based oxides  250
vapour phase deposition
   atomic layer deposition  2–3
   chemical  2
   metal oxide applications
      dielectric oxide materials  18–27
      ferroelectric materials  4–5
      ferroic thin films  5–18
   methodology  1
   precursors and chemistry  3–4
vapour phase synthesis methods, of spinel nanomaterials
   atomic layer deposition  138–140
   chemical vapor deposition  138
   laser pyrolysis process  141–142
   plasma methods  142–143
   spray pyrolysis  140
visible light  319, 329
visible-light photocatalytic activity  295
volatile organic compounds (VOCs)  439
volcano plot  342
Volmer-Heyrovsky mechanism  341
Volmer-Tafel mechanism  341
V$_2$O$_5$ with vertically aligned arrays of carbon nanotubes (V$_2$O$_5$/VACNT) composites  93, 96, 97

## W

water-splitting process   339–340, 344, 346
wet chemistry, for $TiO_2$/CNT hybrid material preparation   100
$W_{18}O_{49}$-based sensor   70
$W_{18}O_{49}$ nanostructures   68
 gas sensing properties   68
$WO_{3-\delta}$ nanoparticles   69
$WO_3$/$TiO_2$ composites, for De-$NO_x$ processes   233
working electrode (WE)   343

## Z

zeolite L   122
zeolites   112
zero polarization   406
zinc ferrites   249
zinc oxide   291
$Zn_3Al$-$CO_3$ photocatalyst   245
$Zn_xCo_{3-x}O_4$ polyhedral spinel material   167
Zn(hfa)$_2$TMEDA   119, 120
$ZnAl_2O_4$ spinel monolith synthesis   143, 144
$ZnAl_2O_4$ spinel nanotubes   156
ZnAl-LDH semiconductor   245
$ZnCo_2O_4$ nano-composites   250
Z-scheme heterojunction   242–244
Z-scheme photocatalytic systems   295